Architecture and Urbanism: A Smart Outlook

Shaimaa Kamel · Hanan Sabry ·
Ghada F. Hassan · Mostafa Refat ·
Abeer Elshater · Ahmed S. Abd Elrahman ·
Doaa K. Hassan · Rowaida Rashed
Editors

Architecture and Urbanism: A Smart Outlook

Proceedings of the 3rd International Conference on Architecture and Urban Planning, Cairo, Egypt

Springer

Editors
Shaimaa Kamel
Ain Shams University
Cairo, Egypt

Hanan Sabry
Ain Shams University
Cairo, Egypt

Ghada F. Hassan
Ain Shams University
Cairo, Egypt

Mostafa Refat
Ain Shams University
Cairo, Egypt

Abeer Elshater
Ain Shams University
Cairo, Egypt

Ahmed S. Abd Elrahman
Ain Shams University
Cairo, Egypt

Doaa K. Hassan
Ain Shams University
Cairo, Egypt

Rowaida Rashed
Ain Shams University
Cairo, Egypt

ISBN 978-3-030-52583-5 ISBN 978-3-030-52584-2 (eBook)
https://doi.org/10.1007/978-3-030-52584-2

This Springer imprint is published by the registered company Springer Nature Switzerland AG
The registered company address is: Gewerbestrasse 11, 6330 Cham, Switzerland

Preface

With the advent of the new millennium, it became obvious that rapid urbanisation has serious repercussions on the world's ecosystems. As innovative solutions are needed to confront new challenges, the Faculty of Engineering at Ain Shams University invited renowned international multi-disciplinary keynote speakers, along with researchers and practical exemplars from the fields of architecture and urban planning, to discuss both innovation and technological opportunities and challenges for the built environment. The aim was to depict how smart approaches and prospects could transform future buildings and urban areas into places of enhanced value and quality.

The event, which took place in October 2019, created an open and dynamic space for profound exchange over the course of three days. The speakers shared different philosophies and experiences, while revisiting the conventional model of design production. Moreover, speakers applied smart solutions to a variety of design challenges touching on urban regeneration and development, contextuality, human-centred design, integrative design, adaptive buildings, innovative technology-based materials, and product design. The research papers collectively intersected with a smart outlook as a common paradigm.

The editors of this book were committed to selecting outstanding leading research papers in an attempt to encapsulate the different concepts and approaches that would help shape a holistic vision towards a smart sustainable built environment.

The book comprises 34 chapters grouped under five main areas: smart architectural design education and practice, smart environmental solutions, smart cultural heritage, smart public spaces, and climate-smart cities.

Part I: Smart Architectural Design Education and Practice

Architecture today is fundamentally different from the time when Vitruvius wrote his books on architecture. Nonetheless, technology and architecture remain epistemologically inseparable. With the increasing importance of technology in the built world, it becomes obvious that technology plays an influential role shaping the architect's artistic and spatial considerations. Today, the role of the architect has developed with the imperative of needing to embrace technology without compromising the artistic integrity. New technologies have always driven innovation in building design, while providing the profession with a myriad of design tools that could take building design further.

The chapters to follow are concerned with architectural education and design. Five chapters tackle enhancing practice-based education pedagogy by adopting concepts that would help students develop their competences and skills for professional practice and adapt to the rapidly evolving future: creativity, ambiguity tolerance, nature imitation, sustainability, and learning algorithms, whereas the last three chapters in this part address the enhancement of building design through the development of three tools and approaches: swarm intelligence, biomimicry, and multisensory design.

Going in depth through Chap. 1, Doaa Hassan investigates the effect of improvisation on students' mindsets and creative climate, then consequently on creative problem solving (CPS) while engaging in an architectural design task. The author develops an educational experiment on two similar brainstorming sessions, where the second session starts with improvisation activities to explore the influence of training in such skill. These results go in line with Chap. 2, where Meral Omar et al. proposed a framework of criteria for adopting biomimicry in architectural design education. The authors provide a comparative analysis of current architectural design programs and experimental studies incorporating biomimetic design into their curricula in junior through senior levels. These studies are reviewed and analysed in terms of their impact on students' creativity.

In Chap. 3, Laila Khodeir et al. assume that teaching sustainable design courses (SDC) in early academic years can have a significant influence on students' works in design studios. The awareness of current sustainable issues could enhance students' design concepts via learning how to integrate different ecological, social, and economic attributes of sustainability in their designs. In Chap. 4, Ayman A. Farid offers a critical analysis of possible gaps in meeting sustainable architecture design training and education in Egypt. He provides a case study that held 90 students from the faculty of engineering at Ain Shams University, who were introduced to multiple architectural short-training modules to test their response to each module. The chapter provides a quantitative analysis for further improvements. Nevertheless, Chap. 5—written by Omar O. Elrawy et al.—tries to develop a computational model that evaluates architectural designs using machine learning algorithms to fulfil the project goal of investigating the computer's ability to generate architectural designs. The results recommend adopting three evaluation criteria related to spaces orientation, functional zoning, and usability of the area.

The next chapters discuss the enhancement of building design through the development of three tools and approaches. Chapter 6 introduces the application of swarm intelligence in architectural design. Algorithms in architectural design have been able to transcend their role as frameworks of formalisation and abstraction. The algorithms' output could now directly visualise the digital fabrication methods that the architect could use to build the production. This argument opens up a new role for algorithms as a design tool. Swarm intelligence represents one of these algorithms. The application of swarm intelligence in architecture ranges from visualisation and self-organisation of a multi-agent system to the design of urban forms. Hagar Magdy and Hazem Eldaly analyse swarm intelligence with their different types and examples in architecture, suggesting three different methods for applying swarm intelligence in architectural design. Those methods include achieving design aims along with optimisation algorithms, applying the strategy of learning and responding, and generating design alternatives.

Furthermore, Sara Samy et al. tackle the biomimicry-based design process in Chap. 7. The biomimicry approach seeks sustainable solutions to human challenges through inspiration from nature, its forms, systems, and procedures. It could be highly beneficial in shaping sustainable design and innovation in architecture. This chapter claims that biomimicry is still an emerging field in architectural design due to the absence of a systematic design process model to serve as a guide throughout the design process. The concluding remarks attempt to bridge this gap, aiming at correlating biomimicry theories and the architectural design process. This theoretical base achieves, through a comprehensive critical analysis of biomimicry definitions, concepts, approaches, levels, initiatives to incorporate these attributes into an integrated architectural design process, and analysis of case studies of biomimicry-based architecture. The results end with a proposal conceptual framework for biomimicry-based architectural practices marking the shift from theory to practice.

For architecture in the digital age, domination of technology implementation usually overshadows multi-sensory design. The challenge is further emphasised by the technology-saturated state that characterises contemporary societies. Architects, now more than ever, should become more sensitive to sensorial experiences as a design tool to enhance the experiential qualities of the built environment. Dina Bakir et al., in Chap. 8, explore a phenomenological investigation of vivid experiences. This chapter is a step forward to achieve a design that creates a truly memorable multi-sensorial experience of place. The authors here investigate the phenomenology of multi-sensory perception of architecture. A phenomenological research approach serves the aim of contributing to a better understanding of the relationship between spatial qualities and the multi-sensorial knowledge of the built environment. Through a qualitative interpretive analysis, different vivid environmental experiences characterised by specific ambiences are explored and analysed. This chapter determines the spatial qualities of the built environment that could provoke more holistic and more productive experiences, moving forward to developing multi-sensory design approaches in architecture.

Part II: Smart Environmental Solutions

A group of researchers discuss sustainable architecture through the usage of design strategies that reduce the negative environmental impact of built environment and ensure an efficient building performance quantitatively. This group discusses the implementation of these strategies in buildings to achieve energy saving. They include building shape, orientation, and façade skin. Generally speaking, researchers worldwide investigate the conjunction of renewable energy systems with smart strategies. Further implementation of these approaches goes forward to concentrate on the concept of net zero-energy building. This implementation is considered the future international goal for saving energy. The results emphasise the importance of investigating the combination of passive design strategies and energy generation systems in residential housing as means to achieving energy saving.

In this domain, one of the traditional passive strategies is the courtyard investigated in Chap. 9 by Mohammed Abd El-Samea et al. The focus is to define the optimum ratio that achieves thermal comfort and efficient energy performance in a school located in Upper Egypt. This study simulates 20 different design scenarios using DesignBuilder. The results show that better thermal conditions are achieved by reducing courtyard width and increasing building height. The results also illustrate that the deep courtyard is more effective and provides better thermal conditions than the shallow one. Moreover, the concluding remarks show that an increase in the courtyard width and building height leads to higher annual school energy consumption, especially when the courtyard width exceeds 10 m due to the façade exposure to solar radiation.

In Chap. 10, Rana El-Dabaa et al. discuss soft, adaptive building skins for energy-efficient architecture. The adaptive façades require complex mechanical systems that control their motion. Programmable materials are increasingly replacing mechanical systems by encoding their inherent properties to achieve passive controlled motion response to external stimuli, thus acting as zero-energy adaptive systems. This chapter identifies design parameters and motion grammars of HMTM, a hygromorphic–thermobimetal laminated composite that consists of wood. This chapter uses a computational approach to capture, track, and programme its response behaviour over time. The results show that the HMTM motion framework pushes the limits of low-tech and low-cost programmable materials to replace mechanical systems with passive, responsive actuators, as well as its applicability in different climatic zones.

Bassent Adely et al. in Chap. 11 evaluate the potential of applying passive techniques and PV panels as a means for reaching nearly/net zero-energy residential housing in an existing gated community located in Greater Cairo. This chapter uses simulation analysis by Design-Builder to evaluate the feasibility of integrating two strategies in a single-family house. As an annual result, there was nearly zero energy consumption, which is quite a reduction for the whole residential buildings of the community. Additionally, in Chap. 12, Hussein Hamza et al. emphasise the balance between energy consumption and production to reach the net plus energy house design. For the case study, the authors here select a three-bedroom apartment prototype set in three different climatic zones in Egypt. These locations are Cairo, Alexandria, and Aswan. IES-VE software presently uses energy simulation analysis. The results indicate that the significance of applying these strategies is not the same for all climatic conditions. Theoretically, the results find that it is possible to achieve net plus-energy housing in Egypt, where energy production from PV cell arrays exceeds the total annual primary energy consumption.

Qualitative approaches in research have always been a trusted tool when it comes to user experiences, as they directly encapsulate the perceptive dimension into conclusions, In Chap. 13, Dalia Hafiz and Vrushali Mhatre investigate the dimension of well-being through qualitative indoor environments via a comprehensive scope, including lighting, acoustics, and thermal comfort. Their findings and proposed recommendations could help in promoting occupant comfort, increasing productivity, and enhancing occupant quality of life. In the reflection towards a broader life spectrum, researchers pay attention to spaces occupied by juniors. Schools are among buildings not only influencing instantaneous occupants' satisfaction but also severely reflecting educational quality and consequently propagating its effect onto future expectations.

Meanwhile, Chap. 14 by Sherif Abdelmohsen et al. experiments a computational approach for the efficient management of school-shared facilities in Egypt. The purpose is to maximise the benefits of integrating BIM and GIS into a comprehensive vision for various stakeholders. In Chap. 15, Alaa El-Telwany et al. explore via a research framework the quality of outdoor environments in primary schools. The authors assess the Egyptian and UK guidelines in terms of values supporting outdoor educational environments. The resulting structure could be incorporated into design guidelines and requirements for education.

Chapter 16, written by Bassant Sorrour et al., concentrates on post-occupancy evaluation (POE) of the emergency department in healthcare facilities. Both qualitative and quantitative methods are applied to define architectural design guidelines that could enhance the performance of the emergency department based on users' experience. The aim is to minimise the gap between the design of the department and its operation phase.

Part III: Smart Cultural Heritage

Architecture, as a cultural discipline, cannot be understood without addressing the gradual build up of the natural and the social sciences applied in technologies and concepts; also history cannot be ignored. In the architectural field, historical values are of importance, given that a building can be envisioned as the outcome of a continuous process of modification and stratification that took place during its life. At the outset of the new millennium, the advent of modern technologies into the cities that are rich in architecture, identity, art, and history, as well as the interaction of the architecture with ever-growing technologies, are significant and noteworthy. The chapters in this part enrich the heritage conservation field through approaching it from various distances and different angles of experience and knowledge.

Chapter 17, by Noran Elbaz et al., investigates system-based knowledge using Heritage Building Information Modelling (H-BIM). They focus on the utilisation of modern technologies in architecture and its integration with traditional and ancient structures and elements. The main conclusion of their work confirms the essential role of cultural values and the principles of local and traditional architecture. A future direction of research could investigate new additions to the architect's toolbox that could change the way of heritage preservation. The universal value of cultural heritage puts a massive demand for applying such technology in this field, which requires an innovative system in data management for the documentation of historical buildings. Modelling historical buildings stands on heterogeneous datasets and could be interoperable with different software tools. The author outlines the importance of interoperability in optimising the usage of data by reducing wasted time and effort in gathering, translating, and integrating data. Besides, it allows for broader dissemination, which permits public users to benefit from the documentation and conservation process for a deeper understanding of heritage sites.

In Chap. 18, Rasha Abd El-Lateef et al. investigate the Christianisation of ancient Egyptian cosmology with an in-depth review of Philae in late antiquity. This chapter focuses on the Ptolemaic temple of Isis on Philae island as the last temple built in that era. The authors here begin by setting the intrigue of historical settings of the regional situation politically and culturally. In doing so, they attempt to interpret the temple's transformation as a unique phenomenon reflecting the public ontological understanding of the sacred power. The concluding remarks stress that the cosmic value of ancient religious buildings should be adopted rather than the traditional absolute practical vision.

Chapter 19 by Sara Adam et al. deals with the concept of Historic Urban Landscape (HUL) that was introduced by the UNESCO as an approach for the identification, preservation, and valorisation of cultural heritage. The authors develop a comprehensive urban regeneration framework through exploring the potentials of the HUL approach and applying it to Rosetta City, which is situated in a unique coastal position in Egypt. In doing so, they give a brief overview of urban regeneration policies and previous heritage conservation theories, emphasising the change from place-oriented to people-oriented approaches, while illustrating the best practices of the HUL approach and learning from their experiences. The authors conclude that the effectiveness of HUL lies in the fact that it is an unconventional adaptive approach that believes in the improvement of lives and knowledge as the genesis of preservation. They emphasise the importance of integrating the conservation of urban heritage with socio-economic development based on a balanced and sustainable relationship between the built and natural environments.

Additionally, the contributing authors have put forward three chapters regarding the rehabilitation of contemporary cities and regenerative urbanism. According to them, the adaptive reuse of buildings is an effective strategy for improving the optimisation of a particular structure to suit another function different than the one it was initially designed for. This process involves multiple challenges through the decision-making phase, becoming an active part of urban regeneration strategies and sustainable development. Besides, the authors emphasise the restorative characteristics of regenerative cities through moving on to circular urban metabolism, eventually leading to the decrease of cities' ecological footprints. The concept of the "Ecopolis" is established in a regenerative city, which supports the ecosystem restoration process.

Chapter 20, Yosra Elnaggar et al., discusses the wall, as an increasing phenomenon in Cairo, materialising in many different ways within the urban context. The authors identify the implications of the wall on the urban environment, regarding urban space vitality, right to the city, social cohesion, crime prevention, and the image of the city. They define the wall as being the stand-alone structure that divides, segregates, differentiates, screens, protects, prevents, and sometimes guides and redirects. Taking Cairo as a case study, the wall is being examined with its increasing manifestations in the urban context: walling public buildings, privatising public spaces, acting as security barriers and encircling gated communities to mention a few examples. The authors indicate that the main reasons behind the growing wall phenomenon in Cairo are mainly represented in the declining sense of security. They conclude that the wall has a widespread impact on the urban environment from the physical, social and visual aspects. They additionally highlight the fact that the wall's intended purpose of securing, banning, isolating, or controlling has failed in most cases. This finding means that more innovative solutions should be found to achieve the same goals with the least negative implications on the city.

Chapter 21 by Nada Hassan et al. investigates the adaptive reuse framework to be applied to the abandoned governmental buildings of Downtown Cairo after the relocation of the administrative centre to the New Administrative Capital. The chapter highlights the fact that such a relocation shall have multiple effects on spatial occupancy and activities, including traffic patterns and unused spaces and buildings. The chapter studies the suggested framework of the relocation process, through two main steps: the relocation of the workforce of each governmental building and the design and construction of the new government buildings in the New Administrative Capital, while putting a specific reuse plan for the abandoned buildings. The authors conclude that adaptive reuse should not only be addressed as an intervention on the building scale, but also on the city scale, as a means of heritage and identity conservation. They emphasise the fact that conducting a comprehensive assessment of the needs and possibilities of Downtown Cairo is essential to allocate the uses accordingly and to avoid failures.

Chapter 22 by Lina Fayed et al. examines aspects of regenerative cities that ensure a robust and intact world for future generations. This chapter highlights the fact that regenerative practices adopt the shift from merely addressing issues such as climate change, biodiversity, ecosystems, and urban development in isolation, to working with them simultaneously and Consequently, enabling urban wealth generated from natural capital in the region and sustaining value in the area with local people, and improving the ability of ecosystems to regenerate resources. The main concluding remark emphasises the importance of regenerative cities, emerging as a solution to the ecological challenges arising from rapid urbanisation, which aims at the establishment of a symbiotic relationship between the urban areas of cities and their hinterlands. This ultimately requires cities to call for holistic strategic choices and long-term planning, not temporary compromises like most of the political decision-making systems in many governments.

Moreover, the last three chapters in this part deal with the challenges that today's cities may have in terms of human behaviour and the surrounding context that have changed rapidly in the age of information and smart technology. Meanwhile, the human behavioural aspects are addressed here by diverse researchers. The concluding remarks, figured out from these pieces of research, highlight the significance of active node, for instance, through driving parameters of behaviour patterns for ever-shifting users. Considering the changes happening on the urban places over time and the study of social practices, informality reveals a realistic primary image of the built environment that should be built with art in urban developers' minds. In this domain, Chap. 23, authored by Khushbu Soni et al., attempts to bridge the gap between collecting information and the behavioural patterns of a pedestrian in outer spaces. Their contribution to this work, based on the spatial analysis of the Kala-Ghoda Festival, is to conclude by drawing parallels for user response accumulatively. The consideration pays attention to the variables of personal, situational, and environmental factors.

Chapter 24—written by Aliaa Ahmad et al.—provides theoretical ground for the quality of life in low-income neighbourhoods. The results of this study figure out the primary role of decision-makers in providing action plans towards the inclusion of people's voices in the development process. Conclusively, Chap. 25—written by Nadine Hindi—addresses the outlook considering the natural recovery phenomenon as a threat instead of an opportunity for thinking paradigms for urban regeneration. The main conclusion in her work about the Badawi neighbourhood highlights the social and construction trespasses. These trespasses are similar essential spaces in this neighbourhood as opportunities in their broader context. A future area of research in this domain could provide resilient solutions to abandoned spaces that consider heritage aspects and avoid social gentrification results.

Part IV: Smart Public Spaces

The quality of the occupied space has always been a concern and a recurrent research topic for architects and design practitioners. This need has continuously been fuelled up via extended time patterns and elongated periods, during which attendees occupy these spaces. Advancement in communication technologies and connectivity allow diversities of tasks to get accomplished within the same physical boundaries, extending the users stay, for maybe numerous daily cycles. In return, these user patterns push out the need to bridge the gap these physical boundaries create for human beings, disconnecting a living creature from its nature. In this part, the researchers end with a discussion about cities' spatial characteristics and the relation between spaces and inhabitants. The authors in this section tackle the topic using different theoretical approaches and methods to investigate the potentials of cities' spaces. Nowadays, small cities and towns around the world have started developing several projects that intend to support the economic welfare of local communities, the importance of cultural assets, community contribution, and small-scale green tourism. In Chap. 26, Ahmed S. Abd Rahman introduces the concept of the Cittaslow movement and tries to apply its principles within a selected Egyptian context. He also suggests an alternative form of delicate tourism, in which tourists abandon modernist fascinations in favour of entrepreneurial ideologies.

From another perspective, improving cities' performance is related to understanding cities as competitive corporates. In Chap. 27, Mahmoud Elnably et al. discuss the competing forces of globalisation that attract investments, as well as people to sustain real estate capital. The authors analyse the competitive advantages of the cities through the concept of city branding, with a direct application on El-Shorouk City using Porter's Competitive Diamond for Nations and its five forces model of competition.

Alongside this direction, cities often face transformation in their physical structures, especially big cities. These transformations tend to increase connectivity for the pedestrian and consider the problems that occur with crossing routes. In Chap. 28, Walid Abdeldayem and Tamer El-Khouly analyse the impact of highways and arterial roads of new urban settlements on the integration and connectivity of the urban structure. The results are achieved using space syntax software. The main conclusion provides the approach to overcome the shortage in the city's network system.

Finally, this part addresses the cities' spaces to accommodate different types of social activities. These spaces range widely in types of activities that occur at a particular place and time. Furthermore, Chap. 29—written by Mohamed Ezzeldin and Ayman Assem—discusses the social production of space. This study tries to determine the social characteristics that are essential in upgrading and development of public spaces that host different events. The results show that using GIS in creating the spatio-temporal analytical maps could help clear the definition of public urban spaces and determine their social characteristics. Consequently, this computational approach should be considered when constructing visions for upgrading and development of urban public spaces. Chapter 30—written by Elaf Raslan et al.—investigates the possibility of developing an evidence-based strategic plan. Qualitatively and quantitatively, this investigation in the case study of Cairo's Downtown focuses on arts and culture activities. The results conclude the essential need for strategic plans that take into consideration arts and culture events as a tool for urban development.

Part V: Climate-Smart Cities

Going through four chapters, the additional challenge here tackles the concept of enhancing the urban ecology of cities. These pieces of research address this challenge by exploring the effects and threats of climate change on the built-up areas. This group of research also goes in-depth to highlight 'light pollution' and the rarely tackled environmental urban issues by showing concepts, approaches, and innovations. This part also shows the enhancement of resilience and sustainability in future cities. In Chap. 31, Esraa Elmarakby et al. explore the relationship between the differences in the morphological structure of four residential areas in Cairo and urban heat island (UHI) intensity at the local scale. This chapter analyses the temporal and spatial change of UHI using remote sensing and GIS applications. It reports that there are considerable discussions among scholars pertaining to urban morphology. Their findings show that urban morphology is the crucial aspect that influences the UHI presence in cities.

Likewise, in Chap. 32, Noha Shaheen et al. attempt to investigate the appropriation of environmental risk assessment towards guiding for eventual future efforts to promote resilience to climate change in historical buildings. This chapter investigates by arguing the built cultural heritage perception and examining the risk assessment based-mitigation, adaptation of strategies, and approaches for implementation. The findings develop a detailed matrix to combine the risk assessment techniques and their related climate change threats hence acting as an assessment quantitative tool to ensure effective responses at all levels.

In Chap. 33, Ashraf A. Nessim analyses and evaluates the environmental impact of light pollution in Egypt resulting from outdoor lighting. An intensive literature review of recent papers discussing the problem has been performed and followed by the assessment of the possible techniques that could be used to limit light pollution. These techniques could help those involved in decision-making and setting legislation to limit light pollution with its impact on humans and nocturnal living organisms.

Eventually, Chap. 34—written by Ahmed K. Abd El-Hameed—highlights the role of artificial intelligence in mitigating climate change in cities, including applications in transportation, urban energy, water use, and waste management. His review proves that these applications could impact city planning by reshaping the built environment. In this chapter, the author shows that implementing such applications could improve the sustainability of future cities.

Cairo, Egypt

Shaimaa Kamel
Hanan Sabry
Ghada F. Hassan
Mostafa Refat
Abeer Elshater
Ahmed S. Abd Elrahman
Doaa K. Hassan
Rowaida Rashed

Acknowledgements

The editors of the presented work would like to acknowledge Ain Shams University for providing the scientific platform of AR_UP 2019 conference, titled: "Architecture & Urbanism … A Smart Outlook" that took place in October 2019, in Cairo, and was chaired by Prof. Dr. Ayman Ashour, Deputy Minister of Higher Education and Scientific Research; and co-chaired by Prof. Dr. Yasser Mansour, Professor and Former Chair of the Department of Architecture, Ain Shams University and Prof. Dr. Tamer Elkhorazaty, Professor of Urban Planning, Ain Shams University. The platform has helped inform the audience of innovative views and exciting arguments in smart architecture and urbanism. The editors would also like to thank the members of the scientific committee for their constructive recommendations in the reviewing process. They would also like to extend their appreciation to the authors for their enthusiastic efforts and to everyone who has participated in the progress of this publication.

Contents

Part I Smart Architectural Design Education and Practice

1 **Improvisation Impact on Creativity Strands in Architecture Education** 3
Doaa Hassan

2 **Biomimetic Approaches in Architectural Design Education** 19
Meral Omar, Shaimaa Kamel, Doaa Hassan, and Sherif Abdelmohsen

3 **Tracing the Impact of Teaching Sustainable Design on Design Projects in Architectural Freshman Level** 33
Laila Khodeir, Ashraf Nessim, and Dalya Maguid

4 **Gaps Facing Sustainable Architectural Design Training and Education in Egypt** 49
Ayman A. Farid

5 **Evaluating the Architectural Designs Using Machine Learning: The Case of Two Modes** 65
Omar O. Elrawy, Tamer El-Khouly, and Omar Abdallah

6 **Applying Swarm Intelligence in Architectural Design** 77
Hager Magdy and Hazem Eldaly

7 **Biomimicry in Architecture: The Potential Shift from Theory to Practice** 89
Sara Samy, Wafaa Nadim, and Morad Abdelkader

8 **Towards Multisensory Architecture: A Phenomenological Investigation of Vivid Experiences** 105
Dina Bakir, Yasser Mansour, Shaimaa Kamel, Yasser Moustafa, and Marwa H. Khalil

Part II Smart Environmental Solutions

9 The Effect of Courtyard Ratio on Energy Consumption and Thermal Comfort in a Primary Governmental School in New Assiut City, Egypt 121
Ahmed Mohammed Abd El-Samea, Nouby Mohammed Hassan, and Amr Sayed Hassan Abdallah

10 Encoding Latent Properties of Hygromorphic-Thermobimetal Composites as a Passive Mechanism for Adaptive Building Skins 133
Rana El-Dabaa, Sherif Abdelmohsen, and Yasser Mansour

11 Retrofit as a Means for Reaching Net-Zero Energy Residential Housing in Greater Cairo 147
Bassent Adly, Hanan Sabry, Ahmed Faggal, and Mahmoud Abd Elrazik

12 Integrating Passive Design Strategies and Energy Generation Systems for Net Plus-Energy Housing in Egypt 159
Hussein Hamza, Mohamed Ayman Ashour, Hanan Mostafa Sabry, and Mostafa Refat Ismail

13 Enhancing Occupants' Well-Being Through Qualitative Indoor Environments 173
Dalia Hafiz and Vrushali Mhatre

14 School2Share (S2S): A Computational Approach for the Efficient Management of Shared Facilities in Egypt 185
Sherif Abdelmohsen, Ayman Assem, and Mohamed Ezzeldin

15 Evaluating the Quality of the Outdoor Environment of Primary School 199
Alaa EL-Telwany, Muhammed Gabr, Ayman Assem, and Tamer Samir

16 Post-occupancy Evaluation of the Emergency Departments: A Holistic Approach to Promote the Healthcare Architecture Design Standards 215
Bassant Sorrour, Akram Farouk, and Mostafa Refat

Part III Smart Cultural Heritage

17 Heritage Building Information Modelling: Towards a New Era of Interoperability 231
Noran Elbaz, Shaimaa Kamel, and Sherif Abdelmohsen

18 The Christianization of Ancient Egyptian Cosmology: Philae in Late Antiquity 241
Rasha Said Abd el-Lateef, Yasser Mansour, Shaimaa Kamel, and Ruby Morcos

19 Historic Urban Landscape Approach for Historic Coastal Cities Regeneration: The Case of Rosetta, Egypt 253
Sara Adam, Marwa Khalifa, and Amr Attia

20 The Wall, An Increasing Phenomenon in Cairo: Main Reasons and Potential Implications 269
Yosra G. Elnaggar, Yasser M. Moustafa, and Ayman F. Wanas

21 Adaptive Reuse as a Revitalization Tool: The Case of Downtown Cairo Post Relocation of Governmental Buildings to the New Administrative Capital 289
Nada Hassan, Marwa A. Khalifa, and Johannes Hamhaber

22 Aspects of Regenerative Cities 303
Lina Fayed, Abeer Elshater, and Rowaida Rashed

23 Behavioural Pattern of Pedestrian Negotiation in Public Spaces 321
Khushbu Soni, Ritesh Agrawal, and Sheetal Pikle

24 Quality of Life and the Architecture of Low-income Neighborhoods 341
Aliaa Ahmad, Yasser Mansour, Muhammad Gabr, and Marwa H. Khalil

25 The Last Train Traces in Beirut: Towards an Ecological Urbanism Perspective 357
Nadine Hindi

Part IV Smart Public Spaces

26 Cittaslow and Applicability in the Egyptian Context: The Case of El-Fayoum 371
Ahmed S. Abd Elrahman

27 City Branding Through Porter's Competitive Models: Investigating El Shorouk City Market Position 387
Mahmoud Elnably, Yasser Mansour, Ayman Assem, and Aya Elsisy

28 Investigating the Urban Structure of Newly Planned Cities in Egypt: The Case Study of New Cairo City 401
Walid S. Abdeldayem and Tamir El-Khouly

29 **GIS-Based Spatio-temporal Analysis for Social Events in Urban Public Spaces** 411
Mohamed Ezzeldin and Ayman Assem

30 **Mapping Arts and Urban Development in Cairo's Downtown** 425
Elaf Raslan, Marwa Khalifa, and Abeer Elshater

Part V Climate-Smart Cities

31 **Spatial Morphology and Urban Heat Island: Comparative Case Studies** 441
Esraa Elmarakby, Marwa Khalifa, Abeer Elshater, and Samy Afifi

32 **Environmental Risk Assessment as an Approach of Resilience to Climate Change in Architectural Built Cultural Heritage** 455
Noha A. Shaheen, Ayman A. E. Othman, and Mostafa Rifat Ismail

33 **Visual Pollution: An Approach to Reduce the Environmental Impact of Light Pollution in Egypt** 469
Ashraf A. Nessim

34 **Artificial Intelligence Shaping Sustainable Cities for Climate Change Mitigation: A Review of Literature** 483
Ahmed K. Abd El-Hameed

Index 497

About the Editors

Shaimaa Kamel is the current head of the Architectural Engineering Department, at the Faculty of Engineering, Ain Shams University. Besides, she is a teaching professor at the Faculty of Engineering, Ain Shams University and a part-time faculty at the American University in Cairo. Since 1992, she has covered numerous teaching courses, developed course syllabi and supervised graduate theses. In her tenure, she has published research papers and participated in organising international conferences. She is a member in the committee of excellence for Architecture at the Supreme Council for Culture. Professor Shaimaa is part of the general assembly within the General Authority for Educational Buildings. Professionally, she has designed several buildings within private firms. She holds a Ph.D. in Architecture from Ain Shams University.

Hanan Sabry is a professor of design and environmental control at the Faculty of Engineering, Ain Shams University, Cairo. She is a specialist in daylighting studies. She participated in many projects as a daylighting consultant. She was also an active participant of a team working in several research projects; one was in the area of integrated desert building technologies as a collaboration between King Abdullah University of Science and Technology (KAUST) and American University in Cairo (AUC). Another project addresses the daylighting and energy performance in intensive care units and patient rooms supported by a research grant from AUC. Starting last year, she has been a part of a team working in a research project about Monitoring Object and Visitor Environments (MOVE) in museums collaboration between Ain Shams University and University of Salford, UK, funded by STDF and Newton Mosharafa. Hanan also contributed to numerous research articles published in scientific journals and international conferences.

Ghada F. Hassan, Ph.D. is a professor and Head of the Department of Urban Planning at Ain Shams University. She served as a member of the Higher Committee of Promoting Professors in Egypt. She was chairman of the Department of Architecture at the French University of Egypt. Ghada had been GM of Technical Office at GOPP, Ministry of Housing Utilities and Urban Communities.

She has more than ten articles published in a scientific journal and one book in the field of urban ecology, landscaping, and housing. She had been organised and participated in more than 15 international and local conferences (H-index 4) and more than citation 9. She is currently a consultant at Informal Settlement Development Funds, UN-HABITAT. She had conducted many terms of references and implemented projects towards slum upgrading, the strategic urban development plan (SUP) for cities, and rural strategic plans sponsored by GOPP. Moreover, she was participating in several coastal tourist development projects at NWC, Syria, and Saudi Arabia.

Mostafa Refat, Ph.D. is the Director of International Credit Hours Programmes and professor of Architecture at the Faculty of Engineering, Ain Shams University. He holds his Ph.D. from the University of Liverpool, where he worked as a part lecturer and researcher from 1998 until 2006. The Institute of Acoustic, UK, certifies him as CCWNRA and CCENM for workplace noise assessment. He worked in EPSRC-funded projects in the UK, and worldwide. In 2009, he was responsible for the Technical Affairs, NUCA, Ministry of Housing and Urban Development, Egypt. He also developed curricula for the Architecture Engineering Department, British University in Egypt, with the University of Loughborough. He is a principal investigator to numerous funded research projects with partners from the UK, France, Italy, and Germany. He practices as well as Architecture Design, Acoustic consultations and building services via his founded firm in Egypt. Refat is a reviewer for several international and national journals.

Abeer Elshater is a full professor of urban morphology in the Department of Urban Planning, Faculty of Engineering, Ain Shams University (ASU), Cairo, teaching and supervising multidisciplinary topics in the ideology of urban design with a focus on urban morphology. She has worked on international research projects. Elshater is an author and co-author of seven books as well as more than 50 manuscripts in scientific journals and conference proceedings. She has given several speeches as an invited speaker in conferences in the UK, Italy, Turkey, and Malaysia. She is a personal investigator of a project funded by Science, Technology Development Fund (STDF), Egypt, about urban morphology and urban heat islands. She served as vice-director of the Continuum Improvement and Quality Assurance Unit, ASU from 2015 to 2018. Recently, she acts as an ambassador of the Regional Studies Association, UK, aside to her position as a consultant to the deputy minister of Higher Education and Scientific Research Affairs, Egypt.

Ahmed S. Abd Elrahman is an associate professor in the Urban Design and Planning Department, Faculty of Engineering, Ain Shams University (ASU). Currently, he is the coordinator of the Architecture Program, ASU. Ahmed is involved in teaching Architecture Design and waterfront development modules in Ain Shams University and Integrated Urbanism and Sustainable Design (IUSD) as a dual programme with his university and Stuttgart University. Moreover, he is the coordinator of the dual degree programme between Landscape Architecture, ASU, and the Architecture Program at Calabria University (UNIRC), Italy. Ahmed has extensive experience with particular emphasis on large-scale residential and waterfront projects. He has such experience during his work in Planning and Urban Development (PUD) firm, Cairo, until he established his office Design and Build Studio (DBS), Cairo. He has experience with international organisations, e.g. UN-HABITAT and national institutions, e.g. Urban Harmony (NOAH), Cairo. Abd Elrahman has several local and international publications.

Doaa K. Hassan is an associate professor in the Architectural Engineering Department, Faculty of Engineering, Ain Shams University. She is currently the academic coordinator of Landscape Architecture Program at Ain Shams University. A focused intention on humanities in both architecture education and professional practices can be seen in her publications. One of her area of interest tackles the creativity studies about teaching and learning methods. Meanwhile, another track adopted the impact of human behaviour in urban landscaping architecture. She is a scientific reviewer in many international journals. She participated in several workshops and conferences nationally and internationally. Doaa has a track record in administrative and professional services. The general authority has consulted her for educational buildings (GOPA) worldwide consultants. She is a member of Continuous Improvement and Qualification Assurance Unit at the same faculty she affiliated. Doaa was also a previous member of the American Society of Landscape Architects (ASLA).

Rowaida Rashed, Ph.D. is an associate professor in the Department of Urban Planning, Faculty of Engineering, Ain Shams University. She earned her doctorate degree in 2007 from the Institute of Urbanism, Graz University of Technology, Austria; with a dissertation titled: "A Framework of Urban Morphological Analysis: The Physical Transformations of Three Districts in Cairo City". Her teaching and research both focus on ecological urban design and regenerative practices. From 2010 till 2015, she has taken part in the establishment of the Architecture and Urban Design Program at the German University in Cairo (GUC), and in her capacity as the faculty coordinator for student affairs and curriculum, she has been responsible for its development, implementation, and advancement. Currently, she is also an associate partner and director of business development at Ökoplan Engineering Consultations, where, over the past 24 years, she has worked on numerous city planning, urban design and architectural design projects inside and outside Egypt.

Part I
Smart Architectural Design Education and Practice

Chapter 1
Improvisation Impact on Creativity Strands in Architecture Education

Doaa Hassan

Abstract Design studio is the core of the architectural education, where students understand how creative solutions are generated, developed and implemented. Design deals with many dynamic factors which demand creativity to cope with this ongoing change. In this sense, divergent and convergent thinking are equally significant to define creativity. However, divergence is associated with ideation. Many studies tackled the concept of improvisation as a crucial power to enhance divergence. Improvisation is to do something without any preparation. Recent studies show that an exposure to improvisation games as a kind of play may create a joyful climate. This in turn helps to release our minds from the preconstructed thinking pattern. Thus, this research aims at investigating the effect of improvisation on students' mindsets and creative climate and consequently on the creative problem solving (CPS) while engaging in an architectural design task. The research depends on an educational experiment that was developed on two similar brainstorming sessions with one exception where the second session started with improvisation activities. The experiment evaluation counted on scores of the divergence components, fluency, flexibility and originality-appropriateness, in addition to questionnaire and observation. The study ended by explaining the pros of improvisation in terms of students' mindsets, creative press and consequently on the ideation experience.

Keywords Improvisation · Architecture design · Creative problem solving (CPS) · Divergent thinking

1.1 Introduction

In architecture education, creativity studies are crucial in developing the learning of how to design. In this respect, it is significant to understand that "design involves a highly organized mental process capable of manipulating many kinds of information, blending them all into a coherent set of ideas and finally generating some realization

D. Hassan (✉)
Architectural Engineering Department, Faculty of Engineering,
Ain Shams University, Cairo, Egypt
e-mail: doaa.kamal@eng.asu.edu.eg

Architecture and Urbanism: A Smart Outlook,
https://doi.org/10.1007/978-3-030-52584-2_1

of those ideas" (Lawson 2014, 6). Thus, design deals with many dynamic factors and accordingly demand creativity to cope with this ongoing change. Igniting creativity and digging into students' inborn powers to discover their potentials should be considered as a core value of design studios (Sidawi 2013). In this sense, it is important to understand what creativity is. In fact, it is difficult to restrict creativity to one universal definition or explanation. Nevertheless, Mel Rhodes adopted more holistic approach. He described creativity with the "Four Ps," creative person, creative press/climate, creative process and creative product (Miller et al. 2011). "The definitions form four strands. Each strand has a unique identity academically but only in unity do the four strands operate functionally" (Rhodes 1961, 307).

Tied in the concept of creative person/mindset, both divergent and convergent thinking are equally significant to reach creative ideas. However, divergence comes first because it is associated with idea generation. Many studies tackled the concept of improvisation as a crucial power or skill to enhance this divergence and hence creativity. To clear up, as stated in Macmillan Dictionary, "to improvise" is to do something without any preparation, often because the situation does not provide any chance to do so. Nevertheless, improvisation is not a matter of intuition action but a skill that can be acquired. In design, improvisation can be stated as the process of generating ideas in the heart of the moment without any setting up (Sowden et al. 2015). In this sense, improvisation cannot be understood as a completely individual activity but should be grasped as a mutually collective action. Individuals are affected by each other's views and actions in the present moment (Larsen and Bogers 2014). Briefly, improvisation could be a tool that helps in developing people's creativity to thrive in manifold and unforeseeable situations (Ross 2013). It is one of "Creativity Trilateral Dynamics", the status that starts with play, then leads to mindfulness and finally reach improvisation (Hassan 2019).

On the practical level, improvisation depends on the concept of "deferring judgment" that is one of the most significant tools to boost fluency, flexibility and originality, the basic components of divergence. Therefore, improvisation has cognitive benefits that would add value if it is introduced into architectural education in order to encourage students to think in more diverse ways (Lewis and Lovatt 2013). Then, the query is about how to potentiate improvisation? Recent researches show that a kind of exposure to improvisation activities can enhance divergent skills (Sowden et al. 2015). Play through specific exercises and games is a core action to learn and practice improvisation virtues. On the other hand, improvisation activities as a kind of play create a joyful press "to improvise" without overthinking as it happens in reality (McWaters 2012). Accordingly, it is crucial to differentiate between two types of press, the physical press which appears in space decorations and furniture, and the psychological press which is the concern of this research as it will be explored (Miller et al. 2011). Hence, this research aims at investigating the effect of improvisation on students' mindsets and creative press and consequently on the experience of creative problem solving (CPS)/creative process. The focus will be only on the production of the ideation phase in terms of divergence components of fluency, flexibility and originality while engaging in an architectural design task.

1.2 Methodology

This research depends on an educational experiment, where the influence of improvisation on students' performance in design was tested. The experiment was developed on two similar brainstorming sessions with one exception where the second session started with improvisation activities. The experiment evaluation mostly counted on scores of the three components/skills of divergent thinking, fluency, flexibility and originality, in addition to questionnaire and observation results.

1.2.1 Experiment Design

The experiment medium was creativity sessions aiming to produce ideas and concepts for specific design project. This experiment is a large-scale study that was carried out on 180 junior students in a selected school of architecture. Students built common knowledge via passing different courses related to architecture studies, such as visual design, freehand drawing, theory and history of architecture, principles of building construction and design studios. The experiment was implemented as two tutorials under the umbrella of a module of an architectural course called "Theory in Architecture-2", and this module is themed as problem solving: Creativity and Criticism. The study was implemented in Fall 2017 and supervised by an associate professor specialized in architectural education, together with qualified teaching assistants.

The experiment tackled improvisation as a type of play that may affect positively on generating concepts and ideas. This experiment depended on carrying out two experiences of brainstorming sessions during two consecutive tutorials. They were almost similar experiences that aimed at generating concepts via brainstorming for a specific design problem. The design task was similar in both experiences to decrease the influence of any other factor. During the first experience, students were asked to form groups (5–6 students each including both males and females) to accomplish a 15-min brainstorming session to generate ideas for an intended focus. In the second experience, the experiment included two phases; the first phase was a play session where students were asked to play specific improvisation activities for 30–45 min. Then, the same groups as formed in the first experience started their brainstorming session for another similar focus. It was essential not to explain the purpose of the play in order to avoid any psychological effect if students know the positive impact of such activities on idea generation process.

Playfulness emerged from improvisation activities can inspire minds and social possibilities which in turn tease out creative ideas (Shyba 2007). Improvisation activities are classified into specific ordered levels of preparing participants' minds. These levels are named: icebreakers, warm-ups and spontaneity. Icebreakers were excluded since participants know each other. Warm-ups are games designed to get participants in a carefree temper and sharpen concentration. On the other hand, spontaneity is

Table 1.1 Improvisation activities/games: warm-ups and spontaneity (Improv Encyclopedia 2017; Farmer 2012)

Warm-ups	
Accepting circle	Participants stand in a circle. One of the participants can start by making a gesture, maybe with a sound. The next tries to mimic the previous action and so on. Despite we expect the gesture/sound not to change, it will
Alphabet letters	One of the participants can start by calling a letter, then each player or pairs forms it with their body parts
Alphabet challenge	Participants stand in a circle. One can start by saying a word that starts with the letter "A", and points at another player who must quickly say a word that starts with the letter "B" and so on
Spontaneity	
What are you doing?	Participants stand in a line. One can begin miming a specific action. (Example: buttering bread). The next participant steps forward and asks "What are you doing?" The first participant replies with something different (Example: "I am climbing a mountain."). The second participant starts to mime the action the first participant said and so on
Barney	Participants stand in a circle, one of them acts as a "caller" in the middle. Caller yells a letter to one of the participants; that participant should think of a name of a person, an object or a service that can be sold, and a location, all starting with the letter given. Example would be "B" which leads to "Barney sells Bread in Bulgaria"
Oracle	This game requires three participants act as one entity (Oracle), a "handler", and audience. The handler introduces the Oracle and proposes a topic. The oracle participants answer in order with only one word each. Once that question is answered, the handle asks the audience for another question. Example: Audience Member: Oracle, why are eggs so good? Player 1: Eggs Player 2: are Player 3: good Player 1: because Player 2: they Player 3: are Player 1: bad

very crucial in our experiment. Human mind is hypothesis-driven and our perception, observation, and understanding of the world are extremely figured by preconceptions. This "top-down" principle is biologically driven and hinders spontaneity. Spontaneity implicates "bottom-up" information processing (Yaniv 2018). Therefore, spontaneity games help in blanking your mind and train players to recognize the issues they subconsciously think about. Accordingly, six games were assigned as shown in (Table 1.1) (Improv Encyclopedia 2017; Farmer 2012). Students were asked to play at least two games in each category. No evidence that games in the same category have different effects.

1.2.2 Data Gathering Tools

Data gathering and assessment tools were assigned to reach the study goal as follows: brainstorming sessions to test divergence components by measuring fluency, flexibility and originality skills using paired sample t-Test, questionnaire and observation.

Brainstorming sessions (divergence skills). Measuring skills of fluency, flexibility and originality were pursued following the content analysis of related studies in how to test divergent thinking. Calculations of these skills were accomplished as assessed with the "Unusual Uses Test" (UUT) or "Alternative Uses Test" developed by Guilford (Vidal 2006; Hamza and Hassan 2016), as follows:

- *Fluency* which is the total number of ideas generated.
- *Flexibility* is the ability to traverse boundaries and make far associations. This is measured by the number of different classes or categories of the generated ideas.
- *Originality* measures how different and new the ideas are, in comparison with the results of other groups. However, it is important to be cautious that novel idea may be ridiculous. Thus, the idea must be appropriate, valuable in some way to some community (Sawyer 2006). *Originality* is crucial but should be complemented by fit, namely, *originality* and appropriateness (Runco and Jaeger 2012).

Despite considering elaboration one of the divergence components, this study excluded this skill since it measures the amount of the details associated with the idea for further harvesting and treatment (developing phase). In this experiment, the focus was only on ideation phase of the CPS and did not extend to developing and implementation phases (Zeng et al. 2011).

Questionnaire. After completing the two experiences, students were required to complete a questionnaire asking about the effect on their performance in the second experiment when a play session took place. The questionnaire asked about rating the efficiency of playing the assigned games on fluency, flexibility and originality-appropriateness in terms of five-point Likert scale (to clarify, for fluency the question is as follows: Play has an impact on your fluency: 5 strongly agree, 4 agree, 3 neutral, 2 disagree, 1 strongly disagree).

Observation. Observation notes were written while following-up during the sessions. This kind of researches could not be understood as completely quantitative but also as qualitative. Research in human science always has increasing challenge in explaining human understanding and behavior in simply measurable terms (e.g., "How often" versus "why" do people understand and behave in a specific manner) (Berg and Lune 2001). Thus, the author was eager to write down all comments regarding students' behavior and reactions during the whole experiment.

1.2.3 Data Analysis Method

This study introduced "descriptive statistics." It is a case study research with a qualitative intention that aims at exploring the impact of improvisation activities on ideation. The question of whether or not the implementation of the same method is predicted to lead to the same results in future cases, does of course require quantitative analysis by comparing samples of a number of classes rather than focusing on one class, in one academic year, in one school of architecture.

1.3 Findings

The findings present the result of investigating the model Brainstorming-Improvisation in terms of the data gathering tools. The evaluation of improvisation activities as a preliminary phase of creativity sessions on fluency, flexibility and originality-appropriateness reveals the following results.

1.3.1 Divergence Components

A statistical inference method called "hypothesis test" is applied on the samples' mean of the data. If the difference between two mean scores is significant, then there is strong evidence of the variation. For paired t-Test, a sample size of thirty groups ($n = 30$) was formed to examine whether or not there has been significant change in its mean scores before and after improvisation activities for fluency, flexibility, and originality-appropriateness. To proceed with "hypothesis test," it is important to carry out sampling shape test. By checking histograms for the collected data for the randomly formed sample of fluency, flexibility and originality-appropriateness before and after improvisation activities, the result showed bell-shaped distribution curve which is a credibility condition to carry out the t-Test.

In spite of the positive sign when embedding improvisation activities in brainstorming sessions in the calculated mean (M), standard deviation (SD), variance (V), coefficient of variation (CV), in the three cases before and after (Tables 1.2 and 1.3), the variation in observed data may be simply due to a chance alone. Therefore, t-distribution test is used for assessing statistical differences between groups to determine if they came from the same population (null hypothesis) by constructing a confidence interval for the true mean to determine if the null hypothesis should be rejected.

Table 1.2 Fluency, flexibility and originality before improvisation

	Fluency	Flexibility	Originality
μ	22.63	8.60	3.87
SD	9.69	2.18	2.00
V	93.90	4.73	3.98
CV	0.43	0.25	0.52

Table 1.3 Fluency, flexibility and originality after improvisation

	Fluency	Flexibility	Originality
μ	24.67	10.03	5.37
SD	10.88	2.97	1.87
V	118.44	8.79	3.48
CV	0.44	0.30	0.35

Paired t-Test. In a paired sample t-Test, each entity is measured two times (before and after), resulting in pairs of observations. In this test, there are two competing hypotheses, the null hypothesis and the alternative hypothesis (research hypothesis). The null hypothesis and the alternative hypothesis are two statements used to express whether or not the different samples are coming from the same population parameter (Levine and Stephan 2014).

The null hypothesis assumes that the true mean difference between the paired samples is zero. Under this model, observed differences are due to chance and random variability as long as the hypothesized population mean difference ($\mu 0 = 0$) falls within the μd sample confidence interval (CI). Interchangeably, the alternative hypothesis proposes that the true mean difference between the paired samples is not equal to zero. With a specific degree of certainty, the hypothesized population mean difference ($\mu 0 = 0$) is not included within the μd sample confidence interval estimate (CI). The alternative hypothesis can adopt one of many forms depending on the expected result. If the direction of the difference does not matter, then a two-tailed hypothesis is used. Otherwise, an upper-tailed or lower-tailed hypothesis can be applied to escalate the power of the test. The null hypothesis stands the same for each type of alternative hypothesis. The paired sample t-Test hypotheses are specified as follows (Levine and Stephan 2014):

- The null hypothesis (H0) assumes that the true mean difference (μd) is equal to zero.
- The two-tailed alternative hypothesis (Ha) assumes that μd is not equal to zero.
- The upper-tailed alternative hypothesis (Ha) assumes that μd is greater than zero.
- The lower-tailed alternative hypothesis (Ha) assumes that μd is less than zero.

In this research, the last case was used (left-tail test) as follows:

Research hypothesis (alternative Hypothesis): improvisation increases divergence components scores
$H0 : \mu d \geq 0\ (\mu 0)$
$Ha : \mu d < 0\ (\mu 0)$
Where, μd = (Sample before − Sample after), $\mu 0$ = hypothesized population mean.
Using t-Test in Excel, variable 1 = sample before & variable 2 = sample after.

Using t-Test in Excel, variable 1 = sample before & variable 2 = sample after.

However, in this test, it is essential to determine significance level alpha "α" that indicates the probability that we wrongly reject the null hypothesis (type I error). Alpha value in this research is 0.05 (default value); this means that the confidence level of non-rejection region for null hypothesis (H0) is 95%, and degree of freedom is 29 (df = no. of groups − 1). Accordingly, comparing findings of fluency, flexibility and originality-appropriateness revealed the following results:

Fluency: Because the p-value (observed level of significance) for this test, 0.195, is greater than the defined level of significance $\alpha = 0.05$, the test failed to reject the null hypothesis. Also, the absolute value for t-Stat < critical value of t (t-Stat = |−0.874| < t = 1.699) and the hypothesized population mean ($\mu 0 = 0$) is included within the sample confidence interval (samples difference mean ± Margin of Error|−2.03 ± 4.769), which all emphasized the failure to reject the null hypothesis (H0) that assumes $\mu d \geq 0$ ($\mu 0$). Therefore, the research hypothesis in *fluency* is not accepted (Table 1.4) and (Fig. 1.1).

Flexibility: In this case, the *p*-value for this test, 0.012, is less than the level of significance $\alpha = 0.05$, and thus, we reject the null hypothesis. Also, the absolute value for t-Stat > critical value of t (t-Stat = |−2.366| > t = 1.699) and the hypothesized population mean ($\mu 0 = 0$) is not included within the sample confidence interval (−1.43 ± 1.24), which accentuated the rejection of the null hypothesis (H0) where $\mu d < 0$ ($\mu 0$). Therefore, the research hypothesis in flexibility is accepted (Table 1.5) and (Fig. 1.2).

Originality-Appropriateness: In this case, the *p*-value for this test, 0.001, is less than the level of significance $\alpha = 0.05$, accordingly, the null hypothesis is rejected. Also, the absolute value for t-Stat > critical value of t, (t-Stat = |−3.419| > t = 1.699), and the hypothesized population mean ($\mu 0 = 0$) is not included within the sample confidence interval (−1.5 ± 0.9), then it is clear that the null hypothesis (H0) where $\mu d < 0$ ($\mu 0$) should be rejected. Therefore, the research hypothesis in originality-appropriateness is accepted (Table 1.6) and (Fig. 1.3).

Table 1.4 Fluency: paired t-Test

	Fluency_Before	Fluency_After
Mean	22.633	24.667
Variance	93.895	118.437
Observations	30.000	30.000
Pearson correlation	0.237	
Hypothesized mean difference ($\mu 0$)	0.000	
df	29.000	
t-Stat	−0.874	
P(T <= t) one-tail	0.195	
t Critical one-tail (absolute value)	1.699	
μd Confidence interval (Samples difference mean ± margin of error)	−2.03 ± 4.77	

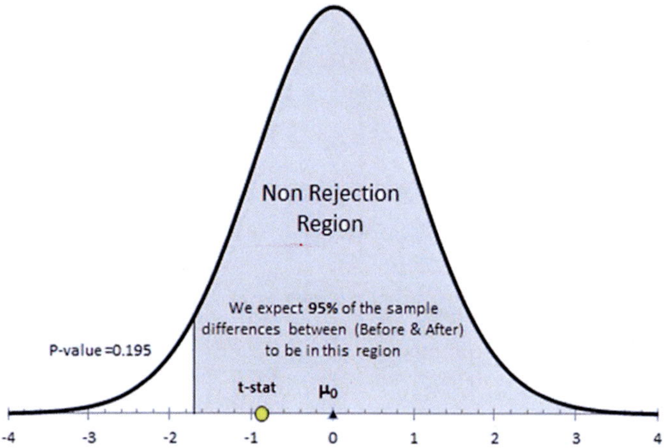

Fig. 1.1 Fluency: paired t-Test

Table 1.5 Flexibility: paired t-Test

	Flexibility_Before	Flexibility_After
Mean	8.600	10.033
Variance	4.731	8.792
Observations	30.000	30.000
Pearson correlation	0.195	
Hypothesized mean difference (μ0)	0.000	
Df	29.000	
t-Stat	−2.366	
P(T <= t) one-tail	0.012	
t Critical one-tail (absolute value)	1.699	
μd Confidence interval (Samples difference mean ± margin of error)	−1.43 ± 1.24	

1.3.2 *Questionnaire*

The questionnaire findings of the 89 participants revealed that students became able to generate more diverse ideas, remote associations and novel production than the first experience (Fig. 1.4).

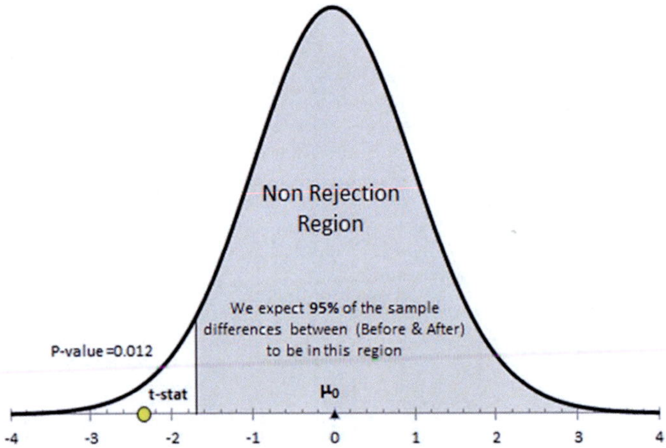

Fig. 1.2 Flexibility: paired t-Test

Table 1.6 Originality: paired t-Test

	Originality_Before	Originality_After
Mean	3.867	5.367
Variance	3.982	3.482
Observations	30.000	30.000
Pearson correlation	0.227	
Hypothesized mean difference (μ0)	0.000	
Df	29.000	
t-Stat	−3.419	
$P(T <= t)$ one-tail	0.001	
t Critical one-tail (absolute value)	1.699	
μd Confidence interval (Samples difference mean ± margin of error)	−1.5 ± 0.9	

1.3.3 Observation

Focusing on students' behavior during and after playing revealed positive clues in specific groups. These clues appeared in their voices' levels and tones, the postures and gestures of their bodies while playing, and when they started their ideation session. Participants of groups, who were vigorously playing, actually showed better response in the ideation session as indicated divergence components scores.

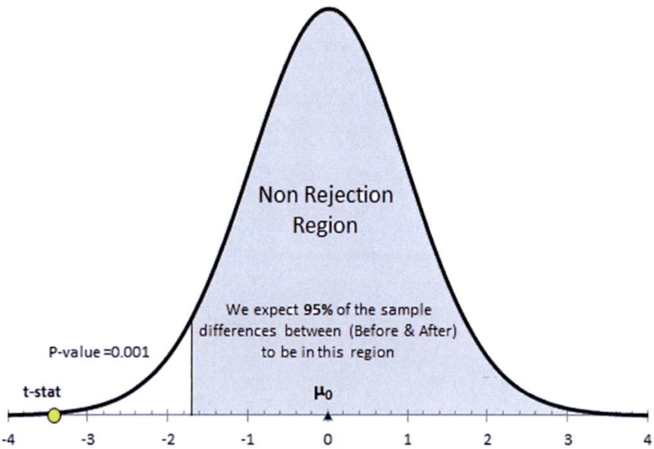

Fig. 1.3 Originality: paired t-Test

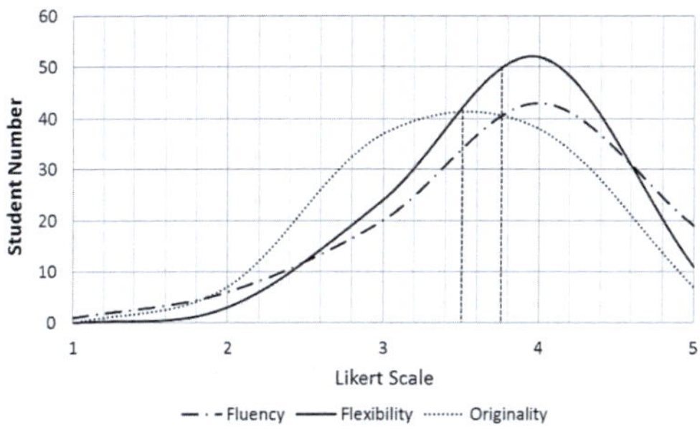

	M	V	SD	CV
Fluency	3.8	0.78	0.88	0.23
Flexibility	3.8	0.48	0.69	0.18
Originality	3.5	0.56	0.75	0.21

M mean
V variance
SD standard deviation
CV coefficient of variation

Fig. 1.4 Questionnaire result of fluency, flexibility and originality-appropriateness

1.4 Discussion

In the light of the literature and research aim, findings will be discussed in terms of the effect of improvisation on creative person/mindset, creative press and creative process/ideation excluding creative product, as it was not required to proceed in development and implementation phases (Fig. 1.5).

1.4.1 Creative Person

Improvisation influence on individuals' minds in terms of enhancing divergence components could be explained following the findings of the paired sample t-Test and questionnaire as follows:

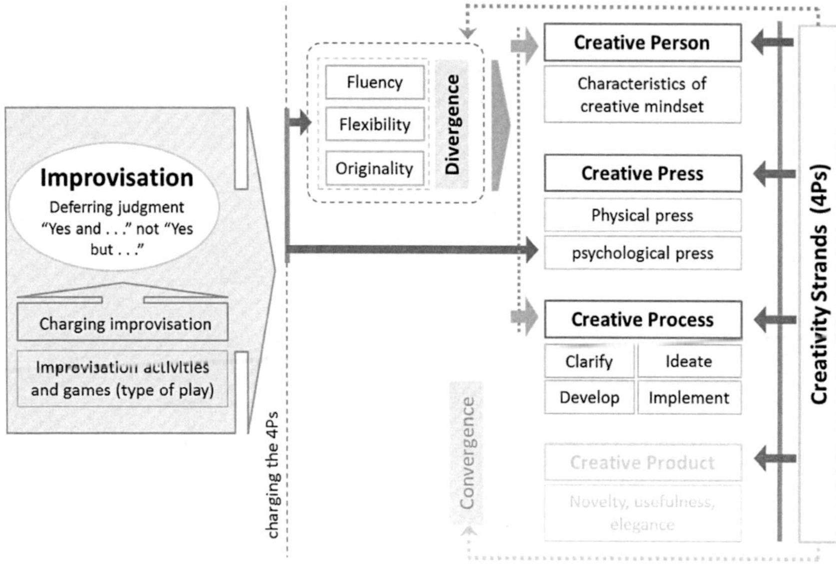

Fig. 1.5 Improvisation impact on the creativity strands (4Ps)

- The results of the t-Test and the questionnaire lead almost to the same findings in both flexibility and originality-appropriateness, and these will be interpreted as follows:
  - Improvisation activities as a play creates a state of mindfulness, where students focus on specific goals within a climate of anticipation and pleasure, i.e., being in the present moment that unfolds its possibilities, far from distractions, which in turn enhances cognitive flexibility.
  - Warm-up and spontaneity activities actually play on the rule of saying "yes"; i.e., agreeing on illogical thoughts. This affects originality in addition to fluency and flexibility. Students commented that "such activities affect crazy ideas to emerge!"
- The results of the t-Test and the questionnaire were different in the case of fluency. While questionnaire results showed positive effect of improvisation on fluency, the findings of the t-Test revealed no evidence of this improvement. However, these findings could be discussed as follows:
  - Improvisation activities open the door for students to improvise by putting individuals in unexpected situations but within a playful press, which gently forces them to immediately concoct ideas to keep playing. This means increasing fluency. This skill of fluency was clearly mentioned in students' comments in the questionnaire. They remarked that such activities affect ideation in terms of speed and quantity. They mentioned they became more open to accept ideas, feeling that there is always a chance to make it real. Contrary to the result

of the questionnaire and conforming to the result of the t-Test, improvisation has no proved impact on fluency. Specific students' comments might bring out explanations to this conflict. They mentioned that despite the similarities of the design task in both experiments, the potentials in each focus were slightly different; then, this might affect the result much more than play. A number of students mentioned that they did not understand the objective of the play session. Concerning the latter reason, hiding the aim was planned to avoid any biased or subjective thoughts or feelings during the experiment and, in turn, on the result. Other students remarked that play consumed their mental power, which in turn affected ideation. These reasons may lead to different results. However, further research could explore more aspects or relations.

1.4.2 Creative Press

As explained previously, the press/climate concept is both physical and the psychological. The latter is the concern of this research. Starting from this point, the findings could be explained as follows:

- Improvisation activities as a kind of play boost the experience of being and immersing in the events with all senses. It creates a clear state of mind far from pre-coined thoughts (Hassan 2019).
- Play experience in terms of improvisation activities creates a climate/press of interest, discovery and pleasure. Namely, improvisation activities create inner inspiring press.

Accordingly, improvisation activities generate a state of mindfulness that affects being in the present moment discovering its potentials, and then, it opens our mind to generate numerous and novel ideas.

1.4.3 Process

As explained earlier, CPS includes four phases, clarification, ideation, development and implementation, but the focus here is only on ideation. In this phase, individuals should come up with as much ideas as possible. Therefore, relevant to discussing findings in terms of the creative mindset in the previous section, these skills of divergent thinking were dramatically increased by improvisation activities. Accordingly, carrying out ideation phase of the CPS was positively affected.

1.5 Conclusion

This research explored the effect of improvisation activities on students' mindsets, creative press and consequently on the experience of the ideation phase of the creative problem solving (CPS)/creative process, while engaging in an architectural design task. Literature in this area showed strong relation between improvisation and creativity. Therefore, this study was an experiment trial to check the improvisation efficiency in architectural design. The study tackled brainstorming sessions, questionnaire and observation as data gathering tools. Findings of this study revealed that improvisation has an impact on divergence skills of fluency, flexibility and originality-appropriateness. The positive impact on these skills assured that improvisation, on one hand, influences creative person/mindset as a result of enhancing divergent thinking, which in turn affects ideation phase of the CPS. On the other hand, improvisation creates psychological press; it develops a state of interest, discovery and pleasure. Reaching such mind status that has powerful impact on creativity is in fact harder to be achieved than setting a creative physical environment. However, in the case of fluency, the results of the brainstorming sessions and questionnaire were not consentient. This was explained by specific but not certain issues. Accordingly, further researches could tackle more investigation via different target groups and using other data gathering tools in order to discover more secrets about improvisation–creativity association.

References

Berg BL, Lune H (2001) Qualitative research methods for the social sciences, 4th edn. Pearson Education, Limited, Massachusetts

Farmer D (2012) 101 More drama games and activities. Drama Resources. ISBN: 978-1-291-02516-3. www.dramaresource.com

Hamza TS, Hassan DK (2016) Consequential creativity: student competency and lateral thinking incorporation in architectural education. Int J Technol Des Educ 26:587. https://doi.org/10.1007/s10798-015-9321-4

Hassan DK (2019) Creativity trilateral dynamics: playfulness, mindfulness, and improvisation. Creativity Stud 12(1):1–14. https://doi.org/10.3846/cs.2019.4313

Improv Encyclopedia (2017) Retrieved from https://improvencyclopedia.org/

Larsen H, Bogers M (2014) Innovation as improvisation 'in the shadow'. Creativity Innov Manage 23(14):386–399. Retrieved from https://doi.org/10.1111/caim.12067

Lawson B (2014) How designers think: the design process demystified, 2nd edn. Butterworth Architecture, Oxford

Levine DM, Stephan DF (2014) Even you can learn statistics and analytics: an easy to understand guide to statistics and analytics, 3rd edn. Pearson FT Press, New Jersey

Lewis C, Lovatt PJ (2013) Breaking away from set patterns of thinking: improvisation and divergent thinking. Think Skills Creativity 9:46–58

McWaters V (2012) Unpacking applied improvisation. Beyond the Edge. Retrieved from https://vivmcwaters.com.au/2012/06/04/unpacking-applied-improv/

Miller B, Vehar J, Firestien R, Thurber S, Nielsen D (2011) Creative climate-creativity unbound: an introduction to creative problem solving. Innovation Systems Group, Williamsville, NY

Rhodes M (1961) An analysis of creativity. The Phi Delta Kappan 42:305–310

Ross K (2013) All about Improv, applied Improv, creativity, play, innovation. Spotlight on Marketing. Retrieved from https://www.kayross.com/blog/2013/04/26/all-about-improv-applied-improv-creativity-play-innovation/

Runco MA (2012) The standard definition of creativity. Creativity Res J 24(1):92–96. https://doi.org/10.1080/10400419.2012.650092

Sawyer RK (2006) Explaining creativity: the science of human innovation. Oxford University Press, New York

Shyba LM (2007) The spontaneous playfulness of creativity: lessons from interactive theatre for digital games. In: Situated Play, Proceedings of DiGRA conference, pp 779–782

Sidawi B (2013) Rethinking architectural education: a focus on creativity. AAE conference. Retrieved from https://architecturaleducators.files.wordpress.com/2013/12/sidawi-2013-rethinking-architectural-education-a-focus-on-creativity.pdf

Sowden PT, Clements L, Redlich C, Lewis C (2015) Improvisation facilitates divergent thinking and creativity: realizing a benefit of primary school arts education. In Press: Psychology of Aesthetics, Creativity and the Arts, American Psychological Association APA. Retrieved from https://epubs.surrey.ac.uk/807281/9/Improv%2520in%2520education%2520accepted.pdf

Vidal RVV (2006) Creative and participative problem solving—the art and the science. Informatics and Mathematical Modelling (IMM), Technical University of Denmark

Yaniv D (2018) Trust the process: a new scientific outlook on psychodramatic spontaneity training. Front Psychol 9:2083. https://doi.org/10.3389/fpsyg.2018.02083

Zeng L, Proctor RW, Salvendy G (2011) Can traditional divergent thinking tests be trusted in measuring and predicting real-world creativity? Creativity Res J 23(1):24–37. https://doi.org/10.1080/10400419.2011.545713. Retrieved from https://pdfs.semanticscholar.org/c801/81faad3839f349709a775048bd924e355b2d.pdf

Chapter 2
Biomimetic Approaches in Architectural Design Education

Meral Omar, Shaimaa Kamel, Doaa Hassan, and Sherif Abdelmohsen

Abstract Biomimicry, bio-inspired design, biomimetic design are all terms used to describe the imitation of models, systems, and elements of nature for the purpose of solving complex human problems. Biomimetic approaches have recently been implemented in design thinking methods due to the global need to utilize technological advances and to meet competencies in all design-based disciplines. Many scholars have incorporated biomimetic design in architectural education to meet those competencies especially for enhancing creativity and complex problem solving. Although several studies have explored the adoption of biomimicry in design, and especially in architectural design education, almost none have developed a clear framework for this purpose. The most famous of those studies are the BID class at Georgia Institute of Technology held in 2011 and the ITECH approach at Stuttgart University carried out in 2018. This paper aims at establishing a framework of criteria for adopting biomimicry in architectural design education. This is achieved by means of a comparative analysis of current architectural design programs and experimental studies incorporating biomimetic design into their curricula in junior through senior levels. The previous mentioned experimental studies and other selected studies were reviewed and analysed, and their impact on students' creativity was evaluated. Despite their varied approaches in teaching and varied tools in assessing the output, they all corroborated that biomimetic design functioned as a catalyst for creating a vibrant studio environment for design education and helped students in understanding the process of creative thinking and complex problem solving. The paper concludes that biomimetic design is interdisciplinary in nature and could be integrated in design curricula specifically at both junior and senior levels for the purpose of enhancing creativity. It is expected that the findings of this research would stimulate additional

M. Omar (✉) · S. Kamel · D. Hassan
Faculty of Engineering, Ain Shams University, Cairo, Egypt
e-mail: meral_omar@hotmail.com

S. Kamel
e-mail: shaimaa_kamel@eng.asu.edu.eg

S. Abdelmohsen
The American University in Cairo, Cairo, Egypt
e-mail: sherifmorad@aucegypt.edu

Architecture and Urbanism: A Smart Outlook,
https://doi.org/10.1007/978-3-030-52584-2_2

interest in the area of biomimicry and design thinking and contribute to developing a database of instructional resources, as well as new and effective teaching methods, to enhance the pedagogy of biomimetic design in architectural education.

Keywords Biomimicry · Bioinspired design · Biomimetic design · Architectural design education · Creativity

2.1 Introduction

Creativity and complex problem solving have been identified by many scholars as meta-competencies facing architectural design education (Nagel et al. 2016). That is why many education concepts like thematic teaching, flexible study and learning-through-research have emerged to face these competencies. Biomimetic design courses lie within those education concepts, coupling a highly flexible, self-tailored, research driven and experiment-based educational framework within a truly trans disciplinary environment (Siddique et al. 2012).

In the middle of the twentieth century, the idea that new technologies can benefit from biological knowledge pervaded the scientific community and began to consolidate as a distinct domain of research and application. Biomimicry is an intended emulation of nature life solutions for solving contemporary challenges (Benyus 1997). Analogies are one of the oldest and most efficient ways to enhance innovation. Repeated findings relate analogy to innovation. Breakthrough innovations are more likely to result from far analogies between distant domains. Between-domain analogies yielded more original ideas than within-domain analogies. Based on these findings, analogies used in biomimetic design are far and between fields and therefore have high potential of innovation. Indeed, class studies in bio-inspired design found that innovation improved by analogical reasoning that link biological functions to engineering challenges (Helfman and Reich 2016).

Biomimetic design concepts, methods, and tools have been used by many institutions to educate students on design innovation as another source of design inspirations (Eggermont et al. 2013). Institutions include Georgia Tech (Centre for Bio-inspired Design), Stuttgart university (the ITECH approach), James Madison University, Clemson University, Texas A&M university, Oregon State University, Arizona State University, University of St. Thomas, Duke University, University of Delaware, and others that have been engaged in innovative research and developing educational materials related to biomimicry. The research aims at establishing a framework of criteria for adopting biomimicry in architectural design education by reviewing and analysing those educational experiments.

2.2 Methodology

Architectural design programs—incorporating biomimetic design method into their educational curricula—were reviewed and analysed through a comparative analysis highlighting the most explicit aspects of those programs (Fig. 2.1).

Those aspects include:

- Context.
- Goal.
- Discipline.
- Assessment tools.
- Output/product.

Case studies were selected following a certain criterion where the ITECH and BID programs were the most significant for their interdisciplinary environments and outstanding outcomes. Other case studies were most relevant to our study. They all focused on learning by doing and teaching how to use inspiration by nature in the design field and the features that can be used. The selected case studies are as follows:

- The ITECH program: Stuttgart university
- BID course at Georgia Tech (Centre for Bio-inspired Design)
- An engineering elective course at Texas A&M University
- The "Arthropods-Architectural Articulation" at the department of Architecture, Turkey.

Based on the findings of the case studies, a synthesis methodology will be used to establish a framework for integrating biomimetic design method in architectural education.

2.2.1 *Case Study 1: The ITECH Program: Stuttgart University*

The Master of Science program (ITECH) offered at the University of Stuttgart provides an international educational environment open to students from a variety of

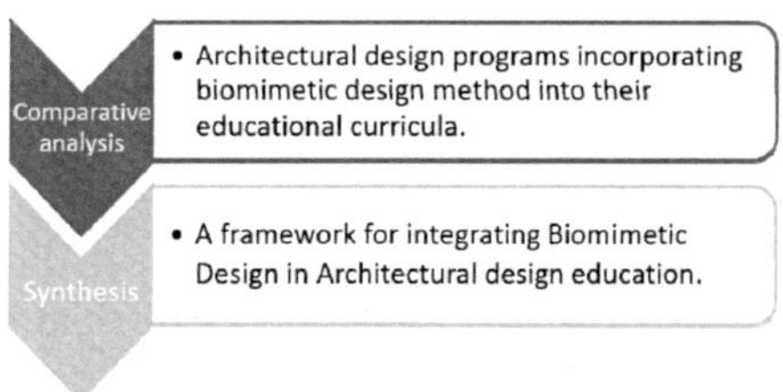

Fig. 2.1 Research methodology

disciplines. ITECH has a strong focus on interdisciplinary skills and offers an integrative approach by merging the fields of architecture, design, engineering, construction, natural sciences, and biology in a research-led teaching environment. The program was first offered in 2013. Since then, 79 students have graduated from the ITECH Program with a Master of Science degree, and three research pavilions have been built through the collaborative efforts of students and researchers.

Course Structure

During the two-year program, the students work together as a group on a collaborative design and build project reinforcing the benefits of collaborative working.

The first year of the ITECH Program is an Integrative Design Research Studio Project, through which students work in teams to solve key technical aspects of the project—under the supervision of ICD/ITKE researchers and Prof. J. Knippers and Prof. A. Menges—through three distinctive phases as follows:

- Ideation and concept development, tightly coupled with a biomimetic seminar co-taught with the department of Evolutionary Biology of Invertebrates.
- Design and development of each project, this phase involves working with experts from several other fields (such as aerospace engineering, biology, geodesy, and textile technologies) to provide technical knowledge and feedback.
- Construction phase, the pavilions are finalized and built between the second and third semester, enabling the students to collectively apply their prototypical designs through technical development to construction of a full-scale prototype.

The second year is structured around the preparation and development of the individual or group Master Thesis Dissertation. In the third semester, students conceptualize their research topic and develop their proposal from both theoretical and technical perspectives. They then consolidate this thesis argument in the fourth semester through the development of their project (Knippers et al. 2018).

Assessment Tools

Assessment tools included classroom observations for the process and Jury evaluation for the final product.

Educational Benefits of the Program

Combining an intensive, critical, and analytical approach to computational design, simulation, and fabrication processes, the ITECH programme is focusing on challenging the design space boundaries of current contemporary architectural and engineering practice. It seeks to provoke a re-examination of techniques, methods, and theories of design in relation to the fields of engineering, robotics, digital manufacturing, material science, and biology.

This unique model forces the students to step out of their comfort zone and forces them to rethink existing methods and theories. The supporting technical and design curriculum of the ITECH Program as well as the increased knowledge and skill sets of students and researchers has enabled the technical scope of the research pavilions to increase over time. Ultimately, the pavilion studio model exemplifies the potentials of learning-through-research and applied making, as well as being a successful combination of research and education. (Knippers et al. 2018).

2.2.2 Case Study 2: BID Course at Georgia Tech (Centre for Bio-inspired Design)

An undergraduate BID class has been offered five times at Georgia Tech (2005–2009) by Jannete Yen and Micheal Helms. The class is a senior-level undergraduate course, taught once a year, and is available to all third, fourth, and fifth year engineering, biology, and biomedical engineering majors (Vattam et al. 2010).

Course Structure

The BID class provides a practical training in methods and techniques that facilitate the identification and translation of biological principles into solutions for human challenges. Their goals in devising this course were related to both developing an approach that will facilitate BID and addressing concerns about the novelty and utility of current practices in engineering design education. The output of the course is a conceptual design that incorporates biological principles into a device or process as well as an account of how the problem was analysed to facilitate the search for useful biological principles. The final project report must contain a detailed quantitative analysis of the relevant governing principles that points to the feasibility of the proposed solution (Yen et al. 2014).

Assessment Tools

Assessment techniques have evolved during the five-year evolution of the course. They included classroom observations analysis of final portfolios (Helms and Goel 2014).

Educational Benefits of the Program

The BID class helped in improving cross-disciplinary interactions and the use of compound analogy enhanced variation in designs. Student comments express greater satisfaction with final designs (Weissburg et al. 2010).

2.2.3 Case Study 3: An Engineering Elective Course at Texas A&M University

An engineering elective course at Texas A&M University explores the effects of five different methods for biomimetic design on the quality and quantity of ideas generated, student self-efficacy, as well as student feedback. These methods include directed, case study, and the use of AskNature.org, BioTRIZ, and functional modelling (Durand et al. 2015).

Course Structure

The course was during two different semesters, one year apart. The participants consisted of 32 students in Semester 1 and 42 students in Semester 2. The experiment took place throughout the course of the semesters. The data collected for this experiment was part of the course homework or in-class assignments.

Assessment Tools

The ideas generated by the students were evaluated using three idea generation metrics: quantity, quality, and number of solutions.

- Quantity of Ideas: determines the number of non-redundant ideas provided by one participant.
- Quality of Solutions: a measure of the feasibility of a solution and how well it meets the problem specifications or customer needs.
- Number of Solutions: This metric shows the number of single product solutions provided by each participant for a given problem.

A survey asking "What methods did you find most useful and why?" each student's answer was read, and the methods mentioned in the answers were recorded and counted as one. The reason "why" they chose the methods was taken into account and used as part of the discussion for the effects of each method. The engineering design self-efficacy was measured through a pre-course and a post-course qualitative questionnaires. The difference between the pre-course and post-course self-efficacy forms was analysed (Glier et al. 2011).

Educational Benefits of the Program

The analysis shows that the methods utilized to teach biomimetic design positively affect the students' design confidence, outcome expectancy, and anxiety while also preserving students' high motivation towards engineering design.

2.2.4 Case Study 4: The "Arthropods-Architectural Articulation" at the Department of Architecture, Uludağ University, Turkey

A Basic Design Studio that is informed by "arthropods", as a rich source of inspiration, was carried out in three consecutive weeks during the 2011–2012 Spring Semester. First-year architecture students were expected to offer a solution to a common human problem through the observation of the forms and behaviour of arthropods.

Course Structure

Students studied arthropods' modes of movement, how they have evolved, and the advantages of their segmented bodies. The "Arthropods-Architectural Articulation" exercises all 48 students taking the Basic Design Studio 2 course participated in the exercise in groups of four. In the first week, concepts relating to the topic were discussed. Design work was carried out in the studio environment during the second week. The third and final week were utilized for classroom evaluation of the finalized projects. Students were asked to pick and observe an arthropod to understand its movement, distinctive behaviour and form, and to then use an analogy relating to this arthropod to design a mobile structure, respond to various stimuli, and strive to solve a problem that the students have identified within their intimate environment. The studio instructor functioned more as a coach than a teacher. Students were guided through a design exercise utilizing a solution-based, bottom-up, "biology influencing design" approach (Yurtkuran et al. 2013).

Assessment Tools

The studio instructor provided subtle critique by asking questions and offering feedback. The critique focused on

- The originality and appropriateness of the digital presentation's conceptual and visual content to the topic at hand,
- The expressive quality of the presentation in terms of graphic representation and the appropriateness of the tools selected to convey the intended message,
- The contribution of group members to the verbal presentation and their use of body language,
- The sufficiency of the research carried out. Group members were expected to critique themselves, as well as reply to questions directed by other participants.

Educational Benefits of the Program

The researchers contend that the goals of the exercise were met where the "Reinterpretation of Arthropods in the Context of Architectural Articulation" exercise aided the process of learning from nature which acted as a design mentor. It allowed the students to abstract new movable structures inspired by the arthropods. Ultimately, students improved and excelled in creative problem solving where the process of thinking deeply into nature and abstracting its forms made their ideas more novel, useful, and their designs appeared more matured.

2.3 Findings

Although the case studies are for different courses in different design-based disciplines for different levels of education, yet they are almost similar in the course structure and assessment tools, they all introduced biomimetic design in a seminar by the studio instructor.

The seminar was carried out by providing hints to the students and engaging them in an inquisitive dialogue. They then have an exercise or a workshop to extract biological concepts and apply them to their design challenges through different tools.

They had a common goal of developing innovative solutions through matching biological analogies to design problems, but they had different tools in achieving that goal.

Some worked in an interdisciplinary context with a strong focus on interdisciplinary skills and offered an integrative approach by merging the fields of architecture, design, engineering, construction, natural sciences, and biology in a research-led teaching environment.

Others worked on introducing students to alternative design methods developing soft skills as problem solving, creativity, and multidimensional thinking in a single discipline. They were different in the scope of the exercise and the required product.

They used different assessment tools to measure the impact of applying the new method on enhancing creativity, these tools included classroom observations, analysis of final portfolios, while others used Jury evaluation and surveys. Table 2.1 shows the goals, assessment tools, focus, pros and cons for the four case studies.

2.4 Discussion

The previous analysis of the four case studies showed that, despite their varied approaches in teaching biomimetic design and varied tools in assessing the output, they all corroborated that biomimetic design functioned as a catalyst for creating a vibrant yet relaxing studio environment for design education.

Biomimetic design helped the students in understanding the process of creative thinking and complex problem solving. Students described their experiences as interesting, informative, and enjoyable. Educational benefits included the introduction of students to alternative design methods and multidimensional thinking specifically related to collaboration, creativity, and innovative solutions to open-ended design problems.

Biomimetic design was introduced as a design philosophy that provides examples of how biological systems were used as inspiration for innovative solutions. Focusing on the content (biology knowledge), students learned that biological systems are surprising complex but have attributes that can easily be applied to design problems. Students learned about the two major paths to a biomimetic design, biology-driven, and problem-driven, as well as how analogies are used to assist with transferring the knowledge from biology to engineering. Both of these themes as well as forming cross-domain linkages link to the competency of critical thinking as students had to analyse the information they were finding and manage their own thinking about the information.

Table 2.1 Biomimetic design courses applied in different universities

University/Institution	Stuttgart ICD/ITKE	Georgia Tech. (Centre for BID)	Texas A&M University	Uludağ University, Turkey
Year	(2013–2018)	(2005–2009)	2015	2012
Context	Master program	Third, fourth, fifth year students	Senior-level elective	First year students
Discipline	Interdisciplinary	Interdisciplinary	Mechanical engineering	Architectural engineering
Sample size	80 students	60 students	32 students	48 students
Product/project	Research pavilion	A conceptual design in a device or a process	Different In-class assignments	A movable structure
Goals	*Developing innovative solutions through matching biological analogies to design problems*			
	Collaboration between disciplines		Exploring the effect of five different methods for biomimetic design	Introducing students to alternative design methods
Assessment tools	*Classroom observations for assessing the process*		*Creativity assessment metrics and surveys*	
	Jury evaluation for the final product	Analysis of final portfolios	Idea generation metrics: Quantity, quality, and number of solutions A survey asking "What methods did you find most useful and why?"	The studio instructor provided subtle critique by asking questions and offering feedback

(continued)

Table 2.1 (continued)

University/Institution	Stuttgart ICD/ITKE	Georgia Tech. (Centre for BID)	Texas A&M University	Uludağ University, Turkey
Focus	*Product-oriented (Focused on developing an innovative product)*		*Process-oriented (Focused on improving students soft skills)*	
Pros	The pavilion studio model exemplifies the potentials of learning-through-research and applied making, as well as being a successful combination of research and education	The output incorporates biological principles into a device or process	The methods utilized to teach biomimetic design positively affect the students' design confidence, outcome expectancy and anxiety, while also preserving students' high motivation towards engineering design	Students improved and excelled in creative problem solving where the process of thinking deeply into nature and abstracting its forms made their ideas more novel, useful, and their designs appeared more matured
Cons	The impact of applying the biomimetic design method was hardly measured as the interdisciplinary context had the higher impact on enhancing creativity and developing innovative solutions		Less innovative or matured products that need more development and implementation due to short period workshops focusing more on soft skills	

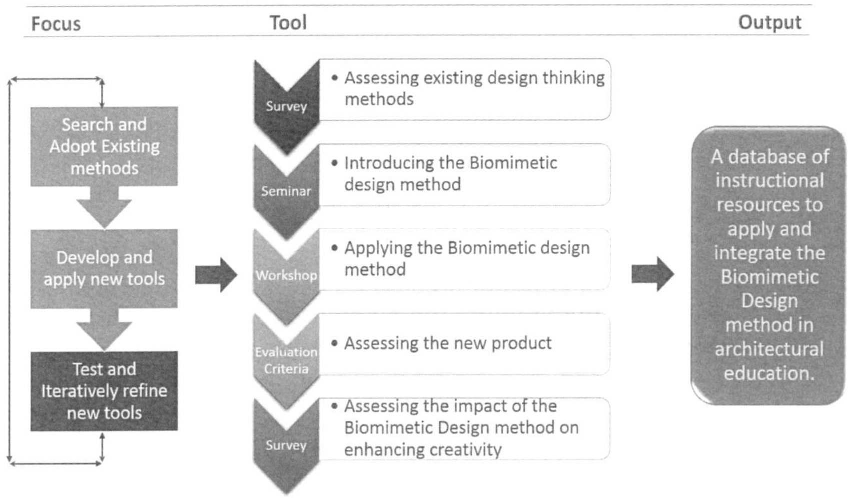

Fig. 2.2 Framework for integrating biomimetic design method in architectural education

Communication across disciplines helped students learning about biological systems by engaging with scholarly resources through independent research, yet the impact of applying the biomimetic design method was hardly measured as the interdisciplinary context had the higher impact on enhancing creativity and developing innovative solutions.

This unanticipated result points towards the significance of teaching biomimetic design in an engineering curriculum. After analysing the previous case studies, we were able to establish a framework for integrating biomimetic design method in architectural education. The framework could be integrated both in the design theory curriculum and in the design studio in order to stimulate students' creative thinking and their ability to solve complex design problems as shown in Fig. 2.2.

The framework would also work as a guide for teaching biomimetic design offering teachers a way to inspire students of all ages by blending biology, creative problem solving, and systems thinking. Teaching materials were also designed to guide students in gradually developing their ideas while completing the initial product design assignment.

The framework presented here will be verified through an empirical case study where architectural design students will be enrolled in an experiment, and their work will be evaluated using a developed assessment criteria.

2.5 Conclusion

It was evident that the biomimetic design had been integrated in architectural design education at various universities and institutions in both graduate and undergraduate curricula. The study showed its significance in exposing students to multiple skill sets, specifically related to collaboration, creativity, and innovative solutions to open-ended design problems. Biomimetic design method proved to achieve various educational benefits as it trains students to transfer knowledge and collaborate across technical and non-technical boundaries while solving engineering challenges. The case studies proved that the biomimetic design is interdisciplinary in nature and could be tailored to fit in many design-based curricula at both junior and senior levels, provided the intensity, scope, and difficulty of the exercise is varied. It is expected that the findings of this research would contribute to developing a database for instructional resources which will enhance the pedagogy of design methods in architectural education.

References

Durand F, Helms M et al (2015) Teaching students to innovate: evaluating methods for bioinspired design and their impact on design self efficacy. In: Proceedings of the ASME 2015 international design engineering technical conferences & computers and information in engineering conference, Boston, Massachusetts, USA

Eggermont M, McNamara C, Nagel JKS (2013) Can biomimicry enhance engineering education? In: 7th Annual biomimicry education summit and 1st global conference. Boston, MA

Glier MW, McAdams DA, Linsey JS (2011) Concepts in biomimetic design: methods and tools to incorporate into a biomimetic design course. In: Proceedings of ASME 2011 international design engineering technical conferences and computers and information in engineering conference, American Society of Mechanical Engineers, pp 655–660

Helms M, Goel A (2014) The four-box method of analogy evaluation in biologically inspired design. In: ASME 2014, International design engineering technical conferences & computers and information in engineering conference. Buffalo, NY

Knippers J, Menges A, Dahy H et al (2018) The ITECH approach: building(s) to learn. In: Proceedings of the IASS symposium 2018 creativity in structural design. MIT, Boston, USA

Nagel J, Pidaparti M et al (2016) Significance, prevalence and implications for bio-inspired design courses in the undergraduate engineering curriculum. In: Proceedings of the ASME 2015 international design engineering technical conferences & computers and information in engineering conference. Charlotte, North Carolina, USA

Siddique Z et al (2012) Competencies for innovating in the 21st century. In: ASME international conference on design education, DETC2012-71170. Chicago, IL

Vattam S, Wiltgen B, Helms M, Goel AK, Yen J (2010) DANE: fostering creativity in and through biologically inspired design. In: Proceedings of 1st international conference on design creativity. Kobe, Japan

Weissburg M, Tovey C, Yen J (2010) Enhancing innovation through biologically inspired design. Adv Nat Sci

Yen J et al (2014) Adaptive evolution of teaching practices in biologically inspired design. In: Goel AK, McAdams DA, Stone RB (eds) Biologically inspired design: computational methods and tools. Springer, New York

Yurtkuran S, Kırlı G, Taneli Y (2013) Learning from nature: biomimetic design in architectural education. In: 2nd Cyprus international conference on educational research (CY-ICER)

Chapter 3
Tracing the Impact of Teaching Sustainable Design on Design Projects in Architectural Freshman Level

Laila Khodeir, Ashraf Nessim, and Dalya Maguid

Abstract Architecture education nowadays is facing changes due to technological developments and the call for more environmentally certified buildings. These changes urge architectural instructors to make alterations either in the contents of their exiting courses, teaching approaches, or methodologies. Among these changes is the introduction of sustainable design courses (SDC) to architecture students (AS). This paper assumes that teaching SDC early on to AS can have a positive impact in terms of leveraging their awareness towards environmental design concepts and integrating such concepts in architecture projects. Thus, this paper aims to investigate the impact of teaching SDCs early on to architecture freshmen students and trace its impact on the output of their design projects. This was achieved through a qualitative case study analysis on a course titled 'Building Services 1: Environmental Control and Design' that is offered to AS in the British University in Egypt. Teaching methods applied were discussed in detail. Furthermore, the responses of the students were collected and analysed through two survey questionnaires. Findings of this paper affirm that introducing SDCs to architecture students in their freshmen year could have a profound impact on their output. Students who actively attended the investigated course were able to implement sustainable design strategies in their design work throughout the studied module and throughout other design modules. Findings of this research can be of great value with respect to designing architecture curriculums in Egyptian universities and can help integrate sustainability concepts early on in students' mindset.

Keywords Architecture students (AS) · Freshmen students · Sustainable design course (SDC) · Architectural design projects

L. Khodeir · A. Nessim (✉)
Faculty of Engineering, Ain Shams University, Cairo, Egypt
e-mail: a.nessim@eng.asu.edu.eg; ashraf.nessim@bue.edu.eg

L. Khodeir
e-mail: drlailakhodeir80@gmail.com; laila.khodeir@bue.edu.eg; leila.mohammed@eng.asu.edu.eg

L. Khodeir · A. Nessim · D. Maguid
Faculty of Engineering, The British University in Egypt, Cairo, Egypt
e-mail: dalya.maguid@bue.edu.eg

Architecture and Urbanism: A Smart Outlook,
https://doi.org/10.1007/978-3-030-52584-2_3

3.1 Introduction and Problem Definition

Since the architectural profession shapes the built environment that we live in (Antonara et al. 2013), the role of architectural education is vital in preparing students on the design of meaningful environments that adhere to users' needs (Salama 2007). However, architectural education nowadays is facing several changes due to technological developments, the call for more environmentally certified buildings, and the globalization of the profession. Thus, it is becoming quite imperative to create awareness amongst future generations of young architects about the importance of sustainable design and the impact of buildings on our environment and on human behaviour (Ismail et al. 2017). Furthermore, integrating sustainability into education helps train students and provide them with the necessary ecological and technological knowledge that is required to create sustainable architectural development (Altomonte 2009; Gucyeter 2016). In addition, there is an increasing demand in the building industry worldwide, for architecture graduates that are proficient in sustainable practices, and commited to their implementation (Torres-Ramírez et al. 2014; Ismail et al. 2017).

While much research has called for the integration of sustainability into architectural curricula, several barriers still exist (Gucyeter 2016). This is due to the lack of interest, awareness, and compatibility in knowledge (Guy and Farmer 2001; Altomonte 2009). Furthermore, there are different views amongst architectural institutions on how and when sustainability should be integrated (Stasinopoulos 2005). Many instructors are concerned that without going in-depth into sustainable principles, architecture schools leave students with broader yet superficial information (Wyckmans 2008). In addition, research indicates that only few schools were able to adapt their curricula to create teaching methods suitable for sustainable design (Elnokaly and Elseragy 2009). Other issues that face the integration of sustainability into architectural education include the fact that there is generally an unclear definition and confusion over the meaning of sustainable architecture, and there is a lack of experts in this subject within the field of education (Gencosmanoglua and Nezor 2010; Taleghani et al. 2011). It was also noted that the teaching content in many schools was directed more towards formal aspects of design than towards environmental problems (Kowaltowski et al. 2005). Furthermore, there is currently an absence of a clear framework that details how sustainability can be integrated into curricula in a way that ensures that students can transfer the knowledge learned into practice (Ismail et al. 2017).

3.1.1 Aim

The main objective of this study is to investigate the impact of teaching courses related to sustainable design principles early on to architecture freshmen students and trace its impact on the output of their following design projects.

3.1.2 Methodology

This study performed a qualitative case study analysis of a course titled 'Building Services 1' that is offered to architecture freshman students in the British University in Egypt. The impact of the course on students' performance was analysed through analysing samples of the students' output in the course and in their following architecture design projects. Furthermore, two survey questionnaires were conducted amongst students who attended the course in order to evaluate how the course affected them. Then, samples of design projects from students who had previously attended the BS (1) course were analysed over the past three years. Figure 3.1 summarizes the research methods applied.

3.2 Teaching Methods in Architecture Education

Research indicates that there are various different teaching pedagogies that can be employed by educators. These can be divided into two main groups; 'student oriented' and 'teacher oriented' approaches, each of which involves several methods and techniques. For example, teaching methods can include 'narrative lectures', 'discussions', 'asking questions', 'sample cases', 'showing samples', and 'problem solving' methods. While teaching techniques can involve 'brainstorming', 'question and answer', 'mimesis', 'pair and group work', 'role playing', and 'creative drama' techniques (Yildirim et al. 2012). Another teaching method that has been widely discussed within the architectural education discipline is 'collaborative learning'

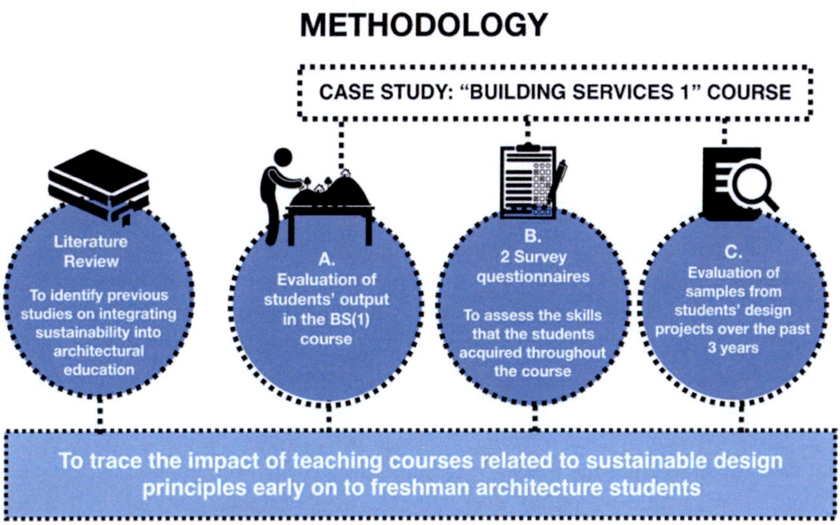

Fig. 3.1 Research methodologies. *Source* Authors

(Emam et al. 2018). Collaborative learning is a teaching strategy where small teams of students work together to finish a task, create a product or design, or solve a problem (Akinciturk 2011). Through this approach, students learn to collaborate together and communicate with each other and with their instructors (Emam et al. 2018). Furthermore, this teaching strategy helps transform students from passive to active participants, allowing them to learn from each other (Salama 2010). In addition, research indicates that students have difficulty integrating things that are taught passively in theoretical and studio modules and that when dealing with sustainability principles, a more active learning and hands-on experimentation approach is recommended (Yuksek 2013).

3.3 Integrating Sustainability into Architectural Education

Recently, sustainable design has been identified as an important aspect that should be integrated into architectural curriculum. Literature suggests that several different approaches of integrating sustainability in architectural education have been applied worldwide, tackling the topic from different aspects (see Yüksek 2013; Altomonte et al. 2014; Karaosman 2011). An example can be seen in a study which presented the experience of a bioclimatic design studio in the University of Campinas, Brazil. In this example, design studio activities were combined with thermal comfort experiments. According to Kowaltowski et al. (2005), environmental comfort is taught mostly in three design courses combining both theory and design practice. An applied physics course is first given to students along with introductory design, followed by discussions on thermal comfort problems. Bioclimatic design however is introduced later on to students starting from the fifth semester of their six-year architecture program. The course studied in this case study however introduced a different approach, where theoretical aspects of thermal comfort were combined with design studio activities, design discussions, lab activities, calculations, case studies, and site visits to sites and buildings. Key conclusions from this study suggest the importance of using models as a source of information regarding assessing the visual and thermal comfort (Kowaltowski et al. 2005). The study also suggests the importance of applying different teaching methods and the use of simulation tools in sustainability related courses.

The importance of simulation tools is also shown in a study by Gaulmynn and Dupre (2019) which presents a new simulation tool to evaluate sustainable performance. This tool is called 'Easy Approach for Sustainable and Environmental Design (EASED)'. Testing took place on a number of case studies by applying it on architectural design exercises for Masters students in Architecture under the guidance of the researcher. A feedback stage at the end of each step of the exercise was carried out to adjust and enhance the tool. The researcher assessed the students' engagement and how they used the EASED. Results revealed that the performance of the students improved during group work but was not satisfactory during individual work. Other studies have also discussed the application of sustainability in architectural curricula in Arab countries (see Salama 2007; Elseragy and Elnokaly 2007).

While a study conducted by Elseragy and Elnokaly (2007) in one of the Egyptian universities (AASTMT) indicated that sustainability-related issues were not introduced to students in their first year, but rather in their 'Architectural Design 1 and 2' modules given in the fifth and sixth semesters, while details on passive design strategies are given in a separate course in the sixth semester (Elseragy and Elnokaly 2007). From the literature reviewed, it is apparent that there are different ways for the application of sustainability in architecture education. It is evident however that the application of different teaching methods, techniques, and simulation tools is quite important when teaching sustainability courses in addition to when and how sustainability courses are introduced to AS.

3.4 Case Study Analysis

This paper performed a qualitative case study analysis of a course titled 'Building Services 1: Environmental Control and Design', offered to AS in the British University in Egypt. First a brief introduction to the course is discussed. This is followed by an outline on the various teaching methods implemented and an analysis for samples of the students' output in the BS (1) course. The BS (1) course aims to introduce students to basic principles and theories that are used and applied for climate control in buildings. This course is taught in the first semester of the first year of architecture, in the British University in Egypt, allowing students to understand sustainability issues and passive design strategies early on. The expected learning outcomes of the course include allowing students to recognize the basic principles of human comfort in buildings, demonstrating principles of passive systems and green architecture, preparing and applying measurements, providing detailed analysis of environmental factors in buildings, and highlighting the importance of climate and its impact on architectural design.

3.4.1 Teaching Methods

Throughout the BS (1) course, multiple teaching methods were employed. These were divided to three main aspects, lectures and in-class sessions, a group model assignment, and a field trip. Lectures and in-class sessions involved weekly classes during which students were introduced to key passive design strategies. Teaching techniques used during the lectures included brainstorming, question and answer, discussions, and showing multiple case studies for each passive strategy discussed. Short videos were also shown during class to help engage the students more and help provide information in an easy to consume manner.

The group model assignment allowed students to practice and implement key passive strategies studied through the design and implementation of a physical model. Students were required to design a prototype building unit that reflected the different passive design strategies taught in class. The model-making assignment involved a collaborative learning approach, where students worked together in class and at home, throughout the duration of the course, to develop their final outputs (Fig. 3.2). Each group was composed of about six students that met with staff members on a weekly basis during tutorial sessions. This allowed students to receive direct one on one feedback on their models and receive possible suggestions on how to enhance their work more. In addition, a one-day model-making workshop was done where students were divided to three groups, and each group worked on designing and completing a model for one of three buildings either a library, a community centre, or a gymnasium.

Lastly, the field trip aimed to show students real-life examples of the strategies discussed in class. The trip involved a site visit to the Sekem Sustainable Community in Cairo. The Sekem community was founded in 1977 with the aim of adapting the vision of sustainable human development. In the field trip, students saw several examples of heat gain prevention strategies, shading devices, courtyards, daylighting strategies, building form and layout, double glass windows, the use of natural building materials, and examples of water features and landscape design elements (Fig. 3.3).

Fig. 3.2 Groups of students working together in class during tutorial sessions. *Source* Authors

Fig. 3.3 Field trip to Sekem Sustainable Community. *Source* Authors

3.4.2 Output of the BS (1) Course

A sample of some of the students' model outcomes for three consecutive academic years is analysed in this section. Students were required to create a 1:50 physical model with the application of various passive strategies. During the academic year of 2016/2017, students were asked to redesign the nursery located in the British University in Egypt. While for the academic year of 2017/2018, students were required to design a prototype of a house unit. For the academic year of 2018/2019, the green school theme was adopted. In all years, the students successfully managed to apply different strategies in a hands-on and creative manner. Examples of strategies implemented included the application of double walls, courtyards, vaults, double glazed windows, and different landscape features, (Fig. 3.4). Examples of products in the academic year 2017/2018 and 2018/2019 are shown in Figs. 3.5 and 3.6. Strategies implemented included the application of heat gain prevention strategies such as double walls with insulating material in between, double glazed windows, the design and correct allocation of shading devices, building layout and form, and landscape elements and water features. Strategies for natural ventilation were also implemented such as cross ventilation, correct allocation of openings, use of wing walls, and the use of landscape features to help direct wind movement (Fig. 3.5). In addition, multiple roof options were also developed including the use of vaults, domes, courtyards,

Fig. 3.4 Samples of students' model outcomes for the academic year of 2016/2017. *Source* Authors

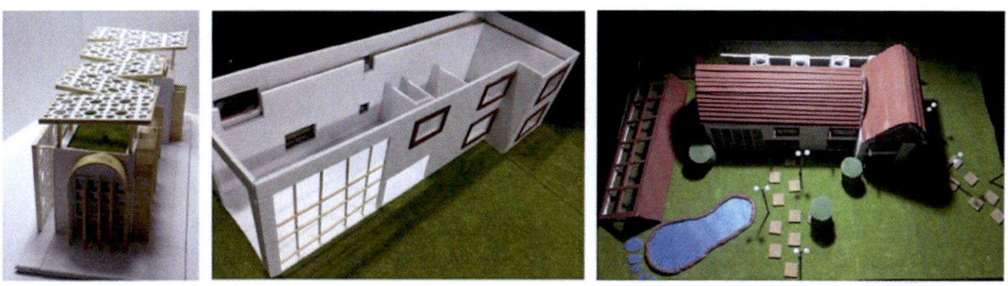

Fig. 3.5 Samples of students' model outcomes for the academic year of 2017/2018. *Source* Authors

Fig. 3.6 Samples of students' model outcomes for the academic year of 2018/2019. *Source* Authors

skylights, and green roofs, and the application of different vertical and horizontal shading devices on the building and in the landscape and surrounding site is shown in Fig. 3.6.

3.5 Discussion and Findings

Based on the literature review, it is evident that there is a clear connection between the skills that students acquire from sustainability courses and their output in design projects. This section of the paper will evaluate the impact of integrating sustainability courses early on to AS in terms of the students' perspectives and in terms of the application of the teaching methods previously mentioned. This will be discussed through two parts; first, an analysis is done for the two survey questionnaires that were conducted to identify students' perspectives. The first questionnaire was distributed amongst first year students who had just attended the course. The second questionnaire was distributed amongst students who had previously attended the course over the past four years. Second, the impact of the course on students' design project outputs was traced through analysing samples of students' design projects. The projects were analysed in terms of the presence and correct application of sustainable design strategies that were taught during the course.

3.5.1 Evaluating the Course's Impact from Students' Perspective

Students' perspectives were evaluated through two survey questionnaires that aimed to assess the course contents, methodology, and teaching methods. The questionnaires also aimed to evaluate the impact of the course on the students' perception regarding environmental design issues. The first questionnaire was sent to DY (1) cohort with a total of 86 students and a response rate of 35%. While the second questionnaire targeted DY (2, 3, and 4) cohorts with a total of 82 responses at almost 33.33% response rate each. In terms of the students' perspective about the course in general, results indicated that 66% of the respondents from the first questionnaire were satisfied about the fact that the course matches well with the nature of the architectural department. In addition, 89% of the respondents from subsequent years asserted that they thought the course should be taught in the early years, in DY (1). The early introduction of the course also proved to be of value to the respondents as 42% of DY (1) respondents approved that they would use the strategies learned in their future projects (Fig. 3.7). A total of 64% of the respondents from the second questionnaire indicated that they actually recalled a lot of the information that they took in the course at 64%.

Regarding teaching methods, 80% of the participants in the first questionnaire were satisfied with the methods of teaching used. Meanwhile, 46% of the participants were satisfied with the methods of practice used during the tutorials. Finally, 53% were completely satisfied with the field trip. Table 3.1 summarizes the students' evaluation of the applied teaching methods. The two questionnaires also addressed the various skills that the students acquired from the course, where 50% of DY (1) students agreed that the BS (1) course helped improve their model-making skills, while 46% agreed that it improved the way they thought about buildings, and 43% agreed that it helped improve their team work skills. Participants were divided equally on the course's impact on developing their communication skills (Table 3.2). Partic-

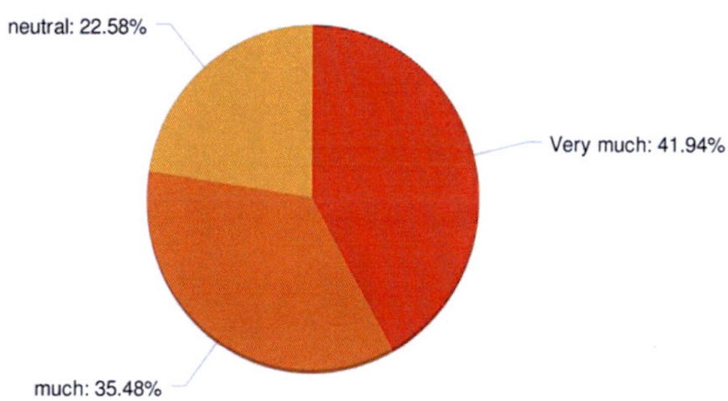

Fig. 3.7 Participants' opinion regarding the possibility of using the learned passive design strategies in architecture design projects (Based on the first questionnaire). *Source* Authors

Table 3.1 Students' opinion regarding the applied teaching methods in the BS (1) course

	Completely dissatisfied (1)		Dissatisfied (2)		Neutral (3)		Satisfied (4)		Completely satisfied (5)	
	Σ	%	Σ	%	Σ	%	Σ	%	Σ	%
Methods of teaching used in lectures	–	–	3x	9.38	3x	9.38	13x	40.63	13x	40.63
Methods of practice used in tutorials	–	–	2x	6.25	5x	15.63	15x	46.88	10x	31.25
Students' poster presentation	2x	6.25	2x	6.25	13x	40.63	8x	25.00	7x	21.88
SEKEM Trip	–	–	1x	3.13	6x	18.75	8x	25.00	17x	53.13

Source Authors

Table 3.2 Students' opinion regarding the impact of BS (1) course on different acquired skills (Based on the first questionnaire)

	Not at all (1)		Little (2)		Neutral (3)		Much (4)		Very much (5)	
	Σ	%	Σ	%	Σ	%	Σ	%	Σ	%
Model making	–	–	1x	3.13	5x	15.63	10x	31.25	16x	50.00
Team work	1x	3.13	2x	6.25	5x	15.63	14x	43.75	10x	31.25
Way of thinking about buildings	–	–	–	–	2x	6.25	15x	46.88	15x	46.88
Presentation skills	5x	15.63	1x	3.13	13x	40.63	8x	25.00	5x	15.63
Communication skills	2x	6.25	2x	6.25	10x	31.25	8x	25.00	10x	31.25

Source Authors

ipants in the second survey indicated that what they recalled in particular from the course was the model making at 65% and lecture content at 19% (Figs. 3.8 and 3.9).

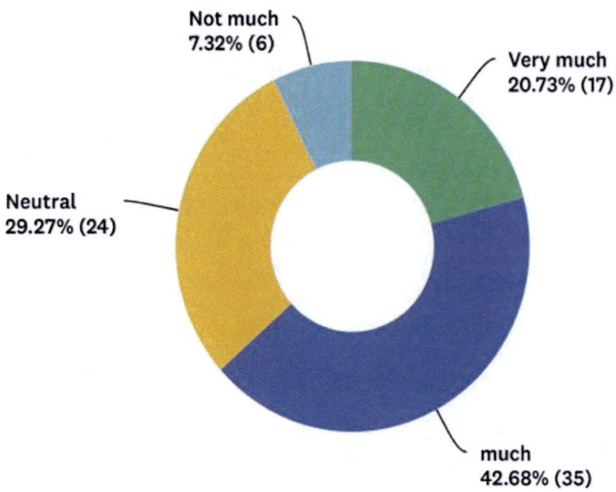

Fig. 3.8 Participants' feedback regarding the level of knowledge and information that they can recall from the BS (1) course (Based on the second questionnaire). *Source* Authors

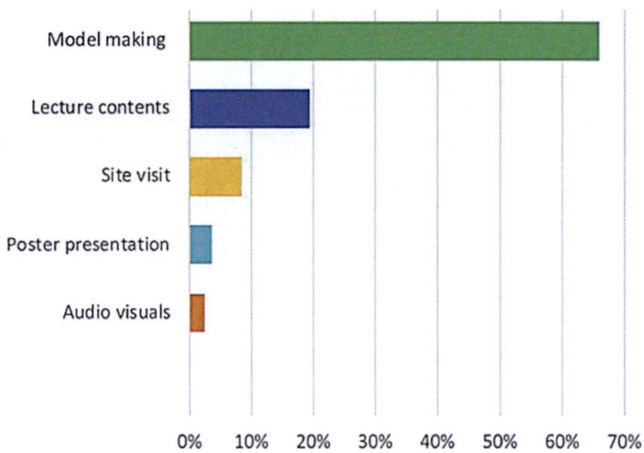

Fig. 3.9 The level of knowledge that students can recall from the course (Based on the second questionnaire). *Source* Authors

Table 3.3 Students' opinion regarding the topics taught in the BS (1) course (Based on the first questionnaire)

	Weak (1)		Fair (2)		Good (3)		Very good (4)		Excellent (5)	
	Σ	%	Σ	%	Σ	%	Σ	%	Σ	%
Identifying green, eco-friendly, sustainable buildings	2x	6.06	2x	6.06	7x	21.21	8x	24.24	14x	42.42
Thermal insulation	1x	3.03	1x	3.03	6x	18.18	10x	30.30	15x	45.45
Design of shading devices	-	-	1x	3.03	6x	18.18	14x	42.42	12x	36.36
Thermal comfort and heat transfer	-	-	2x	6.06	9x	27.27	6x	18.18	16x	48.48
Natural ventilation in buildings	-	-	2x	6.06	9x	27.27	11x	33.33	11x	33.33
Active cooling and HVAC systems	4x	12.12	2x	6.06	11x	33.33	8x	24.24	8x	24.24
Daylighting techniques	4x	12.12	1x	3.03	10x	30.30	9x	27.27	9x	27.27

Source Authors

Concerning the topics and passive strategies that were covered in the BS (1) course, DY (1) students ranked the topics of thermal comfort and heat transfer, thermal insulation in buildings, and identifying environmentally friendly buildings as excellent topics at 48%, 45%, and 42% consecutively. As for the evaluation of active cooling and HVAC systems and daylighting techniques, students ranked them as good with 33% and 30% consecutively. Table 3.3 summarizes the students' evaluation of the taught topics in terms of their efficiency and in terms of students' satisfaction. Participants in the second survey on the other hand indicated that they actually did use many of the passive strategies learned in their design projects. Strategies that were most used included shading devices, daylighting techniques, courtyards, and green walls/roofs at 57%, 50%, 42%, and 40%, respectively. Table 3.4 summarizes the participants' responses.

The two questionnaires also addressed the various skills that the students acquired from this course. 50% of DY (1) students agreed that the course helped improve their model-making skills 'very much', while 46% agreed that it 'much' improved the way they thought about buildings, and 43% agreed that it helped improve their teamwork skills. Participants were divided equally on the course's impact on developing their communication skills, with 31% feeling neutral and 31% feeling that it developed their communication skills very much (Table 3.2). Participants in the second survey indicated that what they recalled in particular from the BS (1) course was the model making at 65% and lecture content at 19% (Fig. 3.9). This asserted that teaching methods that involved hands on and more interactive approaches did indeed leave a greater impact on students.

Table 3.4 Students' feedback regarding the different design features that they used in their projects (Based on the second questionnaire)

	WEAK (1)	FAIR (2)	GOOD (3)	VERY GOOD (4)	EXCELLENT (5)
Double walls/ roofs	12.86% 9	8.57% 6	31.43% 22	31.43% 22	15.71% 11
Green walls/ roofs	1.43% 1	8.57% 6	25.71% 18	24.29% 17	40.00% 28
Coartyards	12.86% 9	8.57% 6	17.14% 12	18.57% 13	42.86% 30
Atria	25.71% 18	15.71% 11	34.29% 24	10.00% 7	14.29% 10
Wind towers	25.71% 18	30.00% 21	18.57% 13	18.57% 13	7.14% 5
Cross ventilation	5.71% 4	11.43% 8	21.43% 15	32.86% 23	28.57% 20
Night ventillation	25.71% 18	20.00% 14	28.57% 20	18.57% 13	7.14% 5
Earth berming	41.43% 29	20.00% 14	22.86% 16	11.43% 8	4.29% 3
Daylighting techniques	4.29% 3	7.14% 5	15.71% 11	22.86% 16	50.00% 35
Shading devices	2.86% 2	5.71% 4	14.29% 10	20.00% 14	57.14% 40

Source Authors

3.5.2 Evaluation of Course Impact on Students' Design Projects

To identify the impact of the BS (1) course on students' design projects, the design output of students who took the building services course during the academic year of 2016/2017 was analysed throughout the subsequent design courses following the module. Figures 3.10, 3.11, 3.12, and 3.13 show examples of the students' design projects in various design courses following the BS (1) course. Many of the passive strategies studied in BS (1) were found to be implemented in the projects shown. For example, the use of courtyards was depicted in the residential villa project as shown in Fig. 3.10a, as well as the use of shading devices and building protrusions and recesses

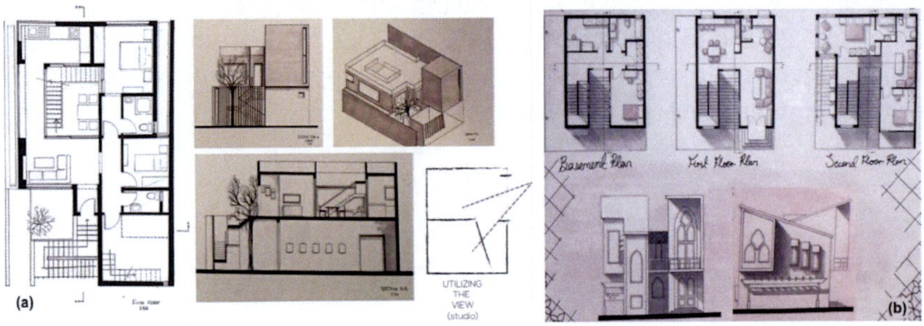

Fig. 3.10 Samples of students' design (1) projects. *Source* Authors

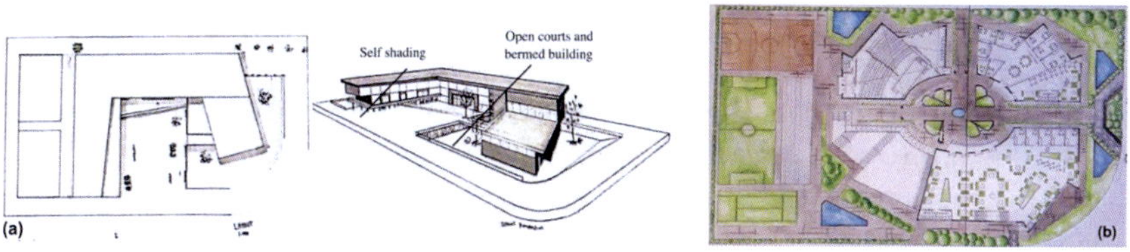

Fig. 3.11 Samples of students' design (2) projects. *Source* Authors

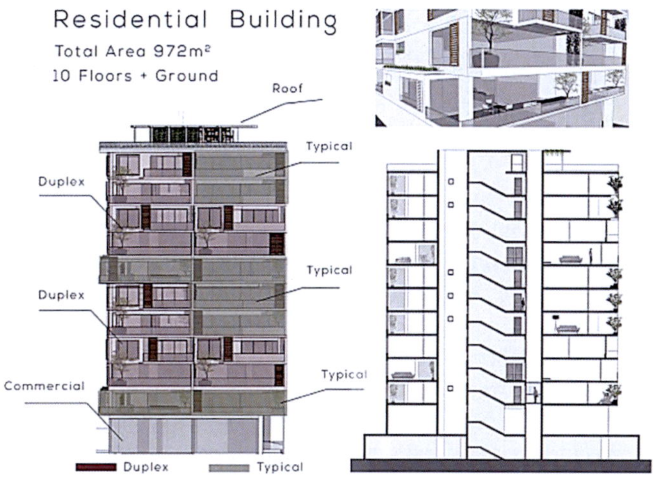

Fig. 3.12 A sample of a design (3) project. *Source* Authors

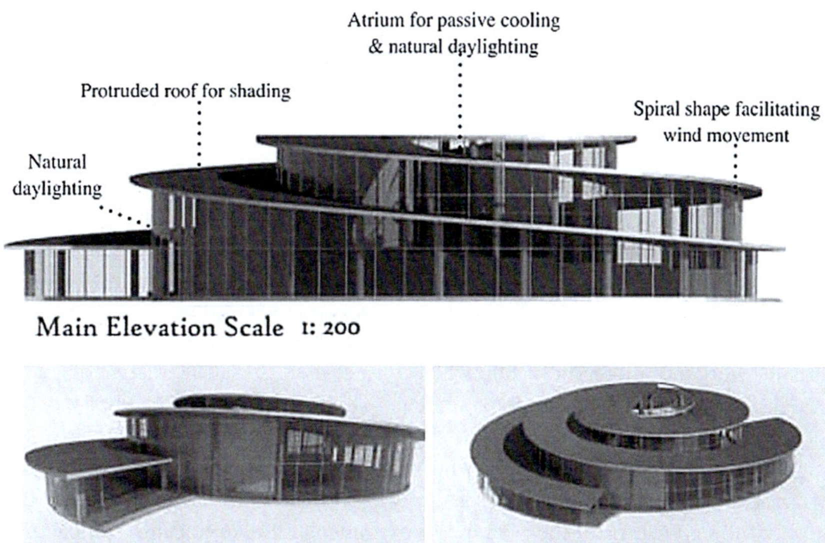

Fig. 3.13 A sample of a design (4) project. *Source* Authors

to create self-shading (Fig. 3.10a, b, respectively). Allocation of windows and openings was also placed to make use of the surrounding views and maximize air flow and ventilation (Fig. 3.10a). Samples of student's design projects for the following semester are shown in Fig. 3.11 depicting samples for the design of a 'Student Centre' project. Examples of passive strategies implemented included the utilization of the buildings' form and layout to create self-shading and utilize the ground space. Roof shading and earth-berming were also implemented (Fig. 3.11a). In addition, other examples of passive strategies implemented in students' design projects include; the use of courtyards, and water features as shown in Fig. 3.11b.

3.6 Conclusions

Achieving the aim of this paper has been fulfilled through first, analysing the literature sources that dealt with the integration of sustainable design courses within the architecture design education in several universities. Second, the application of qualitative analysis of the case study course titled 'Building Services 1: Environmental Control and Design' has been performed. The analysis of the case study offered clear evidence on the impact of teaching such courses on the degree of students' recognition and on their awareness towards the application of suitable design strategies in architecture design projects. Regarding the early integration of sustainability courses to architecture students, the study was able to assert that this was quite beneficial to students in terms of the skills that they acquire and continue to use in their architecture studies. These skills included model making, teamwork, and communication and presentation skills. Particular passive strategies learned in the course also seemed to resonate with students in their coming years. This was evident as they did in fact use the strategies learned in their design projects. Questionnaire findings indicated that the most commonly used strategies included shading devices, daylighting techniques, courtyards and green walls/roofs.

In the end, several suggestions were mentioned by students on how to further improve the course, and that could in turn be tailored into an action plan for improving the teaching process for such courses. Many students suggested that the course could include not only model making but also more practice on the implementation of their adopted strategies in plan and elevation drawings. This could even prepare them more for their coming design modules. Other students suggested for the course to be integrated with all other design modules. Students also requested more practical and hands-on work and more site visits, which further asserts that interactive teaching methods are more successful amongst students. This research traced the impact of the BS (1) course on students who have just attended it, before attending any architecture design modules, and other students who have attended it over the past four years and already produced from 1 to 6 design projects. In addition to evaluating the impact of the course from the perception of students, it also traced its impact on actual design projects of some of the students . Thus, findings of this research have a great value to decision makers in charge of designing architecture curriculums in Egyptian universities and can help integrate sustainability concepts early on in students' mindset.

References

Akinciturk N (2011) Cooperative learning in an architectural design studio. Uludag Univ J Faculty Eng 16(2):35–43

Altomonte S (2009) Environmental education for sustainable architecture. Rev Eur Stud 1(2):12–21. https://doi.org/10.5539/res.v1n2p12

Altomonte S, Rutherford P, Wilson R (2014) Mapping the way forward: Education for sustainability in architecture and urban design. Corp Soc Respons Environ Manage 21(3):143–154. https://doi.org/10.1002/csr.1311

Antonara E, Georgi J, Lianos N (2013) Sustainable architecture through an environmental education program. In: 4th International conference on renewable energy sources and energy efficiency—new challenges. Nicosia, pp 179–188

Elnokaly A, Elseragy A (2009) Amalgamating sustainable design strategies into architectural curricula. International J Environ Cultural Econ Soc Sustain 5

Elseragy A, Elnokaly A (2007) The influence of self-appraisal and international boards validation process upon the architectural education in the Arab Region. In: 24th engineering conference, engineering education in the Arab world, 14–17 May, Jordan, Amman. ISSN 1832-2077

Emam M, Taha D, ElSayad Z (2019) Collaborative pedagogy in architectural design studio: a case study in applying collaborative design. Alexandria Eng J 58(1):163–170. https://doi.org/10.1016/j.aej.2018.03.005

Gaulmynn C, Dupre K (2019) Teaching sustainable design in architecture education: critical review of easy approach for sustainable and environmental design (EASED). Front Archit Res 8(2):238–260. https://doi.org/10.1016/j.foar.2019.03.001

Gencosmanoglua A, Nezor S (2010) Criticizing architecture education through abstraction. Procedia-Soc Behav Sci 9(2010):1335–1341. https://doi.org/10.1016/j.sbspro.2010.12.330

Gucyeter B (2016) The place of sustainability in architectural education: discussion and suggestions. Athens J Archit 2(3):237–256. https://doi.org/10.30958/aja.2-3-4

Guy S, Farmer G (2001) Reinterpreting sustainable architecture: the place of technology. J Archit Educ 54(3):140–148. https://doi.org/10.1162/10464880152632451

Ismail MA, Keumala N, Dabdoob RM (2017) Review on integrating sustainability knowledge into architectural education: practice in the UK and the US. J Cleaner Prod 140((Part 3)):1542–1552. https://doi.org/10.1016/j.jclepro.2016.09.219

Karaosman SK (2011) Options to necessity in architectural education; ecological design. South Archit J 6:63–67

Kowaltowski D, Labaki L, Pina S, Gutierres G, Gomes da Silva V (2005) The challenges of teaching bioclimatic architectural design. In: International conference:passive and low energy cooling for the built environment, May 2005, Santorini, Greece, pp 327–332

Salama AM (2007) An exploratory investigation into the impact of international paradigmatic trends on Arab architectural education. GBER-Glob Built Environ Rev 6(1):31–43

Salama AM (2010) Delivering theory courses in architecture: inquiry based, active, and experiential learning integrated. ArchNet-IJAR: Int J Archit Res 4(2–3):78–295

Stasinopoulos T (2005) Sustainable architecture teaching in non-sustainable societies. In: PLEA 2005—The 22nd conference on passive and low energy architecture, 13 Nov 2005. Beirut, pp 13–16

Taleghani M, Ansari HR, Jennings P (2011) Sustainability in architectural education: a comparison of Iran and Australia. Renew Energy 36(7):2021–2025. https://doi.org/10.1016/j.renene.2010.11.024

Torres-Ramírez M, García-Domingo B, Aguilera J, De la Casa J (2014) Video-sharing educational tool applied to the teaching in renewable energy subjects. Comput Educ 73(2014):160–177. https://doi.org/10.1016/j.compedu.2013.12.014

Wyckmans A (2008) Environmental learning in architecture from individual choice to collective responsibility. Nordic J Archit Res 20(3):73–87. https://arkitekturforskning.net/na/article/viewFile/102/73

Yildirim T, Yavuz AO, Kirci N (2012) Experience of traditional teaching methods in architectural design education: "mimesis technique." Procedia—Soc Behav Sci 51(2012):234–238. https://doi.org/10.1016/j.sbspro.2012.08.151

Yüksek I (2013) The evaluation of architectural education in the scope of sustainable architecture. Procedia—Soc Behav Sci 89(2013):496–508. https://doi.org/10.1016/j.sbspro.2013.08.883

Chapter 4
Gaps Facing Sustainable Architectural Design Training and Education in Egypt

Ayman A. Farid

Abstract Sustainable architecture has become an essential approach for future education especially with the calls for protecting the environment from degradation and resources from consumption. This research offers a critical analysis for possible gaps meeting sustainable architecture design training and education in Egypt. The research offers a case study held among 90 students from the Faculty of Engineering at Ain Shams University who were introduced to multiple architectural short training modules to test their response to each module and provide a quantitative result for further improvements. The training was held inside the faculty of engineering between July and August 2018. The research has tested seven different training modules related to sustainable architecture and design and has conducted a complete set of evaluation for each module based upon its content's facilities and response. The importance of the research is that it provides quantitative data concerning student's response to the different training modules ILOS upon which further enhancement for similar training can be accomplished.

Keywords Sustainable architecture · Training-sustainable design · Sustainability and education

4.1 Introduction

This research work is one of the outputs resulted from BC_SDBE project between Faculty of Engineering at Ain Shams University and the University of East London. The project was funded by the British council under the Newton Institutional Links fund. This research aimed at providing a survey outputs which resulted from a short training held and hosted by the Faculty of Engineering at Ain Shams University. The survey tried to figure out the possible gaps meeting sustainable architecture training, especially in architecture education.

A. A. Farid (✉)
Faculty of Engineering, Ain Shams University, Cairo, Egypt
e-mail: aymanfaridafh@eng.asu.edu.eg

Architecture and Urbanism: A Smart Outlook,
https://doi.org/10.1007/978-3-030-52584-2_4

Sustainability is movement that is widely considered through three major pillars, the environmental, economic and social ones (Altamonte et al. 2014; Murphy 2012; University 2011). Understanding what is sustainability does not mean being sustainable. There are major differences between sustainability as a philosophy and as practice. In order to be truly sustainable, the mankind practice, culture education and behavior should understand the transition from a consumption-based society to a restorative one ("Toward a Restorative Economy with Paul Hawken," n.d.) For that reason, sustainable development should start with proper education for sustainability. There are acceptable amount of concerns and efforts held by educational entities to encompass sustainable development ideals and concepts into their educational programs (Buckler et al. 2014; Lozano et al. 2015). The global effort for improving education for sustainability is led by the European region (Buckler et al. 2014). The Middle East has witnessed in the last decade a considerable development in improving education for sustainable development although it has been more limited to environmental issues rather than social and economic ones (Buckler et al. 2014).

4.2 Architecture Practice and Sustainable Development

The second half of the twentieth century has witnessed the early environmental degradation and lack of resources concerns. There have been significant writings dealing with the concepts of the limited availability of resources and the continual growth of needs. Earth spaceship 1966 was among the early warnings for the massive consumption behavior of mankind dealing with earth resources (Boulding 1966). The early 1970s witnessed more progressive writings that formulated a relationship between rates of consumption and population growth in terms of limits and needs. "Zero Growth Formula" and "Limits to Growth" were among the significant writings dealing with this subject (Meadows et al. 1972). The writings assumed that the irresponsible consumption behavior would lead to an ecosystem misbalance.

After the oil crises, many steps were taken to formulate a common understanding for the meaning of sustainability. The work of the world commission on environment and development led by Gro Harlem Brundtland resulted in coining the term "sustainable development." The term was published in bundled report at "Our Common future" in 1987. This was considered as the spark for the movement toward sustainability. Sustainable development was defined as: "*Sustainable development is the development that meets the needs of the present without compromising the ability of the future generations to meet their own needs*" (Brundtland 1987).

In architecture field, design represents the major step toward creating architectural in the built environment. For that reason, sustainability has to become an essential part of architecture design education. One of the important aspects of sustainable practice is sustainable design as it is considered the action leading to a sustainable method for creating a sustainable end out. Sustainable design is considered to be a respectful process of thinking that encompasses a holistic approach for creating a sustainable product with a zero-net impact on the environment. According to **Van Der Ryn** in his book "ecological design" he stated: "*In many ways, the environmental crisis is a design crisis, it is a consequence of how things are made, buildings are constructed and landscape are used. Design manifests culture and culture rests firmly on the foundation of what we believe to be true about the world*" (Van der Ryn and Cowan 2007).

Sustainable design in architecture is critically important as building represents one of the major elements that consume energy; where residential and commercial buildings only may consume up to 40% of the total energy consumption (Bcapcodes.org 2017). For that reason, understanding sustainable design is critically important for more sustainable practice. In 2004, Jason F. McLennan offered more clear understanding to the term sustainable design where he stated that "*Sustainable design is the philosophy basis of growing movement and individuals and organizations that literally seeks to redefine how buildings are designed, built and operated to be more responsible to the environment responsive to people*" for more explanation, he mentioned that "*Sustainable design is a design philosophy that seeks to maximize the quality of the built environment, while minimizing or eliminating negative impact to the natural environment* "Also he considered sustainable design as a process of environmental healing." *The sustainable design is the approach that looks to the design process to heal exactly as it has damaged*" (McLennan 2004).

The role of architects is greatly essential to lower down the damage happening to the ecosystem and to provide regenerative solutions that does not only have no net impact on the environment or that only heals what has been damaged but also that can add values to the existing environment through a regenerative design mode (Mang and Reed 2012). Architecture design education in terms of sustainability focuses on technology as a tool to achieve better performance and better sustainable solutions to the built environment with more rely on low energy design. Also, it is expected that sustainability can be automatically included within the design; thus, it is not any more an add-on to a given design, but instead, it is embedded in our student's back minds as essential part of the project design as well as a client requirement priority.

4.3 Education for Sustainability

Educational for sustainability is more than a curriculum to be taught. It is more about how to integrate sustainability into the design process discourse (Altomonte et al. 2014). Where successful education would focus on how to interrelate pillars of sustainability into the body of education; where some curricula my only focus on surface learning approaches the emphasis knowledge transfer with less regard to devolving the capabilities for critical or reflective understanding (i.e., transmissive learning) (Altomonte et al. 2014). The actual need in education for sustainability is to provide a deep learning approach which can provide both knowledge acquisition supported with critical thinking and problem-solving techniques. For that reason, solutions and concepts have to be driven as a response to real-life design problems.

According to Cotton and Winter (2010), there should be a support to shift sustainable education from being transmissive to transformative, where courses are structured to provide methods and learning outcomes that focus on deep learning approaches.

As stated by UNESCO "*There is now an increased recognition at the international policy level that education is essential to the advancement of sustainable development.*" (Buckler et al. 2014). In terms of higher education (HE) and according to Sterling (2001), education for sustainability would aim at providing the students with knowledge and skills that helps them providing sustainable solutions for current and future problem (Sterling 2001; Holdsworth and Sandri 2014). Education for sustainability in terms of the built environment discipline probably focuses on teaching students how to think carefully and critically of technology design, production and application; this helps students understanding process of technological change in terms of sustainable development (Linda 2006).

2030 Egypt vision focuses on sustainable development in terms of three goals to be achieved economic development, competitiveness of markets and human capital (SDS Egypt 2030 2016). This makes education for sustainability essential. In Egypt, 2030 development plan the building sectors is considered as one of the largest industries with almost 34% of the National Gross Income. There are several projects held by the Egyptian government in the building sectors; for example the Suez Canal axis comprehensive development, the New Capital project in addition to the new comprehensive city development at Al-Alamain city, New Galala city and others. On housing sector, there are projects targeting social housing like the million social housing units (SDS Egypt 2030 2016). This requires to seriously focus encompassing sustainability in building sector education especially in architecture.

Education for sustainability in Egypt requires a change in the education culture from being perceived as an add-on to a new unique paradigm and concept of education. This new paradigm challenges the current conventional one calls for a new methods and moods for education and learning; such methods can be defined as integrative learning (Barth and Michelsen 2013).

The research work concerning with what should be done in terms of design education for sustainable development is very few, still it is not clear what components should be integrated or how should it be delivered to students. Sobaihi (2017) stated that literature concerning education for sustainability has pointed to three main concerns (Sobaihi 2017).

1. What ideas and concepts should be delivered in a curriculum content seeking sustainable education.
2. In what sequence or distribution should such content be delivered to students.
3. Through which method the curriculum content should be delivered as to be efficient.

Concerning the **curriculum content** and through available research, it seems that most of the curriculum content are resting on the environmental pillar rather than the other two pillars. It focuses on performance, energy reduction, resource preservation and water protection; still there are few curriculums that focus upon economic and social sustainability (Álvarez et al. 2016). There are several recommendations that education for sustainability should develop a framework that gives the chance for interdisciplinary in content (Viegas et al. 2016; Marcus et al. 2015). Also, there has been suggestions that education for sustainability has to encompass inherited knowledge about social and economic aspects (Viegas et al. 2016).

Concerning the **curriculum content** sequence and distribution, there are still different concerns about the time for delivering a certain content related to sustainability, at which sequence and in what level of complicity it should be delivered. Lozano and Lozano identified four different ways to incorporate sustainability in to educational programs (Lozano and Lozano 2014) which are as follows:

1. To incorporate contents on a limited capacity to existing programs
2. To deliver a specific course with such contents within the curriculum
3. To provide sustainability concepts and theories intertwined with disciplinary courses on a regular basis.
4. A clear specialization content for sustainability in each discipline or faculty.

They mentioned that the most spreading type is 1 and 2, while type 4 is considered to be rare. Also, type 3 is the most efficient and beneficial type among all as it allows students to incorporate sustainability concepts smoothly within their study time.

On a similar ground and through reviewing various curriculums for over four hundred architectural degrees and training programs all around Europe, they identified five different types to incorporate sustainability into educational programs (Altomonte 2012).

1. The parallel type: where sustainability and discipline contents run in a parallel way.
2. The partially integrated type: where sustainability contents are integrated within some courses.
3. The fully integrated type: where sustainability content is totally integrated in most courses of a certain curriculum.
4. The interactive type: where sustainability content is based on student interaction during their program study.
5. The elective type: where sustainability content is offered through some elective courses within a certain curriculum.

Type 3 was also regarded as a rare one although it might offer the best scenario, but it needs more awareness knowledge and efforts to be well established and integrated. Also, there has been many concerns about integration sustainability content all along the entire education life and not only the higher education (Buckler et al. 2014).

Concerning **the curriculum content delivery method**, there has been an agreement that education for sustainability seeks to provide students with knowledge and tools that serves them understand, solve and deal with complex design problems. The delivery method should focus on students output rather than teachers and staff inputs (Marcus et al. 2015). Also, there has been a review on how to deliver such content between "sequential learning" that is based on providing lectures and theoretical basis about the topic then having the implementation in a successive step and the "learning by doing" where student practically learns how to solve problems based on given contents on spot. Learning by doing method is backed by multiple reviewers as being the better way for education for sustainability (Viegas et al. 2016; Marcus et al. 2015). Where "sequential learning" is criticized as students are supposed to become creative based on lecturing them with contents without a real doing actions practically (Altomonte 2012), so teaching architecture content usually relies on "learning by doing" approach as the center of curriculum content activities. For that reason, architecture design studio has showed and proven efficiency in delivering a better means for design problems solving as it improves both knowledge and skills.

Design studio is considered the backbone for most design degrees. Based on the following, offering sustainable design contents through a design studio would be an effective method in teaching sustainable architecture design (Walker and Seymour 2008). In 2002, Siddiqi described architecture studio-based learning as follow "*It is a process, a way of thinking during which the many elements, possibilities and constraints of architectural knowledge are integrated. Design studio sequence provides the connective tissue that brings together, progressively, the many elements of architectural education*" (Siddiqi 2002).

One of the other observed phenomena in architecture education is that the content is treated though two separated island; the lecture based for theoretical curriculum material content and the design studio even if sometimes it is integrated within the design studio it is treated as an add-on materials to the original content (Dowdle et al. 2006).

Teaching and learning in higher education intend to provide a certain curriculum content through knowledge, skills and dispositions. In terms of architecture education, it would go for three equivalent themes which are intellectual, technical and intuitive (Biggs and Tang 2011; Angélil and Uziyel 2003).

Unfortunately, Egypt is having limited professional training and education for sustainable development if it is being compared to Europe and USA although Egypt is in great need for applying a sustainable development on different sectors as mentioned above. This results in a lack of training programs and curricula that focus on education for sustainability or that integrate in some means academics and researched, professional practitioners and stakeholders through one whole process for sustainable development.

4.4 Method

This research is based on a recent training case study held within BC_SDBE joint project between Ain Shams University and University of East London where a training survey was held to measure students response to knowledge and education methods concerning sustainability.

4.5 "BC-SDBE" Project Overview

In 2016, the Faculty of Engineering at Ain Shams University has participated in BC_SDBE project that was funded by the British council under the Newton Institutional Links fund. The project was held with the University of East London. It aimed at learning from the UK experience in the field of education for sustainability. The project focused on how to integrate the philosophy of sustainable in the built environment development culture within the Egyptian context through education, training and practice. There had been multiple activities held within the project context which included.

1. Multiple training programs that were held in UK at the University of East London campus for the Faculty of Engineering Ain Ahams University staff members at architecture and urban planning discipline.
2. Two international conferences for sustainable design of the built environment (SDBE) held in UK for sustainable development in the built environment.
3. A skills-gap survey to reach out to diverse stakeholders within disciplines related to the built environment in order to identify the current skills gaps in sustainable design education, training and practice. The results were published in one of the SDBE conferences.
4. Multiple stakeholder events held in Egypt and hosted by Ain Shams University.

4.5.1 ASU-SDBE Short Training for Sustainability Overview, Aims and Objectives

Along with the research project, Ain Shams University establishes multiple research centers, and SDBE was one of these research centers that focused on sustainable development in the built environment. SDBE in collaboration with environmental architecture and urbanism program offered a short training program for integrating sustainability within architecture design. The training was held between July and August 2018. The training program aimed at exploring the responses of the student to multiple modules contents that helps students to have an overall idea about sustainable design practice. The training program results were discussed along with the last stakeholder event held by the research project and showed some significant results that are explained discussed within this research paper.

Environmental architecture and urbanism program (ENVR) aims to equip graduate architects and urban planners with the essential tools and knowledge to respond in their designs to environmental issues, both on building and urban scales. Graduates can integrate all the aspects dealing with the built environment and how it is planned, designed, used, furnished, landscaped, managed and valued by the society in their creative thinking while dealing with complex architectural and urban environment problems. For that reason, SDBE has selected students from ENVR program to be introduced to the short training program and measure their response to the training delivery methods and contents. The program was designed for students who are enrolled at ENVR program level two. Where students we just introduced to basic introduction about architecture design with the support of some theoretical backgrounds dealing with architecture, history and basics for environmental design.

The reason for the choice of students at that level is that they are newly introduced to the basics of environmental design, and at the same time, they still do not have preconceptions about the taught topics which may lead to biased responses to the training program survey and may cause misleading results.

4.5.2 *The Training Program Case Study*

The training program was designed based on "learning by doing" approach to support students in their next design studio at next level 3 as well as it serves some complimentary course that will be taught at the same level. The training program was designed to serve three targets which are as follows:

1. **Architectural skills**: Students learn about essential architectural skills that help them developing their concepts and ideas in a creative way.
2. **Sustainable architecture focused knowledge**: Student learns about current approaches for sustainable and environmental design and green rating systems through an international certified trainer.
3. **Project managements and field study**: Student learns basics about project management in addition to field work on an existing project.

The training program consisted of seven modules to serve these three targets as follows:

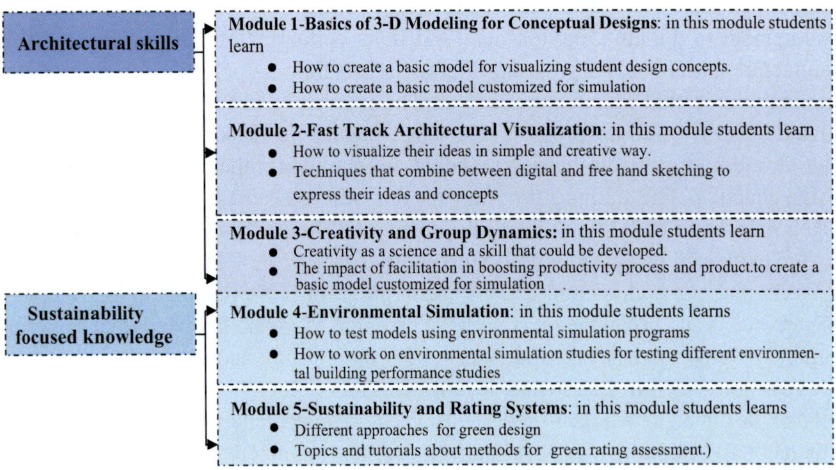

Project managements and field study

Module 6-Basics of Project Management: in this module, students learn
- Basic ideas about steps taken for environmental oriented projects
- About different steps taken in all project stages to ensure that the project meets all possible environmental measures

Module 7-Site Analysis, Case Studies and Field Survey: in this module, students learn
- Students learn how to make onsite studies for a certain project with an environmental perspective

The training program was introduced within 12 days, and student response was measured through a questionnaire held by the training team which consisted of five professors and 13 teaching assistants (Tas). The questionnaire aimed at measuring to three major indicators.

1. **The module satisfaction**: The aim was to measure how much student was satisfied by the module content.
2. **The environment and facility satisfaction**: The aims were to measure the response of students to the training environment settings.
3. **The module independency**: which means to what extent does the student prefer to have the evaluated module to be provided through as a separate training workshop rather than being integrated with other multi-training channels.

More the questionnaire provided the possibility for students to give comments to illustrate their evaluation. Forty students out of 90 responded to the training questionnaire with detailed comments.

4.6 Training Results and Discussion

For **Module 1-Basics of 3-D modeling for conceptual designs**: As seen below (Fig. 4.1), 52.25% of respondents were satisfied with the overall module although 77.5% were not satisfied with the module facilities, and almost 82.5% of the respondents recommended to have this module in a separate training. The respondents

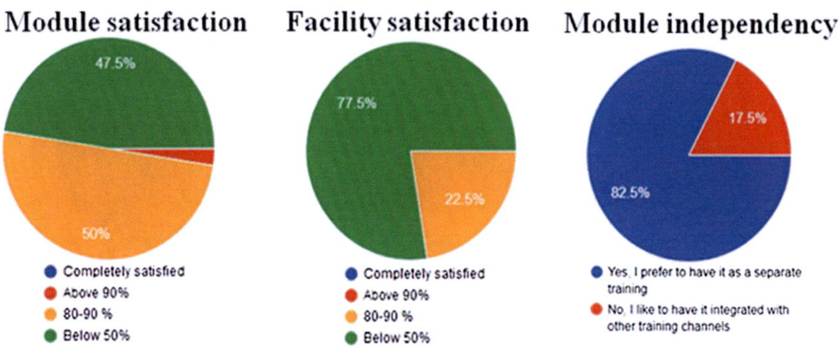

Fig. 4.1 Training results for **Module 1-Basics of 3-D Modeling for conceptual designs**

commented on the importance of useful information and content provided within the module which requires more practice and time. Moreover, the students recommended taking the training sessions out of the University campus which provide an interpretation to the 77.5% dissatisfaction toward modules facilities.

For **Module 2-Fast track architectural visualization**: As seen below (see Fig. 4.2), 72.5% of respondents were satisfied with the overall module, 85.5% were satisfied with the module facilities, while only 40% of the respondents recommended to have this module separately. The respondents recommended to have the training sessions added to module-1 which provide an interpretation to the 40% for module independency. Also students recommended to have more time for this module.

For **Module 3-Creativity and group dynamics**: As seen below (see Fig. 4.3), 72.5% of respondents were satisfied with the overall module with 15% complete satisfaction. 77.5% were satisfied with the module facilities with around 17.5% complete satisfaction, which was the opposite for Module-1; the major difference was that in module, one student needed to be seated individually as in a computer lab, while in Module 3, students are required to work on a dynamic groups with dynamic activities which makes the environment less effective in module 3 compared to module

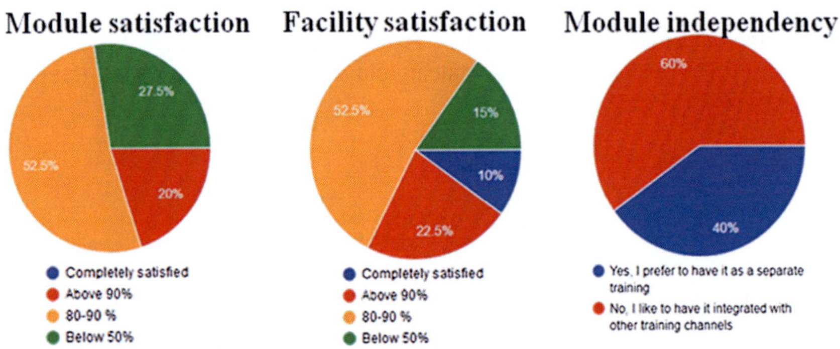

Fig. 4.2 Training results for **Module 2-Fast Track Architectural Visualization**

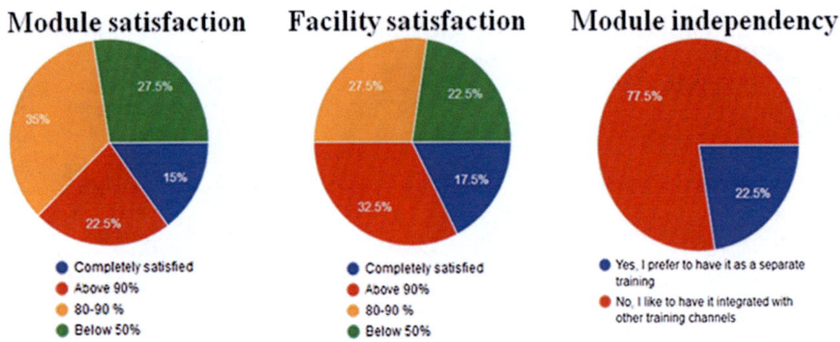

Fig. 4.3 Training results for **Module 3- Creativity and Group Dynamics**

1. Only 22.5% of the respondents recommended to have this module separately, and they preferred to have it integrated within other modules. Also, they recommended to reduce the time dedicated for the module to be added to other modules that needed more time.

For **Module 4-Environmental simulation**: As seen below (see Fig. 4.4), only 30.9% of respondents were satisfied with the overall module, they commented on time limitation and complexity of the subject introduced. Also, 66.7% were not satisfied with the module facilities, which was quite similar to Module-1 which assures that students needed to be seated individually as in a computer lab. This pointed to more careful focus on the teaching settings as well as the content compared to other modules. 77.5% of the respondents recommended to have this module in a separate training.

For **Module 5-Sustainability and rating systems**: As seen below (see Fig. 4.5), 87.1% of respondents were satisfied with the overall module with 46.2% complete satisfaction. They commented on time limitation, and they preferred to have this content on a more interval of time. Also, 89.7% were satisfied with the module facilities although commented on poor projector lighting and acoustics, but still, I

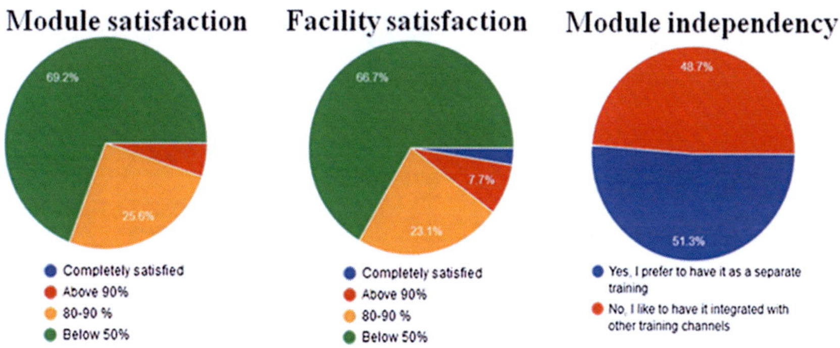

Fig. 4.4 Training results for **Module 4-Environmental Simulation**

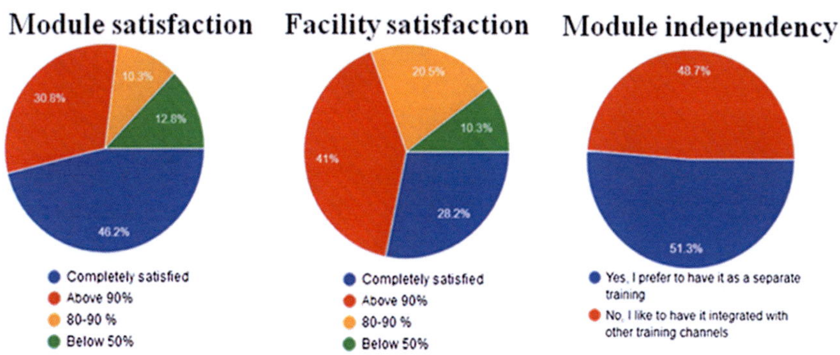

Fig. 4.5 Training results for **Module 5-Sustainability and Rating Systems**

did not affect much their evaluation. 51.3% of the respondents recommended to have this module in a separate training.

For **Module 6-Basics of project management**: As seen below (see Fig. 4.6), 61.5% of respondents were satisfied with the overall module. This shows that the module did not meet the expectations of the students. They commented on the difficulty of the content compared to their knowledge which shows students at that level have less information about project management versus project design. 82.1% were satisfied with the module facilities which showed that the module facilities were not effective on the content evaluation as much as the content difficulty. 69.3% of the respondents recommended to have this module integrated within other training modules.

For **Module 7-Site analysis, case studies and field survey**: As seen below (see Fig. 4.7), 59% of respondents were satisfied with the overall module. They commented on the need to have more in-depth field study compared to what was

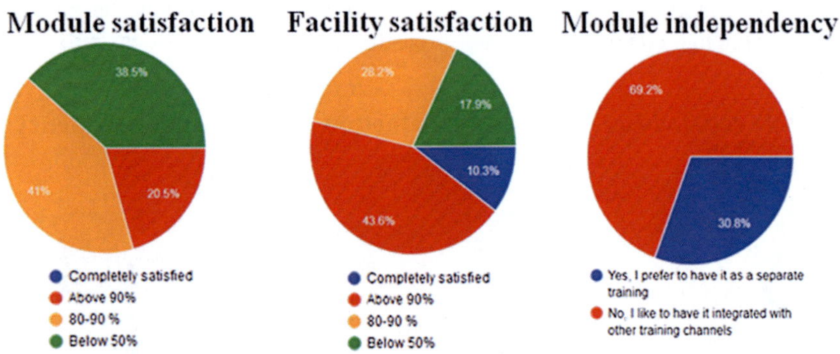

Fig. 4.6 Training results for **Module 6-Basics of Project Management**

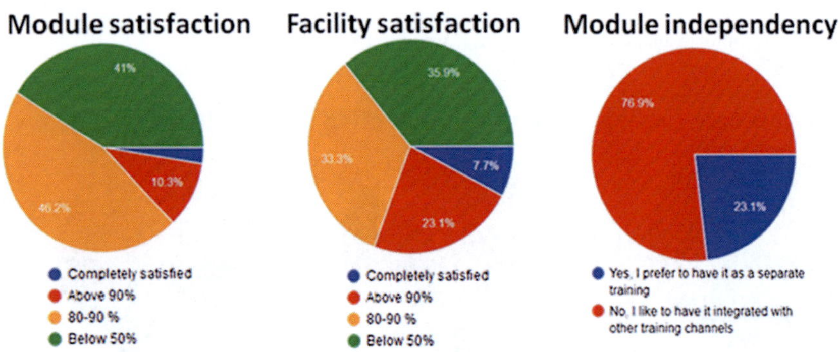

Fig. 4.7 Training results for **Module 7-Site Analysis, Case studies and field survey**

delivered. Also, 64.1% were not satisfied with the module facilities which assure the reason for the module low percentage of satisfaction. 76.9% of the respondents recommended to have this module integrated within other training modules.

4.7 Conclusion

The students provided some significant evaluation for the three major training targets in terms of sustainable design understanding as follows:

- **For student architectural skills**: The training responses showed that students are keen on having a separate content that focuses on 3D architecture modeling with regard to two major points which are 3D modeling skills for quantitative design requirements including environmental simulation as well as qualitative values related to high quality of visualization. The students comments recommended to give more content about 3D modeling that is related to BIM. Also, students refused to have a separate content to empower their ability to creativity and recommended to have them just integrated within any curriculum content. Also, students were keen on having such skills out of the traditional University campus.
- **For sustainable architecture focused knowledge**: The training responses showed that students are keen on having two separate contents that focus on environmental simulations and studies as well as architecture green rating systems. Also, responses showed that knowledge about green rating systems takes the highest level of students attention.
- **For project managements and field study**: The training responses showed that students recommended to have more focus on site visits and requested to focus on sites that are under constructions or buildings that has been awarded a green certificate. Also, students were not familiar with project management philosophy or terminology, and they considered the subject contents a complicated one, which shows that this content needs some more curricula backgrounds before being directly introduced to the students.

Generally, student responses assured on preferring "learning by doing approach" rather than "sequential learning" where most general comments recommended to include more practical work and less lecture. Also, students recommended to have an ongoing training with weekly basis rather than having it condensed within couple of weeks.

Acknowledgements The author would like to acknowledge Prof Ayman Ashour the dean of the faculty of engineering as well as CHEP admin for providing all facilities to accomplish this training. Also, the author would like to acknowledge Prof. Mostafa Refat and all team members of the Center for Sustainable Development of the Built Environment (SDBE) at Ain Shams University for their help and support during all training stages and modules.

References

Altomonte S (ed) (2012) EDUCATE—criteria for professional qualification
Altomonte S, Rutherford P, Wilson R (2014) Mapping the way forward: education for sustainability in architecture and urban design. Corp Soc Responsib Environ Manag 21(3):143–154. https://doi.org/10.1002/csr.1311
Álvarez SP, Lee K, Park J, Rieh S-Y (2016) A comparative study on sustainability in architectural education in Asia. Retrieved from file:///C:/Users/Demo/Downloads/sustainability-08-00290.pdf
Angélil MM, Uziyel L (2003) Inchoate an experiment in architectural education, 565 S
Barth M, Michelsen G (2013) Learning for change: an educational contribution to sustainability science. Sustain Sci 8(1):103–119. https://doi.org/10.1007/s11625-012-0181-5
Bcapcodes.org (2017) Climate change | The building codes assistance project. Retrieved September 25, 2017, from https://bcapcodes.org/topics/climate-change/
Biggs J, Tang C (2011) Train-the-trainers: implementing outcomes- based teaching and learning in malaysian higher education. Malays J Learn Instr 8:1–19. Retrieved from https://mjli.uum.edu.my/images/pdf/8mjli/mjli81train.pdf
Boulding K (1966) Earth as a spaceship. Washington State University Committee on Space. Retrieved from https://earthmind.net/earthmind/docs/boulding-1965.pdf
Brundtland G (1987) Report of the world commission on environment and development: our common future. Oxford Paperbacks, Report of 400. https://doi.org/10.2307/2621529
Buckler C, Creech H (2014) UN Decade of Education for Sustainable Development (2005–2014) FINAL REPORT SUMMARY Shaping the Future We Want UN Decade of Education for Sustainable Development (Shaping the Future We Want Japan Funds-in-Trust United Nations Educational, Scientific and Cultural Organization. Retrieved from https://www.unesco.org/open-access/
Cotton D, Winter JCN (2010) It's not just bits of paper and light bulbs. A review of sustainability pedagogies and their potential for use in higher education. Sustain Educ Perspect Pract Higher Educ 2010:39–54
Dowdle D, Ahmed V, Bouchlaghem D (2006) Teaching and learning building design and construction. Earthscan.
Holdsworth S, Sandri O (2014) Sustainability education and the built environment: experiences from the classroom. J Educ Built Environ 9(1):48–68. https://doi.org/10.11120/jebe.2014.00011
Linda K (2006) Engineering education in sustainable development at Delft University of Technology. J Cleaner Prod 14(9–11):928–931. Retrieved from https://www.sciencedirect.com/science/article/pii/S0959652606000308
Lozano FJ, Lozano R (2014) Developing the curriculum for a new bachelor's degree in Engineering for Sustainable Development. J Cleaner Prod 64:136–146. https://doi.org/10.1016/j.jclepro.2013.08.022
Lozano R, Ceulemans K, Alonso-Almeida M, Huisingh D, Lozano FJ, Waas T, Lambrechts W, Lukman R, Hugé J (2015) A review of commitment and implementation of sustainable development in higher education: results from a worldwide survey. J Cleaner Prod 108:1–18. https://doi.org/10.1016/j.jclepro.2014.09.048
Mang P, Reed B (2012) Regenerative development and design. Encycl Sustain Sci Technol, 1–44. https://doi.org/10.1007/978-1-4614-5828-9_303
Marcus J, Coops NC, Ellis S, Robinson J (2015) Embedding sustainability learning pathways across the university. Curr Opin Environ Sustain 16:7–13. https://doi.org/10.1016/j.cosust.2015.07.012
McLennan JF (2004) The philosophy of sustainable design: the future of architecture. Retrieved from https://www.google.com/books?id=-Qjadh_0IeMC
Meadows D, Meadows G, Randers J, Behrens III W (1972) The limits to growth
Murphy K (2012) The social pillar of sustainable development: a literature review and framework for policy analysis. Sustain Sci Pract Policy 8(1):15–29. https://doi.org/10.1007/978-94-007-4854-5

SDS Egypt 2030 (2016) Reports. Retrieved from Sustainable Development Strategy: Egypt Vision 2030: https://sdsegypt2030.com/wp-content/uploads/2016/08/English-Booklet-2030.compressed.pdf

Siddiqi AA (2002) Architectural design studio projects and the charades of curriculum. Engineering, 1 (December)

Sobaihi M (2017) Establishing an approach to develop a curriculum of sustainability in landscape architecture in Saudi Arabia. In: International conference for sustainable design of the built environment. SDBE, London, pp 9–16

Sterling S (2001) Sustainable education: re-visioning learning and change. Education, 1–94

Toward a Restorative Economy With Paul Hawken (n.d.) Retrieved June 18, 2019, from https://www.life.ca/naturallife/9508/hawken.htm

University T (2011) EFQM framework for sustainability. Retrieved from https://cataleg.uji.es/record=b1417341~S1*cat

Van der Ryn S, Cowan S (2007) Ecological design. Island Press

Viegas C, Bond A, Vaz C, Borchardt M, Pereira GM, Selig PM, Varvakis G (2016) Critical attributes of sustainability in higher education. J Cleaner Prod 126:260–276

Walker JB, Seymour MW (2008) Utilizing the design charrette for teaching sustainability. Int J Sustain High Educ 9(2):157–169. https://doi.org/10.1108/14676370810856305

Chapter 5
Evaluating the Architectural Designs Using Machine Learning: The Case of Two Modes

Omar O. Elrawy, Tamer El-Khouly, and Omar Abdallah

Abstract Machine Learning (ML) is one significant subfield of Artificial Intelligence (AI) that impacts most of today's industries. This research is part of a broader research project that investigates enquiring creativity in machine learning versus the human design processes. The research aims at developing a computational model that evaluates architectural designs using machine learning algorithms, to fulfill project goal of investigating the computer's ability to generate architectural designs. Two ML algorithms were trained with a set of design drawings along with their own evaluation. Three evaluation criteria are identified; Spaces' orientation, functional zoning, and usability of area. Each design case is accompanied by a given evaluation; the evaluation is set by practitioner architects per each criterion. To train the ML algorithm, a custom-built drafting tool was developed, the drafting tool accepts a given architectural plan design drawn on an evaluation grid. Each architectural plan drawing is identified in terms of classes that represent space uses. The tool then scans the grid in order to extract the features related to each of the evaluation criteria. Within this research paper the two algorithms were trained with a dataset of 15 architectural designs, and then tested with five test cases that include the five possible grades. Results showed that both ML algorithms have learning curves, and that one algorithm is significantly better than the other in learning the three grading criteria. This research serves a long-term objective to implement ML in generating designs with specified qualities.

Keywords Artificial intelligence · Architectural design process · Reinforcement learning · Design cognition · Architectural design cognition

O. O. Elrawy (✉) · T. El-Khouly · O. Abdallah
The American University in Cairo, Cairo, Egypt
e-mail: omarelrawy@aucegypt.edu

T. El-Khouly
Ain Shams University, Cairo, Egypt

Architecture and Urbanism: A Smart Outlook,
https://doi.org/10.1007/978-3-030-52584-2_5

5.1 Introduction

Machine learning is one very common subfield of AI that impacted most of today's disciplines and jobs. The origins of machine learning gets back to 1950, when Alan Turing created the Turing test, which was placed to test if the machine can substitute human thinking, and to pass the test, a computer must be able to fool a human into believing it is also human (Marr 2016). Machine learning exceeds the mainstream computation by the fact that it does not only perform a given task under defined rules; but additionally it adapts and improves its behaviour over time while interacting with its environment.

Machine learning encompasses various algorithms, to relate machine learning to the applied algorithms; (Frommberger 2010) stated that: "machine learning is concerned with the desire to build algorithms that enable a computer not only to fulfill a given task according to predefined rules, but also to adapt and improve its behaviour over time while interacting with the environment", he then categorized machine learning in general into three learning cases: supervised learning, unsupervised learning, and reinforcement learning. Under these three categories fall the vast and various types of algorithms. An important difference between reinforcement learning and both supervised and unsupervised learning is that in the beginning of the reinforcement learning process no data is available at all, so the learning literally starts from scratch (Frommberger 2010). Within this research reinforcement learning was used.

5.2 Research Goal

The research eventual goal is to develop a computational model that evaluates architectural designs using two ML algorithms. Design case studies are classified into two sets: the training set and the testing set. Firstly, the training set is evaluated by professional practitioners -according to the highlighted criteria- and implemented to train each algorithm, and then the testing set is applied to each algorithm to test its learning curve. Hence, a comparison between both algorithms is taken to see which of the two is close to the human-being evaluation.

This eventual goal, Fig. 5.1, is part of a broader research, which aims at investigating the computer's ability to generate architectural designs, and to examine whether the computer will follow the same process to generate architectural designs like human beings, or will follow a different process.

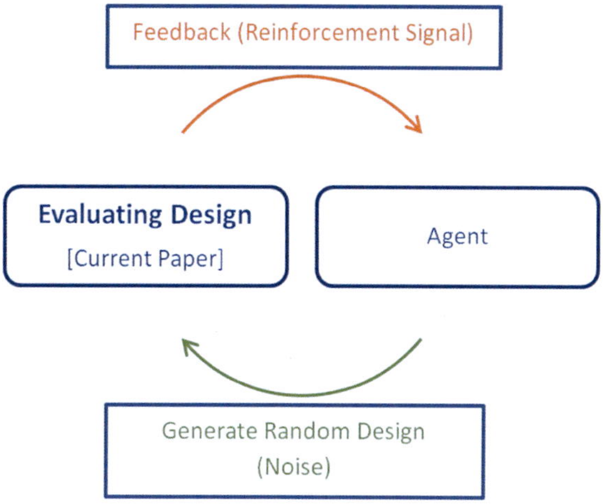

Fig. 5.1 Research plan

5.3 Hypothesis

ML algorithms can be used to evaluate given architectural designs, but some will learn faster/better than the others. The ability of each algorithm to learn architectural design is examined within this research paper.

5.4 Methodology

The research follows inductive reasoning. By creating various ML training cases, and observing the ML algorithms' ability to learn from those cases and self-evaluate; the study can then investigate the potential of Machine Learning (ML) in evaluating subjective aspects of a given architectural design. This is done by building two software tools, and testing several design cases using the developed tools. The research conceptual framework is shown in Fig. 5.2.

5.5 Implementation Method

A computational designing tool is developed to develop designs basically using three main criterions: orientation, adjacency matrix between functional spaces, and best use of area (usability). The tool is developed mainly to design a multiple activity center where there are three classes of spaces: Activity spaces, services (i.e. storage, toilets), and circulation routes (i.e. corridors, entrance lobby and staircases). The tool investigates each space's four coordination vertices, captures any movement or

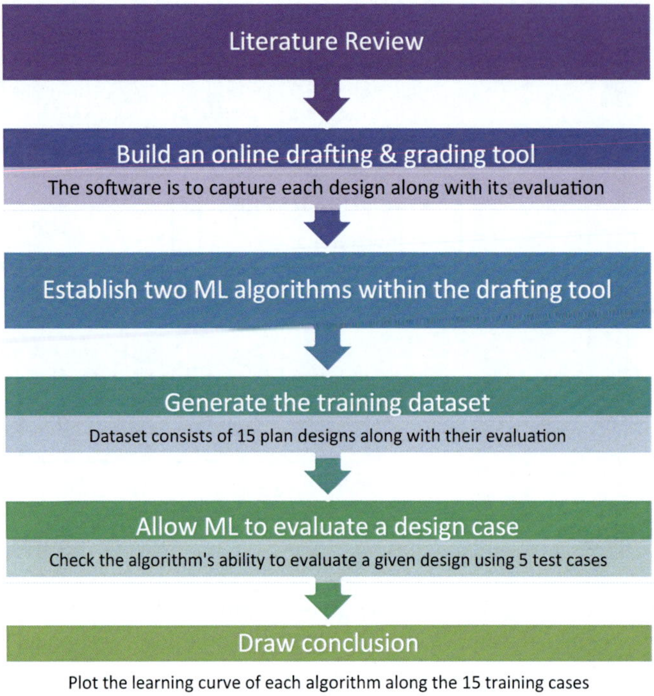

Fig. 5.2 Study framework

amendment then relates each space to the other to examine their adjacency relationship, orientation towards North direction and whether there is a wasted use of area or best usability. Hence, participants are invited to conduct the experiment using this developed tool; keep modifying the spaces to reach the best design. Designs are evaluated by a committee of two expert architectural designers and evaluation criteria are inputted to the ML algorithm in the training set.

The tool will be given five training cases followed by one testing case. This process is repeated three times to comprise the whole 15 training cases. The aim of this process is to investigate the tool's ability to evaluate architectural designs, and to evaluate tool's overall ability to develop a learning curve based on the given training cases.

The architectural design cases represent a hypothetical design of a multiple activity center with the following functional program requirements to be split between two floors:

- A multipurpose room
- Two activity rooms
- One storage room

- An Entrance Lobby
- An outdoor terrace
- Toilets for both genders.

5.5.1 An Online Designing Tool

To prepare the datasets needed for the ML algorithm; an online drafting tool was created for data collection, feature extraction and model training and evaluation, Fig. 5.3. This consists of a drafting grid on which architectural plan drawings are grafted and identified in terms of space uses (Living room, kitchen, … etc.), and identified in terms of their orientation as well.

The drawing interface is developed using Javascript and PaperJs; an open source vector graphics scripting framework that runs on top of HTML5 Canvas. The server backend application is written in Python, which facilitates making use of Python's extensive machine learning capabilities.

The interface consists of a 0.5 × 0.5 m grid canvas, and drawing tools that include: the ability to draw a room with orthogonal walls, draw a door possibly connecting two rooms, draw stairs, move objects around the canvas and give rooms a label out of six possible labels (e.g. multipurpose or terrace). A single drawing may consist of multiple plans. Drawings are saved to the server for grading and further data processing. A drawing is given three grades: An orientation grade, a zoning grade, and area usability grade, in addition to one overall grade. Each of the three grading criteria is defined to the tool in terms of drawing features, as shown in Fig. 5.4.

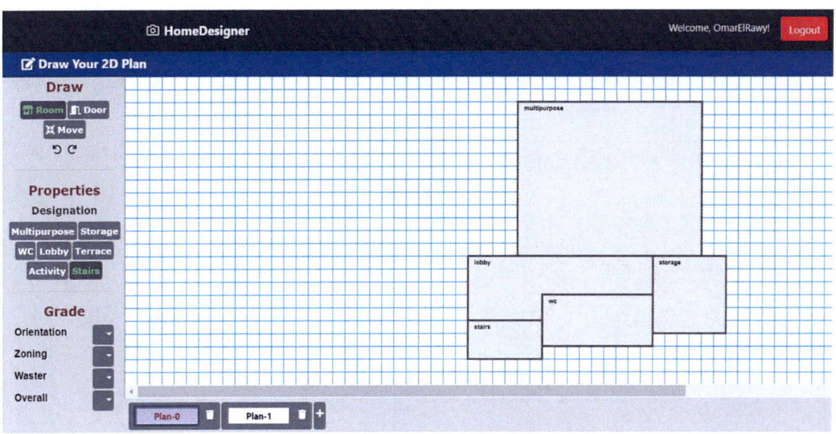

Fig. 5.3 A Screenshot from the developed tool. (Developed by the research team: Omar Abdullah, Omar Elrawy, Mohamed Abdullah and Tamir El-Khouly)

	Criteria	Initials		Criteria	Initials		Criteria	Initials
	Orientation			**Functional Zoning**			**Area Usability**	
Features	Orientation of a given zone compared to North	**Or**		Relation between two zones; i.e., is there a linking Door (D)	**Fz**		Number of Cells occupied by a given zone	**Ua**

Fig. 5.4 The grading features (By research team)

A total of 33 features are extracted from the saved data, which are:

- Orientation (six features): One feature per room label, calculated as the ratio of the cumulative length of rooms with that label facing north to the total length facing north.
- Functional zoning (twenty one features): considered as the cumulative number of door connections from one room label to another (e.g. multipurpose to WC connections). Given six possible room labels we get = 21 total features.
- Wasted area (six features): One feature per room label, calculated as the ratio of the cumulative area of rooms with that label to the total area.

5.5.2 *The Machine Learning (ML) Tool*

First, an ML tool is developed, which follows reinforcement learning technique; it is being trained using a dataset. The dataset consists of several design cases along with their own evaluation, and the evaluation is already set by architectural professors/professionals. The evaluation is set by grading the designs as: (A, A-, B, B- … etc.), instead of marks/values (96%, 88%, 67% … etc.). The grading method is selected based on the selected reinforcement learning algorithm, which is a discrete/classification ML, instead of a regression ML algorithm, Fig. 5.5. The

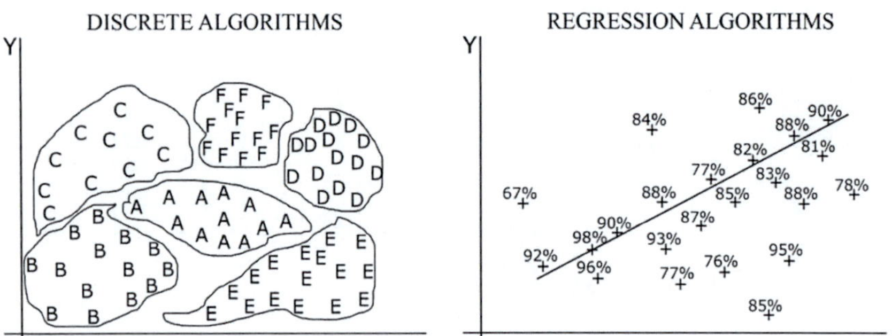

Fig. 5.5 Discrete training algorithm versus Regression training algorithm. *Note* Each point on both graphs represents one data point, i.e. a design along with its evaluation

discrete/classification ML was selected as it is expected to acquire smaller datasets to reach the self-dependent evaluation level.

To obtain useful models for automatic grading of drawings, we use scikit-learn, an open source python machine learning library with simple and efficient tools for data mining and data analysis. It includes implementations of many ML learning models and optimization algorithms.

Two popular models were used for learning: Logistic Regression and Kernel SVM, Appendix A, and data obtained from the drawing tool is normalized and input to each of the two algorithms. Logistic Regression and Kernel SVM algorithms were selected based on the following criteria:

1. Both algorithms are implemented in scikit-learn—"Scikit Learn is built on top of several common data and math Python libraries. Such a design makes it super easy to integrate between them all" (Seif 2018).
2. Both are open-source codes.
3. Both fit the subject problem, where we already have the features extracted, and want to create a predictive model that learns a grading pattern.

5.6 The Training Dataset

The training dataset, Fig. 5.6, contains 15 training cases. The set is created in such a manner to encompass all the possible five grades (A, B, C, D, and F) for the three design criteria: (Orientation, Zoning, and Area usability). The 15 training cases are fed to the ML algorithm in a five by five manner, so that we can test the algorithms learning capability after each five cases. The cases were created to ensure maximum diversity in in each five training cases.

5.7 The Testing Dataset

The testing dataset, Fig. 5.7, contains five testing cases. The cases are created in such a manner to encompass all the possible five grades in all three design criterions; Orientation, Zoning, and Area usability.

The correct/true evaluation of the testing set is inserted to the algorithm so it does calculate the error only, while it's kept blind from the algorithm during the testing process.

5.8 Results

The study results are twofold; detailed results, Table 5.1, and summary graph results, shown in Table 5.2, and Figs. 5.8, 5.9 and 5.10. The detailed results, Table 5.1, shows

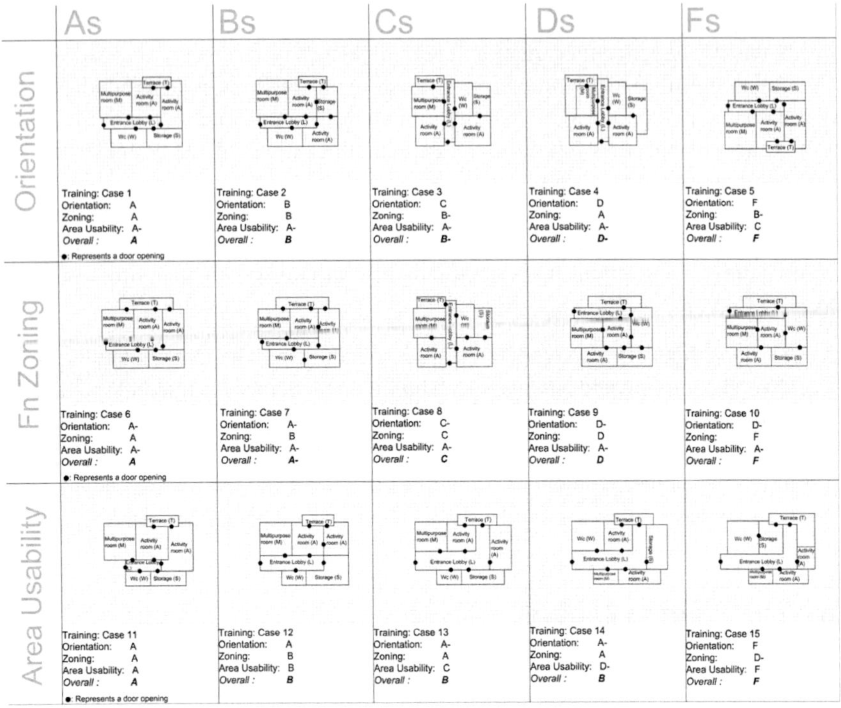

Fig. 5.6 The training dataset (By research team)

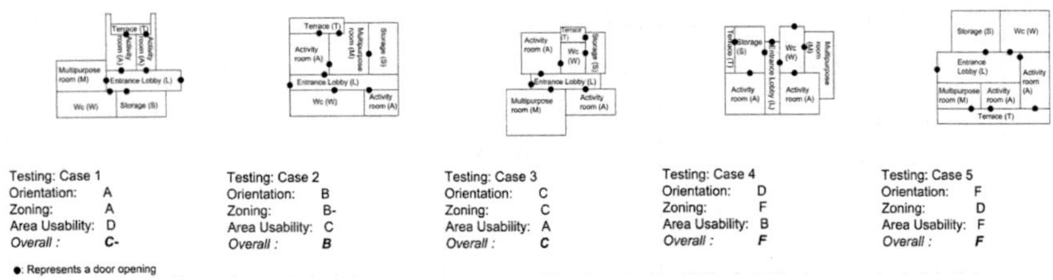

Fig. 5.7 The testing dataset (By research team)

how each algorithm did behave with each testing case after each 5 learning cases, in terms of each grading criteria; Orientation, Zoning, and Area usability.

Under each of the five test cases there are three columns; "Human Grade", "Logistic Regression", and "Kernel SVM Model". The first column entitled "Human Grade" resembles the design case's true grade, which is given by architectural professional(s), that's why the grades under this column is one constant grade, and do not vary along with training steps progression.

Table 5.1 ML training steps and tool's outcome (By research team)

Orientation															
	Test case 1			Test case 2			Test case 3			Test case 4			Test case 5		
Training steps	Human grade	Logistic regression	Kernel SVM model	Human grade	Logistic regression	Kernel SVM model	Human grade	Logistic regression	Kernel SVM model	Human grade	Logistic regression	Kernel SVM model	Human grade	Logistic regression	Kernel SVM model
5	**A**	A	A	**B**	A	A	**C**	B	B	**D**	D	D	**F**	F	*F*
10		C	A-		F	A-		A	B		B	C-		F	*A-*
15		A	A		A-	A-		A	B		C-	C-		D	*A-*
Zoning															
Training steps	Human grade	Regression model	Kernel SVM model	Human grade	Regression model	Kernel SVM model	Human grade	Regression model	Kernel SVM model	Human grade	Regression model	Kernel SVM model	Human grade	Regression model	Kernel SVM model
5	**A**	A	A	**B-**	A	A	**C**	B	A	**F**	A	A	**D**	B-	A
10		A	A		A	B-		B	B		A	B		B-	B
15		A	A		A	A		B	B		C	B		A	A
Area usability															
Training steps	Human grade	Logistic regression	Kernel SVM model	Human grade	Logistic regression	Kernel SVM model	Human grade	Logistic regression	Kernel SVM model	Human grade	Logistic regression	Kernel SVM model	Human grade	Logistic regression	Kernel SVM model
5	**D**	A-	A-	**C**	A-	A-	**A**	A-	A-	**B**	A-	A-	**F**	C	A-
10		A-	A-		A-	A-		A-	A-		A-	A-		C	A-
15		A-	A-		A-	A-		A	**A-**		A-	**A-**		A-	**C**

Table 5.2 Normalized sum of Manhattan distance

	Normalized sum from the 5 test cases					
	Orientation		Zoning		Area usability	
Training steps	Logistic regression	Kernel SVM	Logistic regression	Kernel SVM	Logistic regression	Kernel SVM
5	0.153	0.125	0.222	0.292	0.208	0.250
10	0.417	0.333	0.125	0.167	0.208	0.250
15	0.208	0.306	0.208	0.236	0.236	0.208

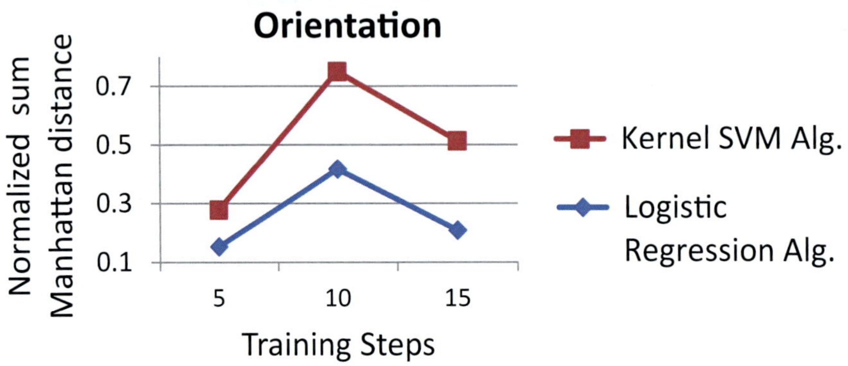

Fig. 5.8 ML learning curve in terms of zones' orientation (By research team)

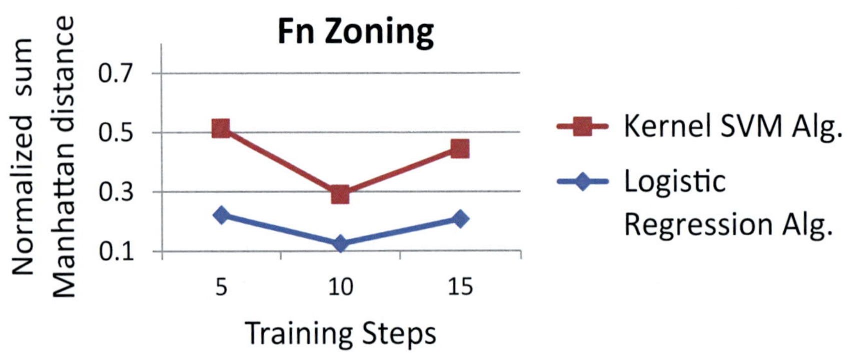

Fig. 5.9 ML learning curve in terms of zones' functional zoning (By research team)

Finally, to detect and evaluate the learning curve of each algorithm: the normalized sum of Manhattan distance values in Table 5.2 is plotted for each of the three grading criteria; Orientation, Fig. 5.8, Zoning, Fig. 5.9, and Area Usability, Fig. 5.10.

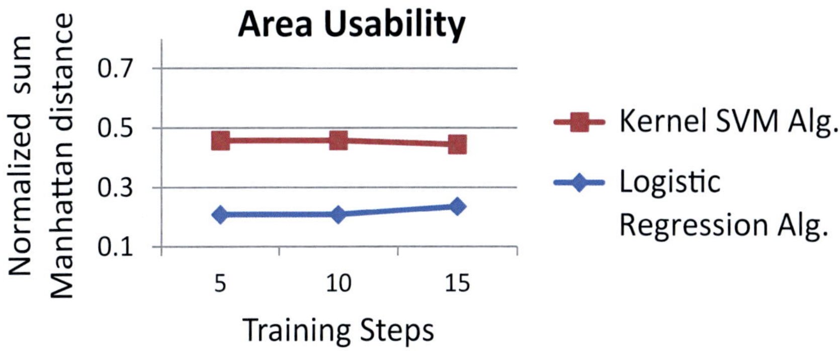

Fig. 5.10 ML learning curve in terms of zones' Area usability (By research team)

5.9 Discussion

There is a lot of potential in today's AI technology in evaluating architectural designs. All those witnessed and upcoming developments are expected to achieve computer-generated architectural designs; however, computer-generated designs' creativity and organicity are still in question—the term "Organicity" is used by (Chaillou 2019) to express non-autonomous designs.

The study results showed that the computer's ability to learn each grading criteria is different from the other. The computer's learning curve experienced high fluctuations in terms of designs' Orientation and Zoning, while in case of Area Usability, the learning curve didn't fluctuate much. This is deemed by the research team to be due to the features' complexity within each criterion, to evaluate a criterion like Area Usability the features of a design are simplified in terms of grids occupied by each zone, while to evaluate another criterion like orientation or zoning the computer needs to understand more complex features, with larger variability.

The study also showed that algorithms learn from design cases almost with the same learning pattern, yet, some algorithms have less error than the others. In the current case, the "Logistic Regression" algorithm showed much less gap/error than the "Kernel SVM algorithm"; hence, algorithm selection is a crucial step while integrating AI in architectural design evaluation process.

The proven learning ability within this study showed that the computer can generate acceptable designs using reinforcement learning, Fig. 5.1. As the reinforcement signal, which is the designs' evaluation, is proven to be possible; this paves the way for the design generation phase, which is the broader objective of the current research project.

It is recommended for future similar research to maximize the number of training cases, since that the plotted learning curves still show error in both algorithms' grades after the 15 training cases.

It's recommended for future research to use Logistic regression algorithms, as the study showed how those algorithms provide less error value, though, both algorithms followed the same learning curve/slope.

The current study used two algorithms that works after extracting pre-defined features from the given design cases. This method was used as it works well will limited datasets; hence, it's recommended for future research to generate huge datasets, as an attempt to test algorithms that do extract the features themselves, this is a significant approach to discover computer's ability to learn "creativity".

It is suggested also for future similar research, not only to compare computer generated designs against human designs, but to track the process followed by the computer to reach a specific design. This also includes extracting plan drawings and relations directly from given designs using Generative Adversarial Networks (GANs)—"Generative modelling is an unsupervised learning task in machine learning that involves automatically discovering and learning the regularities or patterns in input data in such a way that the model can be used to generate or output new examples that plausibly could have been drawn from the original dataset." (Brownlee 2019). This shall eliminate the need for a drafting grid, since that, within this study; the drafting grid was limiting the designs' variability.

Acknowledgements This research has been supported by the Faculty Support Grant at The American University in Cairo, Nov. 2017. The authors would like to thank the empirical studies associated to this research for their valuable participation and contribution.

References

Brownlee J (2019) A gentle introduction to Generative Adversarial Networks (GANs). Retrieved June 17, 2019, from Machine Learning Mastery website: https://machinelearningmastery.com/what-are-generative-adversarial-networks-gans/

Chaillou S (2019) AI & architecture. Retrieved May 5, 2019, from Towards Data Science website: https://towardsdatascience.com/ai-architecture-f9d78c6958e0

Frommberger L (2010) Qualitative spatial abstraction in reinforcement learning. Springer, Germany

Marr B (2016) A short history of machine learning—every manager should read. Forbes. Retrieved from https://www.forbes.com/sites/bernardmarr/2016/02/19/a-short-history-of-machine-learning-every-manager-should-read/

Seif G (2018) An introduction to Scikit Learn: the gold standard of Python machine learning. Retrieved June 17, 2019, from Towards Data Science website: https://towardsdatascience.com/an-introduction-to-scikit-learn-the-gold-standard-of-python-machine-learning-e2b9238a98ab

Chapter 6
Applying Swarm Intelligence in Architectural Design

Hager Magdy and Hazem Eldaly

Abstract During the previous decade, many of architectural design movements are based upon applying computer algorithms. Swarm intelligence represents one of these algorithms, but it is less common to architects. Swarm intelligence can be found in many examples in nature. Merging between computer sciences and swarm intelligence concept results in fruitful applications in architecture. Based on natural swarms of ants, birds, and fish, architects can have unpredictable ideas to achieve their design visions. Swarm intelligence can be applied in architecture in form finding, urban design, user's circulations, structures, etc. In this research, three different methods for applying swarm intelligence in architectural design are proposed. The research will disclose to swarm applications in architectural design to find a guide for architectural designers. Through analyzing the swarm intelligence applications in architecture, different directions are proposed. These directions are classified as, achieving architectural requirements, self-learning architecture, and making process for generating design alternatives.

Keywords Swarm intelligence · Architectural design · Computational design

6.1 Introduction

According to Keerthi et al. (2015), swarm intelligence is a set of similar singular agents that interact with each other and with the environment. The behavior of swarm intelligence is based on the local interactions between the swarm agents which produces a comprehensive behavior between the agents and between them and their environment. Swarm intelligent concept is based on controlling these interactions.

According to Chen (2015), swarm intelligence is a mutual behavior between individuals that have the same size and interacts with each other in groups. It is an

H. Magdy (✉) · H. Eldaly
Faculty of Engineering, Ain Shams University, Cairo, Egypt
e-mail: hager.magdy.ismail@gmail.com

H. Eldaly
e-mail: hazemeldaly@yahoo.com

Architecture and Urbanism: A Smart Outlook,
https://doi.org/10.1007/978-3-030-52584-2_6

expression set to all the individuals that behave as a swarm. This behavior can be applied to systems that are natural or artificial. It can be set to self-organized systems or decentralized systems.

Swarm intelligence agents are working and living together according to a set of rules. These rules control the way they interact with each other. Designers try to understand these rules, to have different applications based on swarm concept. In architecture, swarm rules can be applied in form finding, structure, circulation, urban design, etc.

In architectural design, different applications of artificial intelligence are starting to be common in it. But, swarm intelligence is considered as a metaphor which is still a black box for design. The following research shows the applications of this metaphor with their classifications.

6.2 Swarm Intelligence

Swarm intelligence is an artificial intelligence based on the collective behavior of decentralized, self-organized systems. It is typically made up of a group of simple agents interacting locally and with environment. Despite there is no centralized control structure dictating how individual agents should behave, local interactions between such agents lead to the emergence of global behavior.

Swarm intelligence is found in many examples in nature such as ants, birds, and fish. These examples are smart systems, as they solve problems to reach efficient solutions. Swarm techniques in nature can be understood from studying them. For example, studing swarm technique to find food as swarm agents communicate with each other, to find the best route for food and the best food quality (see Fig. 6.1).

Fig. 6.1 Swarms of birds and fish

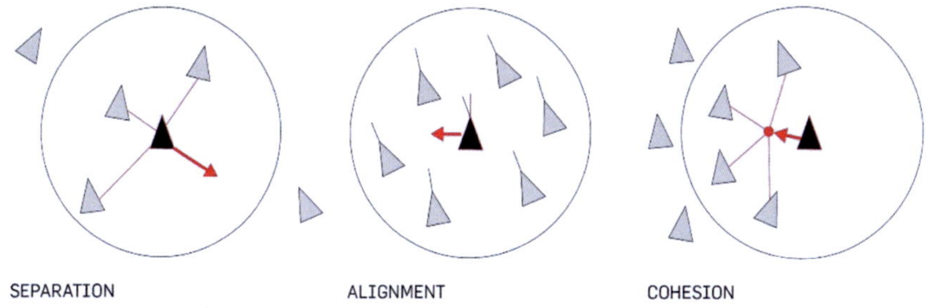

Fig. 6.2 Swarm mathematical model (Yuxing 2015)

6.3 Swarm Model

Through the interpretation of movements of group of creatures, the swarm intelligence is translated into algorithmic steps. And then, a computer-based mathematical model is designed, so it can be applied in simulating swarms. The following Fig. (6.2) shows the basic rules for swarm intelligence. A simple swarm algorithm can be summarized in the following steps:

(1) Agents move together (separation)
(2) They move in the same direction (alignment)
(3) They avoid crushing while moving (cohesion) (Yuxing 2015).

6.4 Case Studies of Applying Swarm Intelligence in Architectural Design

6.4.1 Housing Units

The first example is a project of housing apartments, which considered the building as modules with different functions (Fig. 6.3). It is considered as a computational application of swarm intelligence in architectural design.

The swarm intelligence rules are applied into architecture to achieve different architectural requirements such as the best order of rooms, open spaces, directions, view, privacy and modules configuration. These rules are applied into the 2d and 3d architectural design as they are applied in plan, layout and 3d model.

6.4.1.1 Design Method Based on Swarms

1. The project is considered as multi-agents system, as it consists of units which called voxels (Fig. 6.4). Each voxel is considered as an apartment.
2. These voxels are placed in three-dimensional space that is considered as a matrix.

Fig. 6.3 Simulation engine for evolutionary optimized cellular housing design (Sebastian et al. 2016)

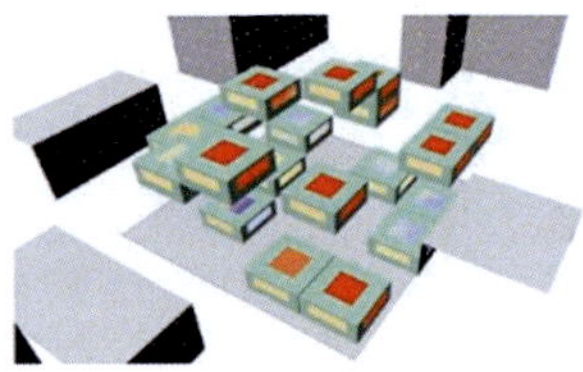

Fig. 6.4 Representation of the voxel in the urban environment (Philippe et al. 2012)

3. The designer simulates voxels movement that is controlled according to swarm mathematical rules to achieve different architectural requirements such as open spaces, directions, view, privacy and modules configuration.
4. These voxels make different alternatives according to these rules.
5. These alternatives are analysed and evaluated to choose the most optimum result from it. A genetic algorithm runs after the swarm intelligence, and it starts to generate offspring based on mutating genes, to choose the optimum alternative.

6.4.1.2 Interactive Genetic Algorithm

In this proposal, rules are set, and based on a genetic algorithm, the selection process runs to generate final form automatically. This algorithm is a "gene class" that controls the voxel, its condition, and its index in the matrix. The voxel condition could take into consideration the programmatic concept of different functions (Philippe et al. 2012).

Fig. 6.5 Folding structure of multi-agents that is adaptable (Sebastian et al. 2016)

6.4.2 Folding Agents' Structure

Swarm intelligence is applied in this application to make an interactive structure that interacts with the environment to reach the optimum performance based on learning from weather conditions (Fig. 6.5). This structure interacts with environmental conditions to gain solar radiation and convert it to power in case of cold weather. In case of hot weather, it can provide cool air inside the building by controlling the airflow according to the wind and also achieve the maximum shading. This structure is designed depending on two natural rules:

(1) Folding rule which is inspired from animals and plants and their ability to fold in a flexible way.
(2) Swarm intelligence rules that controls swarms' movement.

The designer integrates the swarm intelligence rules that control agents' movement in the design of the structure. This technique depends on learning from feedback as it learns from interacting with the environment and that allows the structure to respond in different manners. The structure consists of irregular shapes that fold with different angles to respond to different conditions, and that allows the structure to make different compositions to respond to different environmental conditions.

6.4.2.1 Design Method Based on Swarms

The system consists of triangular agents that generate a free form structure. These agents are robots changing direction and position to change the form of the whole structure to adapt to different environmental conditions. This mechanism produces different forms according to different environmental conditions and designer vision.

1. The project is considered as multi-agents system, it consists of 22 folding elements, and each element is considered as an agent.
2. Each agent connects with sensor and actuator that enables it to rotate with different angles.
3. Based on applying swarm behavior, agents interact with each other and with the environment, so they change position and their folding angles are modified.
4. Agents interact with each other and with the environment to achieve the best performance. The performance is tested on real-time related to the air flow, the shading, sunlight, sun heat, and wind.
5. Each agent is programmed to rotate independently but according to the composition that the whole structure seeks to reach, by learning from the feedback and responding to it.

6.4.2.2 Mechanism

The mechanism contains a spindle that allows the agents to rotate toward or away from each other. It contains 22 inner edges. Mechanism dimensions are 1.10 m × 1.40 m × 0.50 m. Figure 6.6 shows the basic idea of the mechanism. The main aim for the mechanism is to reshape from the initial composition to adapt to weather conditions (Fig. 6.6). As each agent reshapes to reach his target shape according to the whole structure composition (Benjamin et al. 2014).

6.4.3 Flying Architecture Agents

The designer tries to reach a self-organising flying architecture based on swarm intelligence concept. He uses components, which is considered as swarm agents that interact with each other and with the environment. When the interactions between these components reach a high number, these components will self-organise to make the whole architectural composition. The concept of flying architecture is inspired from the flying cities of Gregory Krutikov as you can see Fig. 6.7.

The main concept for the flying cities depends on swarm intelligence agents and mainly on the flocking of birds. It uses the Mascarillons which is developed in the University of Quebec as the main component. They apply the swarm intelligence rules on the components and discover its possibilities in self-organizing systems. It

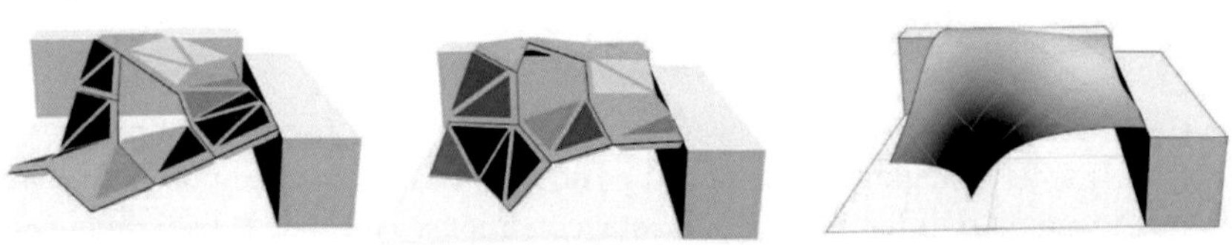

Fig. 6.6 Base idea for structural frame (Benjamin et al. 2014)

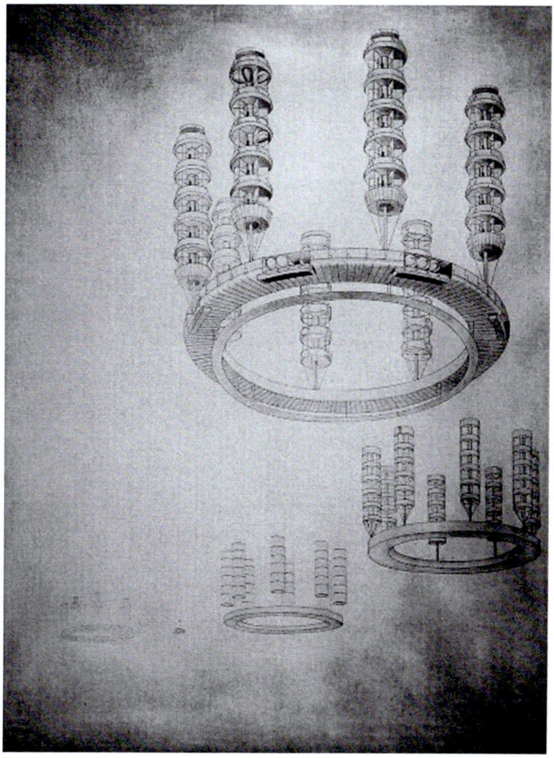

Fig. 6.7 Figure represents flying cities (Julien et al. 2005)

is about 20 cubes that can fly. They are put into semi-spherical environment. These cubes interact with each other and with the environment. The cubes will move to make a pattern as you can see in Fig. 6.8. A structure is produced from their interactions.

Fig. 6.8 Figure represents the Mascarillons in semi-spherical environment (Julien et al. 2005)

Fig. 6.9 Figure represents the real Mascarillon (Julien et al. 2005)

This example tries to reach an architectural form consists of components based on the swarm intelligence rules (Julien et al. 2005).

The flying units are built as cubes that are reduced in size and weight (Fig. 6.9). The cube side is 1.65 m. Each one is considered as a lightweight structure that contains helium. The cube consists of frame and envelope that is tightened to the frame. The cube structure is rigid, so the membrane can be tightened to it and at the same time that structure is light for the loads. Helium exists inside the cube envelope that is made of a polymer material that can expand and shrink according to heat that membrane makes the faces of the cube flat (Julien et al. 2005).

6.4.3.1 Design Methodology

1. Each flying cube contains sensor, actuator, wireless network, and a motor.
2. The flying cubes used the swarm rules (cohesion, alignment, and separation) to control the movement.
3. Based on swam intelligence, boxes are able to fly together in a swarm like birds and make a specific composition (Fig. 6.10).
4. They interact with each other and with the environment as they interact with the color and luminosity of the environment.
5. They make different patterns according to their interactions.

6.5 Analyzing the Three Case Studies

Each of these examples applies swarm intelligence in architectural design differently. The following are three concepts of applying swarm in design.

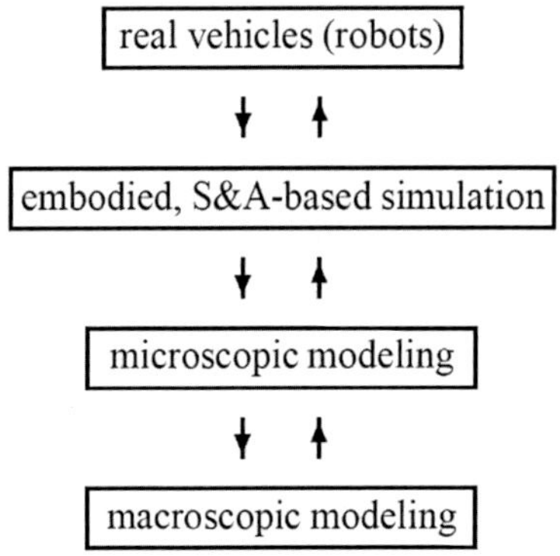

Fig. 6.10 Schematic view of different levels of interaction (Julien et al. 2005)

6.5.1 First Concept: Achieving Previously Set Aims for Design

In this concept, the best solution is selected to achieve specific aims for design. When setting aim for design, applying swarm intelligence has to be accompanied with other optimization algorithms. In the shown example of the housing units, the aim is set to achieve best privacy, view, and zoning.

6.5.2 Second Concept: Applying the Strategy of Learn and Respond to Provide Alternatives to Respond to Unpredicted Conditions

In this concept, swarm is applied by the concept of learn and respond. Swarms and their reactions are linked with real-time updates of design and external conditions.

In previously discussed folding agents' example, swarm makes responsive structure that can respond to different environmental circumstances, to improve the performance. That structure provide more shaded areas and increase the surface exposed to the sun. It uses irregular folding elements which make the structure able to increase the range of produced solutions. It could control the air flow through the building according to wind condition to provide fresh and cold air.

6.5.3 Third Concept: Generating Alternatives for a Design by Making the Design Process

In this concept, swarms are used to inspect alternatives for the design. Based on the designer point of view, the final design is selected. The rules of interaction of swarm agents manage the process that create the design instead of thinking about the final product.

In the discussed flying architecture agent example, the cubes are programmed to interact with each other to generate a form that is self-organizing depending on the interactions of the agent members. The flying structure consists of cubes that interact with each other in different colors and luminosity conditions in an indoor environment to produce the whole architectural composition depending on the number of interactions. If these interactions increased, the produced structure becomes more complex. This example transformed the process of architectural design into rules that control the components of the design.

6.6 Conclusion

Swarm intelligence applications in architectural design are still limited, due to its form complications so it is applied in research more than real world.

To apply swarm in architectural design, design elements should be translated to agents that interact with each other and with the environment.

It can be applied in three concepts, achieving design aims (when used with optimization algorithms), applying the strategy of learn and respond to provide alternatives, and generating alternatives for a design by managing the design process.

The most beneficial use of swarms is to make self-learning architectural product that learns from feedback and responds to different conditions. In this case, the outcome could be solving crucial design problems, rather than form finding.

References

Beekman M, Sword GA, Simpson SJ (2008) Biological foundations of swarm intelligence In: Blum C, Merkle D (eds). Springer, Berlin

Benjamin F, Daniel L, Jörg R, Sebastian W (2014) Experiments with a folding multi-agent system in the design of triangle mesh structures. In: 16th international conference on geometry and graphics (ICGG)

Danial Y, Sarvenaz S-I, Donya Y, Alireza S-M, Mohammad RM (2005) Fish swarm search algorithm: a new algorithm for global optimization. Int JArtif Intell

Girdhar A, Himani (2015) Swarm intelligence and flocking behavior. In: International conference on advancements in engineering and technology

Jan C, Kacper R (2018) Artificial intelligence aided architectural design. In: AI for design and build environment

Julien N, Nicolas R, Eric P, Alcherio M, Alan W (2005) MASCARILLONS : flying swarm intelligence for architectural research. Universit´e du Qu´ebec ‘a Montr´eal

Keerthi S, Ashwini K, Vijaykumar MV (2015) Swarm intelligence. Int J Comput Appl

Pham DT, Ghanbarzadeh A, Koç E, Otri S, Rahim S, Zaidi M (2006) *The Bees algorithm: a novel tool for complex optimisation problems*. Manufacturing Engineering Centre. Cardiff University, UK

Philippe M, Xavier M, Renato S, Gilles D (2012) Creativity with the help of evolutionary design tool. School of Architecture of Lyon, France

Sebastian W, Andreas W, Joerg RN (2016) Swarm intelligence in architectural design. Laboratory of Knowledge Architecture, Faculty of Architecture. TU Dresden

Yuxing C (2015) Swarm intelligence in architectural design. University of California, USA

Chapter 7
Biomimicry in Architecture: The Potential Shift from Theory to Practice

Sara Samy, Wafaa Nadim, and Morad Abdelkader

Abstract The contemporary mode of life has been significantly modified as a logical response to the environmental challenges. This can be seen in the gradual transition from the carbon-intensive age to the ecological age in architectural practices through more ecological solutions and more sustainable framework such as vernacular architecture, zero energy buildings, sustainable architecture, and green architecture. Although the importance of nature has remarkably increased in these approaches, the severe environmental problems need more than a "green" approach to overcome their consequences. Nature provides a large database of strategies and mechanisms that can be applied in different design disciplines. Biomimicry is a rapidly growing engineering design discipline based on abstraction of good design from nature; however, biomimicry is still an emerging field in architectural design due to the absence of a systematic design process model to guide throughout the design process. Different trials have been developed to bridge this gap by providing a systematic way of incorporating biomimicry into architecture. Each of the developed trials focused on some aspects and neglected others. The fragmented outcome of these trials needs to be consolidated toward a comprehensive application of biomimicry in architectural practices. This paper represents the first phase of an ongoing research. The paper aims at correlating biomimicry theories and architectural design process; this is achieved via a comprehensive critical analysis of biomimicry definitions, concepts, approaches, levels, initiatives to incorporate these into an integrated architectural design process and case studies of biomimicry-based architecture. A proposed conceptual framework is then developed to inform the second phase of the research which aims to develop a design process model for biomimicry-based architectural practices to shift from theory to practice.

S. Samy (✉) · W. Nadim
Architecture and Urban Design Program, German University in Cairo, Cairo, Egypt
e-mail: sarasamy_hms@hotmail.com

W. Nadim
e-mail: wafaa.nadim@guc.edu.eg

M. Abdelkader
Faculty of Engineering, Ain Shams University, Cairo, Egypt
e-mail: moradabdelkader@gmail.com

Architecture and Urbanism: A Smart Outlook,
https://doi.org/10.1007/978-3-030-52584-2_7

Keywords Biomimicry · Architectural design · Design process model

7.1 Introduction

For thousands of years, human beings used to fulfill their needs by observing natural systems and getting inspired from them. Nature always provides a large database of various adaptation strategies and mechanisms which is used as a problem solving guide where one can find different solutions for specific problem by imitating nature. One famous example of this process is developing the first flying machine by Wilbur and Orville Wright by imitating the ability of birds to fly.

In the last decade, the contemporary mode of life has been changed gradually from the carbon-intensive age to more ecological age; hence, more sustainable approaches have been emerged in different design disciplines. Many studies have been conducted to define how to extract good design solutions from natural systems, abstract them, and then apply these to a man-made activity.

In 1997, Janine Benyus—a writer and a naturalist—has first coined the term "Biomimcry" in her book "Biomimicry: Innovation Inspired by Nature" to describe this process (Benyus 1997). The term biomimicry comes from a combination of the Greek roots bios (life) and mimesis (imitation).

Different definitions have been introduced to describe the term "biomimicry." It is defined as "learning from and then emulating natural forms, processes, and ecosystems to create more sustainable designs" (Benyus 1997). Vincent et al. (2006) defined it as "abstraction of good design from nature." Janine Benyus introduced a comprehensive, yet brief, definition of biomimicry as "the conscious emulation of nature's genius" (Benyus 1997).

Inspite of the widespread applications of biomimetic design as a design method, it is still an emerging field on the level of architectural design. This is due to the absence of a systematic design process model that enables identifying the relevant natural systems and then abstracting their strategies and mechanisms. This paper introduces a comprehensive critical analysis of biomimicry definitions, principles, approaches, levels, and initiatives to determine how to incorporate them into architectural design practices.

7.2 Aspects of Biomimicry-Based Architecture

7.2.1 Principles

Benyus (1997) defined a set of life's principles in her book to set measures or laws that define different levels of biomimicry design practices. These life's principles are nature runs on sunlight, nature uses only the energy it needs, nature fits form to function, nature recycles everything, nature rewards cooperation, nature banks on diversity, nature demands local expertise, nature curbs excesses from within, nature taps the power of limits.

In 2015, biomimicry design lens have been released which translates the life's principles introduced by Benyus J. into a set of applicable principles for a biomimicry-based design as follows: Adapt to changing conditions, be locally attuned and responsive, use life-friendly chemistry, be resource efficient, integrate development with growth, and evolve to survive (Biomimicry 3.8 2016).

7.2.2 Approaches

To find a solution to any design problem through imitating nature, two approaches are commonly used:

- *Problem-based approach* (Vattam et al. 2009) also referred to as "design looking to biology" (Biomimicry Guild 2007), "top-down approach" (Milwich et al. 2008), or "challenge to biology approach" (Baumeister 2012). In this approach, the design problem is defined, then the organism with similar problem is identified, and the solution is examined and transferred to an applicable strategy.
- *Solution-based approach* (Vattam et al. 2009) also referred to as "biology influencing design" (Biomimicry Guild 2007), "bottom-up approach" (Milwich et al. 2008), or "biology to design approach" (Baumeister 2012). In this approach, an organism's characteristic is observed, then an abstraction of this characteristic is defined, a similar design problem is defined, and finally, the characteristic is transferred to an applicable solution.

7.2.3 Levels

Mimicking nature's solutions can be carried out on different levels. Benyus (1997) defined three main levels from the simplest to the most comprehensive as: form, process, and ecosystem.

On the other hand, Zari (2007) proposed similar levels yet with different terminologies. The proposed levels are organism level (instead of form), behavioral level (instead of process), while they both agreed on the title of the third level "ecosystem." She also investigated the three levels further and produced five tactics that should be addressed in each level which are form, material, construction, process, and function.

The form (organism) level entails mimicking the form or the characteristic of a certain living organism. The process (Behavior) level means to mimic a whole process or behavior of a specific organism to adapt within nature. While the ecosystem level mimics the different processes interaction and integration to achieve the balance state of an ecosystem.

7.2.4 *Building Systems*

In this research, the term "building systems" refers to the different systems of the building that can utilize biomimicry concepts toward an enhanced performance such as ventilation, lighting, and resources consumption.

Biomimicry-based architecture should consider and imply biomimicry concepts in its building systems. Nature's solutions should be the main reference to find the optimum way to ventilate or daylit a building. Nature can also guide how the different systems can work together and how to resolve conflicts between the different systems.

7.3 Initiatives of Incorporating Biomimicry into Architecture

As mentioned above, biomimicry has been adopted as a sustainable tool to design more efficient designs in different design disciplines. However, on the architectural design level, there is arguably still a gap between biomimicry theories and their applications.

Various initiatives have been developed to bridge this gap and to incorporate biomimicry into architecture through providing a framework that links the biomimicry theories and architectural design practices. Eight initiatives have been analyzed in this study which are biomimicry theoretical framework (Gamage and Hyde 2012), the living envelope methodology (Badarnah 2012), biomimetics design methodology (Fornies and Muro 2012), AskNature (AskNature 2018), biomimicry approach model to architectural design (Zejnilović and Husukić 2015), biomimicry application framework (Zari 2007), Deign Spiral (Biomimicry 3.8 2016), and American National Standards Institute Integrative Process Model (ANSI IP©) (ANSI 2012).

In spite of the positive influence of these initiatives on biomimicry-based architectural practices, each of them focused on one or more aspects rather than considering a comprehensive holistic design process. The following part investigates these initiatives in terms of biomimicry-based architecture aspects in an attempt to identify their characteristics, addressed aspects, similarities, and differences.

7.3.1 *Biomimicry Theoretical Framework (BTF)*

Gamage and Hyde (2012) investigated biomimicry as a potential approach to integrate ecology to architectural design through imitating the natural systems processes. The study proposes a biomimicry theoretical framework that addresses the three main levels of biomimicry (organism, behavior, and ecosystem). It also highlights that these levels can be applied through two main biomimicry approaches (problem-based and solution-based) including four stages which are innovation in form, environmental adaptation, functional integration, and categorization.

7.3.2 The Living Envelope Methodology (LEM)

Badarnah (2012) proposed a strategic methodology to help in utilizing biomimicry concepts in the building envelope. The living envelope methodology (LEM) aims at generating design concepts for a "living envelope" based on strategies found in nature. The proposed methodology is based on two main biomimicry approaches (problem-based and solution-based). Both approaches follow a design process of three phases—with different flow—which are challenge, solution, and application. (Badarnah 2012, 2017).

7.3.3 Biomimetics Design Methodology (BDM)

Fornies and Muro (2012) also proposed a biomimetics design process to incorporate biomimicry into industrial design. The proposed process agrees with the previous initiatives in terms of design process approaches (problem-based and solution-based approaches) and design process phases (challenge, solution, and application).

The proposed biomimetics design process investigated further the different stages under each phase, so the challenge phase is divided into two stages: "setting the framework" and "key function definitions," then the solution phase is divided into "biological research" and "translate biological success into key functions," and finally, the application phase is composed of "conceptual design" and "detailed design."

7.3.4 AskNature

AskNature is an online database which facilitates incorporating nature solutions into the design process. The Web site classifies the natural models through two different classification schemes. In the first scheme, the natural organisms are classified by the different functions they use to adapt to the local context such as break down, get, store, distribute resources, maintain community, make, modify, move or stay put, process information, and protect from physical harm. In this scheme, the design team is facing a specific problem or challenge and looking for relevant solution for the problem; hence, it is using a problem-based approach.

The second scheme categorizes the natural models according to the different types of the living systems including plants, mammals, human, bacteria, insects, western honeybee, aves, butterflies and moths, animalia, and ants. The second scheme flows from exploring different natural solutions, thinking of relevant design problem, and applying the solution to a design problem, so it follows a solution-based approach (AskNature 2018).

7.3.5 Biomimicry Approach Model to Architectural Design (BAMAD)

Zejnilović and Husukić (2015) suggested a model to approach biomimicry in architectural design. The model considers environmental, social, and economic aspects of biomimicry-based architecture to create buildings that consider occupants comfort.

The model agreed with the previous initiatives on the two main biomimicry approaches (problem-based and solution-based approaches) and further investigated the three main aspects (environmental, social, economic aspects) to determine the factors affecting the biomimicry-based architecture.

7.3.6 Biomimicry Application Framework (BAF)

Zari (2007) proposed a framework for biomimicry applications based on two main aspects: biomimicry approaches and levels. In the proposed framework, biomimicry applications are achieved through the two main approaches (problem-based and solutions-based) using three levels (organism, behavior, and ecosystem). Five main tactics have been determined to be addressed in each level which are form, material, construction, process, and functions.

7.3.7 Design Spiral

In 2017, Biomimicry Institute has developed a tool that provides specific steps toward a successful biomimetic design. The developed design spiral divides the design process into three main phases: problem, solution, and application. It also divides these three phases into six main stages which are identify challenge, biologize function and context, discover biological strategies, abstract design strategies, emulate nature's lessons, and evaluate fit and function.

7.3.8 American National Standards Institute Integrative Process Model (ANSI IP©)

American National Standards Institute (ANSI) has developed an integrative process model standard (ANSI IP©) for design and construction of sustainable buildings and communities which is used as a basis for the Leadership in Energy and Environmental Design (LEED) green building rating system.

Although this initiative is not directly oriented to biomimicry-based architecture, it addresses a crucial aspect of biomimicry-based architecture which is the integrative design process. For that reason, ANSI IP© has been added to the selected initiatives in this study.

ANSI IP© divides the design process into two main phases (discovery and design and construction) which includes stages (preparation, evaluation, conceptual design, schematic design, design development, and construction documents). It also addresses six main building systems which are material, water, energy, habitat, light, and air. (ANSI 2012).

Different aspects of biomimicry-based architecture have been addressed in the analyzed initiatives. Table 7.1 summarizes the analysis of the selected initiatives.

The analysis clearly shows the existing gap in the developed initiatives. Some aspects are addressed and agreed on in the analyzed initiatives, and others are not addressed at all or not agreed on. To develop a defined framework for biomimicry-based architecture, the analyzed aspects will be further investigated in the following part through selected case studies.

7.4 Case Studies of Biomimicry-Based Architecture

The following part analyzes three selected case studies of biomimicry-based architecture in terms of the previously mentioned aspects.

7.4.1 The Ministry of Municipal Affairs and Agriculture, Doha, Qatar

The form of the building is inspired from the cactus plant that is found in the harsh desert climates similar to the climate of Qatar.

Different references claimed that the building is designed to mimic the ability of cactus to survive in the harsh climate by performing transpiration activity at night rather than at day time to conserve its water from getting evaporated. This is done through designing the windows with movable shades that opens and close according to sunlight intensity (Inhabitat n.d.; Mansour 2010; Anous 2015; Dharek et al. 2017; Aesthetics Architect 2019).

Although movable windows are efficient feature to adopt in harsh climates, it is irrelevant to the functions the cactus applies to adapt such as night transpiration and self-shading spines.

7.4.2 Hydrological Center for the University of Namibia, Namibia

The design team focused on the local priorities of this context which is scarcity of water. The design solution is based on the ability of the Namibian beetle to harvest water from the foggy weather.

Table 7.1 Initiatives of incorporating biomimicry into architecture

Aspects	BTF	LEM	BDM	AskNature	BAMAD	BAF	Design Spiral	ANSI IP©
Principles	Adapt to changing conditions	Adapt to changing conditions	Adapt to changing conditions	Adapt to changing conditions	Adapt to changing conditions	Adapt to changing conditions	Adapt to changing conditions	–
	Be locally attuned and responsive	Be locally attuned and responsive	Be locally attuned and responsive	Be locally attuned and responsive	Be locally attuned and responsive	Be locally attuned and responsive	Be locally attuned and responsive	
	Use life-friendly chemistry	Use life-friendly chemistry	Use life-friendly chemistry	Use life-friendly chemistry	Use life-friendly chemistry	Use life-friendly chemistry	Use life-friendly chemistry	
	Be resource efficient	Be resource efficient	Be resource efficient	Be resource efficient	Be resource efficient	Be resource efficient	Be resource efficient	
	Integrate development with growth	Integrate development with growth	Integrate development with growth	Integrate development with growth	Integrate development with growth	Integrate development with growth	Integrate development with growth	
	Evolve to survive	Evolve to survive	Evolve to survive	Evolve to survive	Evolve to survive	Evolve to survive	Evolve to survive	
Levels	Organism	–	–	–	Organism	–	–	–
	Behavior				Behavior			
	Ecosystem				Ecosystem			

(continued)

Table 7.1 (continued)

Aspects	BTF	LEM	BDM	AskNature	BAMAD	BAF	Design Spiral	ANSI IP©
Building Systems	• Spatial and temporal patterns	Morphology • Form • Structure	–	–	–	• Form • Material • Construction • Process • Function	–	• Material • Water • Energy • Habitat • Daylight • Natural ventilation
	• Processes	Processes • Water • Heat • Light • Air	–	–	–	–	–	–
Approaches	Problem-based	Problem-based	Problem-based	Problem-based	Problem-based	Problem-based	Problem-based	–
	Solution-based	Solution-based	Solution-based	Solution-based	Solution-based	Solution-based	Solution-based	
Phases	–	Challenge	Challenge	–	Challenge	Challenge	–	Discovery
		Solution	Solution		Solution	Solution		Design and construction
		Application	Application		Application	Application		

(continued)

Table 7.1 (continued)

Aspects	BTF	LEM	BDM	AskNature	BAMAD	BAF	Design Spiral	ANSI IP©
Stages	• Innovation of Form	• Identify	• Setting the framework	–	Problem-based • Design problem • Biological analogies • Identification of appropriate principles • Biological model • Analysis and feedback • Design solution	–	• Define Challenges	• Preparation
	• Environmental Adaptation	• Explore					• Analyze Function and Context	• Evaluation
	• Functional Integration	• Discover and Extract	• Key functions definitions				• Discover Biological Strategies	• Conceptual design
	• Categorization	• Elaborate and Analyse						• Schematic design
		• Classify	• Biological research				• Abstract Design strategies	• Design development
		• Abstract	• Translating biological success into key functions				• Emulate Nature's Lessons	• Construction documents
		• Concept					• Evaluate Fit and Function	
		• Transform and Evaluate	• Conceptual design		Solution-based • Biological research • Understanding the principles • Detachment from model • Technical implementation • Design problem			
		• Prototype and Validate	• Detailed design					

The design team applied the same technique by designing a wall that imitates the shape and the characteristics of the Namibian beetle's back, and they designed a water harvesting system based on a curved nylon net wall which faces the ocean and condensates the moisture of the ocean breeze. The designed system integrates with an infrastructure to collect, guide, and store the water in an underground water tank (Maglic 2014; Zari 2007; El Nahas 2014).

7.4.3 Zira Island Masterplan, Baku, Azerbaijan

The design proposal for Zira Island is based on mimicking the natural landscape of Azerbaijan as it is known by its seven peaks. The design does not only imitate the iconic morphology of the seven peaks but also an actual living model of the integrated ecosystem of Azerbaijan's mountains, where the flow of air, water, heat, and energy are handled the same way the nature does.

This was achieved through different strategies such as utilizing the tidal power of the Caspian sea to heat and cool the island buildings, integrating photovoltaic panels in the architecture and wind turbines to supply the energy needed in the buildings, treating wastewater and storm water and reusing them in irrigation, composting solid waste to be used as a fertilized top soil which promotes the existing green cover of the tropical island with minimal footprint (Dharek et al. 2017; ArchDaily 2018).

Table 7.2 summarizes the analysis of the selected case studies in terms of biomimicry-based architecture aspects.

The analysis of the initiatives and case studies has shown the need for a comprehensive framework which consolidates the fragmented aspects of a successful biomimicry-based architecture.

7.5 The Proposed Framework of Biomimicry-Based Architecture

Based on the previous analysis, a comprehensive framework is proposed to bridge the existing gap between biomimicry theories and practices in the field of architectural design. The proposed framework consolidates the outcomes of the analyzed initiatives in an attempt to develop a holistic framework which addresses all aspects of a biomimicry-based architecture. Figure 7.1 represents the proposed framework of biomimicry-based architecture.

Some aspects have been agreed in most of the initiatives such as biomimicry principles, levels, and approaches, while other aspects have only been addressed in one or two initiatives such as building systems. In that case, the most holistic approach is adopted in the proposed framework of this study. For example, since the building systems addressed by ANSI IP© are the most comprehensive, they have been adopted in the proposed framework. Also, for the stages of the design process,

Table 7.2 Analyzing the selected case studies in terms of aspects of biomimicry-based architecture

Aspects	Case Study 01	Case Study 02	Case Study 03
Principles	Adapt to changing conditions	Adapt to changing conditions	Adapt to changing conditions
	Be locally attuned and responsive	Be locally attuned and responsive	Be locally attuned and responsive
	Use life-friendly chemistry	Use life-friendly chemistry	Use life-friendly chemistry
	Be resource efficient	Be resource efficient	Be resource efficient
		Evolve to survive	Evolve to survive
			Integrate development with growth
Levels	Organism	Organism	Organism
		Behavior	Behavior
		Ecosystem	Ecosystem
Building Systems	Form	Form	Form
		Water	Water
			Energy
			Habitat
			Natural ventilation
Approaches	Problem-based	Problem-based	Problem-based
Phases	Challenge	Challenge	Challenge
	Solution	Solution	Solution
	Application	Application	Application
Stages	Define challenges	Define challenges	Define challenges
	Analyze context and functions	Analyze context and functions	Analyze context and functions
	Discover biological strategies	Discover biological strategies	Discover biological strategies
	Abstract design strategies	Abstract design strategies	Abstract design strategies
	Emulate nature's lesson	Emulate nature's lesson	Emulate nature's lesson
	Evaluate fit and function	Evaluate fit and function	Evaluate fit and function

different sets of stages have been introduced in the analyzed initiatives; yet, the design spiral has introduced the most comprehensive set of stages that is relevant to biomimicry approaches and phases. Hence, the design spiral's set of stages has been adopted in the proposed framework.

In the framework, the concluded aspects are arranged in a conceptual order based on how they are used in an actual design process. Guiding arrows have been added to show the flow from one aspect to another. The framework starts with biomimicry

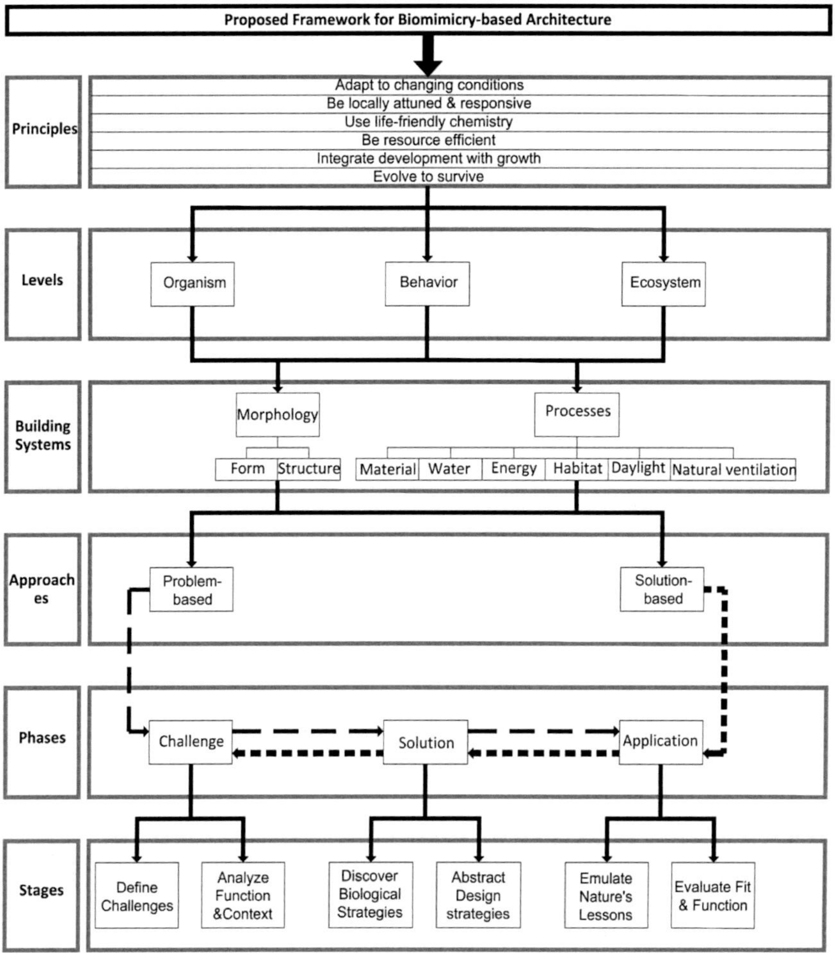

Fig. 7.1 Proposed framework for biomimicry-based architecture

principles, levels, building systems, approaches, phases, and ending with stages. The applied biomimicry principles are defined followed by biomimicry levels in which the design team should determine which level will be implied in the project.

Based on the defined level, the biomimicry-based building systems should be defined.

A biomimicry approach should be then selected to define the sequence of the design process phases and stages. In the problem-based approach, the design process starts with defining challenges, searching for nature's solutions, and applying the optimum solution, while in the solution-based approach, the design process starts with exploring the natural strategies and applications, selecting relevant strategies (solutions), and applying the selected solution on a relevant design problem.

7.6 Conclusions

This paper tackled the existing gap between biomimicry theories and practices in the field of architectural design through a literature study of biomimicry and its relation to architectural design process. It discussed the different definitions, principles, approaches, and levels, and building systems of biomimicry-based architecture.

Different initiatives of incorporating biomimicry into architecture have been also analyzed to define the main aspects that should be addressed in biomimicry-based architectural design process. In addition, selected case studies of biomimicry-based architecture have been investigated in terms of the concluded aspects.

A comprehensive holistic framework has been proposed based on the previous analysis in an attempt to consolidate the existing fragmented trials of incorporating biomimicry into architecture.

The proposed framework organizes the different aspects of a biomimicry-based architecture and shows their correlation to eachs other. It also sets the base for developing a more comprehensive design process model of biomimicry-based architecture in the second phase of the research.

References

Aesthetics Architect (2019) Retrieved 02 14, 2019, from Aesthetics Architect GO Group., Ltd. Website: https://www.aesarchitects.com/page.php?doc=profile&cat=3&id=59
Anous I (2015) Biomimicry: innovative approach in interior design for increased sustainability. AIJRFANS 15(217):18–27
ANSI (2012) Integrative process (IP) ANSI consensus standard guide 2.0 for design and construction of sustainable buildings and communities: American National Standards Institute
ArchDaily (2018) Retrieved 02 15, 2019, from ArchDaily Website: https://www.archdaily.com/12956/zira-island-carbon-neutral-master-plan-big-architects
AskNature (2018) Retrieved 06 18, 2019, from AskNature Website: https://asknature.org/?s=&p=0&hFR%5Bpost_type_label%5D%5B0%5D=Biological%20Strategies
Badarnah L (2012) Towards the LIVING envelope: biomimetics for building envelope adaptation. Doctoral dissertation, Delft University of Technology
Badarnah L (2017) Form follows environment: biomimetic approaches to building envelope design for environmental adaptation. Buildings 7(2):40
Baumeister D, Tocke R, Dwyer J, Ritter S, Benyus J (2012) Biomimicry resource handbook. A seed
Benyus JM (1997) Biomimicry: innovation inspired by nature
Biomimicry 3.8 (2016) Retrieved 06 18, 2019, from Biomimicry 3.8 Website: https://biomimicry.net/the-buzz/resources/designlens-lifes-principles/
Biomimicry Guild (2007) Innovation inspired by nature work book
Dharek M et al (2017) Application of biomimicry towards sustainability of buildings. IJSER 8(10):23–26
El Nahas MMH (2014) Enhancinthe efficiency of natural ventilation systems by biomimicry to achieve sustainability in office buildings. Int J Acad Res 6(5)
FORNIÉS IL, Muro LB (2012) A top-down biomimetic design process for product concept generation. Int J Des Nat Ecodyn 7(1):27–48
Gamage A, Hyde R (2012) A model based on Biomimicry to enhance ecologically sustainable design. Archit Sci Rev 55(3):224–235

Inhabitat. (n.d.). Retrieved 02 15, 2019, form Inhabitat Website: https://inhabitat.com/qatar-cactus-office-building/
Maglic JM (2014) Biomimicry: using nature as a model for design. Masters Dissertation, The University of Massachusetts Amherst
Mansour H (2010) Biomimicry A 21st century design strategy integrating with nature in a sustainable way. BUE, FISC 12
Milwich M, Planck H, Speck T, Speck O (2008) The role of plant stems in providing biomimetic solutions for innovative textiles in composites. In: Biologically inspired textiles. Woodhead Publishing, pp 168–192
Vattam SS, Helms ME, Goel AK (2009) Nature of creative analogies in biologically inspired innovative design. In: Proceedings of the seventh ACM conference on Creativity and cognition. ACM, pp 255–264
Vincent JF, Bogatyreva OA, Bogatyrev NR, Bowyer A, Pahl AK (2006) Biomimetics: its practice and theory. J Roy Soc Interface 3(9):471–482
Zari MP (2007) Biomimetic approaches to architectural design for increased sustainability. In: The SB07 NZ sustainable building conference
Zejnilović E, Husukić E (2015) Biomimicry in Architecture. Int J Eng Res Dev 11(07):78

Chapter 8
Towards Multisensory Architecture: A Phenomenological Investigation of Vivid Experiences

Dina Bakir, Yasser Mansour, Shaimaa Kamel, Yasser Moustafa, and Marwa H. Khalil

Abstract Architecture has long focused on the visual sense with the neglect of other human senses. However, due to the influence of phenomenological writings, there is now a growing notion that our perception of the world is actually multisensory in nature, involving all our senses. By engaging all the senses, the built environment can help provide users with richer experiences. This paper investigates the phenomenology of multisensory perception of architecture. The aim is to contribute to a better understanding of the relationship between spatial qualities and the multisensorial perception of the built environment. The objective is to analyse different vivid experiences. Vivid experiences are experiences characterized by powerful feelings or clear images in the mind. Through the exploration of such vivid environmental experiences, the paper intends to determine the spatial qualities of the built environment that can induce an unforgettable experience. A phenomenological research approach is adopted. The study relies on two informants' vivid experiences. These experiences were documented through personal diaries. To complement findings from the diaries, in-depth semi-structured interviews were conducted. A qualitative interpretive analysis is carried out. The study identifies themes describing the ambiences associated to informants' vivid experiences. Among those ambiences are tenseness, vibrant, ambiguity, attractiveness, tranquility, and predictability. Yet, each experience is shaped by a core ambience. Hence, this paper will focus on the core ambiences: tenseness ambience and vibrant ambience. The findings are discussed

D. Bakir (✉) · Y. Mansour · S. Kamel
Faculty of Engineering, Ain Shams University, Cairo, Egypt
e-mail: Dina_bakir@aast.edu

Y. Mansour
e-mail: yasser_mansour@eng.asu.edu.eg

S. Kamel
e-mail: shaimaa_kamel@eng.asu.edu.eg

D. Bakir · Y. Moustafa · M. H. Khalil
Arab Academy for Science, Technology and Maritime Transport, Cairo, Egypt
e-mail: yasser_moustafa@aast.edu

M. H. Khalil
e-mail: marwakhalil@aast.edu

Architecture and Urbanism: A Smart Outlook,
https://doi.org/10.1007/978-3-030-52584-2_8

in relation to multisensory spatial qualities that have stimulated informants' senses, tying the outcomes to findings from previous research and literature.

Keywords Multisensorial architecture · Ambience · Vivid experiences · Spatial qualities

8.1 Introduction

Architecture has long focused on the visual sense with the neglect of the other human senses, to a degree that architects' design processes attribute a visual bias. As Juhani Pallasmaa states;

> Architecture at large has become an art of the printed image fixed by the hurried eye of the camera. The gaze itself tends to flatten into a picture and lose its plasticity; instead of experiencing out being in the world, we behold it from outside as spectators of images projected on the surface of the retina. (Holl et al. 1994, p. 30)

After highlighting that architecture nowadays is dominated by the visual sense, the phenomenological approach results in the consideration of architectural experience as an embodied experience that is generated through sensory perception. Consequently, many architects and theorists called for a connection with a built space that engage much more than the sighted related experience. Holl's experience of the Ryoan-ji Temple went beyond the appreciation of the space's visual beauty. The space entered his body not only through his eyes, but also through his feet, his ears, and his nostrils (Holl et al. 1994). As he describes: "*…realized how cold it was as moved through the paper-screened walkway and felt the cold boards through thin cotton socks*" (Holl et al. 1994, p. 122). The experience of architecture is a perceptual process, in which the quality of space, matter, and scale is assessed through the combination of multiple senses (Rasmussen 2001; Bloomer and Moore 1977). Architectural design should provide a multisensory experience for people in order to enhance their holistic experience. Emerging research in a variety of fields suggests that vivid experiences are more likely to occur when different senses are stimulated. Through the exploration of such vivid environmental experiences, **the paper intends to determine the spatial qualities of the built environment that can induce a holistic and unforgettable experience**.

8.2 Nature of Vivid Experiences

Vivid experiences, either positive or negative, are experiences characterized by powerful feelings or clear images in the mind. Vividity in cognition theory refers to how easily recalled and convincing a message is (Plous 1993). Additionally, it refers to the sensory components of memory on the past experience (Rubin et al. 2003), suggesting that vivid experiences occur when different senses are stimulated. "The more senses an experience engages, the more effective and memorable it can be" (Pine and Gilmore 1998, p. 104). Similarly, early childhood sensory experiences of a specific place form a vivid and deeply felt attachment of such places (O'neill 2001). Furthermore, sensory stimulation has shown to have positive effect on people with dementia, showing that patients with memory problem, remembered what happened in a previous day, for the very first time, just because of a great sensory stimulation, such as by offering them to make their own ice cream from scratch (Drønen 2013).

Vivid experience research has contributed significantly in fields such as economic business and marketing (Charles et al. 2014). Vividness is particularly important to tourism studies, which has long had the goal of creating vivid memories in the mind of visitors—memories that they will remember for the rest of their lives. The previous research exploring vivid touristic experiences extracted a set of themes as antecedents of memorable tourism among those themes—perceived novelty and local hospitalities (Chandralal and Valenzuela 2013). After reviewing the existing body of research on vivid experiences, it was found that very few research has been done in the field of architecture. Hence, this study tends to answer the following question: what is the role of the built environment in creating vivid experiences?

8.3 Methodology

8.3.1 Research Design

In order to investigate the possibilities of an architectural multisensory design approach, it seems logical that such investigation could benefit from exploring people's lived experiences. The study relies on the analysis of two vivid experiences accounts, with the aim of determining the spatial qualities of the built environment that can induce such holistic and unforgettable experience. This is a task to which the phenomenological research approach is particularly suited. The phenomenology goal is attaining deeper understanding of how humans experience the world they live in (Broussard 2006; Streubert and Carpenter 2011; Rapport and Wainwright 2006). This approach is successful in the investigation and description of all phenomena in the way a phenomena appear. A significant element of this approach is that informant shares experts about their own experience and can offer the researcher an understanding of their thoughts and feelings through their own words (Reid et al. 2005). Typical sample sizes for phenomenological studies range from one to ten persons (Starks and Trinidad 2007). This research is based on two informants' vivid experiences. Although the study is considered limited in terms of the number of informants and that a larger sample might have provided a broader range from which to extract the essence of the phenomenon, data from only a few individuals who have experienced the phenomenon and are willing to provide a detailed account of their experience, was shown to cater for uncovering the phenomenon core elements (Boddy 2016). Additionally, other research stated that depending on the researcher's goals of the study, one informant (sometimes referred to as a co-researcher) can be sufficient for the research inquiry (Starks and Trinidad 2007). Narrative data collection methods will be adopted with their significant advantage that they capture the experience as it is lived out. A narrative inquiry focuses on people, places, and things, all of which are in intellectual and moral relationship. Also, it is based on people making meaning of their lives by reflecting upon their experiences in form of stories (Franzosi 1998). Therefore, by analysing informants' narratives, the research will answer the following questions that guide the study: **what made both informants' experiences vivid? What are the physical elements in the built environment that stimulated informants' senses and created those vivid experiences?**

8.3.2 Accounts of the Two Experiences

Purposeful sampling is employed, by choosing two informants who perceived themselves as having a personal lived experience that until present is characterized for them by a strong, clear image in their minds—vivid experience. Informants with experiences to a newly visited place are chosen, since human senses tend to be more alert when experiencing novel places (Holahan 1982; Wilkie and Roberson 2012). Moreover, informants who are willing to participate and among literate people, who usually use diaries as a method for capturing their experiences, reflections, and events near the time they occur, are chosen. Both experiences were in winter 2018, reflecting upon their experiences a couple of hours after the experience.

The first informant was contacted because of her work as a young female photographer and her ongoing exploration for abandoned places especially cemeteries that she claims to find as relaxing places. Her experience was a visit to a crematorium complex in Bordeaux, France, for people that died in wars. The complex is composed of several things, stone graves in a garden, a crematorium, a place where a dead person's body is burnt into ashes, and a temple-like building where mourners can have a ceremony by a coffin She personally documented her experience through audio recording, the recording spanned 60 min. The second informant is a young female architect, visiting the "Your rainbow panorama" in Denmark. A glass circular walkway, 150-m-long and three-metre-wide walkway in all the colours of the spectrum. The informant documented her experience and sensations through writing in her personal diary a couple of hours after the visit.

8.3.3 Data Analysis

Written diary and transcription of the audio recording are analysed qualitatively using a phenomenological data analysis method adapted from Diekelamann (2001). The transcription stage is followed by reading and re-reading the reflective diaries and transcripts in order to acquire familiarity and understanding. The third step was about realizing the meanings; it was done through diverse coding cycles. The first main cycle was the extraction of *sensory terms* that the informants mentioned when describing their vivid experiences. The following cycle was extracting sensory cues in relation to the *physical characteristics* of the built environment. In order to complement the findings of the initial data analysis, in-depth semi-structured interviews were conducted. Informants' answers were collected through a digital audio recorder, and both interviews lasted for about 45 min. A thematic analysis of the interview transcripts was conducted. It is a method of analysis that is widely used in qualitative research (Braun and Clarke 2008), themes identifying the important data in relation to the research question proposed. The researcher judgment is important to determine what a theme is (Braun and Clarke 2008, p. 79). In conducting the thematic analysis, it is important to try to read each description with fresh eyes allowing the important moments to step out of the linguistic framework (Barrit et al. 1985, p. 41). Both studies indicated that a vivid experience of the built environment is shaped by a certain ambience. An ambience that allowed their perception to create a mental mapping for the sensorial information then retains it as a strong clear image in their minds. What is referred here as ambience (Kuijsters et al. 2014), was previously termed as atmosphere (Zumthor 2006; Kotler 1974). Yet both of those terms are used to describe the surrounding quality. Informants perceived a characterized ambience of the built environment evoking positive as well as negative feelings. Informants expressed those ambiences through sensory terms such as (*striking, frightening beautiful, peacefulness, calmness, silence, anxious, dynamic, cheerful, explorative, quiet, oppressive, energetic, and varied*). Words that occur frequently are often seen as being salient in the respondent's minds (Ryan and Bernard 2003). Sensory terms were sorted into positive terms and negative terms. And then, they were related to each other and a set of ambiences associated to the studied vivid experiences was derived. Those ambiences were **serene/tranquil ambience** (*peaceful, calm, quiet, silent, stillness*), **tenseness Ambience** (*frightening, anxious, oppressive*), **vibrant ambience** (*varied, dynamic, striking, energetic*), **attractiveness** (*beautiful, cheerful, stimulating*), **ambitious ambience** (*curiosity, discover, exploration*). The senses of the informants that recognized such ambient characters were the three sensory modalities (**hearing, touch/skin, and vision**), and the sense of position in the space and movement through the space is referred to in the sensory studies as **kinesthetic sensations and proprioceptive** cues (Proske and Gandevia 2012, 2018).

8.4 Findings

The following sections present and discuss the emergent themes. Those themes are describing the ambiences associated to the informants' vivid experiences. For the first informant, *tenseness* ambience is the most expressed for her memorable experience, while for the second informant, *vibrant* ambience is core in shaping her vivid experience. Each section will link each ambience to physical aspects of the built environment that stimulated the informants' different sensory channels (see

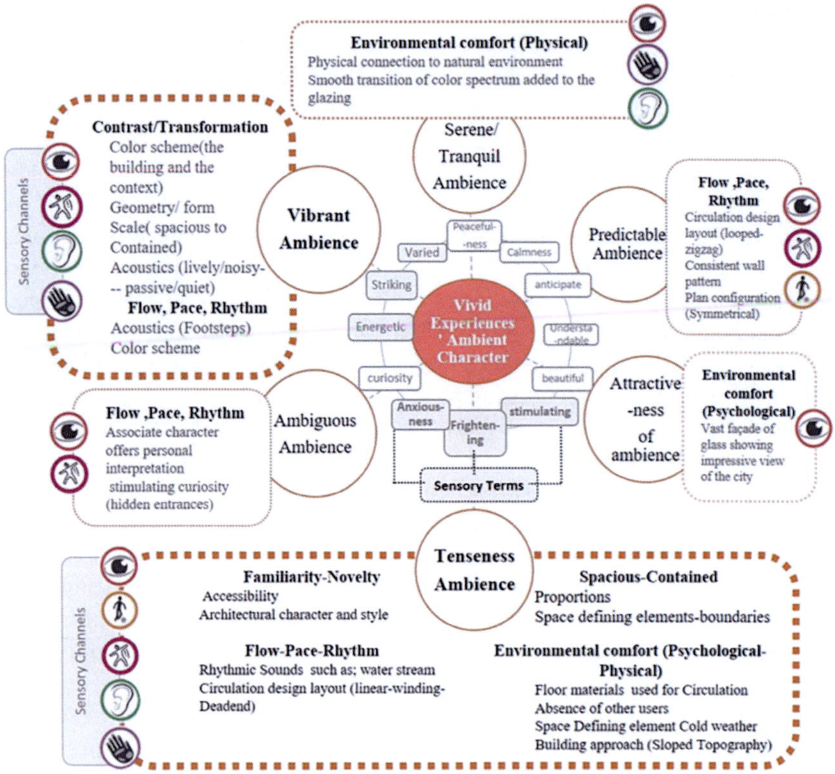

Fig. 8.1 Diagram showing the study findings

Fig. 8.1), presenting them through the illustration of direct quotes from the reflective diaries and interviews and discussing in relation to relevant findings from the previous research and literature.

8.4.1 Tenseness Ambience

Tenseness ambience is the core ambience associated to the first informant's vivid experience. Through her narration, she highlighted that her whole journey was not an easy one, describing it through sensory terms like "*horrible*", "*scary*", "*fearful*", and "*dangerous*". The informant tended to discuss a variety of physical elements that stimulated her senses (sound, vision, touch, and basic orientation sensory systems), hence apprehended such an ambience.

She conveyed that her sense of *novelty* when comparing it to other highways she is used to—*non-familiar*—played a great role in evoking such feeling. As well as the absence *of space defining elements for the sidewalk*; *"the highway was not like one in Egypt …so I felt that it is so dangerous…and there was no pedestrian pathway. I was walking on the edge of the fence and not even a space defined as pedestrian path way"*. After reaching the place, she described the overwhelming *emptiness* that created a *sense of exclusiveness* and that the place is not open to the public, which increased her level of anxiety: "*As I reached the place, at first appeared to be very scary as there was no one…I was filled with anxiety at first*". Hence, the absence of others affected her *psychological comfort*. Additionally, she was reflective of something memorable and lasting, particularly through the utilization of *rhythmic soundscape* such as the sound of water stream and a buzzing sound of a light fixture: "*The building is built on a mountain and underneath it passes a river …so as long as I walked I was hearing a flowing sound of running water, and I felt I was in a scary horror movie ….and I kept walking faster*". In contrast, a recent study exploring spatial experiences without sight concluded that a sound of running water can also produce a restful ambiance (Basyazici-Kulac and Ito-Alpturer 2013).

She highlighted that the change in soundscape from the strong fearful acoustic experience to the silence of quiet natural context can be referred to as a releasing and relaxing experience: "once I was out, I felt that *huuuhh…thank God, I am out*". Her narration conveyed that it was a walking journey with a strong acoustic experience that leads to a powerful vivid experience. Similarly, a recent study revealed the significant impact of soundscape, in urban places, in creating a vivid experience (Wunderlich 2013). Acoustics that are characterized by repetition make the environment recognizable (Augoyard and Torgue 2005). The informant also reflected on a bridge she had to pass that made her pause for a moment commenting on the flooring material used with all its qualities and shapes, connections, and functions: "*it was a very scary steel bridge with overlapping panels… very narrow that only suits one person and when I decided to walk on it, I felt that the floor wasmoving beneath my feet…I was bombed with a lot of ideas that I might fall and it lookedas an unstable structure… so I stood still*". Such feelings evoked by the soundscape or the flooring material and affected the way the informant moved around. Sometimes she was *fast-paced* just to skip this moment, and at other times, she was *slow-paced* or even stood still and avoided passing through. Similarly as mentioned by Emeney (2016), urban design professor Filipa Matos Wunderlick defines this feeling as *place temporality*. Place temporality refers to our movements, encounters, rest periods, and repetitive cycles illustrate specific rhythms we experience when travelling through a space. Wunderlick categorizes place temporality as being dependent on four counterparts: sense of flow, sense of time or pace, soundscape, and rhythmicity (Emeney 2016; Wunderlich 2013). Another issue highlighted by the informant was that while moving within the horizontal winding circulation, she was filled with nervous anticipation of why was the space designed this way and what is she going to see next. "*really the plan was very weird like a zigzag as if you are walking in a star, I felt so anxious, I kept moving wondering why is it designed this way and what is this corridor leading to*". The winding configuration of the circulation created a sense of anxiety and focus. Similarly, a winding configuration of the circulation was previously found to evoke sense of mystery (Franz and Wiener 2008). On top of that, the building had a great sense of flow, discovery by hosting a *symmetric layout* and *consistent wall patterns* throughout, which were both simple and unique to explore. But most importantly, it emphasized important moments: "*After I entered and I reached the middle of the building, I observed that there are two spaces one at the right and the other at the left.... The left half was closed.... And both sides of the narrow corridor, were of the same pattern of wall mounted drawers*". Additionally, a frequently mentioned quality was the sense of *spaciousness.* This notion was often used while she described her experience as anxious and scary. She associated spaciousness to physical elements and the perception of *proportions.* For instance, she narrated how she was taken aback by the vastness of the entrance yard:

> As I reached the place, at first appeared to be very scary and I was so anxious. The first thing I experienced is a vast empty yard then appeared in front of me a building... you enter it from a gate that is too long in height ...the scale of the human to the building is like the scale of an ant toa castle. It was horrible.

Other design elements with direct influence on informant's sense of spaciousness were the sense of openness with no physical boundaries: "*all of it is opened no glass... As I enter, I was having a high level of anxiety... and since the place is very big and opened with no glass... so this made me feel I did not get in a place that can shelter me from the cold windy weather outside*". In a previous research studying the effects of permeability on perceived enclosure and spaciousness, Stamps (2010) highlighted that impressions of enclosure or spaciousness are associated with feelings of safety or danger and are mediated by features of the physical environment such as boundaries or properties of horizontal regions within boundaries. Other than *spaciousness*, the informant added that the *variation and contrast* in several physical elements between the two buildings were part of the experience that left her with a very clear image in mind. *Contrasting styles* between the two buildings since the first building she visited was old, historic, subdued, and all of the *same colour* gave her sense of anxiety. Similarly, the previous research has shown that the lack of colour variety in a hospital can have a negative influence on the state of mind of the already anxious visitor (Hagen et al. 2008). Informant added that other building in the complex appeared to be new and white mass located in the middle of vegetation, which contrasted nicely with the surrounding context: "*it was a completely different building ... A white building surrounded by vegetation... the vegetationcolour was dark so there was a strong contrast between the whitecolour and the surrounding vegetation ... and this building looked new not like the previous one looked too old*". Despite the anxious feelings, the unique *contrasting* styles among the two buildings on the site provided an overall pleasing sensation after exploring both of them. Similarly, *contrast* was utilized through the perception of transition between indoors and outdoors experiences. The indoor spaces were described to be opened with no glass as a physical boundary; however, the informant felt so enclosed and trapped because of the *overwhelming rhythmic soundscape and winding circulation.* The overall surrounding physical elements constituted such a tense ambience: "*regardless the place was considered opened but I was terrified and not feeling comfortable as if I was trapped in a place that I will never get out of*".

The overall analysis strongly identifies how the tenseness of an experience is comprised of rhythms, repetitive and irregular flows as well as *psychological/physical environmental comfort* qualities such as the flooring material that evoked the sense of insecurity and fearfulness, or sometimes evoked by the absence of other users, emptiness, and colour scheme. Sense of *spaciousness* also had an effect on the ambience of the built environment, as well as experiencing *familiarity/novelty*.

8.4.2 Vibrant Ambience

The analysis of the second informant's experience conveyed vibrancy as a strong ambience associated to her vivid experience. She apprehended such an ambience through her senses and expressed it through sensory terms such as striking, dynamic, varied, and energetic.

The most persistent memory of this vibrant ambience was related to visual dimensions of the place visited. Visual dimensions such as colour schemes, illumination, geometry, and form were conveyed by the informant to have a significant effect on how her experience was vivid. The informant highlighted that the staircase *geometry* was very dominant and interesting as she described it as *smooth*. She explained accompanying her answer with a photo she took during her experience (see Fig. 8.2) "*What was interesting is … the smooth massive geometry of the staircase in relation to the geometry of other features… is the most I remember*". This bring to mind

Fig. 8.2 Staircase geometry complementing other features' geometry. *Source* Photograph taken by the informant

the remark by Steen Eiler Rasmussen; when we see a spherical object, we do not simply note its spherical shape. While observing it, we seem to pass our hands over it in order to experience its various characteristics (Rasmussen 2001). The informant added that while she strolled through the space, her impression was altered by variations in the visual field, through *change/transformation* in physical elements such as *colour scheme*, *scale*, and intensity of natural illumination. She described such an experience with sensory terms like *amazing, transformed and varied*: "*As I climbed up the stairs to the glass cylinder, I felt the transformation from the massive white space I have been wandering to a bright colourful containing space… The experience was absolutely amazing and varied*". The panorama located on the rooftop was much more vibrant, with a vast façade of glass allowing great amounts of *natural light* to brighten the space, and transformation from *massive scale* to *contained scale*. Similarly, a recent study confirmed that people's evaluation of architectural spaces is influenced by variations in their physical features including ceiling height and level of enclosure between different spaces (Vartanian et al. 2014). She elaborated more on her colour experience, adding that with the colour spectrum added to the glazing, the natural light provided a glow of smooth transitions between colours from warm to cool shades that evoked different, yet *comfortable* body temperatures for the informant and affected her sense of flow in the place: "*The visual transition between colours was smooth but the physical experience was a bit different. I went for a second round, where I actually felt warmth in the red zone and coolness in the blue zone. The gradience of colour led a very smooth walking experience that was anything but boring*". Such feelings affected the way the informant moved around, making her to go for several more cycles to explore and confirm her sensations. Indeed, the findings of a recent experiment indicated that coloured illumination is merely perceived visually, but it also has the ability to influence human bodily sensations and perception of space (Nielsen et al. 2018). Similarly, a recent study confirmed the notion that colour has an effect on body temperature perception (Ziat et al. 2016). Furthermore, the informant highlighted that the weather that day was rainy and cloudy, but the experience still provided her with an enjoyable sensory experience, regardless of the bad weather. Her diary was accompanied by a group of photographs she took during her experience (see Fig. 8.3):

> The rain stopped and it was really hard to determine the weather from inside, although I knew from the forecast that it was 18 degrees Celsius. At the beginning, I was just watching myself and the people around me smoothly change in colour, from yellow to red, blue, green and then yellow again. The coloured vision was not disturbing at all. It was actually pleasing.

The hearing dimensions was also highlighted to have a significant impact on how vibrant and vivid the experience is. A recent research suggested that there is a direct link between vibrancy and soundscape as well as visual scenery (Aletta and Kang 2018). A sequence of spaces with change in the perceived soundscape, featuring lively/noisy soundscape in spaces and passive/quiet soundscape in others, was highlighted by both informants. They both indicated that the variation in soundscape left them with a strong acoustic experience. The first acoustic experience was through

Fig. 8.3 Vast façade of glass, with the smooth transition in colour spectrum added to the glazing acted as dynamic mediator between informant/herself, informant/other users and the city view. *Source* Photograph taken by the informant

the staircase. Indeed, the staircase was full of footsteps sound that created a repetitive sense of attention and direction. This provided focus to the visitor's experience forgetting about her effort in climbing the staircase. In contrast, when she reached the panorama, she perceived the space as a quiet space and she discovered later that it was not that quiet. Similarly, in the past, the tool of silence has been used to create great atmospheres. Pallasma discussed how one of the most exciting auditory experiences in architecture is tranquillity(Pallasmaa 2005).The informant described that once the space was full of visual cues, her visual sense became busy and she perceived the space as quiet;

> The whole experience was energetic, boosted my visual sense to the extreme. What was really interesting is that when I got back to my recorded video, it was full of footsteps and conversation in the background, I got surprised since I was there, for me the place was very quiet…I left this place feeling HAPPY.

In conclusion, the strongest aspect of this vibrant experience is the variations in physical features, through contrast and transformation in the physical features of sequential spaces such as colour scheme, geometry, and scale. In addition, it is highlighted that the experience is comprised of *rhythms—repetitive* and regular smooth *flows*—that are shaped through smooth colour transition and looped/one-way circulation. The sensation of moving around in a circular path throughout the visit perfectly established a sense of satisfaction for the informant.

8.5 Conclusion

In conclusion, the study confirmed that engaging human senses provides users with more holistic experiences. Additionally, vividness of an initial experience of place stems from the overall perception of the physical characteristics of the built environment, when different senses are stimulated. Humans perceive and recall the environment through their senses. The informants apprehended the sensory cues in the surrounding built environment through several sensory channels: *hearing, touch/skin, vision, and kinaesthetic sensations and proprioceptive cues,* highlighting that movement through the built environment is an important experiencing mode.

Multisensory cues in the built environment stimulate human senses and hence provide users with a perceived ambience. Vivid experiences are characterized by a certain ambience. Ambiences that appeared to have a significant role in creating the two studied vivid experiences are *tenseness ambience* and *vibrant ambience.* Yet, there were other ambiences beside the two core ones that created vivid moments for the two experiences studied such as *ambiguous, attractiveness, tranquil, and predictability.* The perceived ambience is influenced by physical characteristics such as *luminosity, sound, temperature, colour, geometry, materials, boundaries, and circulation.* Attention to spatial qualities such as *rhythm–flow–pace, environmental comfort* (physical and psychological), *spaciousness, familiarity/novelty* and contrast/*variation/transformation* between sequential spaces can contribute to the creation of vivid experiences with characterized ambiences of the built environment.

Methodologically, this study has demonstrated the value of the phenomenological research approach in describing the architectural experience. Narratives through personal audio recording, written diary and the follow in-depth interviews, allowed the consideration of actual users of architectural spaces and to understand their first-hand, personal experiences. The informants' accounts helped in understanding the experiential, multisensory qualities of architectural spaces. This is expected to contribute to an increased awareness of the need for multisensory design approaches in architecture. In turn, a multisensory architecture will provide users with more holistic and richer experiences.

The study is limited in terms of the number of experiences and number of informants; hence, the study findings cannot be generalized and should be confirmed with more analysis of the same type of experiences. However, the findings of the two studies opened a venue for future research such as investigating more vivid experiences, other more sensory channels might be involved in the findings such as taste and smell. Also engaging more data collection sources such as autobiographies for sensory disabled, personal reflections, and documentation of films featuring sensory issues might help in finding shared understanding of vivid experiences from various sources. In addition, the findings of such an exploratory study raise questions such as are all vivid experiences usually associated with a core ambience or a more balanced mixture of several ambiences. Additionally, ambiences such as *ambiguity, attractiveness, tranquillity, and predictability* that emerged from the analysis of both vivid experiences can be further explored for deeper understanding. Further research raising more questions that encompasses such additional phenomenological lived experience investigation may extend upon and bring the current findings further.

References

Aletta F, Kang J (2018) Towards an urban vibrancy model: a soundscape approach. Int J Environ Res Public Health 15(8):1–18

Augoyard J-F, Torgue H (2005) Sonic experience: a guide to everyday sounds. McGill-Queen's University Press, London

Barrit L, Beekman T, Bleeker H, Mukderij (1985) Researching education practice. Grand Forks. University of North Dakota Press, North Dakota

Basyazici-Kulac B, Ito-Alpturer M (2013) A phenomenological study of spatial experiences without sight and critique of visual dominance in architecture. In: Envisioning architecture: design, evaluation, communication conference, pp 167–174

Bloomer K, Moore CW (1977) Body, memory, and architecture. Yale University Press

Boddy CR (2016) Sample size for qualitative research. Qual Market Res Int J 19(4):662–674

Braun V, Clarke V (2008) Using thematic analysis in psychology. Qual Res Psychol 3(2):77–101

Broussard L (2006) Understanding qualitative research: a school nurse perspective. J Sch Nurs 22(4):212–218

Chandralal L, Valenzuela F-R (2013) Exploring memorable tourism experiences: antecedents and behavioural outcomes. J Econ Bus Manag 1(2):177–181

Charles S, Puccinelli NM, Roggeveen DG (2014) Store atmospherics: a multisensory perspective. Psychol Market 31(7):472–488

Diekelamann N (2001). Narrative pedagogy: Heideggerian hermeneutical analyses of lived experiences of students, teachers, andclinicians. Adv Nurs Sci 23(3):53–71

Drønen H (2013) Sensory stimulation: sensory stimulation—a review of how stimulation of the senses can affect the behavior of people with dementia, and inspire the process of designing for them. Retrieved 6 26, 2019, from NTNU : https://www.ntnu.no/documents/10401/1264433962/AshildArtikkel.pdf/8f8ba50d-7425-43c7-9e5f-bfef4380e206

Emeney D (2016) Sensory expereince in space: an analysis of phenomenology and wineries. The University of Utah

Franz G, Wiener JM (2008) From space syntax to space semantics: a behaviorally and perceptually oriented methodology for the efficient description of the geometry and topology of environments. Environ Plan B: Plan Des 35:574–592

Franzosi R (1998) Narrative analysis-or why (and how) sociologists should be interested in narrative. Ann Rev Sociol 24:517–554

Hagen MV, Galetzka M, Pruyn A (2008) Effects of colour and light on customer experience and time perception at a virtual railway station. Noordwijkerhout, The Netherlands

Holahan CJ (1982) Environmental psychology. Random House, New York

Holl S, Pallasmaa J, Pérez-Gómez A (1994) Questions of perception: phenomenology of architecture. Arch Urban (Special Issue):26–37

Kotler P (1974) Atmospherics as a marketing tool. J Market 49(4):48–64

Kuijsters A, Redi J, Ruyter BD, Seuntiens P, Heynderickx I (2014) Affective ambiences created with lighting for older people. Light Res Technol 0:1–17

Nielsen SL, Friberg C, Hansen EK (2018) The ambience potential of coloured illuminations in architecture: a spatial experiment exploring bodily sensations. Int J Sens Environ Archit Urban Space. Retrieved from https://journals.openedition.org/ambiances/1578

O'neill ME (2001) Corporeal experience: a haptic way of knowing. J Archit Educ 55(1):3–12

Pallasmaa J (2005) The eyes of the skin: architecture and the senses. Wiley Ltd., England

Pine J II, Gilmore JH (1998) Welcome to the experience economy. Harvard Bus Rev 76(4):97–105
Plous S (1993) The psychology of judgment and decision making. McGraw-Hill, New York. Retrieved from https://epdf.pub/the-psychology-of-judgment-and-decision-making.html
Proske U, Gandevia SC (2012) The proprioceptive senses: their roles in signaling body shape, body position and movement, and musle force. Am Physiol Soc 92(4):1651–1697
Proske U, Gandevia SC (2018) Kinesthetic senses. Compr Physiol 8:1157–1183
Rapport F, Wainwright P (2006) Phenomenology as a paradigm of movement. Nurs Inq 13(3):228–236
Rasmussen SE (2001) Experiencing architecture. MIT Press, Cambridge, Mass
Reid K, Flowers P, Larikin M (2005) Exploring lived experience. The Psychologist 18(1):20–23
Rubin DC, Schrauf RW, Greenberg DL (2003) Belief and recollection of autobiographical memories. Mem Cogn 31(6):887–901
Ryan GW, Bernard HR (2003) Techniques to identify themes . Field Methods 15(1):85–109
Stamps AE (2010) Effects of permeability on perceived enclosure and spaciousness. Environ Beh 42(6):864–886
Starks H, Trinidad SB (2007) Choose your method: a comparison of phenomenology, discourse analysis, and grounded theory. Qual Health Res 17(10):1372–1380
Streubert HJ, Carpenter DR (2011) Qualitative research in nursing: advancing the humanistic imperative. Lippincott Williams & Wilkins, Philadelphia
Vartanian O, Navarrete G, Chatterjee A, Fich LB (2014) Architectural design and the brain: effects of ceiling height and perceived enclosure on beauty judgments and approach-avoidance decisions. J Environ Psychol 41:10–18
Wilkie R, Roberson GF (2012) Chapter 13: attachment to place. In: Stoltman JP (ed) 21st century geography: a reference handbook. SAGE Publication, Thousand Oaks, pp 134–148
Wunderlich FM (2013) Place-temporality and urban place-rhythms in urban analysis and design: an aesthetic akin to music. J Urban Des 18(3):383–408
Ziat M, Balcer CA, Shirtz A, Rolison T (2016) A century later, the hue-heat hypothesis: does color truly affect temperature perception? Springer, London
Zumthor P (2006) Atmospheres: architectural environments, surrounding objects. Birkhäuser, Boston

Part II
Smart Environmental Solutions

Chapter 9
The Effect of Courtyard Ratio on Energy Consumption and Thermal Comfort in a Primary Governmental School in New Assiut City, Egypt

Ahmed Mohammed Abd El-Samea, Nouby Mohammed Hassan, and Amr Sayed Hassan Abdallah

Abstract The building and spaces design have a great impact on indoor temperature and energy consumption. In Egypt, the general authority of educational buildings (GAEB), which is responsible for designing schools, uses the same typical prototypes across the country despite the diversity of climatic conditions. The main objective of this study is to investigate the effect of courtyard ratio (H/W) on student thermal comfort and school annual energy consumption in order to find out the optimum ratio for school courtyard in one of the typical governmental schools with an E-shaped design. A building simulation model was created using Design Builder software. The model was calibrated using actual measured weather data for the year 2014. A simulation study was carried out for twenty different design scenarios. Two stages for the simulation were performed; the first stage studied the effect of courtyard ratio (H/W) on indoor thermal comfort while using natural ventilation and the second stage studied the influence of using HVAC inside school classrooms on the annual energy consumption. The results showed that high change happened for annual energy consumption when changing the courtyard width from 10 to 5 m, and less change happens when changing the width from 15 to 10 m. The lowest rate of energy consumption is achieved for courtyard ratio which equals 1.5 with annual energy consumption which equals 13.5 kWh/m^2/year. There is significant change in PMV values from 5 m courtyard width to 10 m courtyard, while no change is observed in case of 15, 20, 25 m courtyard width.

Keywords Energy consumption · Thermal comfort · School building simulation

A. M. A. El-Samea (✉) · N. M. Hassan · A. S. H. Abdallah
Faculty of Engineering, Assiut University, Assiut, Egypt
e-mail: ahmedeid2992@gmail.com

Architecture and Urbanism: A Smart Outlook,
https://doi.org/10.1007/978-3-030-52584-2_9

9.1 Introduction

Energy crisis and climate change are considered the greatest environmental threats the world is facing today. The rates of energy consumption have increased significantly, in addition to the negative effects of energy production and consumption on the environment. Also, we can no longer ignore the fact that the world's climate is becoming more extreme and more unpredictable. So, it is important to achieve thermal comfort and energy efficiency in different sectors. The building sector is the most electricity consuming sector in Egypt (Rasazi et al. 2010), especially school buildings, where students—about 16 million students in basic education in Egypt by 2019 (CAPMAS 2019)—spend about one third of their day in classrooms.

The student's density at classrooms in Egypt, especially in basic education, is very high, with an average of 48 students per class (Ministry of Education 2018). Thus, there is an urgent need to build large numbers of schools in order to accommodate the increase in the number of students. According to the Ministry of Education, the number of classes required is estimated to be 200 thousand new classrooms, to reach the number of students in classes to acceptable rates.

The building and spaces design have a great impact on indoor temperature and energy consumption. In Egypt, the general authority of educational buildings (GAEB), which is responsible for designing schools, uses the same typical prototypes (single, double row linear and L-shaped design) across the country despite the diversity of climatic conditions.

Many researches have discussed thermal comfort in (GAEB) typical prototypes. Gado and Mohamed (2009) assessed the thermal comfort of occupants inside nineteen primary schools in Egypt. They found that most of the occupants were thermally discomfort for the majority of time during the academic year. Saleem (2015) performed a simulation study to estimate thermal comfort in different regions in Egypt throughout the year and found that students in Assiut Governorate are expected to spend 40.27% of occupancy hours in classrooms in unsuitable thermal conditions. Wanas (2013) used a thermal simulation program to calculate thermal discomfort amount within classrooms of (GAEB) typical prototypes in Cairo.

Most studies in the literature have focused on (GAEB) typical prototypes, and only few discussed new designs such as the E-shaped design and multi-courtyard design which have recently been used in some new cities in Egypt.

Abdallah (2017) evaluated six different factors to ensure thermal comfort in 13 classrooms in two schools with new designs (E-shaped and multi-courtyard design) during the academic year 2014. The results showed that governmental schools that implemented some passive environmental techniques were more thermally appropriate than the typical prototypes used by GAEB. Taleghani et al. (2014) discussed courtyards as a common passive strategy for buildings in temperate climates. The results showed that adding courtyards, atriums or both helps achieve more energy-efficient and comfortable indoor environments. Kedissa et al. (2017) discussed the effect of courtyard H/W ratio changes on thermal comfort. The results revealed that direct sunlight exposure, temperature and wind speed are affected by changes in courtyard *H*/*W* ratio which plays an important role in obtaining adequate thermal comfort levels. Abdallah (2015) studied the effect of courtyard (height/width) ratio between two buildings (deep and shallow courtyards) in Assiut and New Assiut City. The comparison was based on thermal performance and energy efficiency. It showed a decrease in temperature of the interior spaces in the building with a deep courtyard of 11 °C below the external temperature, while temperature increased in the case of the second shallow courtyard and, thus, caused high energy consumption. Therefore, different factors should be taken into considerations while designing schools, to be consistent with the environmental and climatic conditions of its location and achieve thermal comfort and energy efficiency.

Most of the past literatures that studied the effect of building form and courtyard ratio on indoor student thermal comfort are not sufficiently and need more investigation concerning the optimum ratio for courtyard and its impact on school annual energy consumption and indoor thermal comfort.

The main objective of this study is to investigate the effect of courtyard ratio (*H*/*W*) on student thermal comfort and school annual energy consumption in order to find out the optimum ratio for school courtyard.

9.2 Methodology

A simulation study was conducted for a New Assiut city school building. A model was created using Design Builder software (v5.0.3.007) (D.S.L—D.B 2018) and then calibrated using actual measured weather temperature data for the year 2014 in two classes. A simulation study was carried out for different design scenarios to determine the effect of courtyard *H*/*W* ratio changes on school energy consumption and thermal comfort and to find out the optimum ratio. Thermal comfort was evaluated in classrooms using predicted mean vote (PMV) index.

9.2.1 Case Study Location and Description

The New Assiut city school is located in the eastern side of new Assiut city, the first district, middle of the youth housing service center. Its coordinates are 27° 16′ 12″ N and 31° 17′ 58″ E. The New Assiut city is located in the hot desert climate zone, characterized by high temperature—that varies from 41 to 46 °C as maximum and from 16 to 21 °C as minimum during the summer months—and high solar radiation—from 1000 to 1125 W/m^2 in summer months and 650 to 800 W/m^2 in winter months (Abdallah 2015)—thus increasing the thermal loads on the buildings.

The New Assiut city school was chosen as it is the first school in the New Assiut city with an E-shaped design, as one of the new design styles that have been used recently, by the general authority of educational buildings (GAEB), especially in

Fig. 9.1 New Assiut city school satellite image (left) and the school main entrance (right)

new cities. The school designer tried to apply some passive design techniques such as courtyards, shades and louvers.

The school is surrounded by wide streets and green spaces. Heights of nearby buildings are low, allowing the wind to pass through the building and reduce shadows. The three wings of the school, which include classrooms, were directed to the northwest to take the advantage of natural lighting and wind (see Fig. 9.1).

The school is designed in an E-shaped design. The 12,400 m^2 school consists of three wings surrounding two exposed courtyards and laying perpendicular to the main wing. The main wing contains administrative spaces, teachers' rooms, supervisors, science laboratories, activity rooms, the library, a multi-purpose hall and number of stores and service areas.

The three wings each contain four classrooms on every floor, with a total of 32 classrooms, except for the ground floor in the middle part, which includes the music education room and canteen (see Fig. 9.2).

9.2.2 Modeling and Simulation

School base model. Modeling was carried out for the school building with the same building design and characteristics (see Fig. 9.3). The location and direction of the school were identified. Building materials, openings, lighting, equipment, work schedules and air conditioning systems were added based on field survey for the school**.**

Construction, Material and Glazing. The school building has a reinforced concrete structure, 25 cm-thick exterior walls and 12 cm-thick interior walls from red brick without thermal insulation, with 4 cm-thick finishing of cement plaster and paint. Slabs are made from 12 cm-thick concrete with 10 cm-thickness finishing. Each classroom has six wooden framed windows with 6 mm clear single glazing. It has a horizontal solid shade with 0.9 m depth and a lattice window screen similar to the old arabian mashrabiya used as a shading unit.

Lighting and equipment. The school spaces were divided according to the type of activities in each floor. All details, related to lightning and electrical equipment, were

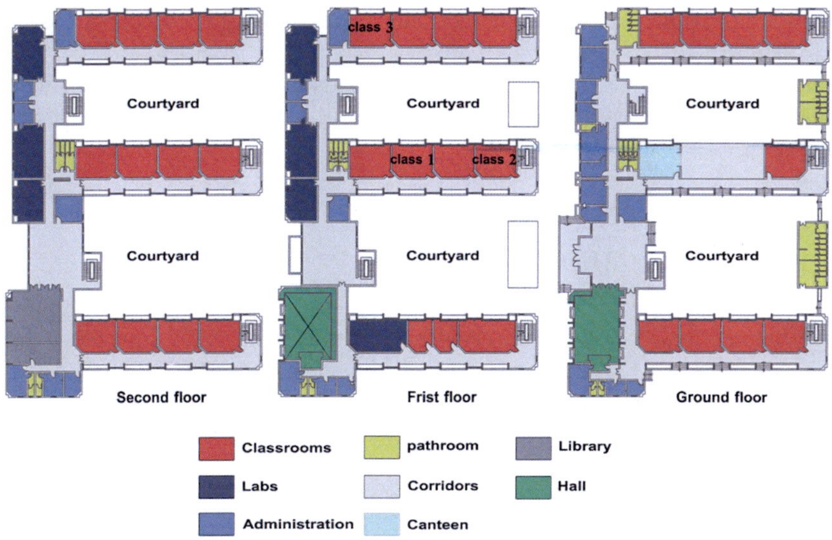

Fig. 9.2 New Assiut city school plans and uses of different spaces and classes 1, 2, 3 position

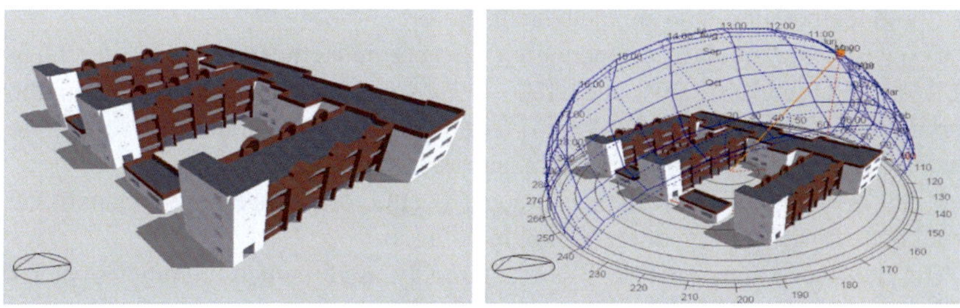

Fig. 9.3 School model made by Design Builder—Sun path diagram at 10:00, May, 15th

added. For example, each classroom has eighteen 40 W fluorescent lamps and a 50 W ceiling fan. The administration rooms and laboratories contain computers and other electrical equipment, taking into consideration that the lightning and equipment are not fully functional during the working hours.

Schedule and Activities. Based on the Egyptian school year which starts in September and ends in June, vacations and working days were identified. The number of working hours during the occupancy days was determined. For example, occupancy hours in classrooms range from 8 am to 2 pm. The occupancy schedules for different spaces such as classrooms, laboratories and administration rooms were also added.

The type of activity affects the temperature of the body and, hence, the room temperature. Activities in the school were determined to be office activities (reading, writing seated) which have a 60 w/m^2 metabolic rate according to ASHRAE Standard 55 (ASHRAE 2010).

HVAC and Ventilation. All classrooms are naturally ventilated, with a single ceiling fan in each class and two fans in each laboratory. Window-to-wall ratio is about 30%.

9.2.3 Validation (Model Calibration)

The validation process checks the accuracy of results obtained by simulation and compares them to the results obtained by measurements. (Andelkovic et al. 2016). It is used to calibrate or verify the models and variables entered into the simulation program by comparing the resulting measurements obtained by simulation with actual measurements and observations (e.g., energy consumption or temperature), so that these actual measurements are approximately equal to the results from the simulation program (Reddy et al. 2006).

Based on the actual measurements for air temperature inside two classes located in the first floor in two different days done by (Abdallah 2017), the base model was calibrated. A simulation study was carried out at the same time to compare actual measured temperature with simulated temperature.

Comparing the simulated temperature with measured temperature showed a 0.07–3.1% variation in classroom 1 and a 0.06–1.9% variation in classroom 2 during the working hour (08:00 am to 13:00 pm) (see Fig. 9.4). The variation is acceptable when the difference is within 5% (Rahman et al. 2008).

It can be concluded that the base model is able to produce approximately the actual operating condition of the school building.

The root mean square (RMSE) is a commonly used method to compare values or variables predicted by the model or simulations and the values actually observed from the real environment (Andelkovic et al. 2016). It can be calculated from the

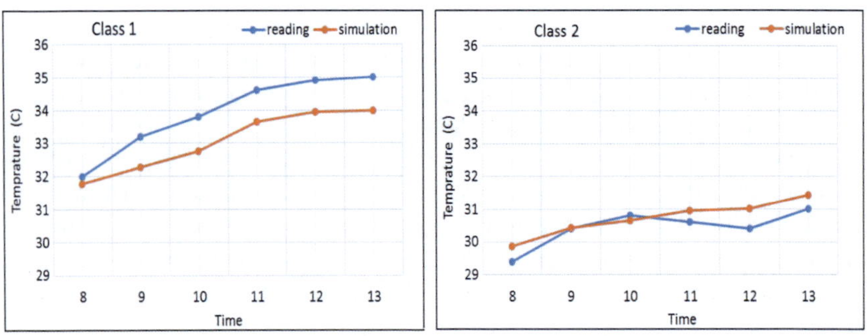

Fig. 9.4 Comparison between measured and simulated temperatures during working hours (8 am: 13 pm) in classroom 1 (on 28 September) and classroom 2 (on 29 September)

following equation:

$$\text{RMSE} = \sqrt{\frac{\sum_{i=1}^{n}(P - O)^2}{n}}$$

where O measured temperature, P simulated temperature and N number of readings.

The value of the root mean square error was calculated as (0.9) in class 1, (0.4) in class 2, less than 2.3, so it lies within the acceptable range.

9.2.4 Simulation Scenarios

Twenty different scenarios were tested for different courtyard H/W ratios. Scenarios were divided into four groups: School height in the first group was fixed as two floors (7.6 m), the second group as three floors (10.9 m), the third group as four floors (14.2 m), the fourth group as five floors (17.5 m) and the courtyard width was variable (5, 10, 15, 20, 25 m) for each group (see Fig. 9.5). A two-stage simulation study was performed. The first was done to determine the effect of these changes on thermal comfort in classroom (2) in case of natural ventilation and the second to determine their influence on school energy consumption after adding HVAC in school classrooms.

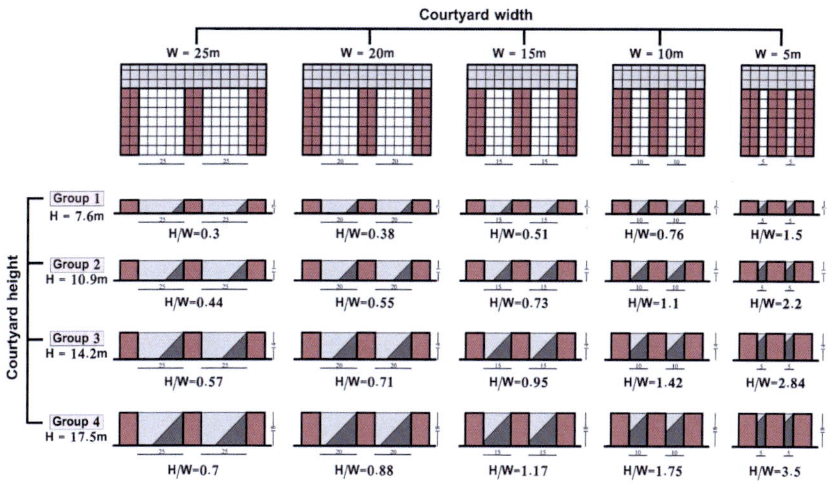

Fig. 9.5 Matrix for different simulation scenarios for school courtyard H/W ratios

9.2.5 Thermal Comfort Evaluation

Indoor thermal comfort is determined by different factors; ASHRAE Standard 55 uses predicted mean vote (PMV) index—which was developed by Fanger in 1970—to assign the requirements for indoor thermal comfort and predict the mean value of the votes for certain combinations of air temperature, metabolic rate, air velocity, humidity and clothing insulation. It evaluates the user's thermal sensation on a seven-point scale from cold (−3) to hot (3), where zero is the typical value with recommended limits between (−0.50) and (0.5) (Energy and Sustainability 2017).

9.3 Results and Discussion

Fanger PMV Simulation results for class no. 2 which located in the first floor—on 30 April—are reported in (Fig. 9.6) based on different floor levels and different courtyard widths (m). The value of PMV gradually increases over time from 8 am to 1 pm. In a single group (same height), it is concluded that there is a significant change in PMV values from 5 m courtyard width to 10 m courtyard, while no change is observed in case of 15, 20, 25 m courtyard width; this is due to when the courtyard

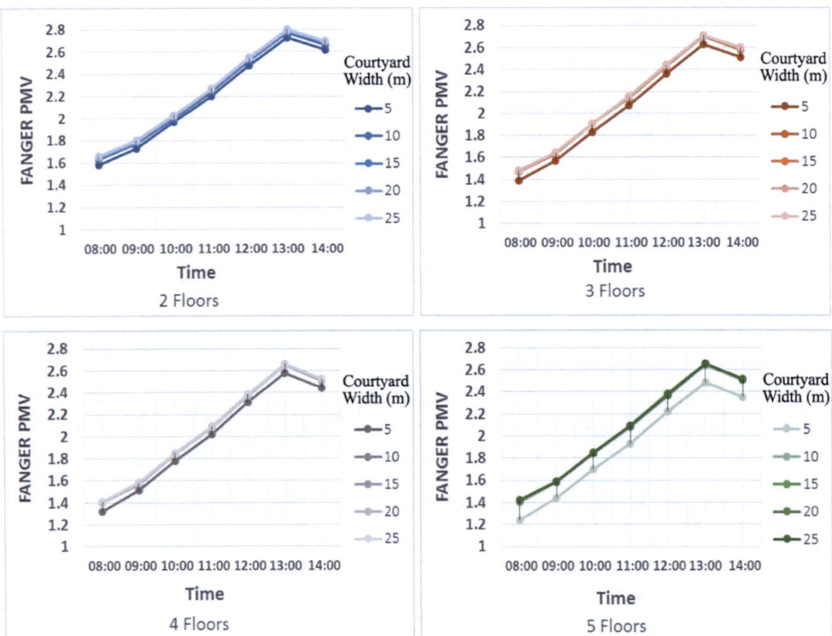

Fig. 9.6 Comparison between simulated PMV results during working hours (8 am: 14 pm) in classroom 2 (on 30 April)

Table 9.1 Annual energy consumption ($kWh/m^2/year$) rates

Group court yard width (m)	25	20	15	10	5
Group 1 (two floors)	13.74	13.72	13.71	13.66	*13.5*
Group 2 (three floors)	14.25	14.24	14.21	14.13	13.8
Group 3 (four floors)	14.35	14.32	14.28	14.19	13.87
Group 4 (five floors)	*14.53*	14.49	14.43	14.28	13.99

width increases more than 10 m, the facades exposure to solar radiation is almost equal. Comparing PMV values in the four groups, there is a gradual decrease from the first group (two floors height) to the fourth group (five floors height).

In spite of changing courtyard width and height in the four groups, all the cases through the working hours are outside the thermal comfort zone. Therefore, a study of using HVAC inside classrooms was done to achieve low energy consumption, with its relation to thermal comfort and courtyard ratio.

Comparing the energy consumption rates in the same group (the same height of the building), it is clear that consumption rates are gradually increasing from the lowest value in case the courtyard which equals 5 m in width to the highest value in case the courtyard which equals 25 m in width. Table 9.1 shows the annual energy consumption rates for each group with relation to courtyard ratio.

It is concluded that the amount of energy saving—between the largest courtyard width equals 25 m and the lowest courtyard width equals 5 m—is 1.8%, 2.7%, 3.4% and 3.7% in groups 1, 2, 3 and 4, respectively. The highest rate of energy consumption is reported in group 4 with five floors, 25-m courtyard width and ratio of 0.7 H/W with annual energy consumption per unit area equals 14.53 $kWh/m^2/year$; this is due to excessive facades exposure to solar radiation and subsequent thermal discomfort and HVAC use. The lowest rate of energy consumption is reported in group 1 with two floors, 5 m courtyard width and ratio of 1.5 H/W with annual energy consumption per unit area equals 13.5 $kWh/m^2/year$; this is due to little facades exposure to solar radiation, which is 7.1% lower than the highest case, so the case with two floors, 5 m courtyard width is considered the most energy-efficient condition.

Therefore, high change happened for annual energy consumption per unit area when changing the courtyard width from 10 to 5 m and less change happens when changing the width from 15 to 10 m, and no change is observed when changing the width from 20 to 15 and from 25 to 20 as the percentage of shading on the facades is almost equal in the three cases (15, 20 and 25 m width).

The relation between the courtyard width and annual energy consumption per unit area for each group (different heights) is represented in four trend curves governed by four equations (see Fig. 9.7).

Those equations give reasonable distribution for energy consumption per unit area over courtyard width. Combining the four equations leads to the main equation that gives the relation between courtyard width and annual energy consumption per unit area according to the number of floors in the school building as follows:

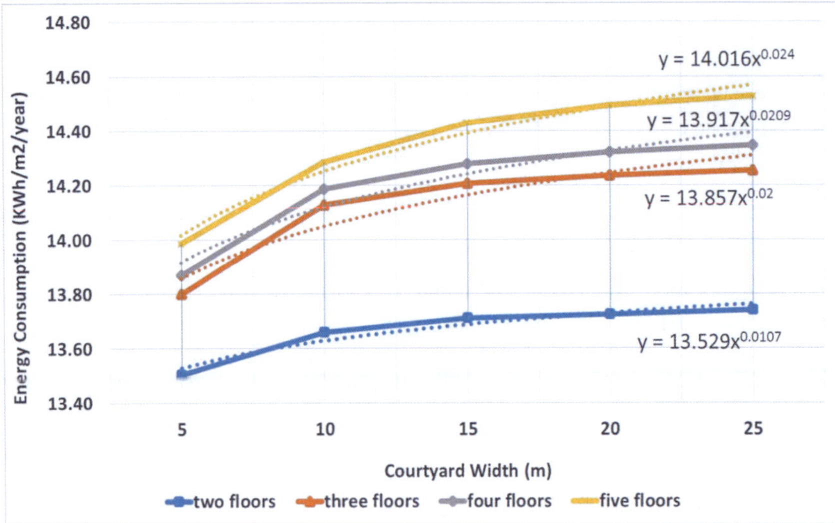

Fig. 9.7 Curves and equations for the relation between courtyard width and annual energy consumption per unit area

$$E = (-0.0572\,n2 + 0.5529\,n + 12.668)\,W^{(-0.0016\,n2 + 0.0149\,n - 0.0124)}$$

where (E) energy consumption, (n) number of floors and (W) courtyard width.

9.4 Conclusion

Schools in Egypt are designed using GAEB typical prototypes which are thermally uncomfortable. Recently, new designs have been used, such as E-shaped and multi-courtyard design which are more thermally comfortable. The effect of H/W courtyard ratio changes on thermal comfort and annual energy consumption was studied in New Assiut city school. The results showed that better thermal conditions are achieved by reducing courtyard width and increasing building height. Thus, deep courtyard (more than 10 m width) is more effective and provides better thermal conditions than shallow courtyard. Also, it is concluded that the more the courtyard width and building height, the higher the annual school energy consumption especially when the courtyard width exceeds 10 m due to excessive facades exposure to solar radiation, with maximum annual energy consumption which equals 14.53 kWh/m^2/year.

Further studies are recommended to examine the relation of courtyard H/W ratio with relation to natural lighting and its influence on energy consumption.

Acknowledgements The research leading to these results used Design Builder license based on DAAD project "High Education Dialogue with Muslim Countries" the cooperation between Assiut University and German partners.

References

Abdallah A (2017) Thermal monitoring and evaluation of indoor CO_2 concentration in classrooms of two primary government schools in New Assiut City, Egypt

Abdallah A (2015) The influence of urban geometry on thermal comfort and energy consumption in residential building of hot arid climate, Assiut, Egypt. Procedia Eng 121:158–166

Andelkovic S, Mujan I, Daki S (2016) Experimental validation of a energyplus model: application of a multi-story naturally ventilated double skin facade. Energy Build 118:27–36

ASHRAE (2010) Thermal environmental conditions for human occupancy. ASHRAE Standing Standard Project Committee 55

Central Agency for Public Mobilization and Statistics (CAPMAS) (2019) Egypt in figures

D.S.L—D.B (2018) Design builder software. Stroud, UK. Retrieved from http://www.designbuilder.co.uk/

Energy and Sustainability (2017). In: Brebbia CA, Sendra JJ (eds) With transactions on ecology and the environment, vol 224. British Library Cataloging-in-Publication Data

Gado T, Mohamed M (2009) Assessment of thermal comfort inside primary governmental classrooms in hot-dry climates Part I—a case study from Egypt

Kedissa C, Outtas S, Belarbi R (2017) The impact of height/width ratio on the microclimate and thermal comfort levels of urban courtyards. Int J Sustain Build Technol Urban Dev 174–183

Ministry of Education (2018) Book of annual statistics. Ministry of Education Information Center

Rahman M, Rasul MMK (2008) Energy conservation measures in an institutional building by dynamic simulation using design builder. In: Conference on energy and environment. University of Cambridge, UK, pp 192–197

Rasazi H, Nzabanita E, Cheikhrouhou H (2010) The arab republic of Egypt-power sector in brief 2010

Reddy T, Maor I, Panjapornpon C (2006) Calibrating detailed building energy simulation programs with measured data part i: general methodology. J HVAC Res 13:221–241

Saleem AA (2015) Experimental and simulation study of the thermal comfort conditions within recent designed governmental primary schools in Egypt. Egypt-Japan University of Science and Technology (unpublished work style), Master of Science

Taleghani M, Tenpierik M, Dobbelsteen A (2014) Energy performance and thermal comfort of courtyard/atrium dwellings in The Netherlands in the light of climate change. J Renew Energy 63:486–497

Wanas O (2013) Assessing thermal comfort in secondary schools in Egypt. Master of science in integrated urbanism and sustainable design. Ain Shams University, Egypt (unpublished work style)

Chapter 10
Encoding Latent Properties of Hygromorphic-Thermobimetal Composites as a Passive Mechanism for Adaptive Building Skins

Rana El-Dabaa, Sherif Abdelmohsen, and Yasser Mansour

Abstract Adaptive facades require complex mechanical systems to control their motion. Programmable materials are increasingly replacing mechanical systems through encoding their latent properties to achieve passive controlled motion response to external stimuli, thus acting as zero-energy adaptive systems. This paper introduces $H_M T_M$, a hygromorphic-thermobimetal laminated composite composed of wood, an anisotropic material whose mechanical properties vary according to fiber direction, and metal; an isotropic material with uniform linear expansion. The composite acts as an embedded sensor and actuator that initiates passive motion in response to temperature variation in hot climates. Physical experiments were conducted to deduce a grammar for the passive motion of the composite. The $H_M T_M$ motion response was captured, analyzed and programmed through a physical-digital interface closed loop using image analysis. The $H_M T_M$ grammar encodes two types of parameters: (1) *embedded parameters* related to latent properties of each of the composite's materials such as dimensional ratio, grain orientation, thickness, lamination and expansion coefficient and (2) *control parameters* related to composite assembly such as fixation position and area isolation. These parameters collectively affect the output motion response in terms of deflection and motion type through semantic rules that define material configuration. The added value in the paper lies in the mutual benefit of integrating hygroscopic and thermal properties, specifically: (1) the extended actuation of the $H_M T_M$ composite in hot arid climates where temperature variation is dominant, as opposed to humid climates only and (2) the extended resulting motion configurations such as sliding, rotation and twisting, as opposed to linear configurations.

R. El-Dabaa
Arab Academy for Science, Technology and Maritime Transport, Cairo, Egypt

R. El-Dabaa · S. Abdelmohsen (✉) · Y. Mansour
Faculty of Engineering, Ain Shams University, Cairo, Egypt
e-mail: sherifmorad@aucegypt.edu

S. Abdelmohsen
The American University in Cairo, Cairo, Egypt

Architecture and Urbanism: A Smart Outlook,
https://doi.org/10.1007/978-3-030-52584-2_10

Keywords Hygroscopic properties of wood · Adaptive façades · Motion parameters · Bimetals · Programmable materials

10.1 Introduction

Adaptive facades are typically characterized by their complex and energy-consuming mechanical systems including sensors, processors and actuators. Fox and Kemp argued for the ability to shift from sophisticated mechanical actuation of kinetic structures to biological passive systems, in what they coined the "End of Mechanisms" (Fox 2016). In parallel, smart material interfaces were developed upon the emergence of ubiquitous computing, which weaves technology into everyday life in an indistinguishable fashion (Nijholt and Minuto 2017). The unique capacity of smart materials in this discourse owes to their ability to mimic biological passive systems through capturing, analyzing and controlling material properties (Kretzer 2017).

Programmable materials are smart materials that have specific fabricated composite layers and are programmed to produce desired and controlled motion in response to stimuli. The programming takes place during the fabrication of the material after studying the properties of each of its layers to control the desired motion. This paper introduces $H_{\mathrm{M}}T_{\mathrm{M}}$ as a novel programmable material that responds to changes in humidity and temperature by utilizing both hygroscopic properties of wood and thermal expansion coefficients of metals. The objective is to employ $H_{\mathrm{M}}T_{\mathrm{M}}$ as an embedded actuation system in responsive and adaptive facades that can control and program its passive motion response to achieve a variety of motion types and configurations. Kinetic motion generally comprises three basic types of motion: sliding, rotation and scaling, in addition to a fourth type which relies on material elastic properties that cause material distortion and complex motion such as twisting, rolling and folding (Moloney 2011).

This paper addresses the following questions: How can integrating different programmable materials allow for the emergence of novel programmable composites that affect and inform adaptive architecture in hot arid climates? How can the latent parameters and properties of materials be stimulated to act as passive motion response with various types of motion and deflection values? How can the response behavior of programmable materials be captured, analyzed and programmed? The paper identifies the design parameters and motion grammars of the proposed $H_{\mathrm{M}}T_{\mathrm{M}}$ composite and introduces a computational approach to capture, track and program its response behavior over time. $H_{\mathrm{M}}T_{\mathrm{M}}$ motion grammars control its response behavior by utilizing the material's parameters with the motion design parameters as discrete events of sequential transformation and targeted output motion.

10.1.1 Utilizing Programmable Materials in Architecture

Relevant studies in developing biological passive systems for adaptive façades have generally addressed three issues: (1) controlling and programming wood motion response through physical experimental approaches (Wood et al. 2016; Correa et al. 2015) using architectural prototypes (Holstov et al. 2015; Wood et al. 2016; Abdelmohsen et al. 2019b), (2) testing material response in real weather conditions (Krieg et al. 2014; Menges and Reichert 2012; Sung 2016) and (3) using numerical simulations to estimate wood response (Rüggeberg and Burgert 2015; Abdelmohsen et al. 2019a).

Experiments at the Institute for Computational Design (ICD) in the University of Stuttgart (Dierichs and Menges 2016) and the Self-Assembly Lab (SAL) at MIT (Correa et al. 2015) were conducted to enhance the ability to control latent material properties of wood and consequently enhance material programmability. Another series of experiments were conducted on wooden configurations in a controlled environment to deduce the effect of hygroscopic design parameters on the motion response of wood (Abdelmohsen et al. 2019a, c). Cross-laminating wood with different materials during composite fabrication was observed to program its motion response and control its activation techniques. 3D printing of nylon and wood was also tested to control the embedded properties of wood and printing it with similar fibers' structures (Correa et al. 2015). Another wooden-based programmable composite was developed by mixing curved wood with shape memory polymers that are activated by heat (Mansoori et al. 2019). Laminating specific parts of wood was also seen to control parts of wood to act as a responsive actuator (Rüggeberg and Burgert 2015).

Several structures utilized hygroscopic design for passive responsive motion that is triggered by variation in humidity levels. The HygroSkin Pavilion uses hygroscopic design properties to achieve passive responsive laminated façades that open in low humidity conditions and close in high humidity conditions (Krieg et al. 2014). A quite contrasting configuration of wood motion was utilized in the HygroScope project whose components open upon increase in humidity level (Reichert et al. 2015). As passive operable façades, hygromorphic skins were also pre-programmed to control aperture size and response speed in response to change in humidity level (Menges and Reichert 2012). Doris Kim-Sung utilized the thermal response of bimetals to achieve self-ventilated and shaded structures (Fox 2016).

10.1.2 Hygroscopic Design Parameters of Wood

Hygroscopic properties of wood are related to its ability to absorb and hold water causing swelling and shrinking upon water release. Hardwood and softwood are different in their biological pores, thus affecting their ability to absorb and handle water within(Forest Products Laboratory 2010). Lamination typically uses layers of softwood and hardwood with different shrinkage values, thus producing passive motion in response to humidity. The previous experiments conducted to identify the effect of each hygroscopic design parameter on the response behavior of wood demonstrated that some embedded parameters like grain orientation, type of wood and thickness and lamination are the main motion drivers in wooden actuators (Abdelmohsen et al. 2019a). It was shown that different lamination configurations control both deflection value and response speed of the material (Abdelmohsen et al. 2019c), in addition to the resultant type of motion especially upon upscaling of the material (Wood et al. 2016).

10.1.3 Thermal Expansion Coefficient of Metals

The passive motion of bimetals relies, like wood, on laminating different types of metals varying in their thermal expansion coefficient when exposed to variation in temperature, leading to uniform linear passive motion. Parameters affecting the deflection value of bimetals, as stated in the Timoshenko formula, are the type of each metal, its linear expansion coefficient, Young's modulus, material thickness and dimensional ratio (Angel and Haritos 2013). Utilizing bimetals in interactive facades has been carried out in various architectural structures to provide shape-shifting structures that allow for either shading or ventilation through the curling process of the bimetals upon temperature increase (Sung 2016).

10.2 Material and Method

This paper introduces $H_M T_M$, a hygromorphic-thermobimetal novel composite and responsive actuation mechanism that relies on merging the latent properties of metal and wood. Metals are isotropic materials that expand linearly, while wood, an anisotropic material, has different mechanical properties according to the orientation of its fibers. This gives an opportunity for different response variations upon change in humidity level. The challenge in controlling the response behavior of programmable materials lies in the ability to extract the design parameters of the material and utilize it in the process of designing and controlling motion. Similar to motion grammars that are extracted from motion maps of robots depending on a predefined series of tasks, the motion programming of $H_M T_M$ relies on controlling the output motion through programming discrete sequential motion by utilizing material latent properties.

One of the principle benefits of adding wood to bimetals is that it supports the coding of the overall type of motion of the composite through hygroscopic properties of wood; however, it is triggered by variation in temperature, as opposed to humidity, thus enabling its implementation in hot climates where temperature variation is dominant. Hygroscopic design parameters of wood are studied and developed as an embedded processor that can be encoded to achieve specific types of motion in building façades, such as sliding, rotation, folding and twisting (Abdelmohsen et al. 2019b).

The essence of the $H_M T_M$ composite is utilizing its latent properties as a tangible interface that encompasses complex mechanisms such as motion sensors, processors and actuators inherently in the material itself. What generates passive motion in the programmable material is the wide array of material behavior in response to differences in both humidity and temperature, depending on which material is triggered by either stimulus. Typically, the active layer is responsible for initiating motion, while the passive layer controls the reversible motion process. However, in the $H_M T_M$ composite, both layers can be interchangeably active or passive according to the type of external stimulus. The metal layer becomes temporarily active when the stimulus is variation in temperature, while the wood layer is temporarily active when the stimulus is fluctuation in humidity. The $H_M T_M$ composite promotes encoding hygroscopic properties of wood for unlimited passive motion types and correspondingly is stimulated by temperature, to extend its usability in hot arid climates such as Egypt.

The paper is divided into two parts. The first part describes a series of physical experiments that were conducted to deduce the necessary $H_M T_M$ design parameters for controlling its response to temperature. Comparing and analyzing the effect of material parameters are used to understand their effect on deflection values to achieve different types of motion. This wood-based actuator plays the main role in achieving basic motion types such as rotation, expansion, sliding and combined motion such as twisting. The second part associates $H_M T_M$ design parameters to motion grammars to deduce generative shape-shifting grammars on linear prototypes. The research framework in Fig. 10.1 illustrates the two parts showing the type of material used, actuation method, analytical approach, control parameters, the enhanced $H_M T_M$ properties to function in specific climates like hot arid climates and the deduced motion grammars to program-specific passive response behaviors in adaptive façades.

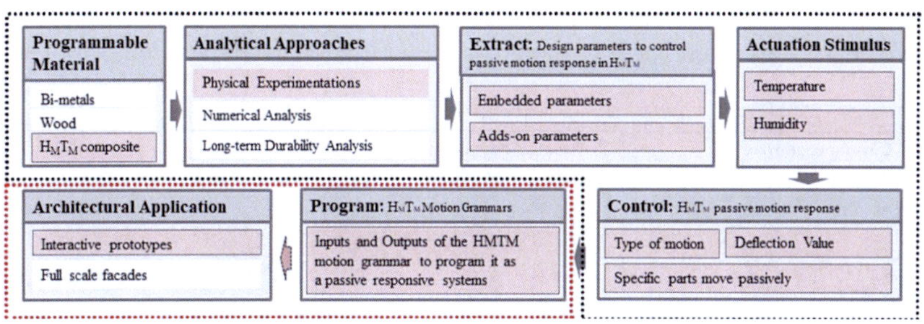

Fig. 10.1 Research framework

10.2.1 Experiment Setup

Encoding the type of response motion of $H_M T_M$ requires controlling the direction of motion and deflection value. Several physical experiments were conducted in a temperature- and humidity-controlled environment to measure the deflection value of $H_M T_M$. Heat is applied by means of an electric oven to reach 40 °C. The deflection of $H_M T_M$ is evaluated by measuring the radius of curvature using image analysis software (Kinovea); a validated motion tracking and analysis method (Puig-Diví et al. 2017; Abdelmohsen et al. 2018). Three markers are located on each side of the sample to denote the start, mid and end point to facilitate tracking. The motion response of $H_M T_M$ results in single and double curved surfaces. The radius of curvature is measured at the highest deflection point reached by single curved samples, while two angular measurements are analyzed in double curved samples; front and back angles (see Fig. 10.2).The results are then exported and analyzed in Microsoft Excel spreadsheets.

10.2.2 Encoding Motion Response of $H_M T_M$ Composite

This paper focuses on aluminum as a metal with high thermal expansion coefficient (23.5 μm/m °C) and beech that has a high shrinkage value of 11.9% in its tangential cut. This aluminum/beech configuration is studied as an $H_M T_M$ base case (parameters shown in Table 10.1). Its response behavior is analyzed and compared to six $H_M T_M$ configurations to deduce motion parameters.

$H_M T_M$ has two fabrication prototypes that differ in their stimulation techniques (Table 10.2). $H_M T_{M1}$ consists of two layers: one wood and one metal layer laminated using silicone-based glue as a thermal- and water-resistant glue. This prototype is triggered by variation in either humidity or temperature, where both layers are active or passive according to the layer responsible for expansion. $H_M T_{M2}$ consists of three laminated layers: one wood layer between two metals with different expansion coefficients. This prototype is triggered only by variation in temperature in bimetals.

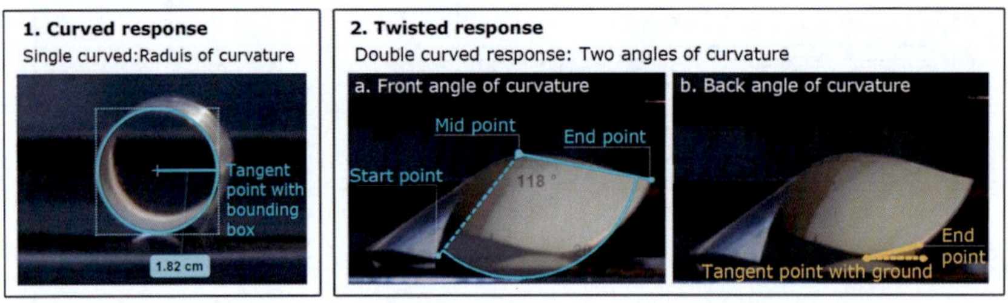

Fig. 10.2 Employed image analysis measurement method in the physical experiments

Table 10.1 Base case $H_M T_M$ composite material parameters

First layer (wood): American beech (hardwood)		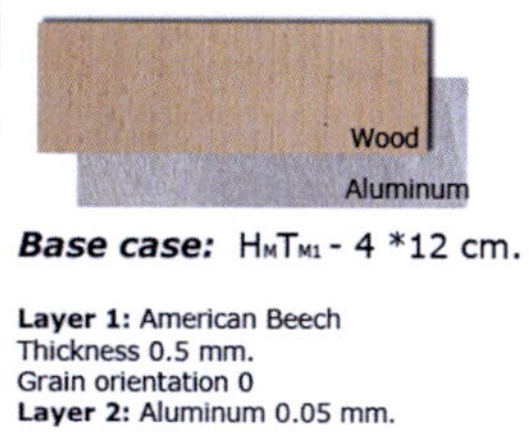
Thickness	0.5 mm	
Grain orientation	Tangential cut with 0° wood grain orientation (fibers are parallel to the short side of the sample)	
Shrinkage value	Tangential: 11.9%	
Second layer (metal): Aluminium		
Thickness	0.05 mm	
Expansion coefficient	23.5 µm/m °C	
Adhesive layer: Silicone-based glue (water- and heat-resistant base)		
Dimensional ratio: 1:3 (4 * 12 cm)		

Table 10.2 Response behavior of $H_M T_M$ prototypes

Prototypes	Stimulus	Motion Mechanism		Radius of curvature
HMTM1 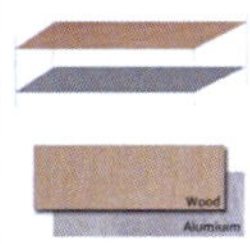	Temperature Humidity Both layers can be active layer according to the stimulus	Temperature increase Humidity increase	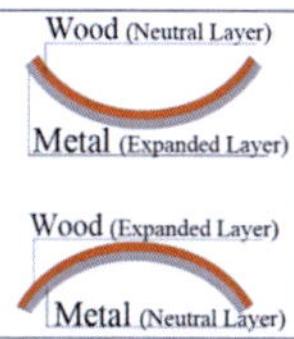	 Average = 1.84 cm.
HMTM2 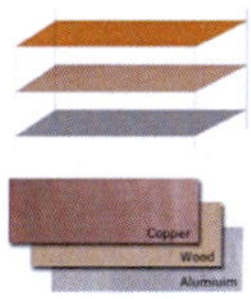	Temperature only Al:Active layer Cu: Passive layer Beech: microcontroller for motion direction	Temperature increase Wood (controls type of motion) Metal 1: metal with high expansion coefficient Metal 2: metal with low expansion coefficient	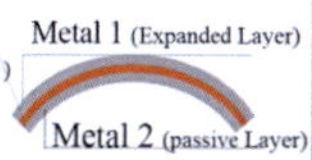	 Average = 25.6 cm.

The wood however is used as an embedded actuator that controls the direction and type of motion by utilizing its hygroscopic design parameters.

$H_M T_{M2}$ can be distinguished from typical bimetal composites due to its ability to achieve a variety of motions other than linear expansion. The difference in their radius of curvature results in different types of motion like rolled and curved samples. By comparing the response behavior of both prototypes upon temperature increase, $H_M T_{M1}$ demonstrates a uniform curved and rolled form with higher deflection and average radius of curvature 1.84 cm, while $H_M T_{M2}$ demonstrates a non-uniform wrinkled curved form with less deflection and radius of curvature 25.6 cm (Table 10.2). In this paper, we focus on $H_M T_{M1}$ design parameters to deduce a generative shape-shifting grammar to control motion response in hot climates.

We identified two types of design parameters for the $H_M T_M$ composite: (1) embedded parameters and (2) control parameters. Embedded design parameters are

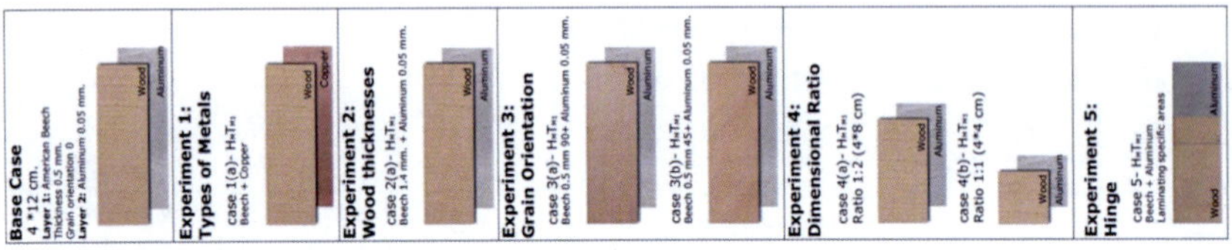

Fig. 10.3 Tested material configurations

related to the latent properties of materials such as (a) type of material, (b) layer thickness, (c) dimensional ratio and (d) grain orientation. Control parameters are related to external parameters that take place during the composite fabrication and assembly such as the fixation position or location of isolated areas in the sample. Five experiments were conducted, in addition to the base case (see Fig. 10.3) to deduce the effect of the different parameters on radius of curvature and motion response. Three identical samples were used in each experiment, and the average result was taken for radii of curvature. Large radius of curvature values denote little deflection response.

Type of Material. The response behavior of $H_M T_M$ varies according to the type of wood and metal used. Wood types vary in shrinkage value, while metals vary in terms of expansion coefficients. The effect of different types of metals was tested on the response behavior of $H_M T_{M1}$ (in this case aluminum with a coefficient of 23 μm/m °C and copper at 17 μm/m °C). Radii of curvature were compared for aluminum/beech and copper/beech composites when exposed to increase in temperature (see Table 10.3). The experiment showed that using metals with high expansion coefficients (aluminum/beech) results in higher deflection values.

Layer Thickness. The thickness of $H_M T_M$ layers affects its response behavior to temperature increase. Two $H_M T_{M1}$ thickness configurations were tested: beech 0.5 mm/aluminum 0.05 mm and beech 1.4 mm/aluminum 0.05 mm. The experiment showed that low-thickness samples have higher deflection values (see Table 10.4).

Dimensional Ratio. This is the ratio of $H_M T_{M1}$ length to width. In this experiment, three ratios were tested with the same material configuration of aluminum/beech (1:3 as the base case, 1:2 and 1:1), with a width of 4 cm. The experiment showed that the larger ratio of 1:3 had the highest deflection value (see Table 10.5).

Table 10.3 Effect of different types of metals on the response behavior of $H_M T_{M1}$

	Aluminum/beech (base case)	Copper/beech (Case 1a)
Maximum deflection		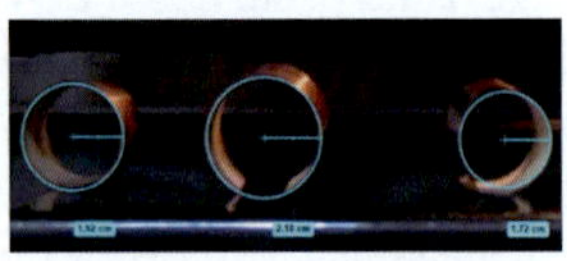
Deflection	Average radius = 1.84 cm	Average radius = 1.94 cm

Table 10.4 Effect of different beech thicknesses on the response behavior of $H_M T_{M1}$

	Beech 0.5 mm/aluminum 0.05 mm (base case)	Beech 1.4 mm/aluminum 0.05 mm (Case 2a)
Maximum deflection		
Deflection	Average radius = 1.84 cm	Average radius = 7.95 cm

Table 10.5 Effect of dimensional ratio on response behavior of $H_M T_{M1}$

	1:3 Ratio (4 × 12 cm) (base case)	1:2 Ratio (4 × 8 cm) (Case 3a)	1:1 Ratio (4 × 4 cm) (Case 3b)
Maximum deflection			
Deflection	Radius = 1.84 cm	Radius = 2.00 cm	Radius = 2.52 cm

Grain Orientation. Three wood grain orientations (parallel to the longitudinal side of the sample at 0°, parallel to the short side at 90° and at 45°) were tested (see Table 10.6). The experiment showed that the 0° and 90° samples exhibited single curved forms, while the 45° samples exhibited a twisted motion behavior. Also, the $H_M T_{M1}$ samples with 0° were shown to have higher deflection values than the 90° samples.

Control Parameters. Control parameters mainly involve parameters not directly related to the latent material properties, such as laminating and isolating specific parts of the sample to control its response behavior. Figure 10.4 shows three different configurations for laminating aluminum and beech: (a) two aluminum pieces covering 50% of the sample on opposite sides of the wood layer, (b) two aluminum pieces covering 50% of the sample at the two ends and same side of the wood sample

Table 10.6 Effect of different wood grain orientations on the response behavior of $H_M T_{M1}$

	Beech 0.5 mm 0°/aluminum 0.05 mm (base case)	Beech 0.5 mm 90°/aluminum 0.05 mm (Case 4a)	Beech 0.5 mm 45°/aluminum 0.05 mm (Case 4b)
Maximum deflection			
Deflection	Radius = 1.84 cm	Radius = 2.26 cm	Twisted angle = 109°

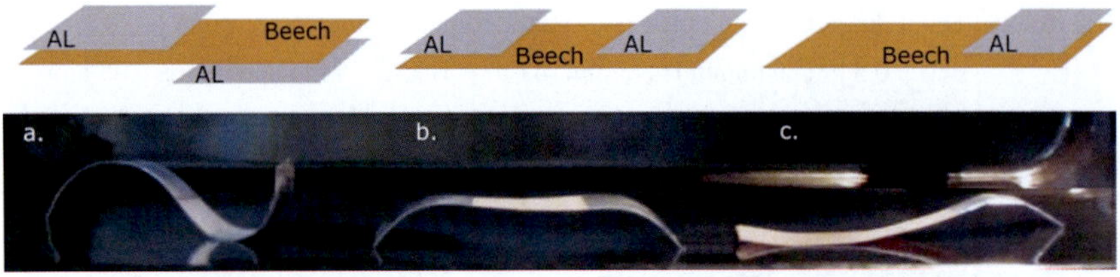

Fig. 10.4 Controlling the motion response behavior of $H_M T_M$ through control parameters

and (c) one aluminum piece covering 25% of the sample at one end. As shown in the figure, response behavior varies according to the control parameter configurations. This can be utilized to achieve different types of complex motion in kinetic façades.

10.3 Grammars

Grammars are the structure of formal languages presented as a set of rules that define the control and dynamics of a system (Dantam et al. 2012). Stiny and Gips introduced shape grammars to describe generative computational approaches through a series of transformation rules applied on initial shape configurations (Knight 2000). Formal grammars were used to understand and analyze the motion of kinetic elements (El-Zanfaly 2011). Motion grammar was developed as a computational approach and linguistics system for designing motion tactics that regulate robotic motion associated to 3D object movement (Bidgoli and Cardoso-Llach 2015; Dantam and Stilman 2012). In complex systems, motion is described as strings with continuous and discrete events (Dantam et al. 2012). Composite motions are derived from a series of rules that are combined and sorted to express all options of change (Bidgoli and Cardoso-Llach 2015).

In motion grammar logic, context-free language defines motion through *sensors*, *output productions*, *tokens*, *tokenizing functions*, *variables*, *semantic rules* and *commands* (Dantam and Stilman 2012; El-Dabaa and Abdelmohsen 2018). Using this logic in the context of a shape-shifting grammar framework using hygroscopic and thermo bimetal design parameters addresses the control, tracking and programming of $H_M T_M$ response for adaptive façades. The framework is based on two types of formal language grammars: shape and motion grammars, where a set of semantic rules and procedures define the process of transforming an initial state with a desired state through several rules and variables. *Sensors* in this context represent the parameters for passive motion response design including embedded and control parameters. *Productions* refer to the resulting output motion behavior which is represented through *tokens* including deflection values and motion types. These are controlled by *semantic rules* which constitute the primary component of the grammar that relates the combination of variables to reach specific productions and describes the ability of the composite to expand and shrink. *Tokenizing functions* denote the change of

Table 10.7 $H_M T_M$ shape-shifting grammar database

Variables (Embedded parameters) — Variables (Controlled parameters) — Production (Output - Type of motion)

Semantics rules with relation to temp and humidity

Material configuration

Tokens

Record

Metal Embedded Parameters		Wood Embedded Parameters				Control Parameters			Output	
						Isolation through lamination wood/metal				
Metal thickness	Metal Type	wood Type	Grain orientation	Dim. Ratio	Thickness	Percentage	Direction	Location	Deflection	Type of motion
thickness 1	Aluminum or Copper	Beech or Spruce	0	1:1	thickness 1	100%	Single side Opposite sides	All covered		
						75%	Single side	End		
								Middle		
								Start		
							Opposite sides	End		
								Middle		
								Start		
						50%	Single side	End		
								Middle		
								Start		
							Opposite sides	End		
								Middle		
								Start		
						25%	Single side	End		
								Middle		
								Start		
							Opposite sides	End		
								Middle		
								Start		
					thickness 2					
				1:2	thickness 1					
					thickness 2					
				1:n	thickness 1					
					thickness 2					
			90	1:1	thickness 1					
					thickness 2					
				1:2	thickness 1					
					thickness 2					
				1:n	thickness 1					
					thickness 2					
			45	1:1	thickness 1					
					thickness 2					
				1:2	thickness 1					
					thickness 2					
				1:n	thickness 1					
					thickness 2					

moisture content in wood in response to humidity or change in thermal response to temperature in metals. *Variables* are the embedded parameters that affect the overall motion response like grain orientation, ratio, thickness and type of wood or metal. *Commands* denote the resulting output motion type such as twisting, bending and rotation. (Abdelmohsen et al. 2019b). The $H_M T_M$ shape-shifting grammar database (see Table 10.7) shows the output motion response for each material configuration through a series of records. Relating and combining the variables with the records reveal semantic rules for each material configuration. This logic is used as a base for the $H_M T_M$ shape-shifting framework.

Using this database, it is possible to program different output tokens including motion types and deflection values using specific semantic rules for material configuration variables. Semantic rules include the variable configuration algorithm for material fabrication and its properties, targeting the ability of the material to handle either percentage of moisture content or expansion in a given sample to reach specific deflection values and motion types. The material lamination configuration addresses most of the design parameters including embedded and control $H_M T_M$ parameters.

The $H_M T_M$ shape-shifting grammar database depends on various records to reach the desired types of motion. This database is flexible enough to encompass future design parameters or more records for the existing design parameters. As an example, the bending command for $H_M T_M$ due to increase in temperature requires the following semantic rules: the expansion of the metal layer more than the wood layer, an increase in the deflection of the sample and an increase in its radius of curvature. The type of deflection depends on the material configuration variables such as the type of layers, grain orientation, dimensional ratio and thickness. The same applies in the case of humidity increase, but with wood acting as the active layer that expands and drives the motion through a different set of semantic rules.

10.4 Conclusion

The $H_M T_M$ motion framework pushes the limits of low-tech and low-cost programmable materials to replace mechanical actuators with passive responsive actuators with applicability in different climatic zones. This paper demonstrated design parameters for a novel programmable wood–metal composite ($H_M T_M$) to generate different motion types including simple motion, bending, rolling, twisting or combined motion types such as double curved or s-shaped motion through controlling embedded and control parameters. Embedded $H_M T_M$ parameters such as grain orientation and ratio were shown to control simple motion types, while control parameters were shown to program more complex motion types. All design parameters were shown to affect deflection value, for example, thick samples tend to have less deflection value than veneer and metal sheets. Future work will study different design configurations to achieve more controlled motion responses within a shape-shifting grammar framework and expand its database of variables, tokens and semantic rules.

Acknowledgements The authors are grateful for the support provided by Bartlett's Fund for Science and Engineering Research Collaboration in supporting the research project "Soft Adaptive Building Skins for Energy Efficient Architecture."

References

Abdelmohsen S, Adriaenssens S, El-Dabaa R, Gabriele S, Olivieri L, Teresi L (2019) A multi-physics approach for modeling hygroscopic behavior in wood low-tech architectural adaptive systems. Comput Aided Des 106:43–53

Abdelmohsen S, Adriaenssens S, Gabriele S, Olivieri L, El-Dabaa R (2019b) Hygroscapes: innovative shape shifting façades. In: Bianconi F, Filippucci M (eds) Digital wood design: innovative techniques of representation in architectural design, pp 675–702

Abdelmohsen S, Massoud P, El-Dabaa R, Ibrahim A, Mokbel T (2018) A computational method for tracking the hygroscopic motion of wood to develop adaptive architectural skins. In: ECAADe 2018: 6th annual conference on education and research in computer aided architectural design in Europe, vol 2. Lodz, Poland, pp 1–9

Abdelmohsen S, Massoud P, El-Dabaa R, Ibrahim A, Mokbel T (2019) The effect of hygroscopic design parameters on the programmability of laminated wood composites for adaptive façades. In: 18th CAAD futures 2019 international conference. Presented at the Daejeon, Korea

Angel GD, Haritos G (2013) An immediate formula for the radius of curvature of a bimetallic strip. Int J Eng Res 2(12):1312–1319

Bidgoli A, Cardoso-Llach D (2015) Towards a motion grammar for robotic stereotomy. In: 20th international conference of the association for computer-aided architectural design research in Asia (CAADRIA), Hong Kong, pp 723–732

Correa D, Papadopoulou A, Guberan C, Jhaveri N, Reichert S, Menges A, Tibbits S (2015) 3D-printed wood: programming hygroscopic material transformations. 3D Print Add Manuf 2(3):106–116
Dantam N, Essa I, Stilman M (2012) Linguistic transfer of human assembly tasks to robots. In: IEEE/RSJ international conference on intelligent robots and systems, pp 237–242
Dantam N, Stilman M (2012) The motion grammar: linguistic perception, planning, and control. In: Robotics: science and systems VII. MIT Press, Cambridge, Mass, pp 49–56
Dierichs K, Menges A (2016) Towards an aggregate architecture: designed granular systems as programmable matter in architecture. Granular Matter 18(2)
El-Dabaa R, Abdelmohsen S (2018) A methodology for evaluating the hygroscopic behavior of wood in adaptive building skins using motion grammar. IOP Conf Ser Mater Sci Eng 362:1–8
El-Zanfaly D (2011) Active shapes: introducing guidelines for designing kinetic architectural structures. In: SIGRADI 2011. Presented at the XV CONGRESO DE LA SOCIEDAD IBEROAMERICANA DE GRÁFICA DIGITAL, Argentina
Forest Products Laboratory (2010) Wood handbook: wood as an engineering material. U.S. Department of Agriculture, Forest Service, Forest Products Laboratory, Madison, Wis.
Fox M (ed) (2016) Interactive architecture: adaptive world, 1st edn. Princeton Architectural Press, New York
Holstov A, Morris P, Farmer G, Bridgens B (2015) Towards sustainable adaptive building skins with embedded hygromorphic responsiveness. Advanced building skins. Presented at the Graz, Graz
Knight T (2000) Introduction to shape grammars. Lecture Notes presented at the MIT, MIT/Miyagi Workshop
Kretzer M (2017) Information materials: Smart materials for adaptive architecture
Krieg OD, Christian Z, Zuluaga DC, Menges A, Reichert S, Rinderspacher K, Schwinn T (2014) HygroSkin–meteorosensitive pavilion. Fabricate 2014:61–67
Mansoori M, Kalantar N, Creasy T, Rybkowski Z (2019) Adaptive wooden architecture. Designing a wood composite with shape-memory behavior. In: Bianconi F., Filippucci M (eds) Digital wood design, vol 24, pp 703–717
Menges A, Reichert S (2012) Material capacity: embedded responsiveness. Archit Des 82(2):52–59
Moloney J (2011) Designing kinetics for architectural facades: state change. Routledge, Abingdon, Oxon, New York
Nijholt A, Minuto A (2017) Smart material interfaces: playful and artistic applications. In: Conference on imaging, vision and pattern recognition (IcIVPR), pp 1–6
Puig-Diví A, Padullés-Riu JM, Busquets-Faciaben A, Padullés-Chando X, Escalona-Marfil C, Marcos-Ruiz D (2017) Validity and reliability of the kinovea program in obtaining angular and distance dimensions. Preprints 1
Reichert S, Menges A, Correa D (2015) Meteorosensitive architecture: biomimetic building skins based on materially embedded and hygroscopically enabled responsiveness. Comput Aided Des 60:50–69
Rüggeberg M, Burgert I (2015) Bio-inspired wooden actuators for large scale applications. PLoS ONE 10(4):1–16
Sung D (2016) Smart geometries for smart materials: taming thermobimetals to behave. J Archit Educ 70(1):96–106
Wood DM, Correa D, Krieg OD, Menges A (2016) Material computation—4D timber construction: towards building-scale hygroscopic actuated, self-constructing timber surfaces. Int J Archit Comput 14(1):49–62

Chapter 11
Retrofit as a Means for Reaching Net-Zero Energy Residential Housing in Greater Cairo

Bassent Adly, Hanan Sabry, Ahmed Faggal, and Mahmoud Abd Elrazik

Abstract The net-zero energy building (NZEB) concept is the future international goal for saving energy. It has led to many countries implementing policies designed to target efficient built environment. In Egypt, the built environment consumes more than 70.1% of total sold energy, with the residential sector consuming 54.9%. According to the Egypt's annual electricity report, "the increase in the rate of sold energy of the residential sector, in comparison with industry and other purposes, is a result of the expansion of residential compounds and new communities." This paper introduces the concept of retrofitting the existing residential gated communities to achieve zero energy residential community (ZEC) and discusses the integration of two strategies: energy efficiency retrofitting techniques to reduce energy demand and renewable ensergy systems to generate sufficient energy, resulting in nearly zero energy consumption on an annual basis. Design Builder is used to analyze the feasibility of integrating these two strategies in a single-family house, in a residential gated community in the eastern part of Greater Cairo. The simulation goes through two stages. The first stage, which itself consists of three phases, is focused on the building scale. During this stage, energy efficiency retrofitting techniques are investigated,and renewable energy (PV panels) is integrated. The second stage is focused on the community scale and involves applying the results on the whole residential community. The results achieved a reduction of 88.68% in energy consumption for the whole compound and proved that NZEB could be met by applying the two strategies. Moreover, a significant reduction in energy is shown when applying (NZEB) retrofit on a community scale.

Keywords Retrofitting gated community · Net-zero energy building (NZEB) · Zero energy community (ZEC) · Energy efficiency · Renewable energy

---

B. Adly (✉) · H. Sabry · A. Faggal
Faculty of Engineering, Ain Shams University, Cairo, Egypt
e-mail: bassentalaa@gmail.com

M. A. Elrazik
Housing and Building National Research Center, Giza, Egypt

Architecture and Urbanism: A Smart Outlook,
https://doi.org/10.1007/978-3-030-52584-2_11

11.1 Introduction

The increasing rate in energy consumptionis considered to be one of the world's most significant challenges today. The building sector is widely known as one of the key energy consumers worldwide. The analysis of the energy performance of the existing building stocks is highly increasing worldwide. (Pombo et al. 2016) The total energy sold in Egypt for the year 2017 reached 174.6 terra watt hours, with an annual average growth rate of 3.7% per year during the period from 2012/2013 to 2016/2017. According to the Egypt's annual electricity and renewable energy consumption report, residential buildings, especially gated communities, consume the largest ratio. (EEHC 2017) Therefore, implementing energy-efficient retrofit on the existing residential compounds in new communities provides opportunities for effectively reducing energy consumption in the residential sector;this represents both the problem and the solution at the same time. (El-Darwish and Gomaa 2017).

One of the most widespread strategies across the globe in the past few years is Net-Zero Energy Building (NZEB). This concept aimed at designing highly sustainable buildings, and it is based on two main aspects: The first is saving energy and the second is producing energy through renewable energy resources. Both parts should be regarded in the Egyptian building context. (Albadry et al. 2017) The main concept of (ZEB) has produced a new concept, namely net-zero energy community (NZEC). This includes communities which have reduced their energy intake through efficient techniques and met remaining energy needs through renewable energy resources. (Carlisle et al. 2009).

Attia and Herde (2009) investigated the potential and impact of retrofitting on two different scales: the building scale and the community scale. They used Transys to evaluate and compare the energy performance of retrofitting a middle-income apartment in a residential community in Cairo and the whole urban community. Envelope retrofitting and internal loads' retrofitting were applied. In addition, domestic water heating, photovoltaic panels and solar thermal air conditioning were integrated. The final result showed significant savings in energy consumption when applied to a community scale rather than on a building scale. (Attia and Herde 2009).

El-Ghetany and Hussein (2016) studied energy and water conservation in a residential compound in Egypt, in order to investigate the impact of retrofitting techniques on energy and water conservation in the compound. RetScreen software was used as the energy simulation program to calculate the effect of retrofitting actions—such as shading device retrofitting, lighting retrofitting, solar water heating and PV system—on the energy performance of the buildings. The final result revealed that by applying the retrofitting process, the compound could save 30% in its energy consumption. (El-Ghetany and Hussein 2016).

Al-Ragom (2002) examined multiple retrofitting techniques in residential buildings in hot arid climates. The study aimed to address the situation in countries with such climates, and to justify the implementation of the retrofitting process. Several retrofitting options were identified and defined, including roof insulation and glazing systems. The methodology relies on performing the analysis using theoretical techniques and modeling different options using DOE-2.1E as the building energy simulation program. The final result showed that significant savings could be achieved at the national level even if the cost of implementation was fully supported by the government (Al-Ragom 2002).

Mourad et al. (2014) applied ZEB techniques on a single-family detached house in a new community in a desert in Egypt. Renewable energy technologies were used to provide the house with electricity, heat and cold air (need in the summertime), specifically by using hybrid photovoltaic thermal solar collectors, PV module and small-scale wind turbines for generating electrical power.

Through simulation, it was shown that by applying ZEB techniques, the house could be efficient and save around 20% of its energy bill. (Mourad et al. 2014).

Within the relevant literature, there are not many publications concerning net-zero energy community (NZEC) in Egypt. Moreover, methodologies usually tackle only one of two basic aspects of the research, either net-zero energy methodology or retrofit methodology. This paper aims to investigate the result of applying a combination of retrofitting and net-zero energy concepts, while striving to reach zero energy community consumption annually in the Egyptian building context.

The main goal of this paper is to evaluate the potential of applying retrofitting techniques and PV panels as a means for reaching net-zero energy residential housing community in Greater Cairo by simulation.

11.2 Methodology

A simulation using Design Builder is applied to a single-family house, in a residential gated community located in the eastern part of Greater Cairo.This is done in order to investigate the effect of applying the energy efficiency retrofitting techniques and (PV panels), on the building scale and the community scale, and to determine whether a zero energy residential gated community can be achieved.

11.2.1 The Case Studies

The selected gated community has a total area of 542,000 m^2. It consists of 253 three story villas and includes a central service area consisting of a sport and social club, a pharmacy, a cash dispenser and a mosque. Each villa consists of three floors, the ground floor, which is 260 m^2, the first floor, which is 275 m^2 and the third floor, which is 260 m^2 as shown in (Fig. 11.1). On average, each villa contains one family of four to six occupants. During weekdays, the occupancy of the house is almost full from 3:00 pm until 8:00 am. As for most weekends, the villa is occupied almost 24 h a day. In addition, there are two workers working six days a week (they take Friday as a day off).

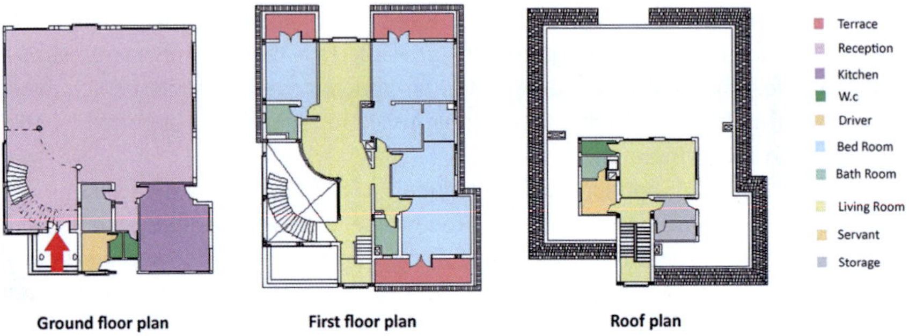

Fig. 11.1 Villa's floor plan

11.2.2 Simulation Phases

This paper represents a practical application of the proposed guideline, by applying a simulation on a single-family house in a residential gated community in the eastern part of Greater Cairo. Design Builder is used as a simulation tool to measure and assess the energy performance and feasibility of applying each strategy, with the simulation analysis divided into the following two consecutive stages:

a) The first stage involves three phases and is applied to the building scale as follows:
- Phaseone: The actual energy performance and the user's electric consumption patterns were determined in four main orientations in the community layout. The objective is to identify the maximum annual energy consumption kWh. This is considered the worst case on which the simulation will be applied in the second phase.
- In phase two: Three energy efficiency retrofitting techniques are applied: insulation and glazing in the building envelop and retrofitting of lighting systems, Aiming to investigate their effect on energy consumption and to reach the maximum energy reduction. As for the techniques that achieved the maximum energy reduction, they are combined and applied in the next phase.
- In the third phase: Renewable energy sources such as PV panels are integrated to produce enough energy to cover the remaining energy from phase two.

b) In the second stage, the simulation results of the first stage are implemented on the entire gated community (the community scale)with the aim of reaching zero energy consumption.

11.3 Phase One

The actual energy performance and the user's electric consumption patterns and peaks were determined in four main orientations in the community layout, as follows:

Table 11.1 Represents a description of the building elements in the current situation

Building envelop elements	Walls	
	Description	R-value
	The building has a concrete structure with single concrete brick walls. There was not any insulation in the exterior wall	0.37
	Openings	
	Description	SHGC
	All windows have a single-blue 6-mm panel, and 25% of the window frames were made of wood and the rest are made of aluminum	0.58
	Ceiling	
	Description	R-value
	The roof is made of cement tiles and is insulated by using (rock wool with 5 cm thickness) as a thermal insulating material	1.89
	Shading device	
	Description	
	42% of the total number of windows has wooden louvers to prevent direct sunlight from entering the interior space	
Internal loads elements	**Lighting**	
	Description	
	Each room has different amount and types of lighting units, including incandescent lamps (40 W), halogen ceiling spotlight lamps (60 W) and led lamps (13 W)	

11.3.1 The Current Situation

For purposes of analyzing the current situation of the building, Table 11.1 represents a description of the building elements before the energy efficiency retrofit process.

11.3.2 Simulation Results of Phase One

Four different orientations (A, B, C and D) which represent the four main directions were selected to investigate the effect of the building orientation on the energy efficiency retrofitting process by simulation. (A) represents the South orientation, (B) represents the West orientation, (C) represents the North orientation and (D) represents the East orientation as shown in (Fig. 11.2).

By analyzing the current energy performance in the four orientations by simulation, the result is as shown in (Fig. 11.3):

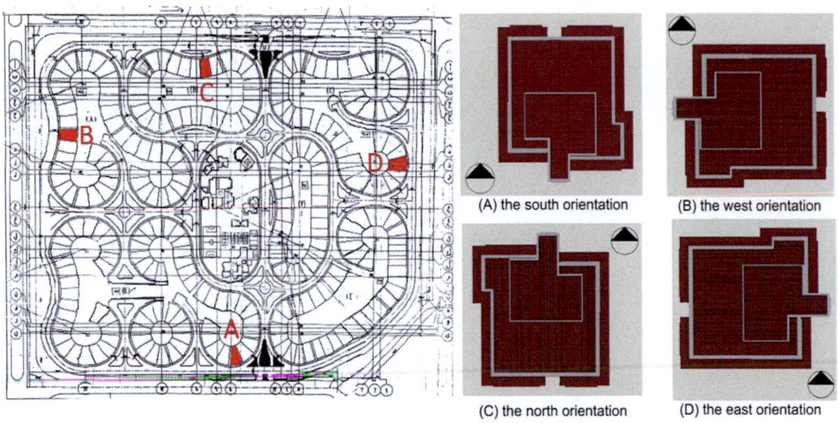

Fig. 11.2 Four orientations A, B, C and D

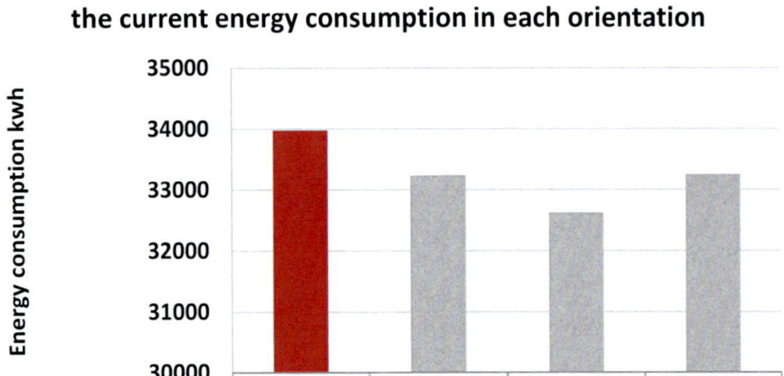

	(A) South	(B) West	(C) North	(D) East
energy consumption	33980.136	33236.846	32622.996	33245.876

Fig. 11.3 Energy consumption results for the four orientations

- There are two types of loads affecting the energy performance of the buildings in any orientation: external thermal loads on the building envelops and internal loads (lighting, equipment) inside the buildings.
- The results show that the internal gains (lighting, equipment) in all the previous cases (A, B, C and D) are constant, but the cooling loads vary due to the change in the orientation and the external loads affecting the building envelop.
- All these loads should be reduced by applying retrofitting techniques which lead to a reduction in consumption, as follows in the next phase.
- By simulation, case (A) has the highest energy consumption compared to the rest. Therefore; it represents the base case for applying the retrofitting techniques in phase two.

11.4 Phase Two

Energy efficiency retrofitting techniques involves buildings envelop and internal loads retrofitting.

11.4.1 Energy Efficiency Retrofitting Techniques

Five types of energy efficiency retrofitting techniques were selected. They include wall, opening, ceiling, shading device and lighting retrofitting. These techniques aim to achieve the maximum energy reduction in every retrofitting type as shown in Table 11.2.

11.4.2 Simulation Results of Phase Two:

The retrofitting actions that achieved the maximum energy reduction in every retrofitting type were applied to the base case. The amount of reduction in energy consumption achieved by every retrofitting type is shown in (Fig. 11.4). And the total reduction in energy consumption achieved when all retrofitting types were combined was 22.6% as shown in (Fig. 11.5).

Applying energy efficiency retrofitting on the building envelop produced the following results:

- The maximum savings in energy consumption are achieved by the application of shading device retrofitting, while the application of ceiling retrofitting achieved the least savings.
- The results of applying opening retrofitting, and shading device retrofitting on the building envelop, are close. Therefore, it is possible to apply only one of them.
- The energy reduction achieved by applying wall retrofitting is less than the reduction achieved by applying shading and opening retrofitting.
- The ceiling retrofitting is not effective because the rate of reduction in energy consumption due to its application is weak and insufficient. Therefore, it cannot be applied.

It is recommended to start the building envelop retrofitting process with opening retrofitting or shading device retrofitting, in order to guarantee the maximum reduction in energy.

Table 11.2 Retrofitting action that achieved the maximum energy reduction in every retrofitting type

Building envelop elements	Wall retrofitting	
	Retrofitting action	R-value
	Increasing the thermal resistance (R-value) of external walls, by applying wall retrofitting using an insulating material including (extruded polystyrene 2 cm, expanded polystyrene 2 cm, fiberglass roll 3 cm, rock wool 2 cm, polyurethane 3 cm, light sand brick 10 cm) that achieved the best energy reduction	1.07
	Opening retrofitting	
	Retrofitting action	SHGC
	Decreasing the solar heat gain coefficient (SHGC) of opening glazing, by applying opening retrofitting using a glazing material including (single electric absorbed colored 6 mm-double ref-B-H clear 6 mm/6 mm air)that achieved the best energy reduction	0.27
	Ceiling retrofitting	
	Retrofitting action	R-value
	Increasing the thermal resistance (R-value) of the ceiling, by applying ceiling retrofitting using an insulating material consisting of (expanded polystyrene layer covered with cement mortar and fiber glass mesh from both sides) that achieved the best reduction	2.7
	Shading device retrofitting	
	Retrofitting action	
	Decreasing the solar heat gain coefficient (SHGC) of building opening glazing, by applying shading device retrofitting using an exterior roller blind that is made of aluminum on all the building openings	
Internal loads elements	**Lighting retrofitting**	
	Retrofitting action	
	Decreasing the internal heat gain in the building, by applying lighting retrofitting,through replacing all the incandescent lamps (40 W) and the halogen ceiling spotlight lamps(60 W), by Led lamps (9, 13 W)	

11.5 Phase Three

PV panels are integrated in the roof of every villa in the compound, and the type of connection chosen is on-grid. It allows for purchasing energy from the grid or feeding in excess energy to the grid. (Marszala et al. 2011).

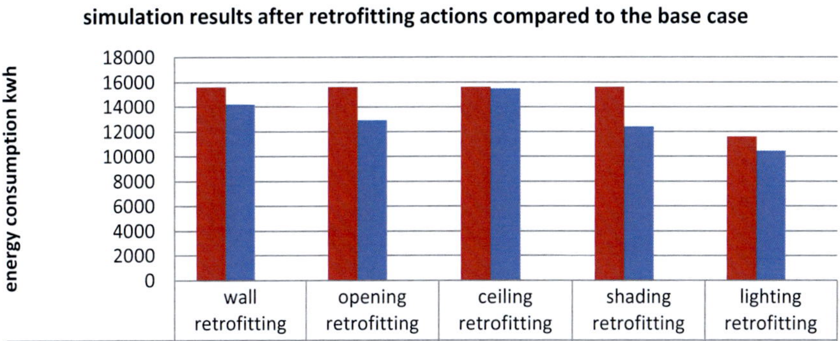

	wall retrofitting	opening retrofitting	ceiling retrofitting	shading retrofitting	lighting retrofitting
base case	15624.3	15624.3	15624.3	15624.3	11606.85
after retrofitting	14221.2	12926.05	15485.98	12392.01	10456.28
reduction %	9.12%	17.26%	1%	20.68%	9%

Fig. 11.4 Amount of reduction in energy consumption that is achieved by every type

Fig. 11.5 Energy consumption after applying all retrofitting actions (building envelops retrofitting and lighting retrofitting) on the base case

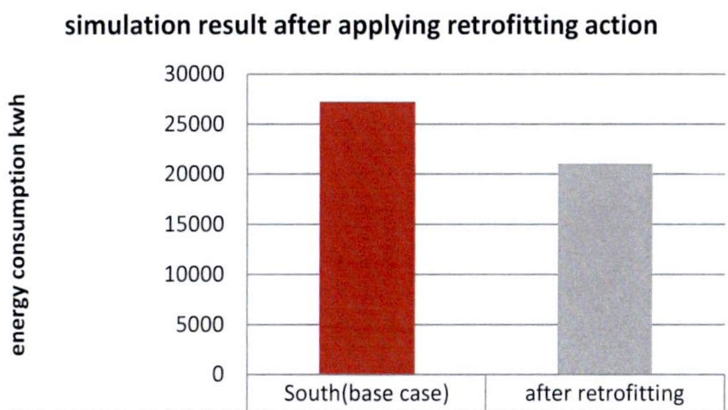

	South(base case)	after retrofitting
energy consumption	27231.136	21081.89

11.5.1 Integrating Renewable Energy Systems (Zero Energy Techniques)

According to Albadry et al. (2017), the polycrystalline panel with 30 degree inclination generates the largest amount of energy. There must be a space of 60 cm between each row of panels and the one following it. (Albadry et al. 2017).

- Many PV panels array had been tested, and the following array shown in (Fig. 11.6) represents the best arrangement due to the maximum generated energy. The data of the panels is as follow:
- The orientation of the panels: It is installed facing the South direction based on the solar data analysis.
- The roof floor area is 260 m^2, and the photovoltaic dimension is 164 × 99 cm. Therefore, the number of photovoltaic that can be installed is 44 photovoltaic cells, with electric power 250 W each.

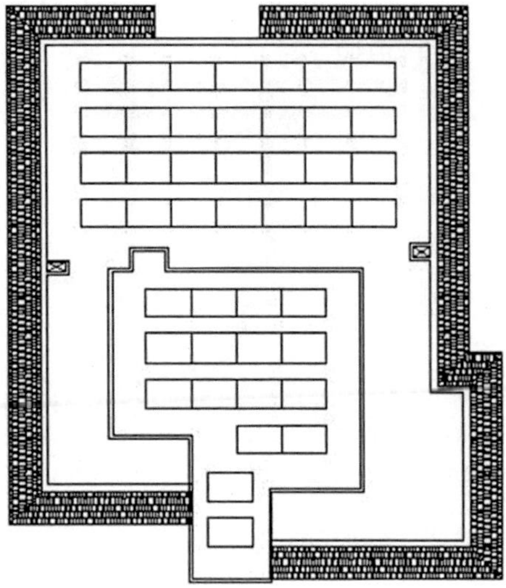

Fig. 11.6 PV panel's numbers and array on every villa's roof

- The average working hours for the photovoltaic is 6 h per day. (Abd elrazik and Selim 2017)

  The total energy produced by PV panels per day is calculated as follows:
  44 Cells * 250 W * 6 h = 66,000 W.H = 66 k.W.H
  The total energy produced by PV panels per year is calculated as follows:
  66 365 day = 24,090 k.w.h per year.

11.5.2 Simulation Results for Phase Three

Based on simulation, the reduction in energy consumption after applying an energy efficiency retrofit and integrating renewable energy techniques is shown in Table 11.3.

11.6 Energy Reduction for The Whole Compound (Community Scale)

After applying energy efficiency retrofit techniques and integrating renewable energy systems on the building scale, all the results are applied to the community scale to investigate the feasibility of achieving zero energy community in Greater Cairo.

The number of villas in each orientation is as follows: South = 70 villas, West = 60 villas, North = 68 villas and East = 55.

Table 11.3 Energy reduction before and after the retrofitting process and integrating the PV panels applied to all orientations

Orientation	Energy consumption before retrofitting k.w.h	Energy consumption after retrofitting k.w.h	Energy consumption after retrofitting and integrating PV panels k.w.h	Energy reduction achieved (%)
South	33980.136	27830.9	3740.9	89
West	33236.846	27740.916	3650.916	89.02
North	32622.996	27914.706	3824.706	88.27
East	33245.876	27947.946	3857.946	88.67

Table 11.4 Total energy consumption of all the entire compound's villas before and after energy-efficient retrofitting and integrating zero energy techniques

Orientations	The total energy consumption for all the villas **before** energy efficiency retrofitting and integrating zero energy techniques k.w.h	The total energy consumption for all the villas **after** energy efficiency retrofitting and integrating zero energy techniques k.w.h
South	70* 33980.136 = 2,378,609.52	3740.9 * 70 = 261,863
West	60* 33236.846 = 1,994,210.76	3650.916 *60 = 219,054.96
North	68* 32622.996 = 2,218,363.728	3824.706 * 68 = 260,080.008
East	55* 33245.876 = 1,828,523.18	3857.946 * 55 = 212,187.03
Total = the sum	8,419,707.188	953,184.998

The total energy consumption of the entire compound's villas before and after energy efficiency retrofitting and integrating zero energy techniques is shown in Table 11.4.

The ratio of energy consumption after applying an energy efficiency retrofit and integrating renewable energy techniques = 953,184.998/8,419,707.188 *100 = 11.32%.

Therefore, by applying an energy efficiency retrofitting and integrating renewable energy techniques, the compound achieved an energy reduction of **88.68%** of the total energy consumption before retrofitting.

11.7 Conclusion

In Egypt, the expansion of residential gated communities is one of the primary causes for the increasing rate of energy consumption in the building sector. Retrofitting existing gated communities into the net-zero energy housing is a feasible solution to solve this problem. This paper shows a practical application of the proposed methodology and guidelines. By applying selected retrofitting techniques—involving building envelop and internal loads retrofitting—as well as integrating PV panels as renewable energy technique on a case study in the eastern part of Greater Cairo. The result shows a reduction in energy consumption reaching 18% after applying energy efficiency retrofit. Moreover, the reduction reached 88.68% for the whole gated community after integrating the PV panels' technique. The result did not achieve zero energy community due to the existence of some obstacles, including human behavior, and the need to leave empty spaces on the roof for maintenance issues. Ultimately, this paper provides a vision and examination for retrofitting gated communities that may soon be implemented.

References

Abd elrazik M, Selim R (2017) B approaching zero energy consumption in administrative building. Hous Build Nat Res Cent (HBRC) 1–11

Albadry S, Tarabieh K, Sewilam H (2017) Achieving net zero energy buildings through retrofitting existing residential building using PV panels. Int Conf—Altern Renew Energy Quest 195–204

Al-Ragom F (2002) Retrofitting residential buildings in hot and arid climates. Elsevier

Amazon. (n.d.). Retrieved from https://www.amazon.co.uk/Silver-Reflective-Window-Control-Privacy/dp/B00OJ87M64

Attia S, De Herde A (2009) Impact and potentials of community scale low-energy retrofit: Case study. In: Cairo. 3rd CIB international conference on smart and sustainable built environment (SASBE) 2009. TU Delft, Netherlands

Carlisle N, Geet OV, Pless S (2009) Definition of a "Zero Net Energy" community. National Renewable Energy Laboratory, United States

EEHC (2017) Retrieved 14 May 2019, from Egyptian electricity holding company: https://www.eehc.gov.eg/eehcportal/Eng/

El-Darwish I, Gomaa M (2017) Retrofitting strategy for building envelopes to achieve energy efficiency. Alexandria Eng J 1–14

El-Ghetany H, Hussein H (2016) Study of energy and water conservation in a residential compound in Egypt. Int J Appl Eng Res

Marszala A, Heiselberga P, Bourrelleb J, Musallc E, Vossc K, Sartori I et al (2011) Zero energy building—A review of definitions and calculation methodologies. Elsevier

Mourad MM, Ali A, Ookawara S, Abdel-Rahman A, Abdelkariem N (2014) An energy-efficient smart home for new cities in Egypt. Eco-Architecture 115–126

Pombo O, Allacker K, Rivela B, Neila J (2016) Sustainability assessment of energy saving measures: a multi-criteria approach for residential buildings retrofitting—A case study of the Spanish housing stock. Elsiveir 1–30

Chapter 12
Integrating Passive Design Strategies and Energy Generation Systems for Net Plus-Energy Housing in Egypt

Hussein Hamza, Mohamed Ayman Ashour, Hanan Mostafa Sabry, and Mostafa Refat Ismail

Abstract Global risks and challenges concerning climatic change, resource depletion and environmental degradation are becoming increasingly obvious. Egypt has recently suffered from fuel shortages combined with rising electricity costs and growing demand for energy. With the residential sector responsible for 50% of the national electricity consumption, net zero and net plus-energy buildings are becoming high priority. This paper investigates the combination of passive design strategies and energy generation systems to achieve compromises between the heating, cooling, lighting, domestic hot water and plug load requirements. The aim of this study is to achieve a balance, on an annual basis, between the energy consumption and production to reach net plus-energy house design. A three-bedroom apartment located in three different climatic zones in Egypt has been selected as a case study. These relevant locations are: Cairo, Alexandria and Aswan. IESVE software is used for energy simulation analysis. Results indicate that the significance for applying these strategies is not the same for all climatic conditions. Theoretically, it was found possible to achieve net plus-energy housing in Egypt where the energy production from PV cells array exceeded the total annual primary energy consumption. Area needed for PV array system in this case study reached 145 m2. Although roof may not provide sufficient space to cover consumptions for all units, additional spaces can be provided on building façades as well as parts of a community renewable energy supply system, thus moving steps forward from net zero-energy balance toward plus-energy performance.

Keywords Net zero-energy building (NZEB) · Net plus-energy building (NPEB) · Renewable energy · Residential · Dynamic thermal simulation

H. Hamza (✉) · M. A. Ashour · H. M. Sabry · M. R. Ismail
Faculty of Engineering, Ain Shams University, 1 Elsarayat St., Abbaseya, Cairo 11517, Egypt
e-mail: hussein.a.faried@eng.asu.edu.eg

Architecture and Urbanism: A Smart Outlook,
https://doi.org/10.1007/978-3-030-52584-2_12

12.1 Introduction

Global risks and challenges concerning climatic change, resource depletion and environmental degradation are becoming increasingly obvious. Energy use in buildings worldwide accounts for over 40% of primary energy use and 24% of greenhouse gas emissions (Garde et al. 2017). Measures and changes in the building modus that can yield substantial savings in energy performance have thus become an essential need. On April 23, 2009, the EU Parliament has requested that by 2019, all new buildings conform to zero energy and emission (Kolokotsa et al. 2011). Recently, Egypt has suffered from fuel shortages combined with rising electricity costs and growing demand for energy. With the residential sector solely responsible for 50% of the national electricity consumption, net zero and net plus-energy residential prototypes are becoming high priority (Renewable-Energy 2017).

A number of previous publications addressed similar research topic. Clarke et al. (2005) applied building simulation to a case study of hybrid renewable energy systems for residential building in Korea. The study aimed at evaluating the feasibility of new technologies using a simulation-based decision support system. Bernier, Ferron and Biaou (2006) also applied building simulation using TRNSYS software to simulate a net zero-energy home in Montreal equipped with photovoltaic (PV) panels and a geothermal heat pump for heating and cooling. Norton and Christensen (2007) chose the three-bedroom Denver zero energy home as a case study, combining envelope efficiency, efficient equipment, appliances and lighting, a photovoltaic system, passive and active solar thermal features. This study shows that it is possible to build efficient affordable zero energy homes in cold climates. Wang et al. (2009) employed EnergyPlus and TRNSYS 16 simulation software tools to investigate the possible solutions for zero energy building design in UK under Cardiff weather conditions. Simulation software (EnergyPlus and TRNSYS 16) are employed in this study, where EnergyPlus considering building materials, window sizes and orientations, renewable electricity, solar hot water system and energy-efficient heating systems under Cardiff weather conditions. Gilijamse (1995) investigated three configurations of zero energy houses in the Netherlands, all using PV electricity. TRNSYS TYPE 56 was used to model the house and TUTSIM to simulate the heat demand with the current best practices and advanced systems. Wang et al. (2009) employed EnergyPlus and TRNSYS 16 to investigated possible solutions for zero energy building design under Cardiff weather conditions in UK.

The above literature review illustrates the fact that many studies have focused on applying passive strategies and energy generation systems to reach net zero and net plus-energy housing prototypes. However, little research work addressed the influence of these strategies and systems under hot climatic conditions.

12.2 Objective

This paper investigates the combination of passive design strategies and on-site renewable energy generation systems. The aim is to achieve compromises (balance) between energy demand for heating, cooling, lighting, domestic hot water and energy supply from renewable resources on an annual basis to reach net plus-energy performance for residential prototyping in Egypt's hot climatic conditions.

A net zero-energy building (NZEB) or a net plus-energy building (NPEB) refers to a building with a zero or negative net energy consumption over a typical year (Kolokotsa et al. 2011). NZEB/NPEB building performance can be measured or evaluated using different indicators, including net primary energy consumption, net energy costs, carbon emissions, etc. However, in this study, net primary energy consumption over a period of one year has been considered. While the energy demand is met on an annual basis from renewable energy supply systems, these systems can either be integrated into the building footprint or can be provided to building, for example, as part of a community renewable energy supply system (Kolokotsa et al. 2011). This study, however, focuses only on renewable energy supply systems integrated into the building footprint.

12.3 Methodology

In order to explore the potential variations in passive strategies and energy generation systems applied to building design, computer dynamic thermal simulation of building systems can serve as a good alternative for the construction and laboratory experiments. It can provide quick predictions for net zero and net plus-energy housing design, and meanwhile, it can avoid large costs and expenses of building construction and field experiments. In this study, a three-bedroom apartment located in three different climatic zones in Egypt has been selected as a case study. These relevant locations are Cairo, Alexandria and Aswan. These geographic locations were chosen to study the influence from the contrasts and similarities across different climatic conditions in Egypt. The *IES*[1]*-Virtual Environment* (VE) software has been used for the energy simulation analysis. IES tools meet the following approved international standards: *(1) ASHRAE 140: 2001, 2004, 2007, 2014, (2) BEST TEST, (3) CIBSE TM33 and (4) EU EN13791: July 2000.* IESVE is fully validated under ASHRAE Standard 140 and has published the results for all versions of ASHRAE Standard 140; 2001, 2004, 2007 and 2014—Heating, Cooling and Envelope (IESVE).

ASHRAE design weather database v6.0 was used as the source for simulation weather data for these three locations where all the four main orientations N, S, W, E were investigated. To decide the percentile of heating and cooling hours taken into account for the design weather, the annual percentile is set to 99.6% for heating loads

[1] **Integrated Environmental Solutions Ltd.** is a software and consultancy company that specializes in building performance analysis.

weather and 0.4% for cooling loads weather as typically mentioned in the ASHREA standard. This neglects the few hours of extreme weather conditions over the year, rather than including them in the calculations. Building energy model components was created using ModelIT program. ApacheSim program was used for the dynamic thermal simulation at sub-hourly time steps. Results from ApacheSim were viewed using VistaPro program.

12.4 Experimentation Approach

This research work was divided into two main stages. The first stage focused on creating building energy model components for a generic three-bedroom apartment—identified in this study as base case—and calculating its total annual energy consumption. The second stage was built on the results of the previous stage and entailed two steps. First, passive treatments are applied to the building's envelope, including both brickwork and glazing, in order to decrease energy demand and minimize total annual energy consumption. Next, on-site renewable energy generation systems are added trying to achieve compromises between energy demand and energy supply from renewable resources to achieve a balance, on an annual basis to reach net plus-energy performance.

Both in the two stages, dynamic thermal simulation was performed for three different locations using the ApacheSim simulation engine. These locations are Aswan (23.96° N, 32.82° E), Alexandria (31.18° N, 29.95° E) and Cairo (30.12° N, 31.41° E). Results from these three locations are compared to study the influence from the contrasts and similarities across different climatic conditions of Egypt in all four main orientations.

12.4.1 *First Stage*

The aim of this stage was to create a simulation 3D energy model for the base case and calculate its total energy consumption in the four main orientations (N, S, W, E) over a period of one year. This included assigning all construction materials to the simulation model and defining their thermal properties (materials' thermal conductivity, density, specific heat capacity and heat resistance). It also included defining daily and weekly occupancy profiles that describe the time variation of thermal input parameters. This process included modulating casual gains and ventilation rates, specifying infiltration rates and defining time-varying set points and supply temperatures for HVAC. These inputs are used in dynamic thermal simulation.

The building energy model used for the dynamic energy simulation is shown in Figs. 12.1 and 12.2. The design of the case study unit was adopted from the housing prototypes that have been recently launched by The New Urban Communities Authority (NUCA) in the New Capital and New Al-Alamain projects. This shall

Fig. 12.1 Case study unit model using modelIT tool

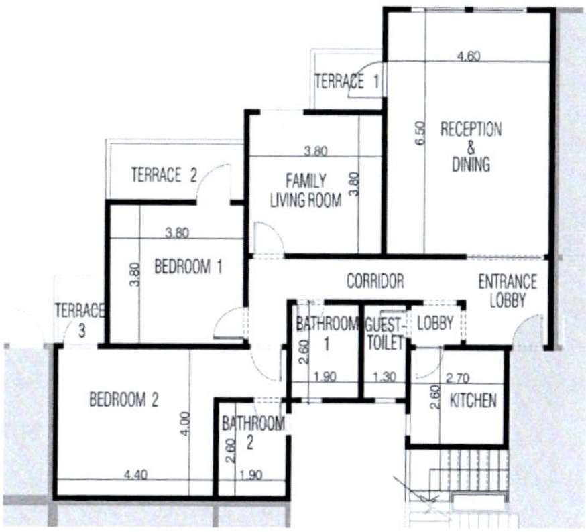

Fig. 12.2 Case study unit floor plan

reflect the typical practice in the domestic mass housing market in Egypt. The unit was assumed to be located in a typical floor, with external adjacency from only one side facing one of the main four orientations. All other sides of the unit will have internal adjacency where any wall in a conditioned space/room in contact to any of these adjacencies will be treated as adiabatic.

For the dynamic thermal simulation, in order to create envelope component sections that are representative of the region, the constructions assigned to the simulation model should accurately reflect the thermal and surface properties of the materials in the Egyptian thermal insulation specification manual. Thermal properties include the materials' thermal conductivity, specific heat capacity and density, while surface properties include emissivity, solar absorptance, visible absorptance, roughness, color and texture. In this study, all thermal and surface properties for building envelope constructions (floor, wall, roof and windows) are derived from *the Database of Egyptian Building Envelopes* (DEBE); an online database presenting envelope materials and constructions of projects and case studies conducted as a part of an ongoing research that aims to describe the influence of building constructions on energy consumption through a survey that is conducted in Cairo and its surrounding residential neighborhoods (Attia and Wanas 2012), together with the thermal conductivity specifications of typical building materials commonly used in Egypt according to EE Building Code Egypt, *Housing & Building National Research Center* HBRC (Visser and Yeretzian 2013). All construction properties for the base case unit envelope are listed in Table 12.1.

For each room, data was assigned to specify heating and cooling, internal gains and ventilation. The value used to calculate the sensible heat gain from occupants, lights and other equipment was specified according to typical values which are listed in CIBSE Guide (A) Table A7.1. The magnitude and types of these casual gains, together with profiles to indicate their time variation, were specified individually for each room as indicated in Table 12.2.

The cooling seasonal EER was set to 4.4, and the delivery efficiency was set to 1.5. Cooling set point was adjusted to 23 °C. DHW efficiency was set to 0.95. Hot water supply temperature was set to 60 °C with a storage volume of 30 L which is considered a typical size for a family. Infiltration maximum flow rate was set to 0.7 ach for all living spaces, and max flow rate for kitchens and toilets was set to 1.4 ach (since their air exchange rate is expected to be higher). To define the pattern of variation for these parameters over a day, modulating daily profiles is used to modulate inputs such as gains, ventilation and infiltration rates and to schedule air-conditioning plant. A daily profile is used as a control switch taking the form of a time series of values in the range 0–1. where values greater than 0.5 are interpreted as 'on' and other values as 'off.' Daily profiles are built into weekly profiles to establish a repeating pattern over the weeks of the year.

To calculate the total annual energy consumption for the base case in the three indicated locations for the main orientations (N, S, W, E), year-round dynamic thermal simulation was performed using ApacheSim. Table 12.3 shows the total monthly energy consumptions for the four orientations of the base case in Aswan, Alex and Cairo, respectively, over a typical.

Overall, it was found that total annual energy consumptions vary for these locations and results ranged from 13.52 MWh to 15.46 MWh. However, where results indicate that southern orientation has the highest annual consumption rate, differences in annual consumptions for the four orientation within one distinctive location are not highly significant. In the following stage, comparative analysis drawn to identify various potentials and limitations of passive techniques and renewable energy sources applied to the base case will be limited to southern orientation only.

12.4.2 Second Stage

The second stage was built on the results of the previous one. This stage entailed two steps. In the first step, passive techniques are applied to the building's envelope in order to decrease the energy demand and thus minimize the total annual energy consumption. Clear single glazed windows used in base case are replaced by double glazed windows with low-E coating. Solar control low-E coating aims at minimizing

Table 12.1 Construction properties for base case unit envelope

Glazing

Glazing type	Thickness mm	Coating yes/no	Conductivity W/(m·K)	Angular dependence	Resistance m^2K/W	Transmittance	Refractive index	Net u value W/(m·K)
Clear glass window	6.0	No	1.0600	Fresnel	0.0057	1.000	1.000	5.9628

External walls

Wall type	Construction layers	Thickness mm	Conductivity W/(m·K)	Density Kg/m^3	Specific heat capacity J/(Kg·K)	Resistance m^2K/W	Vapor resistivity GN·s/(Kg·m)	Net u value $W/(m^2·K)$
Cement hollow brick wall with plaster finish	Cement plaster	40.0	0.7200	1860.0	800.0	0.0556	140.0	2.4415
	Cement/hollow brickwork	250.0	1.6000	1140.0	800.0	0.1563	0.000	
	Cement plaster	40.0	0.7200	1860.0	800.0	0.0556	140.0	

Table 12.2 Internal heat gain assigned for each space

Room/spaces	Gain 1	Gain 2	Gain 3	Gain 4	Gain 5	Total no. of gains
Reception	Fluorescent lighting	People	T.V	–	–	3
Bedrooms	Fluorescent lighting	People	T.V	Computer/laptop	–	4
Bathrooms and guest toilets	Fluorescent lighting	People	Dish washer	–	–	3
Family living	Fluorescent lighting	People	T.V	Computer/laptop		4
Corridors	Fluorescent lighting	–	–	–	–	1
Kitchen	Fluorescent lighting	People	Dish washer	Cooker	Refrigerator	5

the amount of heat that gets absorbed into a living spaces during the time of the day. Also, to increase the thermal insulation of external walls, double brick walls with an inside cavity is employed. Net U-values for glazing and external wall were further improved to 1.5047 and 2.3437 W/(m·K), respectively. Construction properties for these parameters are listed in Table 12.4.

In the southern orientation, the total monthly and annual energy consumptions for the three chosen locations over a typical year after these passive measures have been adopted to building's envelope is shown in Table 12.5. It has been found that total annual energy consumptions were reduced by an average 10% as follows: the total annual energy consumptions were reduced (1) from 15.5307 to 15.0383 MWh in Aswan, (2) from 13.6453 to 13.5368 MWh in Alexandria and (3) from 14.0500 to 13.8000 MWh in Cairo.

In the next step, on-site renewable energy generation systems are added to achieve compromises between energy demand and energy supply from renewable resources. Since photovoltaic (PV) systems enable moving toward NZEB/NPEB performance, using a renewable energy source that has no moving parts and requires minimum maintenance, they are the most deployed system among other renewable energy generation systems (Garde et al. 2017). Therefore, in this study, a roof-mounted PV array system is used to address electric energy requirements assuming that an average 30 m_2 of roof area is allocated for each apartment to install PV cells. Since surrounding topographical shading surfaces are modeled in the virtual environment using ModelIT, shading factor was set to 1.0 (no shading was taken into consideration in addition to shading from surroundings buildings). The PV array parameters used are listed in Table 12.6. Total monthly and energy consumptions as well as energy generated by photovoltaic system are shown in Table 12.7.

Table 12.3 Total annual energy (MWh) for base case for all orientations

	Total energy (MWh)											
	Aswan				Alexandria				Cairo			
Date	East	South	West	North	East	South	West	North	East	South	West	North
Jan 01–31	1.0358	1.0413	1.0330	1.0327	1.0225	1.0223	1.0223	1.0225	1.0220	1.0219	1.0219	1.0219
Feb 01–28	0.9489	0.9796	0.9443	0.9428	0.9152	0.9154	0.9152	0.9152	0.9196	0.9246	0.9190	0.9190
Mar 01–31	1.1811	1.2012	1.1737	1.1687	1.0163	1.0247	1.0158	1.0155	1.0423	1.0570	1.0396	1.0385
Apr 01–30	1.3044	1.3056	1.3004	1.2892	1.0742	1.0802	1.0686	1.0636	1.1357	1.1409	1.1322	1.1245
May 01–31	1.4495	1.4379	1.4454	1.4383	1.2035	1.2009	1.2015	1.1944	1.2726	1.2662	1.2672	1.2607
Jun 01–30	1.4857	1.4680	1.4796	1.4761	1.2402	1.2359	1.2408	1.2354	1.3113	1.2995	1.3066	1.3004
Jul 01–31	1.5315	1.5140	1.5260	1.5197	1.2992	1.2961	1.3000	1.2943	1.3754	1.3646	1.3701	1.3615
Aug 01–31	1.5456	1.5371	1.5405	1.5291	1.3479	1.3498	1.3459	1.3387	1.3849	1.3819	1.3775	1.3670
Sep 01–30	1.3989	1.4097	1.3939	1.3844	1.2322	1.2478	1.2289	1.2244	1.2677	1.2836	1.2594	1.2538
Oct 01–31	1.3624	1.3890	1.3508	1.3444	1.1865	1.2186	1.1768	1.1744	1.2231	1.2472	1.2149	1.2128
Nov 01–30	1.1575	1.1846	1.1476	1.1437	1.0399	1.0356	1.0350	1.0332	1.0464	1.0442	1.0438	1.0425
Dec 01–31	1.0630	1.0626	1.0569	1.0552	1.0187	1.0181	1.0183	1.0182	1.0195	1.0185	1.0186	1.0185
Summed total	15.4643	15.5307	15.3921	15.3243	13.5963	13.6453	13.5690	13.5297	14.0206	14.0500	13.9708	13.9209

Table 12.4 Construction properties for modified building envelope

Glazing

Glazing type	Thickness mm	Coating yes/no	Conductivity W/(m·K)	Angular dependence	Resistance m^2K/W	Transmittance	Refractive index	Net U-value W/(m·K)
Double glazed with low-E coating	6.0 double layer with cavity	No	1.0600	Fresnel	0.0057	0.530-0.780	1.526	2.3437

External walls

Wall type	Construction layers	Thickness mm	Conductivity W/(m·K)	Density Kg/m^3	Specific heat capacity J/(Kg·K)	Resistance m^2K/W	Vapor resistivity GN·s/(Kg·m)	Net u-value $W/(m^2·K)$
Cement hollow brick wall with plaster finish	Cement Plaster	40.0	0.7200	1860.0	800.0	0.0556	140.0	1.5047
	Cement/hollow brickwork	250.0	1.6000	1140.0	800.0	0.1563	0.000	
	Cavity	120	-	-	-	-	-	
	Cement/hollow brickwork	250.0	1.6000	1140.0	800.0	0.1563	0.000	
	Cement plaster	40.0	0.7200	1860.0	800.0	0.0556	140.0	

Table 12.5 Total monthly and annual energy consumptions for modified building envelope

	Total energy (MWh)		
Date	Aswan	Alexandria	Cairo
Jan 01–31	1.0622	1.0380	1.0261
Feb 01–28	0.9724	0.9163	0.9272
Mar 01–31	1.1717	1.0246	1.0569
Apr 01–30	1.2528	1.0731	1.1222
May 01–31	1.3718	1.1821	1.2347
Jun 01–30	1.3927	1.2083	1.2566
Jul 01–31	1.4337	1.2625	1.3124
Aug 01–31	1.4511	1.3061	1.3275
Sep 01–30	1.3365	1.2119	1.2367
Oct 01–31	1.3285	1.1954	1.2219
Nov 01–30	1.1706	1.0806	1.0768
Dec 01–31	1.0943	1.0379	1.0316
Summed total	15.0383	13.5368	13.8000

Table 12.6 PV array parameters installed on building's roof

PV array parameters					
Module nominal efficiency	Nominal cell temperature (NOCT) (°C)	Reference irradiance for NOCT W/(m^2)	Temperature coefficient for module efficiency (1/K)	Degradation factor	Electrical conversion efficiency
0.0700	46.0	800	0.0024	0.9900	0.8500
Area (m^2)	30.0		**Azimuth**	180.0 (clockwise from North)	
Shading factor	1.00		**Inclination**	45.0 (From horizontal)	

Although the total energy consumption was reduced by an average 20%, it can be deduced from the results above that the area allocated on roof for PV system is not enough to reach net plus-energy level. Therefore, energy simulation was repeated after *(1) introducing additional PV solar panels to the external façade with an inclination angle 45 (only 11* m^2 *area was available)* and *(2) increasing the area allocated for PV solar panels on roof to reach 135* m^2. The simulated results in Table 12.8 show that in this case, the amount of energy generated by PV systems could significantly compensate the total energy consumption with an excess ranging from 2.7266 to 3.5817 MWh for the total year (an average 20% annual plus-energy).

Table 12.7 Total monthly and annual energy consumptions after installing roof-mounted PV array system

	PV generated electricity (MWh)	Total energy (MWh)	PV generated electricity (MWh)	Total energy (MWh)	PV generated electricity (MWh)	Total energy (MWh)
Date	Aswan		Alexandria		Cairo	
Jan 01–31	−0.2139	0.8483	−0.1426	0.8954	−0.1370	0.8890
Feb 01–28	−0.2350	0.7375	−0.1489	0.7674	−0.1626	0.7646
Mar 01–31	−0.3123	0.8594	−0.2315	0.7932	−0.2360	0.8209
Apr 01–30	−0.3435	0.9094	−0.2812	0.7919	−0.2930	0.8292
May 01–31	−0.3566	1.0152	−0.3101	0.8720	−0.3345	0.9002
Jun 01–30	−0.3737	1.0190	−0.3039	0.9043	−0.3614	0.8952
Jul 01–31	−0.3800	1.0537	−0.2983	0.9642	−0.3658	0.9466
Aug 01–31	−0.3620	1.0891	−0.2832	1.0229	−−0.3316	0.9960
Sep 01–30	−0.3142	1.0223	−0.2363	0.9756	−0.2833	0.9535
Oct 01–31	−−0.2706	1.0578	−0.2117	0.9837	−0.2028	1.0191
Nov 01–30	−0.2205	0.9501	−0.1578	0.9228	−0.1492	0.9276
Dec 01–31	−0.1995	0.8948	−0.1212	0.9167	−0.1267	0.9049
Summed total	−3.5817	11.4566	−2.7266	10.8102	−2.9839	10.8469

12.5 Conclusion

From this study, through dynamic thermal simulation, it can be concluded that it is theoretically possible to achieve net plus-energy housing in Egypt. However, results showed that applying solar control low-E coating on double glazed windows together with double brick walls with an inside cavity accounted only for an average 10% reduction in the total annual energy consumptions. It can also be concluded from results that for multi-story residential buildings with large number of apartments, roof area is not sufficient to allocate space needed for PV cell arrays to compensate the total annual energy demand. Total needed area in this case study reached 145 m^2. Additional spaces can be provided on building façade as well as parts of a community renewable energy supply system. The annual PV generated electricity is predicted to range from −13.9354 MWh (in Cairo) to −17.2292 MWh (in Aswan). The remaining electricity can either be stored in batteries, used for other functions (i.e. electrical charging of vehicles for transport for instance) or sold back to the national grid.

This study, however, focused on integrating only two specific passive treatments. Further research work should extend the possibilities to more different combinations to study their impact on total annual energy consumption reduction. Further investigations should also consider the potential of applying other renewable energy-generation systems (such as wind energy for instance). It is also important to study to

Table 12.8 Total monthly and annual energy consumptions after increasing PV cells area and installing PV on façade

	PV generated electricity (MWh)	Total energy (MWh)	PV generated electricity (MWh)	Total energy (MWh)	PV generated electricity (MWh)	Total energy (MWh)
Date	Aswan		Alexandria		Cairo	
Jan 01–31	−1.5084	−0.4492	−1.0735	−0.0366	−0.9029	0.1228
Feb 01–28	−1.4444	−0.4769	−0.9313	−0.0153	−0.9617	−0.0354
Mar 01–31	−1.5971	−0.4310	−1.2381	−0.2154	−1.2378	−0.1839
Apr 01–30	−1.4551	−0.2063	−1.2869	−0.2182	−1.3181	−0.2002
May 01–31	−1.2721	0.0986	−1.2451	−0.0657	−1.3135	−0.0811
Jun 01–30	−1.1956	0.1969	−1.1583	0.0483	−1.2974	−0.0422
Jul 01–31	−1.2714	0.1619	−1.1660	0.0946	−1.3694	−0.0590
Aug 01–31	−1.4210	0.0278	−1.2251	0.0774	−1.4087	−0.0854
Sep 01–30	−1.5052	−0.1744	−1.1902	0.0172	−1.4324	−0.2011
Oct 01–31	−1.5789	−0.2561	−1.3168	−0.1261	−1.1682	0.0497
Nov 01–30	−1.5067	−0.3404	−1.1681	−0.0899	−0.9743	0.1008
Dec 01–31	−1.4735	−0.3819	−0.9361	0.1010	−0.8604	0.1708
Summed total	−17.2292	−2.2309	−13.9354	−0.4287	−14.2448	−0.4442

validate the conclusions drawn from this study against measured data from physical models to ensure that the outputs of the simulation are close enough to empirical data.

References

Attia S, Wanas O (2012) The database of Egyptian building envelopes (DEBE): A database for building energy simulations. Proc SimBuild 5(1):96–103

Bernier M, Ferron Y, Biaou A (2006) Simulation of zero net energy homes. Journée thématique du 21

Clarke J, Hong J, Kim J, Strachan P, Hwang I, Lee H (2005) Simulation-based design procedure to evaluate hybrid renewable energy systems for residential buildings in Korea. Paper presented at the proc. ninth int. IBPSA conf., building simulation

Garde F, Ayoub J, Aelenei L, Aelenei D, Scognamiglio A (2017) Solution sets for net zero energy buildings: feedback from 30 buildings worldwide: Wiley

IESVE. Software-validation. Retrieved from 19 June 2019. https://www.iesve.com/software/software-validation

Kolokotsa D, Rovas D, Kosmatopoulos E, Kalaitzakis K (2011) A roadmap towards intelligent net zero-and positive-energy buildings. Sol Energy 85(12):3067–3084. https://doi.org/10.1016/j.solener.2010.09.001

Norton P, Christensen C (2007) Performance results from a cold climate case study for affordable zero energy homes: National renewable energy lab (NREL), Golden, CO (United States)
Renewable-Energy MOE (2017) Egyptian electricity holding company annual report 2016/2017
Visser F, Yeretzian A (2013) Energy efficient building: Guidelines for MENA region: MED-ENEC project office
Wang L, Gwilliam J, Jones P (2009) Case study of zero energy house design in UK. Energy and buildings 41(11):1215–1222

Chapter 13
Enhancing Occupants' Well-Being Through Qualitative Indoor Environments

Dalia Hafiz and Vrushali Mhatre

Abstract In today's competitive business environment, corporations can no longer afford to waste the potential of their workforce as personal costs of salaries and benefits are considered the most expensive items in an office building. There are certain essential factors in the employee's workplace that influence significantly on their level of motivation and performance. The main objective of the study is to understand the relationship between the employee's performance and their physical work environment. Lighting is an effective element in interior design, especially when it comes to deciding between natural and artificial lighting, where several light measurements, calculation, and fixtures types are needed. This paper examines a small design project in which occupants' comfort and indoor environment quality aspects are tested through the collection of occupants' feedback under a variety of conditions: 1. natural light in addition to energy-efficient artificial lighting, 2. acoustics, noise control and acceptable sound level indoors, and 3. thermal comfort through natural and artificial ventilation. The parameters are analyzed and concluded with various alternatives to achieve an optimum solution for better indoor working conditions for the proposed space. Finally, based on the evaluation findings, a set of recommendations is proposed to evaluate indoor working spaces' qualities. The paper findings and proposed recommendations can help promoting a healthy environment, improving the occupant's comfort, increasing workers' performance, reducing stress, and ultimately enhancing the occupant's quality of life.

Keywords Occupants' comfort · Workers' performance · Indoor environment quality

---

D. Hafiz (✉)
College of Architecture and Design, Architecture Department, Al-Ghurair University, Dubai, UAE
e-mail: d.hafiz@agu.ac.ae

V. Mhatre
College of Architecture and Design, Interior Design Department, Al-Ghurair University, Dubai, UAE
e-mail: vmhatre@agu.ac.ae

Architecture and Urbanism: A Smart Outlook,
https://doi.org/10.1007/978-3-030-52584-2_13

13.1 Introduction

The quality of the indoor environment of a work/educational space has a direct relationship with the occupants' satisfaction and well-being. In addition, an increase in performance is expressed in a direct quantified reduction of employees' absenteeism. The loss of productivity is related to absenteeism as well as the hindrance, which is expressed through the presence of employees in workplaces during working hours but without reaching an optimum work concentration (Busan and Korea 2009). This is related to the quality improvements of the indoor environment. There is a number of factors that affect the indoor environment as well as productivity levels. These factors include temperature, draught, noise, lighting, and air pollution. Each factor under certain conditions can be a type of discomfort within the working space. This holds especially true if the temperature and ventilation are only controlled centrally without user control, while other aspects like lighting and noise can be influenced by the users within the building (Michael and Gregoriou 2018). The integration of user influence on the indoor environment is allowed if it only permits any improvements in the indoor air quality as shown in Fig. 13.1.

13.1.1 Interior Design Factors with Impact on Productivity and Performance

Today's workforce believes that a noisy work environment inhibits effective work; statistics from the Data Management Association show that a lack of acoustical privacy in open-plan offices can reduce productivity by as much as 40% and increase errors by as much as 27% (McLaughlin 2000). In recent studies, lighting was found to be the main factor, which has an impact on the productivity of employees in workplaces. The commission for Architecture and the Built Environment and the British Council for Offices carried out a study which revealed that even simple things such as adequate lighting and adequate daylight can decrease absentee rate by 15%. Also, they increase productivity between 28 and 20%. Temperature and humidity can strongly affect how tired someone might feel; although the World Health Organization recommends a maximum working temperature of 24 °C (Leaman 1995), there are still little regulations governing high-temperature levels in the workplace and the responsibility of companies to their employees in this respect.

Thermal Comfort Thermal comfort has a considerable and measurable effect on workers' performance regardless of their activities. It is considered a key aspect in determining the acceptable indoor operative temperatures, as well as identifying passive solutions and strategies for ventilation if the design is being within a hot context. In order to achieve thermal comfort through any design, it is important to know the actual situation of the indoor environment considering temperature and relative humidity. The United Arab Emirates (UAE) has a desert climate, characterized by warm winters and very hot, sunny summers, with higher levels of humidity. Typically, temperature control is needed year-round (Ballarini and Luca 2018).

Indoor Air Quality A strong relationship has been demonstrated between indoor air quality and the productivity of employees. For example, providing a good climate control within the workspace that generate different air qualities can be quantified through the satisfaction of the employees in different percentages. Understanding and controlling common pollutants indoors can increase productivity by decreasing the risks of indoor health concerns within any workspace (Montgomery 2004; Busan and Korea 2009). Fanger (2006), discussed the impact of indoor air quality on productivity in offices, learning in schools, and the growth of allergic and asthmatic diseases. He went on to discuss solutions, including air cleaning, personalized ventilation, and the need for cool (21 °C) dry air.

The Auditory Indoor Environment (Acoustics) The interior acoustics have a strong impact on the indoor environment quality. Different substantial studies have clearly addressed the influence of sound on the human performance. Such studies show that large acoustic differences in experimental situations and conditions prevent the transition of the acoustic result to be realistic. However, whenever people are working indoors, they are continuously subjected to sound; whether they are working in private offices, group offices, or any industrial settings, they will never be subject to a complete absence of sound, which occurs not only due to people but also due to the HVAC and machinery installations that can cause distraction, stress, fatigue, or even hearing loss (Buratti et al. 2018).

Visual Comfort Different hypotheses are considered when there is a need to examine the visual comfort within a workspace, where these hypotheses are helpful to analyze the visual perception and mood responses under different factors such as illuminance conditions and different color temperatures. Among these hypotheses, one that states that visual perception is affected by the color temperatures of light sources, while the second hypothesis indicates that better moods are associated with low color temperatures while are given equal illuminance levels (Roelofsen 2016a, b). On the other hand, visual discomfort is related to different spatial and lighting aspects and conditions, where the dimensions of the space and the lighting settings and fixtures affect the human visual experience and comfort within the workspace. In order to test the occupant's visual comfort in a working space, questionnaires with different types of questions are used including general questions, questions regarding overall comfort, and questions regarding the visual perception of visual stimulation and mood. Some general questions examine age, gender, color blindness, and performance and sensitivity of lighting, while questions regarding comfort include visual responses, visual satisfaction, and visual distraction. Moreover, different questions regarding mood according to visual perception are included to test the visual comfort and annoyance (Michael and Gregoriou 2018).

Artificial and Natural Lighting in the Interior Environment When we think about visual comfort, we think about the integration of artificial and natural lights within the workspace. The importance of daylight is that it creates an ambiance of quiet tranquility and visual comfort within the workspace. In order to achieve the best-expected comfort of natural light, control of glare and illumination is important to avoid the negative aspects of glare and heat. To preserve the importance of daylight within a workspace, it is important to know why and how daylight should be integrated into the indoor environment of a workspace (Babizhayev et al 2003). In reading spaces, for example, different strategies are adopted for daylight integration to lighten the space. The integration between natural and artificial light is essential to create a vision that involves not only physiology but also psychology. That is because visual perception is more than a visual task which is influenced by the psychological process. The psychological levels of comfort address how light can serve as an excellent means for controlling our perception on a psychological level. Different light settings effects can influence cognitive performance and improve human behavior and enhance productivity (Xie and Guo 2017).

13.2 Research Methodology

The research method used in this paper can be subdivided into four sections: A space examination, B-survey questionnaire: defined as "a list of research questions asked to respondents and designed to extract accurate information"(Khanie et al 2015). It aims at (1) collecting the appropriate data, (2) making data comparable to analysis, and (3) minimizing bias in composing and asking questions. The survey collected opinions from a group of employees in an open-plan office in a university building. The survey respondents were asked to evaluate the space acoustical/thermal and lighting conditions. C—lighting distributions, D—survey analysis, and E—observations, conclusions, and recommendations.

13.2.1 Space Examination

The examined office space is located on the south façade (with a south-facing window) of an educational building located in Dubai, UAE. The office is a shared open space, and it has an operable window, one central AC supply outlet and returns as shown in Table 13.1 and Fig. 13.1.

13.2.2 Survey Overview—Perceptual Judgment

Perceptual judgment is defined as the way in which we make judgments primarily based on sensory experiences and is influenced by experiences encountered since early childhood. The previous research suggested that occupants' indoor comfort perception is an active reaction that cannot be isolated to only a single sense organ and should evaluate more than one sense (Leaman 1995). Consequently, perceptual judgment was used in this paper to collect the occupant's responses in three different sections: thermal, acoustical, and lighting/visual comfort. The survey collected employees' feedback over the month of March and April at different times varying from 10 AM to 4 PM.

The employee's feedbacks were analyzed while pursuing two different working activities: 1—typing/reading from their computer screens and 2—reading a printed text. They were asked to express their level of satisfaction in each activity based on the provided conditions. Three cases were explored:

- Case 1: window blind open—natural light only.
- Case 2: window blind closed—electric light (direct LED).
- Case 3: window blind open + direct LED—combination of natural + electric light.

The office included three working tables, each is 140*100 Cm. The survey results were collected based on the employee's table location:

- Table 1: the closest to the window which center is at 216 Cm from the window.
- Table 2: The middle table, center at 400 Cm from the window.
- Table 3: The furthest from the window, center at 600 Cm from the window.

Table 13.1 Space characteristics

Room dimensions	854*616 cm
Ceiling height	313 cm
Window location	South façade
Window size and characteristics	210*245 cm, sill at 95 cm
Shading	Interior white opaque blinds—no exterior shading
Glass properties	Transparent (6 mm glass and 8 mm air gap) double glazing
Electrical lighting characteristics	Two recessed square light (120*120 cm each) (two LED tube lights inside): 16 W (40-W equivalent) 100 cm. 1800 lumens daylight deluxe (6500 K)
Daylight control	Window roller blind (sunscreen, blackout, translucent)
Thermal comfort	Window and A/C
Noise source	Window and A/C

Note window glazing operability was not a reasonable idea since the harsh weather in the UAE does not allow for natural ventilation. Consequently,the air conditioner was active at all times and was considered in the acoustics comfort evaluation

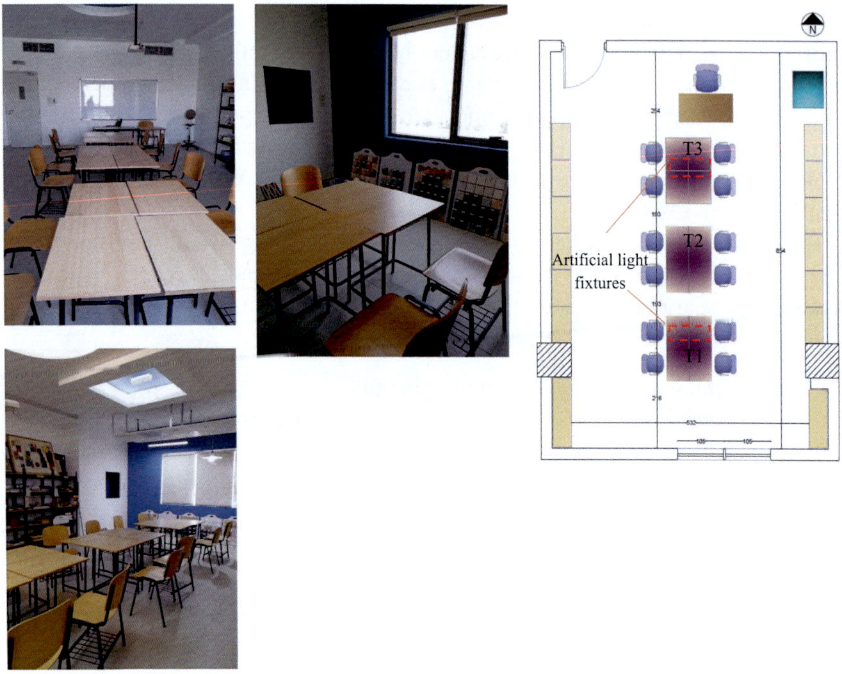

Fig. 13.1 Space images and floor plan

13.2.3 Light Distribution

Prior to the feedback collection, illuminance values were measured using UNI-T UT383 Digital Lux meter in each of the proposed alternatives. The examined region (Dubai, UAE) has a year-round clear sky with sun conditions. For this reason, lighting measurements were recorded over a two weeks period (April 21 to May 5) from 10:00 AM to 4:00 PM only. Average readings were calculated (in Lux) as shown in Table13.2.

The measurements were used to generate some hypothesis when analyzing the data including:

1. Uniform lighting is favorable in electric lighting condition-only case,
2. Less glare is generated in electric lighting condition-only case,

Table 13.2 Average Lux meter calculation at the center of each table (in Lux)

	Case 1 natural light	Case 2 electric light	Case 3 combination
Table 1	320	145	420
Table 2	110	50	160
Table 3	90	150	185

3. Natural lighting-only case has a high level of fluctuation in lighting distribution.

Employees' feedback collected from a set of open-ended questions and was analyzed using thematic methods. Those themes can be summarized as follows (Tables 13.3, 13.4 and 13.5).

13.3 Observations and Discussion

The focus of this survey study is applied to an open-plan office space. Based on the literature review, the factors considered in this study are key factors affecting the indoor air quality of any interior space which in turn affects the employee's performance. Based on the above literature, the conceptual model tested in this paper is presented in Fig. 13.2. The dependent variable in this research is the working environment in which the employees are working which includes the indicators like thermal comfort, acoustics, and lighting and the independent variable is the performance level of employees.

Indoor air quality which consists of the three factors thermal comfort, acoustics, and lighting affects employee performance to a large extent in any work environment and vice versa; employee performance, in turn, is affected by the indoor air quality of office space. The model in Fig. 13.2 indicates clearly the interconnection between IAQ and employee performance:

Table 13.3 Section 1: Thermal comfort

Possible recommendations for thermal satisfaction	Sources of thermal dissatisfaction
– Provide external insulation – Enhance AC performance and add ventilation – Use plants and breathable materials – Use monochrome glass or double skin facade – Apply exterior shading – Open windows – Enhance windows protection from the sun	– AC is not functioning well – Windows are not well protected from the sun

Table 13.4 Section 2: Acoustics comfort

Possible recommendations for acoustical satisfaction	Sources of acoustical dissatisfaction
– Provide acoustical insulating/absorbing materials – Add acoustical panels	– Echo control – Lower the AC noise – External noise

Table 13.5 Section 3: Visual comfort

Possible recommendations for visual satisfaction	Sources visual dissatisfaction
– Use blinds and curtains – Add more electrical lights at the rear areas of the office space	– Glare was detected when the blinds were not pulled down

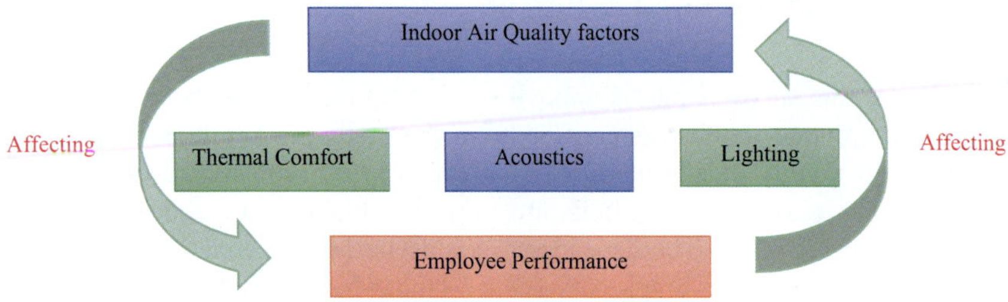

Fig. 13.2 Interconnection between the IAQ and employees' performance and satisfaction

The solution from this experiment is to emphasize the fact that improper IAQ significantly affects the performance of real office work overtime, and making changes in this area can be more profitable to any organization when compared to its labor cost.

Based on the results of the survey, 45 employees were questioned on the three important aspects being thermal, acoustical, and lighting comfort. While conducting the survey, there were certain limitations such as availability of time which was one of the restrictions in collecting data. As observed from the data collected, it seemed that the employees of the organization seemed hesitant to share their real opinions and true feelings. As perceived from Figs. 13.3, 13.4, and 13.5, most of the employee's attitudes were neutral toward the three factors considered thermal, acoustical and lighting conditions in the existing premises.

The indicators are more inclined toward a neutral attitude (not matching the proposed hypothesis and mismatching with the illuminance levels distribution). It was recorded that the workers have been working in the same room for an extended period of time (average of one year). The researchers concluded that with time, the

Fig. 13.3 Thermal comfort analysis

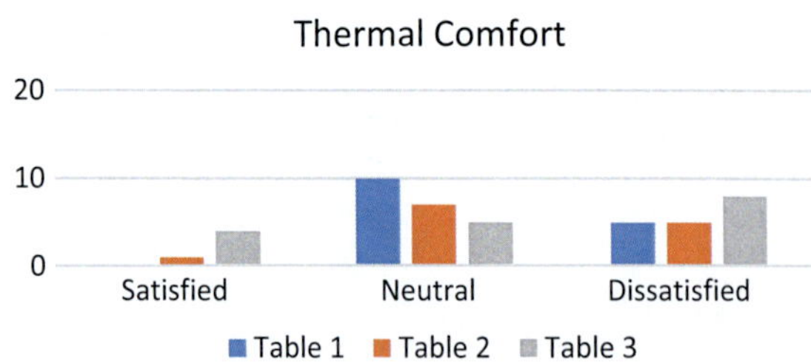

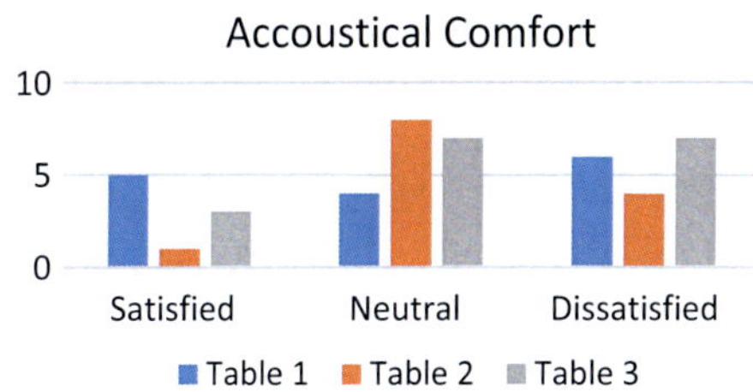

Fig. 13.4 Acoustical comfort analysis

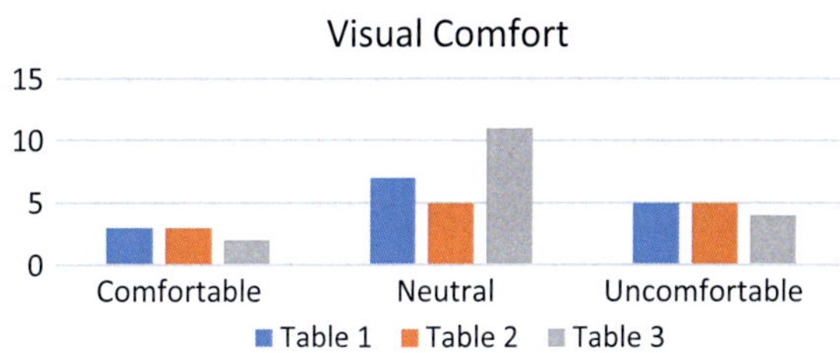

Fig. 13.5 Visual comfort analysis

occupants are adapted to any condition and their feedback becomes neutral. The researchers strongly felt that in such working environments where employees' opinions and views are key design factors, it is highly recommended that occupants' feedback is collected in the early stages of the building occupancy considered to be reflected in the design process. Such reflection can definitely have a positive impact on the performance of employees (as seen in the survey results). This would lead to a high level of employee's satisfaction and productivity for their organizations, thus promoting their respective industries in the years to come.

13.4 Conclusion and Future Research

From the study, it can be concluded that the IAQ of office space has a direct impact on employees' performance. Improper working conditions can impair performance levels to an enormous extent. This research paper highlights the importance of good work conditions imperative for employee satisfaction and performance level. The research paper influences the importance of taking the work environment more seriously within a business model to increase employees' morale and commitment.

The literature review and the analysis of data gathered show that a significant influence of an employee's attitude and comfort level toward the work environment results in a neutral, dissatisfied, or satisfied response toward working conditions.

The study further highlights the importance of the indoor air quality that has a great influence on an employee's behavior, satisfaction level, and finally performance. As stated by (Leaman 1995) in his research, "People who are unhappy with temperature, water quality, lighting and noise conditions in their offices are more likely to say that this affects their concentration at work."

The proposed hypothesis and measured illuminance distributions showed a strong level of dissatisfaction represented in glare, poor illuminance distribution, and lake of uniformity—natural lighting only-case. In addition, poor color rendering in the artificial lighting only-case was detected. However, the results still showed the same levels of satisfaction over all the cases. Consequently,occupants' educations are highly needed on satisfaction indexes and performance enhancement methods and techniques, followed by periodical monitoring of behavior, performance, and satisfaction levels.

Various factors that can essentially influence the employee's performance still remain a huge challenge. However, it can be concluded that not only thermal comfort, lighting, and acoustics but also the office layout, ergonomic furniture, connection to outdoors, and color schemes of office spaces also contribute to employee performance which has to be considered by designers as well as office management.

For future investigations, researchers could use further comparison between the productivity of office employees and their overall office environment taking into consideration smart technologies and parametric design which could enhance productivity and reduce stress levels. The studies perhaps could compare the performance level of the office staff of a well-designed office and another office where attention has not been given to design and IAQ. Finally, it can be concluded that the employee's education on the importance and guidelines of a calm comfortable and employee-friendly work environment is important for the overall mental and physical health of the employees, which would further enable them to concentrate and do their jobs perfectly with ease, increase productivity, and bring in profits not only to the business but also the economy.

References

Babizhayev M, Deyev A, Yermakova V, Davydova N, Kurysheva N, Doroshenko V, Zhukotskii A (2003, 5) Image analysis and glare sensitivity in human age-related cataracts. Clin Exp Optom 86(3):157–172

Ballarini I, Luca G (2018) u. (n.d.). Integration of thermal and visual comfort in the retrofit of existing buildings. ieeexplore.ieee.org

Buratti C, Belloni E, Merli F, Environment P-B (2018) u. (n.d.). A new index combining thermal, acoustic, and visual comfort of moderate environments in temperate climates. Elsevier

Fanger PO (2006) What is IAQ? Indoor Air 16(5):328–334

Khanie M, Wienold J, Andersen M (2015) A sensitivity analysis on glare detection parameters

Leaman A (1995) Dissatisfaction and office productivity. Facilities 13(2):13–19

Michael A, Gregoriou S, Energy S-R (2018) u. (n.d.). Environmental assessment of an integrated adaptive system for the improvement of indoor visual comfort of existing buildings. Elsevier

Montgomery K (2004) Understanding the relationship between the design of the workplace environment and wellness

Proceedings T-R, Busan U, Korea U, (2009) u. (n.d.). The effects of indoor temperature and outdoor air supply rate on learning performance of late-teen students. archi.tohoku.ac.jp
Roelofsen CPG (2016a) Modelling relationships between a comfortable indoor environment, perception and performance change
Roelofsen CPG (2016b, 10 23). A computer model for the assessment of employee performance loss as a function of thermal discomfort or degree of heat stress. Intell Build Int 8(4):195–214
Roelofsen P (2018, 7 3) A new methodology for the evaluation of the perceived air quality depending on the air pollution, caused by human bioeffluents, the temperature, the humidity as well as the air velocity. Intell Build Int 10(3):154–161
Sarey Khanie M, Stoll J, Einhäuser W, Wienold J, Andersen M (2017, 11 30) Gaze and discomfort glare, Part 1: Development of a gaze-driven photometry. Lighting Res Technol 49(7):845–865
Sciences R-Z-P-S (2012) u. (n.d.) The interior design of workplace and its impact on employees' performance: a case study of the private sector corporations in Egypt. Elsevier
Xie X, Gou Z (2017, 1) Building performance simulation as an early intervention or late verification in architectural design: same performance outcome but different design solutions. J Green Build 12(1):45–61

Chapter 14
School2Share (S2S): A Computational Approach for the Efficient Management of Shared Facilities in Egypt

Sherif Abdelmohsen, Ayman Assem, and Mohamed Ezzeldin

Abstract This paper introduces School2Share, a cloud-based system, that utilizes BIM and GIS for the efficient management of shared school facilities. Despite the abundance of schools in developing countries, there remains a significant lack of facilities per school and an underutilization of resources and regulatory mechanisms that are true to children needs. Several efforts have been developed to enhance the management of school facilities in terms of space and asset management, automating labor-intensive tasks, and managing campus facilities and curricula, but none address the efficient management of school networks in relation to funding and government agencies, or underprivileged children in need of available and underutilized school facilities. Our approach links data from different stakeholders using BIM and GIS and addresses the efficient management of the integration and sharing of school facilities. We present the system architecture and technology infrastructure of the proposed system and a working prototype for an existing neighborhood in Cairo, Egypt, demonstrating the main functionalities, workflows, and different portals. As our approach incorporates all stakeholders including children, schools, local authorities, and funding agencies, we argue that it empowers children to have broader access to community-based education at three hierarchical levels: children-to-school interaction, where children identify school facilities that address their needs in a customized fashion; school-to-school interaction, where shared and round-the-clock activities are facilitated through a comprehensive and real-time networked system, and policies and development, where funding agencies identify schools and communities in need of development through an informed system, and authorities monitor effective use of funding.

Keywords Facility management · Building information management · Geographic information systems · School management systems

---

S. Abdelmohsen (✉)
Department of Architecture, The American University in Cairo, Cairo 11835, Egypt
e-mail: sherifmorad@aucegypt.edu

S. Abdelmohsen · A. Assem · M. Ezzeldin
Department of Architecture, Faculty of Engineering, Ain Shams University, Cairo 11535, Egypt

Architecture and Urbanism: A Smart Outlook,
https://doi.org/10.1007/978-3-030-52584-2_14

14.1 Introduction

Developing countries are facing increasing children illiteracy rates as high as 25% (UIS and UNICEF 2015) and out-of-school children rates (9, 16 and 36% in primary, lower secondary and secondary school age respectively) (UNESCO-IIEP, UNICEF and World Bank 2014). With challenges to bring children back into education and over 263 million out-of-school children (UNESCO 2015), access to community-based education is becoming increasingly in demand. Despite the abundance of schools, there is significant lack of basic facilities per school, including labs, libraries, theaters, and sports facilities, leading to underutilized resources and lack of regulatory mechanisms that utilize school facilities efficiently based on children's needs.

Most approaches that address such mechanisms tend to involve localized management of school operations and maintenance, space and asset management, finance and accounting, enhancing facility management systems and inventory tracking, automating labor intensive administrative tasks, providing computer-aided facility management (CAFM) systems, developing methods for energy and cost efficiency of school facilities, or the coordinated management of students, curricula, and campus facilities (Akcamete et al. 2010; Asiabaka 2008; Kassem et al. 2015 Olayinka and Andrew 2018; Volk et al. 2014). None, however, address the efficient management of school networks linked to funding agencies, government organizations, or underprivileged children in need of available and underutilized school facilities.

This paper introduces a computational approach that exploits building information modeling (BIM), geographical and geospatial information systems (GIS), and cloud-based technology to link data from different stakeholders and efficiently manage the integration and sharing of school facilities. We present the system architecture and technology stack of S2S, and a working prototype for an existing neighborhood in Cairo, Egypt, illustrating workflows, user views, and functionalities. We argue that using our approach, children can have wider access to community-based education through three hierarchical levels: (1) children-to-school interaction, where underprivileged children can identify school facilities that address their needs, customized to satisfy age group, social status, and level of education; (2) school-to-school interaction, where the listing and availability of facilities per school allows for shared activities and their effective use throughout times of the day and year; and (3) policies and development, where government agencies and funding organizations can identify schools and communities in need of development based on real-time data of missing facilities or gaps in community service, and regulatory bodies can effectively monitor fund use, accountability, and allocation.

14.2 BIM and GIS in Building Management

The process of utilizing building information modeling (BIM), BIM open standards, data structure specifications, and geographic/geospatial information systems (GIS) technology is being increasingly used to streamline management data pertaining to buildings, neighborhoods, and cities due to the credibility of these systems in terms of exchanging and delivering data, and acting as repositories of project information within operational phases and facility management phases of the lifecycle of a given building or group of buildings (Volk et al. 2014). This has been implemented in a variety of solutions that have explored the role of this technology in decision support systems, knowledge management and acquisition, the support of improved workflow and efficiency, management workflows, and spatial visualization and analysis (Akcamete et al. 2010; Shen et al. 2012; Motawa and Almarshad 2013; Motamedi et al. 2014; Kassem et al. 2015).

This paper exploits the collective added value in BIM and GIS technologies to visualize, manage, and analyze information in the network of schools, educational institutions, and shared communal facilities that comprises demographic data, geographic data, financial data, school resources, facilities, utilities and infrastructure, and building components and systems, etc. This is all developed in a Web-based interactive platform with networked portals that attempts to achieve smooth and easy communication among its participants.

The added value in using BIM and GIS in our approach for efficient management of shared school facilities is multi-faceted. First, the BIM/GIS integration allows for data archival on different scales and levels of detail, including the school network level, the school building level, and the school systems and functional spaces level, in addition to linking this data with other networked entities with different geographical locations and different viewpoints and access privileges, including children, school authorities, government organization, and funding agencies. These are documented in digital repositories with the record history intact. Second, this integration allows for a platform of communication between several stakeholders involved in the management of school shared facilities. These include funding organizations, schools, local municipalities, and governments. Third, it allows for an interactive tool for analysis of what-if scenarios, reporting of available resources, as well as informed decision making and planning for shared school facility management justified with credible data.

14.3 Approach

This paper introduces School2Share (S2S), a cloud-based platform, that links data from stakeholders including schools and school authorities, children, government agencies, and funding organizations, to manage the efficient integration and sharing of school facilities. School2Share allows children to access available shared activities, maximizes using underutilized school facilities, provides live advertising for schools to promote community services, allows real-time monitoring of government agencies and efficient budget allocation to underserved areas, and allows funding organizations to effectively use allocated funds.

School2Share users include a multitude of stakeholders. The first type of users involves children and youth from different age groups, including primary school age, and lower and upper secondary school age. The focus is on underprivileged children who are facing out-of-school and illiteracy challenges in their communities in developing countries, amounting to over 250 million possible outreached children. The second type of users includes schools, and specifically public schools in developing countries, represented in school administrators and boards. The third type involves government agencies, including regulatory bodies at different levels, such as ministries of education and educational governorates that oversee and monitor the operational procedures and financial agenda of school facilities in different districts. The fourth type involves funding organizations, including a variety of national and international sponsoring institutions and entities, in addition to independent donors.

We claim that School2Share provides empowerment for the community and by the community. Using the proposed integrated and comprehensive platform, schools have registered accounts containing data involving available shared facilities, preferred sharing time schedules, and prioritized wish lists for requested facilities. Children can perform facility booking requests and participate in shared communal activities based on social status and geographic location. Schools can coordinate shared uses of facilities by sending and receiving live on-demand requests. Funding agencies can identify schools in immediate need of financial support. Government organizations and local authorities can perform regular monitoring of schools and their corresponding facilities and administrative entities in different districts, and the allocation of funding and effective policies and funding strategies in these facilities through a networked and real-time database.

By means of this approach, we argue that children can have wider access to community-based education, through three hierarchical levels:

1. Children-to-school interaction: Using personal computers or mobile applications, underprivileged children can identify school facilities (laboratories, theaters, libraries, arts and music rooms, football courts and sports facilities, cultural and communal activities, etc.) that address their needs. These are tailored within the logic of the system to satisfy age group, geographical location, demographics, social status, and level of education. Patterns of use and rating of services both contribute to the sustainable development of community facilities.
2. School-to-school interaction: The listing of facilities per school and their availability allows for shared activities through "school-to-school requests" that are sent and received among the networked system of schools within specific geographical locations, in addition to enabling and facilitating their effective use throughout times of the day and year.
3. Policies and development: Government organizations and funding agencies can identify schools and communities in need of development based on real-time data of missing facilities or gaps in community service. In this context, regulatory bodies can effectively monitor the allocation, accountability, and use of funds.

14.4 System Modules

School2Share comprises a wide variety of features based on its array of modules that include the management of shared facilities, task management, project management, and building management, navigation, file management, management of funds, stakeholder management, human resources, data analysis and reporting, and a chatroom/messaging system. This is described as follows:

Management of shared Facilities: This module enables major authorities such as school administrators, government agencies, and local municipalities concerned with the management of school facilities and their shared networks to monitor these efforts at the wider scale of the district, neighbourhood or city, in addition to managing everyday operations pertaining to revenues, requests, expenses, control, and performance. Using the navigation capabilities and database technologies embedded in BIM/GIS systems, the module is appropriated for a wide range of services. Funding agencies and government officials, for example, can browse interactively between GIS maps of shared school facilities in large neighborhoods in developing countries at the macro-level, while school administrators can navigate through specific zones that need their surgical intervention, together with all services, facilities, and necessary assets. Some of the significant features of such a module include an interactive dashboard for the real-time tracking of revenues, budget and expenses, and updated status of community sharing, for navigation in 3D in the shared schools, and monitoring completed and pending tasks, requests and performance of employees in each project.

Building management: This module enables building/facility managers affiliated with municipalities at the local level to manage FM systems within the designated schools with shared facilities. These systems include a wide range of disciplines involving roof and wall systems, insulation, protection control, heating, cooling, air conditioning and ventilation systems, communications, safety and electrical systems, control systems, finishes, circulation systems, equipment, waste disposal, and various other systems. This module allows the managers to navigate easily inside the 3D BIM model representation of the facility by means of a variety of methods, either level by level, or system by system. All data and building systems are represented in real time, in addition to maintaining a record and history of all operations and shared operations and activities in school networks.

Task management and project management: This module supports the assignment of projects and tasks among stakeholders involved in the school sharing process. This is crucial to the planning of operations and the overall lifecycle of the facility. The role of each stakeholder and the project duration and completion status are all reflected in real time. Each task can be edited, including the ability to add comments and documents, and identify and change activities and log hours.

Process management: Once a certain facility or building is selected for the purpose of conducting any given task or operation, a "ticket" is issued for that operation. All metadata including coordinate data, identifiers for building elements, properties, and other data is automatically detected by the system. Other data can be inserted manually by authorized users. This can include descriptions of the request or ticket at hand, allocation of date preferences, duration of operations, in addition to attaching images from the physical setting. The full historical record of these tickets can be monitored and queried by various means including status and type. They can also be visualized, managed, and resolved depending on the privileges and roles of each user. Managers can also dynamically customize and edit "actions" and "steps" within a given ticket workflow and delineate specific permissions, interrelations, and durations.

Navigation module: Different stakeholders can visually navigate through 3D BIM models of a given facility, aided by the integrative capabilities embedded in the system between BIM and GIS technologies. Using these capabilities, they can launch, modify, or manage any given ticket interactively. The benefit of each of these technologies independently as well is related to the scale of intervention and visualization. Using the GIS navigator, different users can track and manage macro-level sites, assets, utilities, and buildings within a given cluster or neighborhood. Using the BIM navigator on the other hand enables visualization of more detailed building components within a given school facility, in addition to free and custom navigation through detailed building systems, components, and floors.

File management: In this module, users can visualize, document, and modify the files, drawings, and documents pertaining to a given project. These include but are not limited to certificates, reports, forms, scanned images, and digital models. Comments and mark-ups can also be inserted on those documents to enable a tailored and annotated collaboration and communication throughout the whole process.

Funding management: This module involves the detailed management of related monetary and fiscal data for a given project. Users, based on permissions and roles, are assigned and designated different access levels and privileges to different databases that are connected to the school project database. These include several issues that a traditional accounting module would include, such as product data, vendor background information, invoice data, budgets, and revenues.

Stakeholders: In this module, the main settings pertaining to stakeholder data are defined and managed. Each of the stakeholders, including managers, administrators, suppliers, or other authorities and organizations, would have a specific portal or view of the project at hand. They would also have varying permissions related to the assignment, issuing, and manipulation of tickets throughout the workflow of the project. A multitude of complex types' relationships, privileges, and roles can be assigned based on the dynamic and flexible database technology of the system. These include the assignment of managers to properties, the assignment of relations between each stakeholder and a specific facility or building, and the assignment of a property or building and an array of building or facility managers.

Reporting and analysis: Using this dynamic module which comprises a reporting engine, users can produce customized reports for a given school project. The types of reports generated are endless and customizable. These include but are not limited to revenue reports, statistical reports concerning project status, ticket status and types, ticket actions and steps, school sharing status, expenses data and statistics, and many others. A summary of the major data outcomes from the embedded reporting engine, in addition to the added layer of analysis and some predictive and evaluative reports, is presented in graphical format on the main dashboard for each of the project stakeholders, based on their assigned roles and privileges. Another type of reporting for instant action is the notification panel which, as opposed to dashboard analytics, demonstrates issues that require the immediate action or at the very least acknowledgement of the user, based on their role and authority in the project.

Human resources: This module is connected to the databases of each of the school projects. Designated personnel can use this module to monitor data related to the background data, recruitment and employment data, and performance of their staff and employees.

Chatroom and messaging: Using this interactive module, smooth communication, exchange of data, and real-time messaging can be enabled between each of the stakeholders in the project. This module also keeps track of specific queries, events, and user inquiries and other technical support issues, in addition to receiving and responding to specific actions and steps using its embedded notification module.

14.5 Technology

The operating system of School2Share employs Linux (CentOS or Ubuntu as primary technologies preferred to run the main server). Apache runs as a main Web server using PHP7. The Laravel Web framework is used to run the core application. Laravel is open source, uses PHP, and is linked through drivers to databases such as MySQL. MySQL can be flexibly replaced through any hooked driver. The main server communicates with BIM servers and GIS servers, so that any specific data requests are processed and analyzed. Its outputs are managed via a API gateway to communicate with various devices and users including desktop devices (using CSS3, HTML, Vue JS, and JavaScript), in addition to mobile applications (iOS/Android) using the Swift programming language. Figure 14.1 illustrates the technology stack diagram used for the proposed S2S school management system.

In this structure, the back-end consists of various databases and cloud servers. Examples are the MySQL database, and Google Maps, which are responsible for navigation in maps. It also includes BIM servers which are responsible for data translation from school BIM models, in addition to GIS servers which are responsible for the launching and management of GIS data. It also includes Web applications pertaining to global school maps. These servers communicate with the core application via an intermediate database. This is done for both Web and software development using PHP7, Java, Python, and CentOS. At the front-end, this communication is handled via a platform for managing the system dashboard.

In terms of interface, each user has specific available views and functionalities based on need and role. Children can simply access, follow, view, and book available facilities (laboratories, theaters, libraries, sports facilities, etc.) based on need, social status, and location. Schools have more detailed functionalities, including basic statistics, locations and 3D model views of school, spatial allocations for each facility,

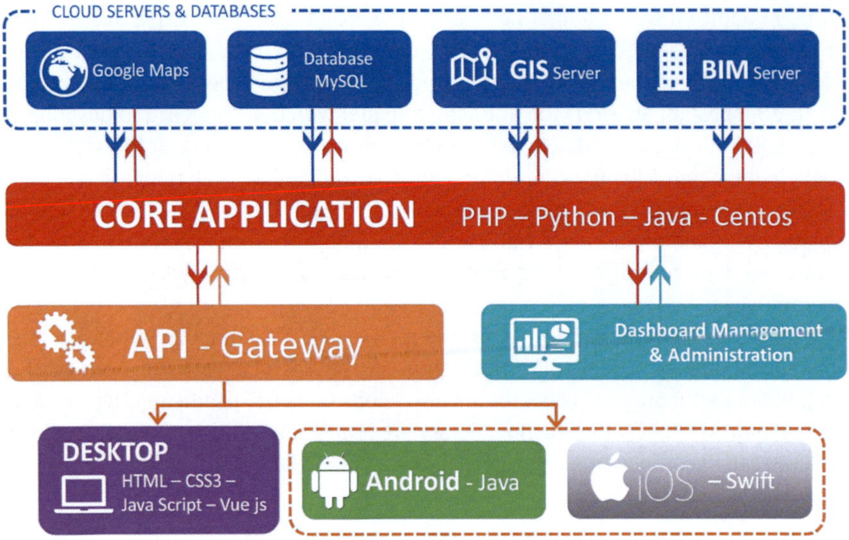

Fig. 14.1. Diagram showing the technology infrastructure of the proposed S2S system

and "sharing manager" to manage facility sharing to nearby schools or children in need. Government agencies have holistic views of schools within the community and their sharing activities. Funding organizations can track shared requests, available budgets, and requested funds via the "funding manager.".

In terms of real-time data utilized by the system, this includes (1) user activity and status, e.g., total number of accounts, number of registered children, schools, government agencies and funding organizations, number of active or inactive users, number of children in need of facilities; (2) facility sharing status, e.g., total number of schools, number of facilities, number of available facilities for sharing, number of shared facilities, number of needed facilities, coverage of children needs, school facility rating, and (3) request status, e.g., number of school-to-school requests, number of children activity and booking requests, number of accomplished requests, number of pending requests, number of funding requests, available budget, requested funds.

14.6 System Overview

The school management system approach goes beyond the micro-level related to the building to include the district and often the city, and extends to comprise different stakeholders, drawings, and documents of existing and new school facilities, funds, resources, and other activities and processes. At the entry level, the system dashboard outlines a map of the whole country with the status of available schools, resources, and events, as well as a fundamental project analysis and a reporting of its metadata. The proposed school management system comprises several portals addressing the needs of each of its stakeholders: (1) children portal, (2) school administrator portal, (3) government organizations portal, and (4) funding agencies portal, as described below.

14.6.1 Children Portal

Figure 14.2 shows a typical dashboard for a children portal, where children from different backgrounds, strata, and geographical locations can browse through maps, events, and image galleries of available school facilities for sharing in their neighborhood. Upon creating their accounts and filling in their basic data, the system recognizes their preferences, locations, and desired activities. Children can identify then available services and facilities in their immediate neighborhood and choose to follow different schools to keep track of upcoming events, or book school facilities based on their preferred date and time.

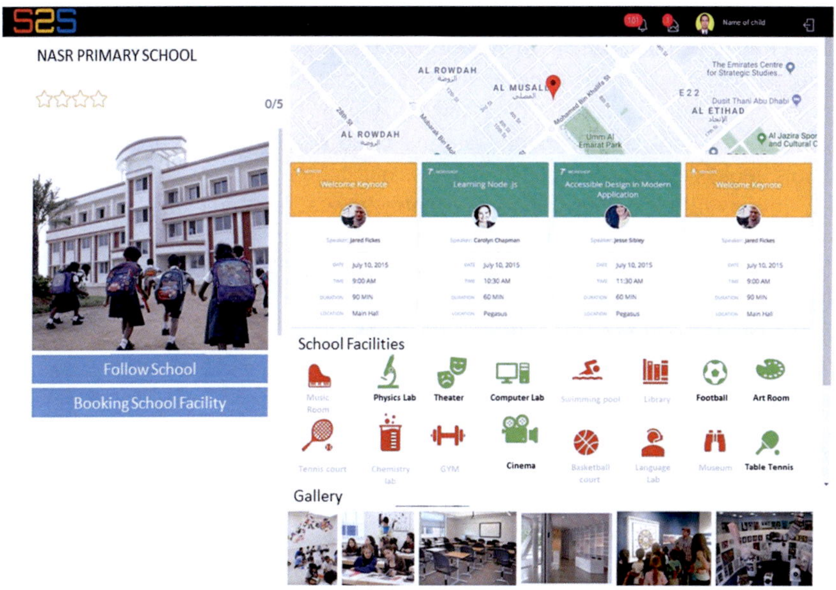

Fig. 14.2. Dashboard for children portal in proposed school management system

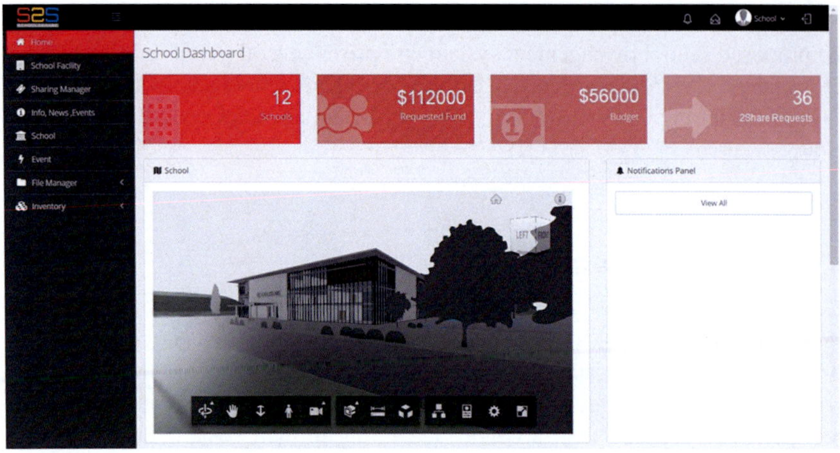

Fig. 14.3. Dashboard for school administrator portal in proposed school management system (3D exterior view)

14.6.2 School Administrator Portal

Another level of management of school facilities is conducted through the school administrator portal, where school authorities can manage the school network sharing process through their customized dashboards. Figure 14.3 illustrates a typical dashboard for a school administrator featuring basic reporting related to number of schools, requested funds, available budgets, and sharing requests.

By means of a "sharing manager," schools have access to detailed databases of school maps, 3D building model views, available facilities for sharing, and spatial allocations for each facility per school (see Fig. 14.4). School administrators can create, edit, or delete shared facility allocations using this interactive dashboard based on children needs and preferences uploaded to the system.

14.6.3 Government Organizations Portal

Regulatory bodies and government agencies can use this portal to oversee and monitor operations within the school management system by means of an interactive map. The dashboard alerts authorities of pressing issues related to lack of funding, allocation of resources, spending patterns, and other imminent issues pertaining to the effective operation of the school management system (see Fig. 14.5).

Fig. 14.4. Dashboard for school administrator portal in proposed school management system (2D interactive plan view)

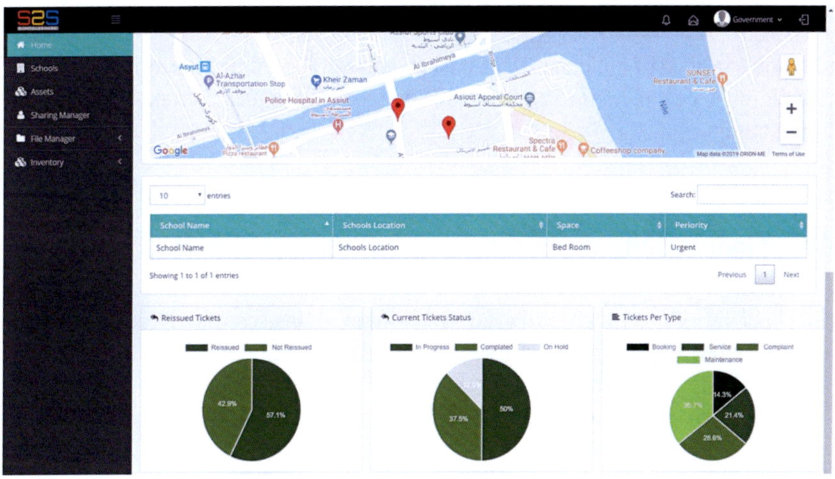

Fig. 14.5. Dashboard for government organizations portal in proposed school management system

14.6.4 Funding Agencies Portal

Funding agencies and sponsoring organizations can use this portal to manage funding and spending on school sharing projects and for addressing urgent resource allocation in different geographical locations (see Fig. 14.6). The dashboard provides funding agencies with macro-level data related to number of cities, neighborhoods,

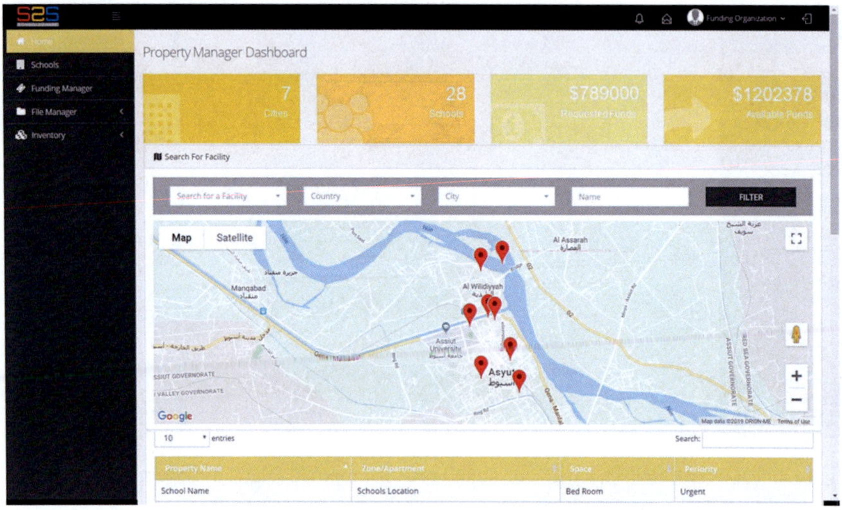

Fig. 14.6. Dashboard for funding agencies portal in proposed school management system

and schools in need of financial attention, categorized according to priority of funding and children needs, in addition to available budgets and requested funds from school administrators.

14.7 Conclusion

This paper introduced a cloud-based system that addresses the effective management of school shared facilities for underprivileged children in developing countries. The system captures the power of GIS and BIM technologies to analyze, manage, and visualize information in an interlinked network of schools and shared communal facilities. The system incorporates all stakeholders including children, schools, local authorities, and funding agencies, with different views and levels of access through their customized portals. We presented the approach, system technology infrastructure, and an overview of system portals and functionalities. We argue that this system empowers children in underserved areas, where they could identify customized school facilities addressing their needs using a comprehensive and real-time networked system. We also claim that this approach provides informed decision making and support to funding agencies and policy-makers, where schools and communities in need of development are identified, and regulatory bodies can oversee wise and effective spending for educational institutions.

References

Asiabaka IP (2008) The need for effective facility management in schools in Nigeria. New York Sci J 1:10–20

Akcamete A, Akinci B, Garrett-Jr J (2010) Potential utilization of building information models for planning maintenance activities. Proceedings of the international conference on computing in civil and building engineering. pp. 151–157

Kassem M, Kelly K, Dawood N, Serginson M, Lockley S (2015) BIM in facilities management applications: a case study of a large university complex. Built Environ Proj Asset Manage 5(3):261–277

Motamedi A, Hammad A, Asen Y (2014) Knowledge-assisted BIM-based visual analytics for failure root cause detection in facilities management. Autom Constr 43:73–83

Motawa I, Almarshad A (2013) A knowledge-based BIM system for building maintenance. Autom Constr 29:173–182

Olayinka O, Andrew S (2018) Achieving sustainability in commercial buildings in Nigeria: The FM approach. J Civil Constr Environ Eng 3(1):1–4

Shen W, Hao Q, Xue Y (2012) A loosely coupled system integration approach for decision support in facility management and maintenance. Autom Constr 25:41–48

UIS and UNICEF (2015) Fixing the broken promise of education for all: findings from the global initiative on out-of-school children. UNESCO Institute for Statistics, Montreal

UNESCO-IIEP, UNICEF and World Bank (2014) Education sector analysis methodological guidelines: sector-wide analysis with emphasis on primary and secondary education. UNESCO IIEP/UNICEF/World Bank, Paris/New York/Washington, DC

UNESCO (2015) Education for All 2000-2015: achievements and challenges; efa global monitoring report. UNESCO Publishing, Paris

Volk R, Stengel J, Schultmann F (2014) Building information modeling (BIM) for existing buildings—literature review and future needs. Autom Constr 38:109–127

Chapter 15
Evaluating the Quality of the Outdoor Environment of Primary School

Alaa EL-Telwany, Muhammed Gabr, Ayman Assem, and Tamer Samir

Abstract Primary education is a key factor in developing children's character; subsequently, the governments and educational organizations pay significant attention to this educational phase. Moreover, the quality of the educational environment impacts educational performance. The Egyptian schools ignore the design quality of the outdoor environment other than the UK's primary schools involved in outdoor education. Hence, this study aims to build a framework to evaluate the quality and design parameters of primary schools' outdoor environments. This framework studies the relationship between the educational settings (based on Bloom's theory) and the characteristics of children's outdoor environment. Our framework classified these characteristics into three categories: the motivations factors, the usability factors, and the legibility factors. In addition, the compatibility of the framework relative to the governmental Egyptian and UK's design guidelines of the primary schools' outdoor environment was analyzed. These guidelines (Egyptian and UK) were assessed based on the framework to review their strengths and weakness in supporting the outdoor educational environment. This led the study to analyze these impacts on real schools' models of both countries. This framework could be utilized to reflect on the design guidelines and educational requirements that led by operators and official agents of the national community.

Keywords Children's outdoor environment · Primary/elementary schools · Outdoor education

---

A. EL-Telwany · M. Gabr · A. Assem (✉) · T. Samir
Faculty of Engineering, Ain Shams University, El-Abaseya, Egypt
e-mail: Ayman.assem@eng.asu.edu.eg

M. Gabr
e-mail: muhammad_gabr@eng.asu.edu.eg

T. Samir
e-mail: tamer_hamza@eng.asu.edu.eg

Architecture and Urbanism: A Smart Outlook,
https://doi.org/10.1007/978-3-030-52584-2_15

15.1 Introduction

Many societies, countries, and educational systems are providing curriculum scenarios and different models of outdoor education within the school's outdoor environment. The school's outdoor environment has educational potentials, where education curriculum can be alive to the pupils. Using the outdoor environment of schools in the educational process does not mean moving the educational activities from the indoor spaces to outdoor spaces, wherever identifying applications that emulate exploratory, experiential, and educational aspects of the primary stage. This type of education has a significant impact on pupils' personal skills and educational performance. In many cases, the design quality of the outdoor environment is absent in the designing of the primary schools. The current state quo in Egypt, the primary school's outdoor environment lacks the necessary design consideration of educational and development needs. The schools' outdoors are just open space for morning line and may include play equipment that is not designed adequately to this age stage, other than the UK, that pays more attention to the outdoor education within schools. The paper emphasizes that the design quality of the school's outdoor environment could reflect the governmental design guidelines of each country. Initially, it impacts the area requirements of the school's outdoor environment as shown in the next table (Department of Education & Education Funding Agency 2014; GAEB 2011).

15.1.1 Problem

Nowadays, primary schools lack the design quality of the outdoor environment. In most cases observed, the school's outdoor area has been designed without consideration of the pupil's personal and educational needs. This design quality is always due to the lack of national design guidelines of schools (Table 15.1).

Table 15.1. Site area guidelines of Egypt and the UK

The guidelines	The UK's design guidelines	The EGYPT'S design guidelines
Total site area	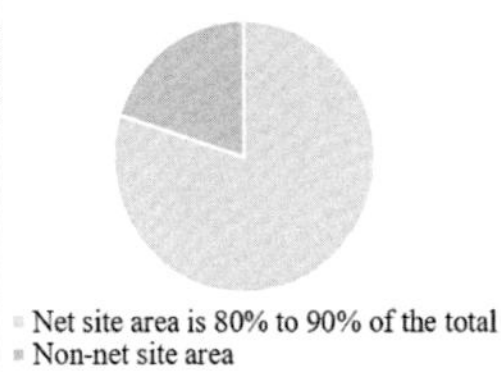	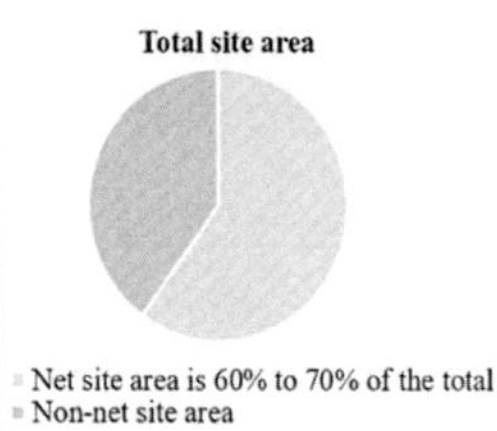
Area/pupil in the total area	42m2	4m2
Area/pupil in the net area	33.3m2	2.5m2

15.1.2 Objective

The paper aims to highlight the design quality of the school's outdoor environment by simulating axes concerning the child (outdoor environment–educational domains). Also, it aims to put design framework of primary school's outdoor environment and producing a set of designing recommendations.

15.1.3 Methodology

In order to achieve the research objective, multilayered methodology will be approached as follows:

- Investigating child–nature environment relationship theories that help in building a framework of the environment.
- Investigating education domains that build integrated educational process for the pupils.
- Comparative analyzing studies of the Egyptian design guidelines and the UK's guidelines and its impact on real schools' models.

15.2 Children–Nature Environment Relationships

The nature experiences are helping the children to develop their cognitive, affective, and evaluative skills (Kellert 2002). Kellert proposed three types of connecting the children to nature experiences ('direct', 'indirect', and 'vicarious' or 'symbolic'). The educational outdoor environments aim to direct the experience of children with nature in educational settings. Based on Kellert (2002), direct experience is direct interaction between the children's contact and natural settings like creatures, plants, and habitat. The capabilities of space are different based on the age of users and the function (Oloumi Alireza et al. 2012). However, the paper will describe the natural environment according to the children's perspective.

15.2.1 Affordance Approach

The concept of 'affordance' (perceived action possibilities) is the main approach of the mutuality between the organisms and the environment (Oloumi et al., 2012, Jenkins, Harold S., 2008). This approach defined by Gibson who adopted formal and visual properties of environmental characteristics. Gibson used this concept to build the framework of interactive relations between the human and the environment used in a specific function or specific activities (Oloumi et al., 2012). He described the affordance concept as "When the constant properties of constant objects are perceived (the shape, size, color, texture, composition, motion, animation, and position relative to other objects), the observer can go on to detect their affordances" (Jenkins, Harold S., 2008).

Then, Harry Heft developed this concept and launched the functional dimension to describe the relations between humans and the environment. Heft's theory moves from the knowledge direction of the environment into the functional direction of the environment. According to Heft's theory, there are a lot of functional considerations that could be paid attention:

- The simulators factors that produced the interaction process.
- The functional significances of the environmental features for an individual description. This approach may be more meaningful from a psychological standpoint than focuses on form.

Based on the previous, in order to describe the environment that requires widely aware of the environment's possibilities and limitations. Based on Heft's theory, the environmental characteristics and features can be designed and described according to the functional manner of the activities that need. Consequently, evaluating the functional possibilities or affordance of a particular space, it must be building on the physical and psychological characteristics of the user.

In this approach, demanding a specific reaction from origin or human in the environment requires recognizing the interaction process between the organisms, the environment's features, and characters. This led us to identify the constraints and opportunities of that environment (Jenkins, Harold S., 2008). For example, the functional requirements for children in the school's educational outdoor environment differ from children in a park of the same age; however, the environments are similar in visual characteristics.

15.2.2 The Educational Domains

The education settings are the main theme of the school's outdoor environment. The main purpose of the outdoor education environment is applying the domains of education. According to Bloom's theory, the educational process mainly focuses on three domains: cognitive, affective, and psychomotor domains.

The cognitive dimension:

According to Kellert (2002) studies, the philosophers and educators set the cognitive domain is the main aim of the educational process and children's development process. The cognitive is an approach to the learning, content, knowledge, and entities' concepts that can be done by the learner and possessed internally. Based on Piaget's theory, the cognitive process is different according to the age stage of the children. The abilities of children within middle childhood (concrete stage) are understanding logic, problem-solving, and understanding reversibility (Kendra, Cherry 2018).

According to Birdsongs, Kistina (2016), natural environments can interact with pupils' cognition in a radically different manner of fascination or "effortless attention". Based on Cachelin, A., Paisly, K., and Blanchard, A. (2007) studies, the cognitive domain gained from outdoor programs observed is greater than the classroom learning. In this approach, the outdoor environment can be related to the curricula process of many subjects that are based on mind and knowledge like the mathematics applications and science applications, dealing with animals and academic tasks.

The affective dimension:

According to Bloom's taxonomy, the affective domain is based on the emotional field and its values like receiving phenomena, then valuing their phenomenon or their behavior finally organizing the values into priorities. Building on Jarvis, P. (2006) studies, the learning through emotions has a great effect on educational process, where emotions can have a considerable effect on our thoughts, motivations, and beliefs, attitudes and values, in addition to area where we can find the vocabulary to help express the expectation" (O'Neil et al. 2010).

Several authors set the affective domain as one dimension of an outdoor educational environment (Waite 2011), by developing the pupils' skills, creative thinking, and problem-solving through the movement, interaction, and harmony with nature. The pupils can express the effective outcome in forms of literature, music, visual art, poetry, independent application, and competitive spirit.

The psychomotor dimension:

The psychomotor domain is the third major category of educational domains that set by Bloom's theory. This domain involves the relationship between the children's body and the surrounding environment. Moreover, it is considered as a result of the pupil's reaction to the other domains (affective–cognitive). Psychomotor objectives are a reflection of discreet physical functions, reflex actions, and interpretive movements. These types of objectives are concerned with the physical encoding of information with movement and/or with activities where the gross and fine muscles are used for expressing or interpreting information or concepts.

Within the educational outdoor environment, the psychomotor domain responses or reflexes about nature activities serve the body of the pupils and educational curricula. This process is done by merging the children's motor abilities to education process activities within the surrounding environment.

15.2.3 Children's Outdoor Environment's Characteristics

The study identified the characteristics of children's environment according to many theories, books, and qualitative studies that described the natural environment. The paper classified these characters into three categories: motivation factors, usability factors, and legibility factors.

The motivation factors:

The outdoor environment could be attractive for children to passionate them to interact. This study emphasizes that simulations of the intellectual possibilities of the children and their sense of designing the environment for their own self are directly related to the design and the kind of environmental features. The paper classified these motivations into three: scales, flexibility, and quality of aesthetic.

Scale: The form specifications of children's features are different from the adults in the same space. Based on (Oloumi et al. 2012) studies, they defined the environment's scale as "proportion with children size". According to Moore et al. (1997), child-scaled environments promote to produce children's feelings of control and self-identity. E. J. Gibson emphasized that the information to specify the utilities of the environment is accompanied by information to specify the observer himself, his body, legs, hands, and mouth (Gibson 2001). In this approach, the paper divided the environment scale into space scale and physical scale.

Flexibility: According to Gibson's theory, the children will change and alter their environment to change their affordances to suit them. From this point, the environment could be organized; however, it could be able to change its structure according to the children's needs. The flexible environment can receptive to the interests and capacities of every pupil and offer different alternative educational activities and different events that support pupil's imagination and creativity (EYLF PLP 2011). These spaces could include performance area, event areas, galleries, and adventure playgrounds.

Quality of aesthetic: Various aesthetic expression values can be used to improve understanding of ecological processes (Wilhelmsso 2012). According to several studies, the children in the middle childhood has attracted to various emotional aspects of the environment (Kellert 2002), and the children pay more attention to shape, color, sound, textures, and smells of the expressing aesthetic (Wilhelmsso 2012). The paper divided aesthetic appeal into a variety of textures and a variety of colors.

The usability factors:

Usability is a term to describe the ability of children to directly interact with the environment without fears or danger. The accessibility and safety of the environment have met the usability of children.

Accessibility: Many scholars proposed that the accessibility has a significant role on building ideal outdoor environment for pupils (Fadzila and Said 2010), as the indoor education and outdoor education are an integrated process (Mozaffar and Mirmoradi 2012). The direct access increases the pupils' awareness of the role of the outdoor environment, being more included and view pupils' abilities (Sandberg and Niklasson 2010), and also impact on their emotional side (Mozaffar and Mirmoradi 2012).

Safety: Increasing the outdoor time for children causes increasing pupils' risks that can be exposed to the children (DCELLS 2009). It does not imply that outdoor education is ought to be risk-free, yet that the risks that can be faced should be addressed and assessed. In this approach, the outdoor environment could provide opportunities and plan strategies to evaluate and manage children's safety, challenge, and risk like observation management, risk considerations, and identifying boundaries.

Legibility factors:

Meaning is reading the function of the environment that can be represented in its features, interaction elements, and terrain. Creating a legible interpretation of outdoor environment information could follow the required function. In the educational outdoor environment, the legibility could follow the educational domains. The paper divided the legibility factors into physical elements and landscape characteristics.

Physical elements: the environment's elements could be indicated to the required function of the environment. According to Kellert 2002; Moore and Marcus (2008) classified the environment's elements into:

- Direct natural elements: the elements that reproduce or keep up a generally wild or natural setting (like woods, bond area, trees, plantings…, etc.).
- Indirect natural elements: the elements that require human upkeep and guided to interaction with environment procedures (like maintenance elements, composting elements, roof garden…, etc.)
- Symbolic natural elements: the elements that represent the environment abstractly without engaging to nature (like models, subculture, signs…, etc.).
- Non-natural elements: the elements that not engage with the natural environment and represent the required function process (like seats, playground equipment…, etc.).

Landscape characteristics: The site topography and layout surfaces provide opportunities to various interactive programs (Fadzila and Said 2010). According to Gibson and Heft studies, the terrain represents the function for the children in contrast with the adults as it represents forms (Fjørtoft and Sageie 2000). This means that the children are predicting an outcome based on the interactive program of the terrain or topography that can be offered within the landscape. According to Fjørtoft and Sageie (2000) noted that the variables habitat is a common way to describe the natural outdoor environment and impact on choosing suitable terrain, habitat, or vegetations according to the case. In this approach, the variable terrain and topography can offer a sensory experience for children to enhance the effective education process. In the context of environment, the variables of the terrain (mounds, terraces, slopes) can offer to provide physical changes to encourage the pupils to climb, roll, and jump (Fjørtoft and Sageie 2000).

15.3 The Difference Between the Design Guidelines of Egypt and UK Primary Schools

Based on the previous study, the school's outdoor environment could satisfy the educational process under the design principles of children's outdoor environment. On the paper view, these could impact on the design guidelines of the school's outdoor area. The next table will review the governmental design guidelines of Egypt's primary schools according to GAEB (2011) (Table 15.2).

The next table will review the governmental design guidelines of the UK's primary schools according to Department of Education and Education Funding Agency (2014) (Table 15.3).

After reviewing both guidelines, the study concludes that the UK's design guidelines appreciate merging the educational process within nature outdoor environment of schools other than Egypt's guidelines to support some aspects of the psychomotor domain. The next table will review the aspects supported by both guidelines (Tables 15.4 and 15.5).

After reviewing the strength and weaknesses of both design guidelines, the study will analyze how it impacts the design of real models of primary schools.

15.3.1 The Case Study

The study analyzes the capability of real schools' models under the principals of the design guidelines by analyzing the site area of South Molton Primary School in UK and EL-Qatamia Primary School in Egypt according to (researcher visits and South Molton Primary School 2018) (Table 15.6).

Based on the previous table, the pupils of EL-Qatamia primary school has the lowest of area per pupil through the outdoor environment. The next table will analyze the education process within the school's outdoor environment characteristics from the view of study (researcher visits and South Molton primary schools 2018) (Tables 15.7, 15.8 and 15.9).

Based on the previous tables, the South Molton Primary School applied most of the design principles of educational outdoor environment in different ways. The South Molton Primary Schools represent the means of learning methods through the

Table 15.2. Egypt's design guidelines for the primary school's outdoor environment

The Egypt design guidelines		
Zone	Description	Objective
Outdoor court	– The double playground. – The shaded area could be 20% of the area	This area is dictated to the morning timeline, break time, and PE activities
Indoor court	– An open area	The guidelines do not define its using

Table 15.3. UK's design guidelines for the primary school's outdoor environment

The UK guidelines			
Zone	A/P	Description	Objective
Soft outdoor PE area	20	– Winter pitches to preferred team games, such as football, rugby, and hockey – Overlapping summer pitches, such as cricket, athletics track. – Trim trails and bike tracks.	– Providing the PE facilities and social events for pupils
Hard outdoor PE area	1.5	– A porous macadam MUGA, with three netball courts, overlaid. – Tennis/netball courts	– Enhancing physical education
Soft informal and social area	2	– Soft surfaced areas, such as grass, sand, bark mulch, grassed banks, or terraces – Shrubs, planted, Meadowland, or woodland	– Providing a safe and shaded area. – Providing playing, large gatherings, and sociability
Hard informal and social area	1	– Hard-surfaced playgrounds. – Furniture created through the introduction of natural materials that blend into the landscaping	– Providing outdoor study groups area to engage pupils in outdoor art, theater, dance, and design activities
Habitat area	.5	– Grounds developed for a range of supervised activities and Meadowland. – Areas condition to be fenced to avoid unsupervised access	– Providing wildlife habitats such as ponds and gardens – providing outdoor science areas to support educational activities

Table 15.4. Comparison between guidelines of Egypt and the UK's educational domains

Educational domains	The UK's principles					Egypt's principles	
	Soft outdoor PE area	Hard outdoor PE area	Soft informal and social area	Hard informal and social area	Habitat area	Outdoor court	Indoor court
Cognitive	•	•	•	✓	✓	•	•
Affective	•	•	✓	✓	✓	•	•
Psychomotor	✓	✓	✓	✓	✓	•	•

Table 15.5. Comparison between the design guidelines of Egypt and the UK's environment characteristics

Characteristics of the outdoor environment		The UK's principles					Egypt's principles	
		Soft outdoor PE area	Hard outdoor PE area	Soft informal and social area	Hard informal and social area	Habitat area	Outdoor court	Indoor court
The motivation	Scale	•	•	•	•	•	•	•
	Flexibility	✓	✓	✓	✓	•	•	•
	Quality of aesthetics	✓	✓	✓	✓	•	•	•
The usability	Accessibility	✓	•	✓	•	✓	•	•
	Safety	•	•	✓	•	✓	✓	✓
The legibility	Physical elements	✓	✓	✓	✓	•	•	•
	Landscape characteristics	✓	✓	✓	✓	•	•	•

Table 15.6. Comparison between two primary schools in Egypt and the UK

	South Molton primary school	El-Qatamia primary school
Overview of school		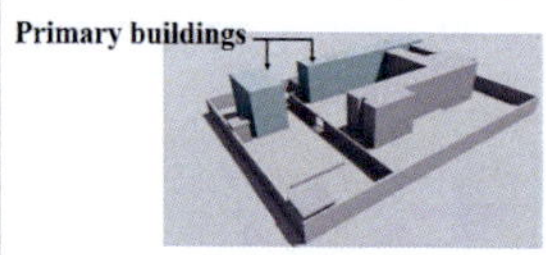
Zoning of the schools	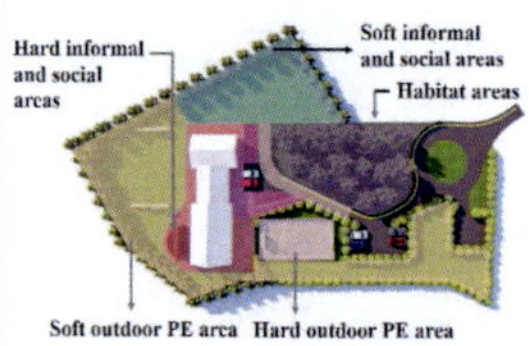	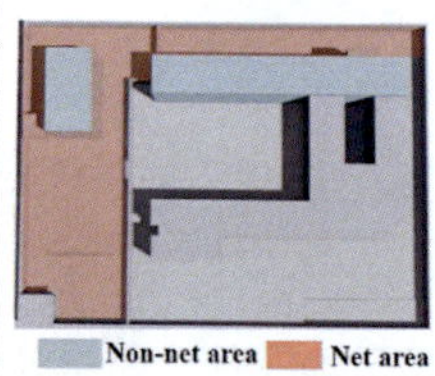
No. of pupils	240	2000
Total site area	9712 m2	2787 m2
Net area	8812 m2	2139 m2
Net area/pupil	40.5	1.06

school's outdoor environment under the constrains of the UK's guidelines. In this approach, the study proved that any shortage in the national guidelines impacts on the design quality of the educational environment.

Table 15.7. Comparison between children's motivations and legibility characteristics of the case study

Category			South Molton Community Primary Schools	El-Qatamia Primary School	
The motivation	Scale	Environment	The forest zone is the most zone represents enclose to the children by using natural elements		The outdoor environment does not provide any enclosure for children
		Physical	Overall in the school's environment, all playing equipment follows the scale of the children		The physical elements lack to appreciate the children's scale
	Flexibility		The environment's zones have the flexibility for children to do more than one function		The environment lacks a variety of functions that can be offered
	Quality of aesthetic	Textures	The zones lack using a variety of materials, wherever the forest zone is, the most zone used a variety of textures like mud, planting, and lake		The environment lacks to use a variety of textures based only on one hard material (interlock)
		Colors	– The forest zone using a variety of colors that appear in equipment, textures, and natural elements		There are no any varieties of colors in the outdoor environment
Legibility	Landscape characteristic		– The terrain used only in wild zone in the forest zone, wherever it lacks identifying the function that required		The terrain never regards in the outdoor environment

(continued)

Table 15.7. (continued)

Category		South Molton Community Primary Schools	El-Qatamia Primary School	
	Physical elements	– The environment used the natural elements and non-natural elements to represent the activity required		The environment using only non-natural elements to represent the required function like seating fixtures

Table 15.8. Comparison between children's usability characteristics of the case study

Category		South Molton Primary School		El-Qatamia Primary School	
Usability	Accessibility		There is direct access between the building and outdoor and direct access between all zones of the outdoor environment		There is direct access between the first building and outdoor area and indirect access between the second building and outdoor area by using the fence
	Safety	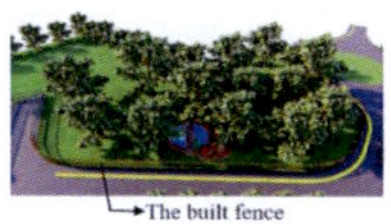	Safety appears only in identifying boundaries by using natural and build a fence		Safety appears only in identifying boundaries by using a build fence

Table 15.9. Comparison between the education domains' achievements of the case study

Education domains	South Molton Primary School		El-Qatamia Primary School
	Zone	Activities	Activities
Cognitive	Forest zone	– Biology class – Hunting class – Dealing with insects – Geographical class – Outdoor classroom	– The school lacks to appreciate cognitive and affective activities
	Soft social and informal	– Math activity – Outdoor classroom	
Affective	Forest zone	– Drawing class – Drawing class – Sculpture class – Arts activities	
Psychomotor	Hard PE area	– Football – Basketball	– Football – Running – Athletics
	Soft PE area	– Football – Athletics – Tag Rugby	

15.4 Conclusion

Providing outdoor education within nature outdoor environment has a significant impact on developing personal skills and educational performance for the pupils. That paper assumed that the schools' national design guidelines of each country reflect its quality of the educational environment. As these design guidelines are a key factor in applying the educational outdoor concept, the paper built a design framework for the school's outdoor environment to identify the strength and weaknesses of the design guidelines. This framework is based on the relationship between the educational settings and the characteristics of children's outdoor environment.

Based on the paper design framework, the outdoor environment could directly relate to applying the educational domains cognitive, affective, and psychomotor domain. The paper highlighted the educational activities and application within the outdoor environment:

- Cognitive: mathematics, science, experimental experience, dealing with animals, and academic tasks.
- Affective: activities of literature, music, visual art, poetry, culture.
- Psychomotor: PE activities and motor reflexes through affective and cognitive activities.

The paper concluded that all of these educational dimensions could follow the principles and design parameters of the outdoor environment of the children in this stage. The paper classified these design parameters and principles into three categories: motivation, usability, and legibility.

- Motivation factors: represents in a scale of space and physical, flexibility of function, and quality of aesthetics.
- Usability factors: represents in accessibility and safety of the environment.
- Legibility factors: represents in reading the function of the environment's elements and landscape characteristics.

After identifying the Egyptian's design guidelines of the schools' outdoor environment, the paper concluded that Egypt ignored the design principles of the educational outdoor environment. This weakness impacts the design quality of the outdoor environment and does not provide outdoor educational process at the Egyptian case study (El-Qatamia Primary School). In contrast with the UK's design guidelines that pay more attention to the quality of the outdoor environment zones, this attention appears in the outdoor educational environment of the UK's primary schools (South Molton Community Primary Schools).

Finally, this study demonstrated that all educational domains and design principles of the outdoor environment could be recommended in the governmental design guidelines to be reflected in the real school's design. This design framework can provide guidance to school operators, architects, landscapers, and policymakers who are interested in creating supportive educational outdoor environments.

References

Birdsongs K (2016) Get outside, it's good for your brain from www.scilearn.com/blog/cognitive-benefits-of-nature. Last Accessed May 2019

Cachelin A, Paisly K, Blanchard A (2007) Using the significant life experience framework to inform program evaluation. J Environ Educ

DCELLS (2009) Foundation phase: outdoor learning handbook. DCELLS

Department of Education and Education Funding Agency (Jun 2014) Area guidelines for mainstream schools. Building Butin (103)

EYLF PLP (2011) The early years learning framework professional learning program. EYLFPLP e-Newsletter No. 11 2011

Fadzila AN, Said I (Dec 2010) Asia pacific international conference on environment. Faculty of built environment, University Teknologi Malaysia

Fjørtoft I, Sageie J (2000) The natural environment as a playground for children: landscape description and analyses of a natural playscape. Landscape and Urban Plann

GAEB (2011) Standards and requirements for the validity of sites and school buildings for primary and secondary schools in the existing villages and cities. GAEB, Egypt

Gibson EJ (2001) Perceiving the affordances: portrait of two psychologists. Mahwah: Lawrence Erlbaum Associates

Jarvis P (2006) Towards a comprehensive theory of human learning. Routledge, London

Jenkins HS (Dec 2008) Gibson's "Affordances": evolution of a pivotal concept. J Sci Psychol

Kellert S (2002) Experiencing nature: affective, cognitive, and evaluative development in children. In: Kahn P, Kellert S (ed) Children and Nature. Psychological, sociocultural, and Evolutionary Investigations, vol. Mass: The MIT Press, Cambridge

Kendra C (2018) Reviewed by a board-certified physician. The concrete operational stage of cognitive development, from the very well mind website: https://www.verywellmind.com. Last Accessed May 2019

Moore R, Marcus CC (2008) Healthy planet, healthy children: designing nature into the daily spaces of childhood. Science and Practice of Bringing Buildings to Life, Biophilic Design

Moore R, Goltsman SM, Iacofano DS (1997) Play for all guidelines: planning, design, and management of outdoor play settings for all children. 2nd edn

Mozaffar F, Mirmoradi SS (2012) Effective use of nature in educational spaces design. Technol Manag Constr Int J

O'Neil G, Feagal M (2010) Guide to taxonomies of learning. UCD teaching and learning, Dublin

Oloumi S, Mahdavinejad M, Namvarrad A (2012) Evaluation of outdoor environment from the viewpoint of children. Procedia—Soc Behav Sci

Sandberg A, Niklasson L (2010) Children and the outdoor environment. Europ Early Child Educ Res J

South Molton primary school (2018) From south molton primary school website: www.southmoltonprimary.devon.sch.uk. Retrieved May 2019

Waite S (2011) Teaching and learning outside the classroom: personal values, alternative pedagogies, and standards. Int J Prim Elementary Early Years Educ

Wilhelmsso B (2012) Teachers' intentions for outdoor education: conceptualizing learning in different domains. Umeå University, Department of Science and Mathematics Education

Chapter 16
Post-occupancy Evaluation of the Emergency Departments: A Holistic Approach to Promote the Healthcare Architecture Design Standards

Bassant Sorrour, Akram Farouk, and Mostafa Refat

Abstract Healthcare facilities (HF) are complex environment to design, build and operate. Design guidelines define the spatial requirements of HF are interfacing stakeholders' demands for satisfaction. Many details are left for HF operators to address causing a gap between designers and operators. For example, emergency department (ED), trends like room saturation, length of stay, etc., tend to re-form the concept design of ED. The research aims to narrow the gap between the designers' knowledge about the built environment (BE) and the stakeholders' claims for efficiency. Both qualitative and quantitative methods are undertaken to generate a checklist to guide designers to enhance the performance of ED based on users' experience. A quantitative analysis is used for post-occupancy evaluation (POE) by means of principles of evidence-based design and survey on users' experience in ED. A qualitative analysis is used for selected case studies using filtered measures from quantitative surveys. The selection of case studies is based on the application of design guidelines of ED. Nationally, ED in Dar Al-Foad Hospital (DAF, Giza) and the extension of DAF (Cairo) were selected. Internationally, the United Kingdom National Healthcare Services (NHS) is utilizing a POE system for Brent Emergency Care Centre. Proportions between smooth flow among rates of admission and reallocation of facilities were observed. This paper is limited to study ED operational demands. Further investigations should proceed concerning departments adjacent to ED, like imaging, operation theatres, etc., applying the feedback from POE to have a holistic approach to the operational design of hospitals.

Keywords Post-occupancy evaluation (POE) · Emergency department ED · Operational efficiency · Patient experience

B. Sorrour (✉) · A. Farouk · M. Refat
Faculty of Engineering, Ain Shams University, Cairo, Egypt
e-mail: Bsorrour@gamil.com

A. Farouk
e-mail: akram_mohamed@eng.asu.edu.eg

M. Refat
e-mail: mostafa_ismail@eng.asu.edu.eg

Architecture and Urbanism: A Smart Outlook,
https://doi.org/10.1007/978-3-030-52584-2_16

16.1 Introduction

A hospital building can be identified as a secondary care building (AIA 2014), where patients streamed to after cases have been stabilized. It is composed of five categories; Impatient unit, diagnostic and treatment department, public services department, support services department and ambulatory department (EHFG 2013). The assorted departments came up with diversity of stakeholders in HF which are (1) patients (inpatients and outpatients) (2) board of trustees and leadership (3) researchers and designers (4) vendors/suppliers (5) caregivers/family/visitors (6) staff (7) community partners (8) community organizations and (9) donors (CHD 2010).

The reason of choosing ED as an exemplar is to magnify the complication operational process in HF, in addition to focusing operational challenges and capability of architectural design to deal with the challenges. The number of inpatient bed saturation in Egyptian HFs is around (1.5–1.7 beds/1000 population), which is significantly lower compared to world global average of 2.7 beds/1000 population (The Pulse 2015). The low impatient bed density reveals no significant relationship with long waiting\staying (+30 min) intervals for patients to be diagnosed in ED (Becker and Douglas 2008).

16.2 Research Problem Definition

Reference to Japan International Cooperation Agency (JICA) report as a technical assistance to Egyptian MOH (Ministry of Health) vision in quality improvement of HCFs concerning emergency medicine departments (JICA 2017), the below challenges facing the transfer of healthcare service in ED were observed (Table 16.1).

In conclusion, the research problem can be summarized in the following points:

- No strategic planning for locations of ER departments all through Egypt that indicate destination that patient should be directed in case of emergency on sudden disasters.
- Long waiting time intervals in treatment rooms as a result to the lack of knowledge of the caregivers about triage rooms that its role is to identify walk-in patient conditions.
- Violence against medical staff and caregivers in ER reception as a result to poor access control and streaming.
- A poor patient privacy and safety in treatment rooms.

Table 16.1 Challenges and reasons of emergency healthcare architecture

Findings	Challenges	Reasons
• *Emergency transfer*	• Paramedic teams lack of experience (JICA 2017) • No data collection and analysis of data • Emergency transfer is not tested in terms of building efficiency (JICA 2017) • Synchronization of hospitals for transferring patients is at its minimum (JICA 2017) • Violence against staff and caregivers • Triage room was not neither found nor utilized among many hospitals in the study (JICA 2017)	• Training curriculum modification • Insufficient continuous education for quality and technical improvement (JICA 2017) • Insufficient data collection on patient • Process for calls to ambulance lacks consistency (JICA 2017) • Insufficient information of bed occupancy rate • Poor access control for entrances and circulations • Insufficient knowledge about the triage room role in patient streaming
• *Emergency nursing*	• Nursing staff is not sufficient to cover a good surveillance for patients (JICA 2017)	• Lack of motivation for patient care
• *Emergency medicine*	• Desynchronization between health facilities in streaming patients (JICA 2017)	• Undeveloped referral system (JICA 2017) • Paramedic teams lack experience about utilizing triage room to fit in dedicated function (JICA 2017) • Encouragements for physicians to join ED are declining (JICA 2017)

Source Japan International Cooperation Agency (JICA) 2017

16.3 Aim

The research aims to fill in the gap between demands of operation and architectural design of physical built environment of ER layout as an approach to solve observed problems.

16.4 Methodology

The research method was based on inducting results from a post-occupancy evaluation (POE) on a currently operated ER department. Both qualitative and quantitative methods were undertaken in this research. Firstly, level of information provided in national and international design guidelines was evaluated, and a quantitative survey was used to identify framework of evidence-based design methods and was fortified with field survey on local stakeholder's experience. Secondly, a qualitative analysis was used through currently operated case studies of ER departments. Two case studies were selected based on (1) the compatibility with design guidelines (2) to have been operated for a quite some time to enable evaluation of post-occupancy aspects, socially, technologically and environmentally. Lastly, a checklist is generated for proposed design leads for improvement of efficiency of built environment of ED.

16.5 The Patient Journey in Emergency Department

The emergency department is considered as a critical patient entrance to the hospital building (AIA et al. 2014). This entrance is responsible to receive the patient (walk-in cases) with conditions that may not be specified unless reaching the ER by means of ambulance where diagnosis is undertaken through (HBN 15 2013). Therefore, patients must path through three types of streaming. Firstly, admission data concerning security reasons for the patient might be convicted for crime. Secondly, the patient must be screened for possible infection condition weather he is infected (+ve cases) or his body is venerable to acquire infection (−ve cases). Lastly, the patient is passed through triage room to be able to identify a primer medical diagnosis to be able to elect the most suitable specialist physician dedicated to treat the patient's case or discharge patient according to his condition urgency; this process must be conducted as short as possible for patient to avoid enduring long waiting intervals (Fig. 16.1).

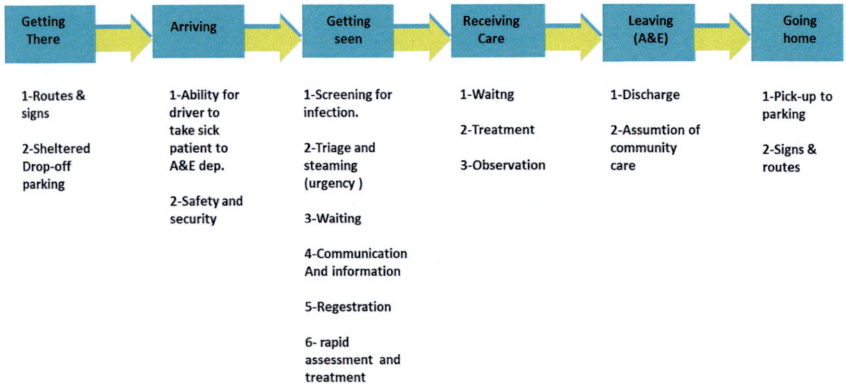

Fig. 16.1 Patients journey through ED. *Source* Health building Notes 15 (HBN 15 2013)

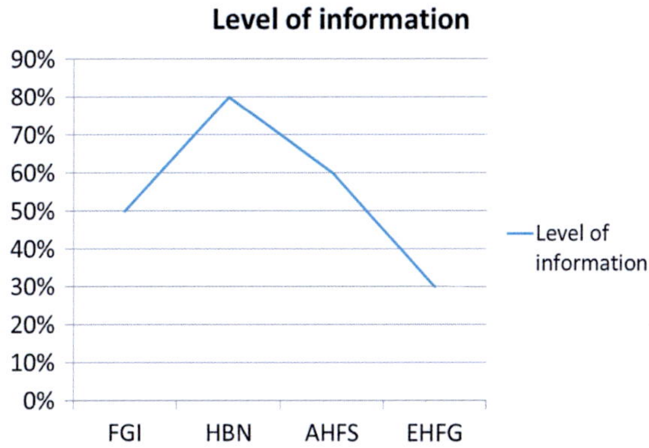

Fig. 16.2 Design guidelines level of information

16.5.1 *The Compatibility with HF Design Guidelines*

In order to process gathered data about the ED being compatible with the country HF design guidelines, the below comparison table is conducted to pinpoint design guidelines of built environment in ED. This comparison is used to define the available level of information in FGI (Facility Guidelines institute 2014, US), HBN 22 (Health building notes 22, 2013, UK), AHFS (Australian Health facility guidelines 2010, Australia) and EHFS (المعايير التصميمية لمبني المستشفي et al. 2010) (Fig. 16.2; Table 16.2).

16.5.2 *The Measures of Operational Efficiency*

The evaluation of level of information in design guidelines directs the research towards more investigations on POE using evidence-based design methods in order to magnify patients and staff outcomes. This method aims to reach a concept for what can be addressed as an efficiently operated built environment of HF. The results show the effects on patients, staff and family from several dimensions of the physical environmental factors of HF as shown (Sigma Theta Tau International, 2010) (Fig. 16.3).

In addition to the principles defined by EBD, the chart below was deduced from a local quantitative survey to investigate assorted users experience including medical staff, operators, caregivers, patients and visitors. The survey target highlights the operational aspects that claim to be fulfilled in built environment in ED and to detect the degree of satisfaction. The chart recorded higher levels of satisfaction for lights 31.58% and relatively lower ones for noise control 2.63% and way finding 2.63% (Fig. 16.4).

Table 16.2 National and international design guidelines comparisons

Point of comparison	FGI 2014	HBN15	AHFS	EHFG
Levels of care	2 Care levels	3 Care levels	6 Care levels	3 Care levels
Number of admissions per year	N\V	20 million new patients attend ED in England per year. Furthermore, a number of attendances are assumed to be raised from 5–7% per year	N\V	N\V
Notices on Area of Treatment room	Area has no significant relation to either level of care or no of attendances	Area is related to number of expected attendees	Area is related levels of care	Area has no significant relation to either level of care or no of attendances
Notices on Area of triage room	Area has no significant relation to either level of care or no of attendances	Area and number of rooms are verified with reference to number of expected attendees	Area and number of rooms are verified with reference to levels of care	N\V
Notices Areas of treatment rooms	Area has no significant relation to either level of care or no of attendances	Area and number of rooms are verified with reference to number of expected attendees	Area and number of rooms are verified with reference to levels of care	N\V
Operational Scenarios	N\V	Verified	Verified	N\V
Circulation	Min. corridor width is 2400 mm	25% of ED total area	N\V	Min. corridor width is 2400 mm

16.5.3 Case Studies

The criteria of selection of case studies were based on the following (1) applying design guidelines, (2) the hospital is operable to deliver healthcare services and (3) a comparative method is applied using a base case.

Case study 1: NHS trust (United Kingdom Department of Health, National Healthcare States); post-occupancy evaluation (POE) (NHS 2013), method was based on

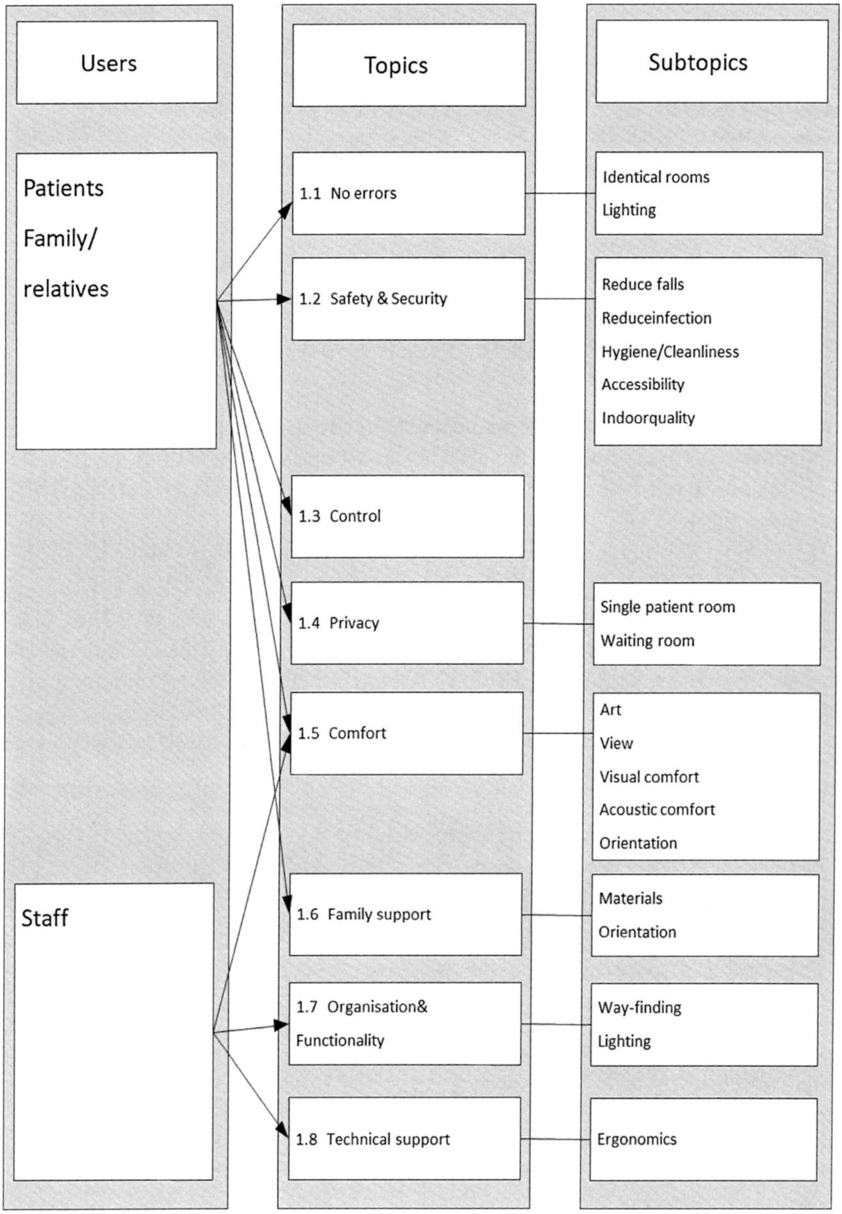

Fig. 16.3 Users influencing physical built environment. *Source* (McCullough 2010)

	0 (1)		1 (2)		2 (3)		3 (4)		4 (5)		5 (6)			
	Σ	%	Σ	%	Σ	%	Σ	%	Σ	%	Σ	%	Ø	±
a) Security and surveillance	3x	7.89	7x	18.42	9x	23.68	12x	31.58	1x	2.63	6x	15.79	3.50	1.47
b) Signs for way finding	1x	2.63	9x	23.68	7x	18.42	8x	21.05	6x	15.79	7x	18.42	3.79	1.51
c) Acoustic control	1x	2.63	14x	36.84	4x	10.53	11x	28.95	2x	5.26	6x	15.79	3.45	1.50
d) Lights	-	-	6x	15.79	3x	7.89	11x	28.95	10x	26.32	8x	21.05	4.29	1.33
e) Interior finishing colors	2x	5.26	12x	31.58	4x	10.53	11x	28.95	4x	10.53	5x	13.16	3.47	1.50
f) Lack of nature scenes (w...	6x	15.79	15x	39.47	4x	10.53	7x	18.42	2x	5.26	4x	10.53	2.89	1.56

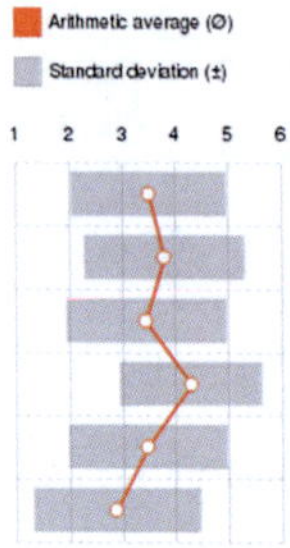

Fig. 16.4 Users feedback on field survey for local hospitals. *Source* conducted by researcher, full statistics about survey can be found at https://www.esurveycreator.com/results/5a8a46d-a03e2a9

emergency department in Middlesex hospital, and an exemplar plan developed as a comparative base case for trust undertaking new build schemes. Both the BECaD and base case layouts were designed in compatible with HBN 22 (2013) (Figs. 16.5, 16.6 and Table 16.3).

Case study 2: Emergency departments in DAF (Dar Al Fouad Hospitals). The old one is located Giza Governorate 6 October district, dissert road; as specialized hospital for cardiac diseases. The new Extension in located in Cairo, Nasr City district, Nasr Road. Both hospitals are compatible with the Egyptian design guidelines for healthcare facilities (EHFG 2006) and accredited Joint Commissioning Institute (JCI 2005) (Figs. 16.7, 16.8 and Table 16.4).

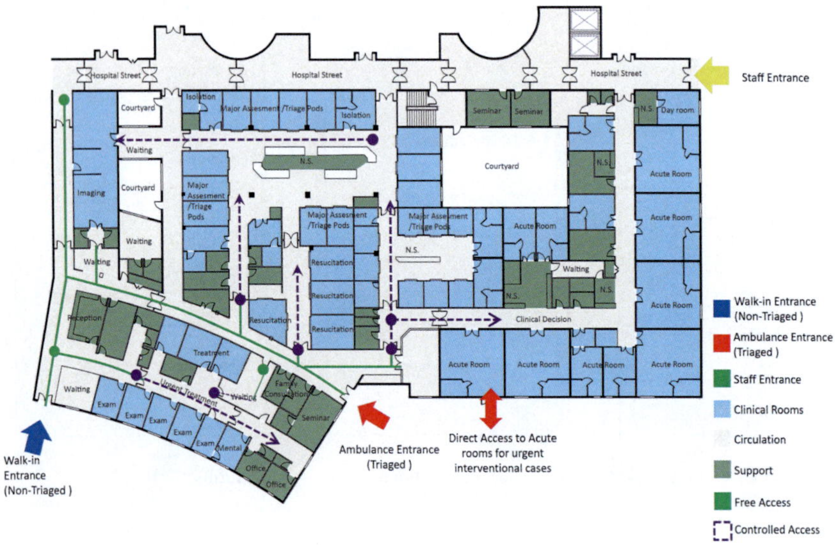

Fig. 16.5 ED in Middlesex Hospital. *Source* (NHS 2004)

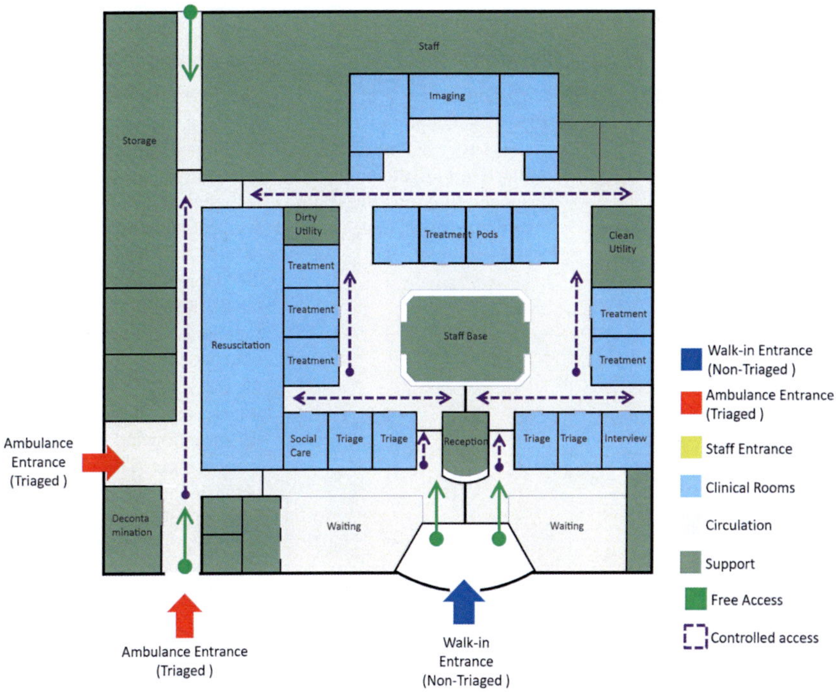

Fig. 16.6 The exemplar plan. *Source* (NHS 2004)

16.6 Generated Checklist for Development and Design Leads

From further analysis on the measures of efficiency by means of both qualitative and quantitative methods of analysis, it can be deduced that the design guidelines are not enough to guarantee operational efficiency. The EBD principles, the survey and the analysis, for case studies, are recording more design leads for designers supported by user claims for greater value of ED (Table 16.5).

16.7 Conclusions

- Post-occupancy evaluation should be applied through repetitive rounds for receiving feedback from wide-ranging stakeholders in order to have an accurate in development plan and measuring the efficiency of "performance of the design" before and after the project.

Table 16.3 Post-occupancy evaluation measures

Main findings	The example base case	Middlesex hospital ED
• Access and circulation	• Circulation areas are represented along two clear routes. Access control is utilized to mitigate public from entering sensitive room functions (NHS 2004)	• The department needs to change the use of entrance to mitigate public surge shown in layout (NHS 2004) • Although is a strong segregation between the "Major Accidents Unit" and the "Urgent Treatment Unit", fitting in their functions, staff or patients who need to travel between the units will be forced to travel a long distance across the reception and main waiting area. A direct link between the two units could solve the observed circulation complexity (NHS 2004)
• Way finding	• The main entrance directs patients straightforward to the front desk. Way finding to treatment rooms is clear and easy (NHS 2004)	• The reception area is well positioned by the main entrance, to support way finding for new arrivals and visitors (NHS 2004)
• Waiting areas	• ED layout design fortifies a strong surveillance in waiting areas (NHS 2004)	• Waiting areas are poorly surveyed, which put ED under the risk of crime especially in medication stores and reception areas • The waiting area seating capacity is beyond departments demand (NHS 2004)
• Triage, assessment and treatment room\s compatibility	• The capacity and flexibility of assessment and treatment rooms in the plan may not be enough to accommodate future demand or changes in use (NHS 2004)	• There is no triage in ED. Instead, a "conductor" will assess and guide patients to either "Major Access "or "Urgent treatment Units" (NHS 2004)

(continued)

- It is important to involve stakeholders since early design process (including strategic, operational and spatial procedures), to ensure correct decisions. The processes should record relevant to the time and detail defined in "the business plan".
- The design team requires detail from a variety of sources at suitable project stages.

Table 16.3 (continued)

Main findings	The example base case	Middlesex hospital ED
• Privacy and dignity	• Treatment rooms are fully enclosed which will provide audio and visual privacy for patients. Moreover, patients' details are taken within this room rather than at reception, this will further enhance the privacy for patients (NHS 2004)	• Surveillance for circulation areas and waiting zones provides privacy in both terms of audio and visual aspects (NHS 2004)

Source conducted by researcher from retrieving (NHS 2004)

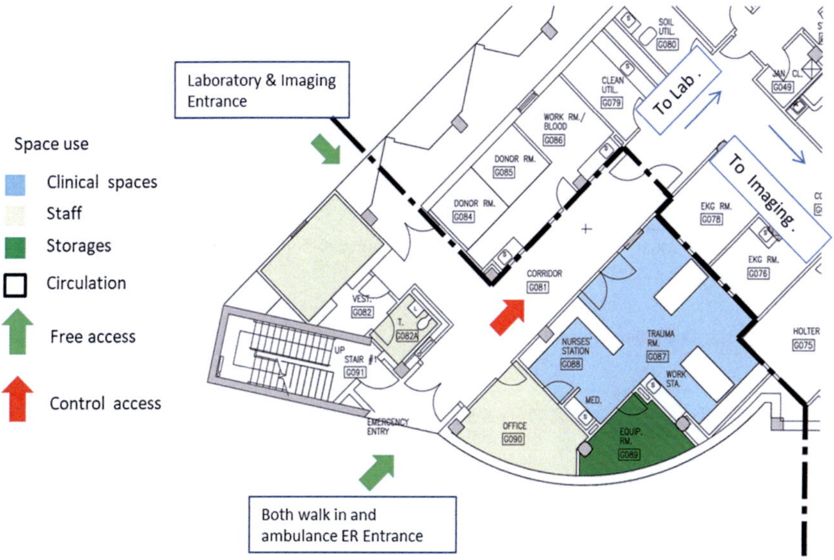

Fig. 16.7 Dar Al Fouad ED 6th October plan view. *Source* Conducted by researcher form main consultant (Dar Al Handasa Shaer and partners)

- Design briefing should be informed by a combination of "quantitative" and "qualitative" information. If that happens, measures that identify the requirements for "accommodation", "adjacencies", "operational processes" and "spatial qualifications" should be sketched reflecting the stakeholders' spectrum.

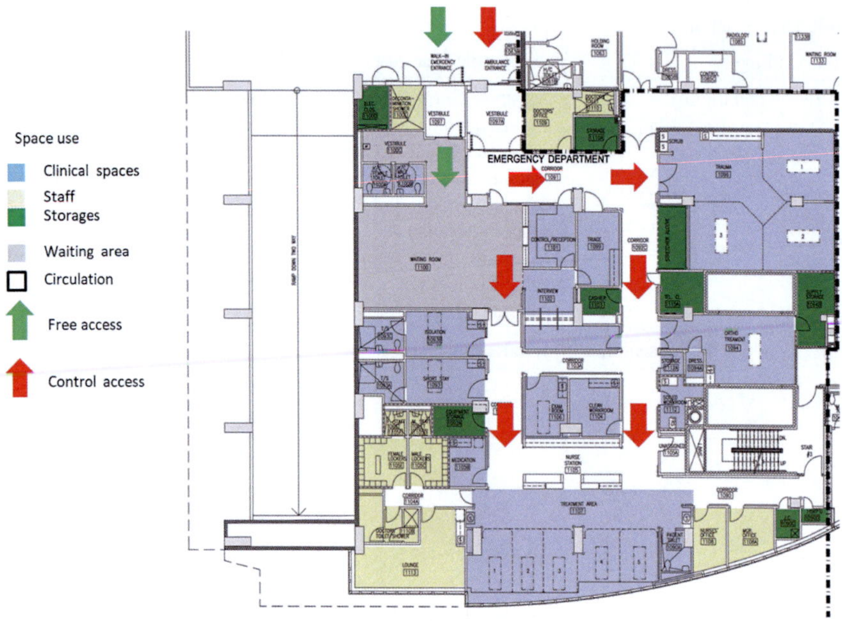

Fig. 16.8 Dar Al Fouad ED Nasr city plan view. *Source* Conducted by researcher form main consultant (Dar Al Handasa Shaer and partners)

Table 16.4 Post-occupancy evaluation measures

Main findings	DAF 6th October branch	DAF NASR city branch
• Access and circulation	• No segregation for walk in or ambulance entrances.Both from same entrance • Crowd on streaming corridor • The entrance area does not support the location of ER facility (on Desert road)	• Segregation of entrances between walk in and ambulance entrances • Yet both entrances are in a very close distance which is negatively affecting access control and cross circulation between highly acute cases entering ER by ambulance and relatively lover acuity walk-in patients
• Way finding	• Having one entrance for all is not supporting clear way finding	• Good placement of main entrance to provide clear way finding especially for first time visitors
• Waiting areas	• N\A	• Strong surveillance of waiting areas and tackling, the risk of crime especially in sensitive locations

(continued)

Table 16.4 (continued)

Main findings	DAF 6th October branch	DAF NASR city branch
• Triage, assessment and treatment room\s compatibility	• N\A (EHFG 2005 does not mention the triage room is obligated.) • Trauma cubicles are used only to stabilize urgent upcoming caser • Trauma rooms are less surveyed by nurse station	• Good surveillance by staff and flexible patient flow through triage and treatment rooms
• Privacy and dignity	• Privacy in trauma is very poor	• Poor privacy of patients in treatment areas by means of curtain cubicles
• Infection control	• No isolation room was observed	• Infection control is achieved via isolation rooms

Source conducted by researcher from site visit

Table 16.5 Front-end planning, design checklist. *Source* Conducted by researcher

Pre-design checklist	Comments
• What are the project performance efficiency benchmarks? Define similar facilities as case comparisons?	
• Will you test the design contribution in shaping to complement the patient's journey within ED?	
• Are the future operational requirements of the project facilities, as well as equipment, technology and infrastructure, measured against an "on field realization plans"?	
• Patient, staff and career experience: is the architect consulting with patient liaison groups?	
• What are the "Activity Analysis" that will demonstrate capacity of high dependency to low dependency causality rooms to flow between acuity? • Will you identify the pattern of "Occupancy Load" for ED per year? As well as identifying pattern of number change per 5, 10 and 25 years?	
Design checklist	Comments
• Are room areas based on detailed information about activity analysis conducted from pre-design stage? • Single-patient accommodation for decision-making (observation rooms) to maintain privacy and noise control • Does the design concept support standardized, acuity and speciality adaptable rooms were utilized to enhance patient experience and raising staff efficiency?	
• Does the design embrace the future changes for expansion within the actual flow, resilience and operational demands?	
• Is the support facilities in the ED included in "design discussions" (e.g. psychological consultation, children care and bariatric requirements)?	
• Does the design concept support "Same Handed rooms" module to reduce nursing medical errors?	

(continued)

Table 16.5 (continued)

Pre-design checklist	Comments
• Do spaces provide sufficient clearance (e.g. wide corridors, space between furniture) for wheelchair use?	
• Are there visible and accessible facilities that support physical activity (e.g. well-maintained walking paths, biking trails, bike racks, exercise stations, playgrounds) in outdoor areas?	
• There are TVs, brochures, art and other visual displays of age appropriate health education materials including healthy lifestyles, healthy food, etc	

Source Conducted by researcher

References

American Institute of Architects (AIA). (2014). Guidelines for Design and Construction of Health Care Facilities. Dallas, TX: Facility Guidelines Institute (FGI).

Australasian Health Infrastructure Alliance AHIA (2010) Australasian health facility guidelines HFG 4, Australia. Retrieved from: https://www.health.nsw.gov.au/utilities/disclaimer.asp

المعايير التصميمية للمستشفيات والمنشآت الصحية ج1(مكونات عامة مركزية - خاصة ومتطلباتها603 / 2010) . المركز القومى لبحوث الاسكان والبناء وزارة الإسكان والمرافق والتنميه العمرانيهEHFG .

Becker F, Jones-Douglass S (2008) The ecology of the patient visit: physical attractiveness, waiting times, and perceived quality of care. J Ambul Care Manage 31(2):124–137

Health building Notes 15 (HBN 15) (2013) Accident and emergency departments. Planning and design guidance, National Healthcare States, United Kingdom

Japan International Cooperation Agency (JICA) (2017) Survey in Arab Republic of Egypt final report, Arab Republic of Egypt ministry of health and population egyptian ambulance organization, health sector in cooperation planning with, Fujita Planning Co., Ltd.

Sigma Theta Tau International (2010) Evidence based design for healthcare facilities. In: McCullough C (ed) RN, MSN, Indianapolis, IN 46202 , US

National health care states NHS (2004) A&E design evaluation; evaluation of two proposed accident and emergency departments: brent emergency care and diagnostic centre at central middlesex hospital, and an exemplar plan, Wales. Retrieved from info@whe.wales.nhs.uk

The Center for Health Design (CHD) (2010) An introduction to evidence-based design, exploring healthcare and design Part 1. Retrieved from www.healthdesign.org/blog/

The Pulse 7th edn (2017) Egypt Healthcare, Colliers International Research. Retrieved from colliers.com/egypt

Part III
Smart Cultural Heritage

Chapter 17
Heritage Building Information Modelling: Towards a New Era of Interoperability

Noran Elbaz, Shaimaa Kamel, and Sherif Abdelmohsen

Abstract Heritage Building Information Modelling (H-BIM) is an application of Building Information Modelling (BIM) in the field of documentation, preservation and managing of historic sites. The universal value of cultural heritage made a huge demand on applying such technology in this field, such important value needs an innovative system in data management for the documentation of historic buildings, where a historic building 3D model stands on heterogeneous data sets and should be interoperable with different software tools in order to handover the information to several users. The interoperability is defined as the ability of two or more systems to exchange information, data and knowledge. This paper outlines the importance of interoperability in optimizing the usage of data; by reducing wasted time and effort in gathering, translating and integration of data, and how it gives a wider dissemination which allows public users to benefit from the documentation and conservation process for a deeper understanding of heritage sites; outlying and reviewing the case of Notre Dame Cathedral in Paris, and how the Heritage Information Modelling could provide a precise restoration documents form; being interoperable with game engines.

Keywords H-BIM · Historic buildings · Interoperability

17.1 Introduction

The cultural heritage configures the collective memory of humanity and relates the past with the present. It acts as a national income source, creating a growing need for

N. Elbaz (✉) · S. Kamel · S. Abdelmohsen
Department of Architecture, Faculty of Engineering, Ain Shams University, Cairo, Egypt
e-mail: elbaz_noran@yahoo.com

S. Kamel
e-mail: shaimaa_kamel@eng.asu.edu.eg

S. Abdelmohsen
e-mail: sherifmorad@aucegypt.edu

S. Abdelmohsen
Department of Architecture, The American University in Cairo, Cairo, Egypt

Architecture and Urbanism: A Smart Outlook,
https://doi.org/10.1007/978-3-030-52584-2_17

optimum documentation and preservation of the heritage sites, using latest technologies. Digital documentation provides a wider dissemination through audio-visual tools, while earlier process of documentation was based on human interpretations that has been wasting time and giving space for human error. This paper highlights the main problem of the lack of synchronization of data which duplicates the effort and multiplies the error (Juan and Zheng 2014). The Heritage Building Information Modelling offers an innovative interoperable platform that provides an upgraded level of data management and data utilization, resulting in a better interactive environment of the historic sites and more advanced and accurate display of the documentation of the historic buildings (Murphy et al. 2017).

17.2 Methodology

This paper outlines the potential of interoperability of H-BIM in the visualization and dissemination of historic buildings and production of restoration documents. Section one gives a literature background on how H-BIM offers a new approach for heritage preservation, while the second section reviews the concept of Heritage Building Information Modelling and its related terminologies. Section three outlines the idea of interoperability and how it could provide wider dissemination. Ending with section four, which thematically analyses the case study of Notre dame Cathedral as an example where the benefits of H-BIM could be implemented in historical restoration and heritage conservation. The findings of this paper elaborate the interoperability benefits in the restoration of remarkable historical landmarks and how we can reverse the conservation process of historic buildings to produce restoration documents from game engine platforms as one of the examples of the potentials of interoperability of H-BIM.

17.3 Heritage Building Information Modelling

The Heritage Building Information Modelling is an innovative environment of managing data of historical building conservation process; it is represented as an intelligent method (Murphy a et al. 2017). The H-BIM model is built by a library of parametric objects that contains the semantic and geometrical information of the historical elements. This intelligent process also has the potential of being interoperable with other digital platforms involved in the building's life cycle (Simeone et al. 2014). Parametric modelling provides instant update of elements with all involved teams in that process. The parametric library is also reusable, which saves time and effort used to be wasted in duplicated work (Murphy et al. 2019). The H-BIM system turns to be a reusable, interoperable database that enables managing, analysing and representing historic buildings. Researchers from different disciplines such as architecture, computational architecture and archaeology are working on optimizing the H-BIM by creating an open H-BIM system, with automated object recognition (Murphy et al. 2019).

17.4 Concept Design and Initial Level of Detail (LOD)

The H-BIM system requires a new definition of progress, as it totally differs from conventional CAD systems. A new term of LOD referring to level of detail is introduced. It classifies the model category based on the amount of detailing the model; for example if LOD 100 is a low detail level, it may include only the masses of the building; however, LOD 400 represents a high level of detail approaching to an as built model (Murphy 2012).

17.4.1 Architectural Libraries

The architectural digital libraries reflect the level of detail of H-BIM model, however, the BIM library is oriented towards the modelling of modern components; however, a new plug-in for BIM was introduced by Murphy for the complicated elements of historic buildings (Murphy a et al. 2017). The coding approach offers accurate mapping onto survey from laser scanning. The detailing can be ranged from architectural pattern books, to laser scanning and point cloud. The H-BIM system offers a deeper documentation and analysis that goes beyond the geometry, to the methods of construction and materials. Figure 17.1 shows the different parametric library objects built from pattern books and shape grammars in GDL (Murphy et al. 2017).

17.4.2 Shape Grammars

Shape grammars applications initiated forty years ago, working on different areas of design, especially architectural design. However, it has been developed in the last ten years to integrate shape representation and computation (Eloy et al. 2018). The shape grammars are sets of rules responsible for the composition of building elements and the relation between them within spaces. In H-BIM, the shape grammar can be based on architectural style, which configures the building's spaces. However, it gives the potential of transforming or deforming space by replacing or changing shapes (Murphy et al. 2019). The semantic aspect of H-BIM model upgrades the model from static representation of buildings to an interactive mode; these interactive results can be traced in a multidisciplinary collaborative environment. This open system saves time, effort and cost (Murphy et al. 2019).

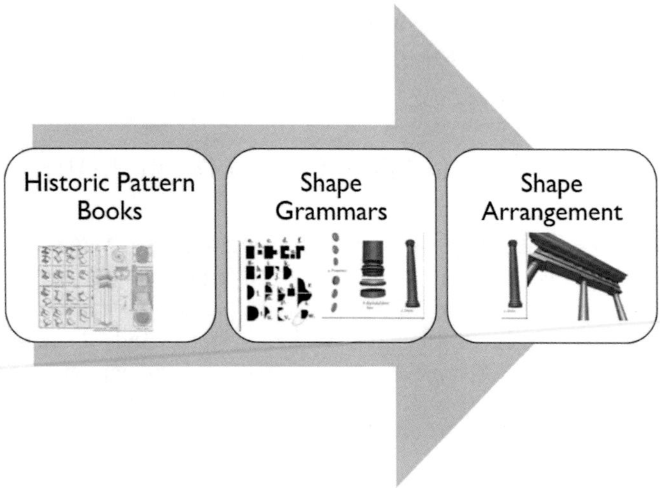

Fig. 17.1 Parametric library objects built from pattern books and shape grammars in GDL (Murphy et al. 2017)

17.4.3 Procedural Modelling

The potential in H-BIM is in creating an automated procedure, in order to generate high-quality 3D models of existing historic buildings to overcome the high cost and effort needed for the modelling and conversion of the semantic aspect of buildings (Murphy et al. 2019). It became the focus of many researches, in which it gave a promising result in extracting basic architectural elements from point cloud data. However, the researchers are still working on complicated historic elements. The procedural modelling refers to the automation of the process of generation and transformation of data from one form to another. For now, the conversion is carried out manually (Albourae et al. 2017).

17.5 H-BIM and Interoperability

Interoperability can be defined as the ability of two or more systems to exchange information (Santos 2009). The upraising potential of H-BIM is opening the system to web-based files which give a wider dissemination of historical buildings documentation. The idea of interoperability became an important research field nowadays (Murphy et al. 2019). Two of the main points in this field are increasing the automation of the process of producing H-BIM model to eliminate human intervention and conversion of semantic H-BIM data into web repository to transform to an interactive, smart and user-friendly interface (Quattrini et al. 2017). Figure 17.2 shows the architecture of the system of Archival and storage repository (Murphy et al. 2019).

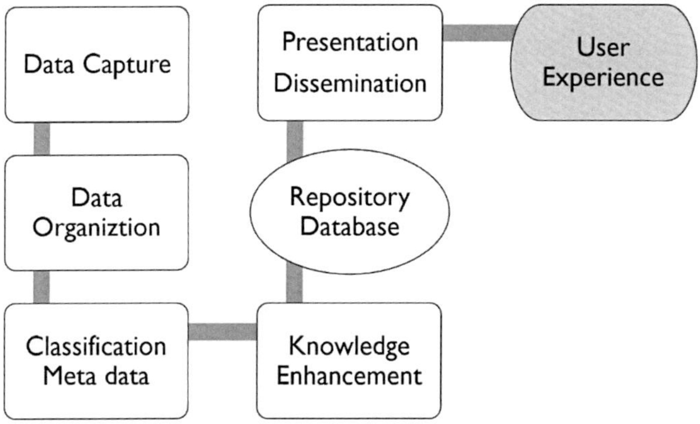

Fig. 17.2 Architecture of the system of archival and storage repository (Murphy et al. 2019)

17.5.1 Towards Wider Dissemination Through H-BIM and Game Engine Integration

Architectural visualization is one of the main outputs of H-BIM. However, it is most likely limited only to specialists (Murphy et al. 2019). Unlike game engine platforms, which are more user-centric, the integration of both Heritage Building Information Modelling and game engines is the upcoming upgrade of the user experience of heritage sites (Albourae et al. 2017). The game engines offer virtual reality (VR) which is a simulated environment that enables users to experience heritage sites without visiting them. While the augmented reality (AR) as a system that allows the integration between reality and computer model aligned together by the Geographic Information System (GIS), this great potential can store and visualize geographic data, in order to experience heritage sites in an interactive way, or virtually reconstruct a specific time in history that helps public users in understanding heritage more in-depth (Albourae et al. 2017), in order to provide a user-centric experience to be used as a learning tool in pedagogical approaches (Wua and Kaushik 2015). The game engines are also used for combing visuals and audio in virtual reality, which riches the experience of spaces (Postma et al. 2016). Figure 17.3 shows interoperability tests for data visualization in AR and virtual reality (VR) (Osello et al. 2018).

17.6 Notre Dame Cathedral Case Study

Notre Dame Cathedral is a famous and remarkable heritage landmark, located in Paris, France. The construction started in 1163 and was not completed until 1345; it was built by Gothic French architectural style. Notre Dame means ‘Our Lady’ in English, which refers to Virgin Mary, the cathedral witnessed most of the epic

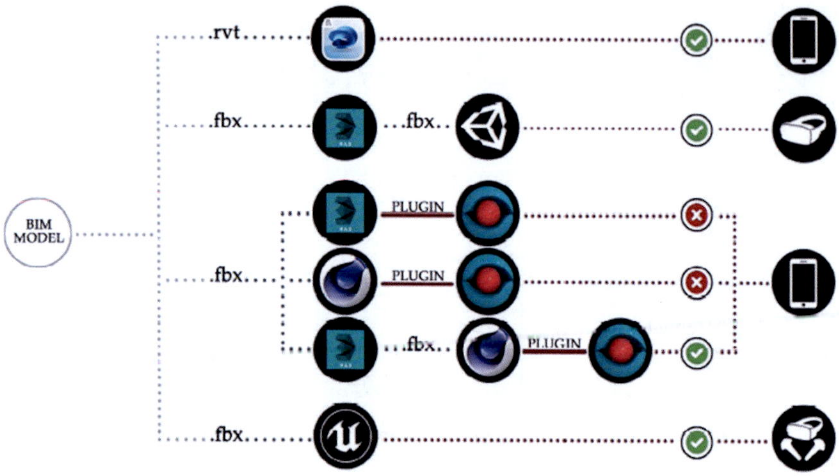

Fig. 17.3 Interoperability tests for data visualization in AR and virtual reality (VR) (Osello et al. 2018)

French history and contains revered relics as well (BBC News 2019) (Notre Dame Cathedral, Paris 2019). The cathedral roof was damaged by fire in April 2019, and restoration action plans of the cathedral became now a main issue in France and a part of the international heritage concern. The common process of restoration from available documents and manuscripts does not provide an accurate restoration document. H-BIM presents another approach based on the interoperability, in integration with game engines; the cathedral was digitally built in a game called Assassin's Creed Unity in 2014 and was modelled over a period of 14 months by Obisoft (BBC News 2019); Fig. 17.4 shows a snapshot from the game, where the Notre Dame Cathedral appears at the background.

Fig. 17.4 Snapshot from the game of Assassin's Creed Unity model showing Notre Dame Cathedral

Fig. 17.5 a Picture of the Notre Dame de Paris Cathedral from the altar towards the organ. b Similar view in the Blender VR model (Postma et al. 2016)

There was another digital representation of the cathedral the Ubisoft's, and yet is not the only one. Another digital representation was made by the Vassar College historian Andrew Tallon, and it was based on more than 50 locations laser scanning that contains more than a billion data points (BBC News 2019). The game engines are used in combining visuals and audio as mentioned before. There is also another model of the cathedral generated on a 3D MAX visualization platform and exported to Blender game engine. It was combined by an acoustical software creating computational acoustic model to study auralization in virtual reality performance, as shown in Fig. 17.5 (Postma et al. 2016).

The relation between Heritage Building Information Modelling and Game Engine is a two-way data transferring channel (Edwards et al. 2015), as shown in Fig. 17.6. Here stands the benefit from the interoperability of H-BIM, integrated with game engine in providing a wider interactive dissemination, with a user-friendly approach; the process could be reversed to create an accurate restoration documents from game engine models. Figure 17.7 shows a concluded summary of the potential of H-BIM, in which it reverses the conservation process of historic buildings to produce restoration documents from game engine platforms.

17.7 Conclusion

The Heritage Building Information Modelling has the capability of being interoperable with different platforms, which saves effort and time and provides wider dissemination through the integration of game engines platforms, through a common data repository that acts as two-way transfer channel. As a result of wider dissemination,

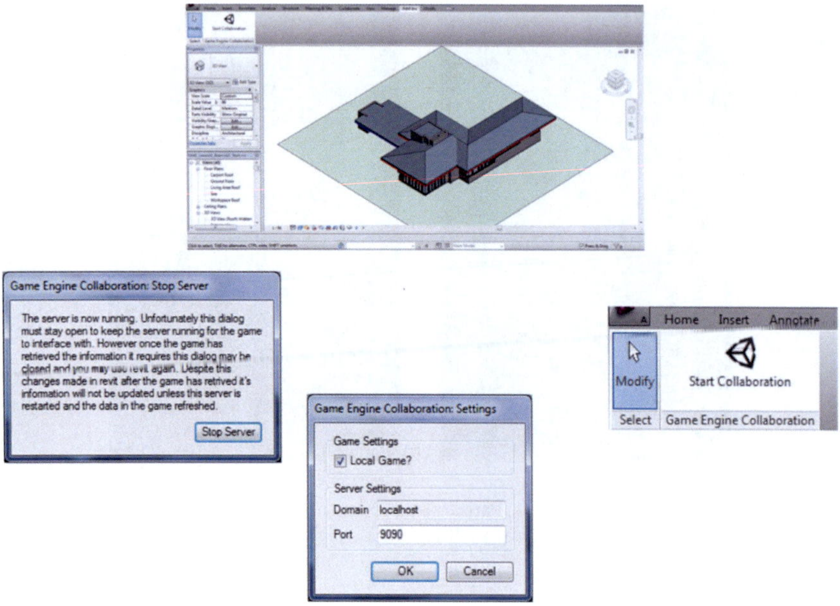

Fig. 17.6 Starting the server to create the link from the game to Revit (Edwards et al. 2015)

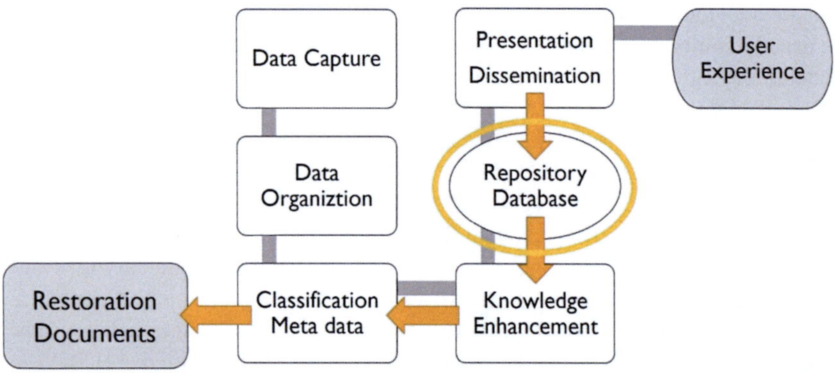

Fig. 17.7 Potential of H-BIM to reverse the conservation process of historic buildings to produce restoration documents from game engine platforms

a better user experience and a deeper understanding of cultural heritage are now available for public users.

The Notre Dame Cathedral as a case study proves that interoperability of H-BIM system is an added value in the conservation of cultural heritage, which in this case could have been lost totally or partially; but with the new conception of H-BIM, the process of conservation could be reversed to start from the game engine platform to extract the restoration documents for the roof of the cathedral in an easy and accurate process. H-BIM is the new era of interoperability of heritage conservation.

References

Albourae AT, Armenakisa C, Kyan M (2017) Architectural heritage visualization using interactive technologies. The international archives of the photogrammetry, remote sensing and spatial information sciences, vol. XLII-2/W5. Ottawa, Canada: 26th International CIPA Symposium 2017

BBC News (2019). Retrieved from https://www.bbc.com/news/world-europe-47946675

Edwards G, Li H, Wang B (2015) BIM based collaborative and interactive design process using computer game engine for general end-users. Vis Eng

Eloy S, Pauwels P, Economou A (2018) Advances in implemented shape grammars: solutions and applications. Artif Intell Eng Des, Analysis and Manufacturing

Juan D, Zheng Q (2014) Cloud and open BIM-based building information interoperability research. J Serv Sci Manag

Murphy M, Corns A, Cahill J, Eliashvili K, Chenau A, Pybus C, Shaw R, Devlin G, Deevy A, Truong-Hong (2017) Developing historic building information modelling guidelines and procedures for architectural heritage in Ireland. Remote sensing and spatial information sciences. 26th international CIPA symposium, Ottawa, Canada

Murphy M, Pavia S, Cahill J, Lenihan S, Corns A (2019) An intial design framework for virtual historic dublin. 2nd international conference of geomatics and restoration, Milan, Italy: The international archives of the photogrammetry, remote sensing and spatial information sciences

Murphy M (2012) Historic building information modelling (HBIM). Dublin university, Dublin

Notre Dame Cathedral Paris (2019). Retrieved from https://notredamecathedralparis.com/

Osello A, Lucibello G, Morgagni F (2018) HBIM and virtual tools: a new chance to preserve architectural heritage. Buildings

Postma BNJ, Poirier-Quinot D, Meyer J, Katz BFG (2016) Virtual reality performance auralization in a calibrated model of notre-dame cathedral. EuroRegio, Porto, Portugal

Quattrini R, Pierdicca R, Morbidoni C (2017) Knowledge-based data enrichment for HBIM: exploring highquality. J Cult Heritage

Santos ET (2009) Building information modeling and interoperability. SIGraDi

Simeone D, Cursi S, Toldo I, Carrara G (2014) BIM and knowledge management for building heritage

Wua W, Kaushik I (2015) Design for sustainable aging: improving design communication through building information modelling and game engine integration. Int Conf on Sustain Des Eng Constr

Chapter 18
The Christianization of Ancient Egyptian Cosmology: Philae in Late Antiquity

Rasha Said Abd el-Lateef, Yasser Mansour, Shaimaa Kamel, and Ruby Morcos

Abstract Abandonment, destruction, being a source for retrieving building materials, and the reuse, these were multiple destinies that ancient traditional temples reached in Late Antiquity, accompanying the religious paradigm shift from polytheism to monotheism. The First Cataract in the frontier zone between Egypt and Nubia has a significant ontological value to ancient inhabitants, studying the fate of its temples cannot be done without approaching the significance of the sacred landscape, a significance that is perceived in the topographical and astronomical features representing the cosmic impact on the Egyptian mythology and theology. This paper will focus on the Ptolemaic temple of *Isis* on Philae Island, as the last temple to be closed in Egypt, declaring the absolute domination of Christianity. Starting by setting the historical background explaining the complexity of the regional situation politically and culturally. Accordingly, trying to interpret the temple transformation as a unique phenomenon reflecting the public ontological understanding of the sacred power, and its points of access, in spite of changes in the perceptions of religion. The objective is to bring to light a new perspective for the transformation of traditional sacred spaces, and how transformation can be a manifestation of the localization of a new faith. An assumption is based on applying Mircea Eliade's phenomenological analysis of the religious experience. The research is using the tools of ethnographic descriptions, iconographic representations, archeoastronomical studies, and archeological examinations, for applying a thematic analysis, within the hermeneutic phenomenology as a methodology.

Keywords Philae · Cosmology · Sacredness · Christianizing · Late Antiquity

R. S. Abd el-Lateef (✉) · Y. Mansour · S. Kamel · R. Morcos
Faculty of Engineering, Ain Shams University, 1 El Sarayat S, Abasseya, Cairo 11535, Egypt
e-mail: arch.rasha.said@gmail.com

Architecture and Urbanism: A Smart Outlook,
https://doi.org/10.1007/978-3-030-52584-2_18

Fig. 18.1 Complex in its new location "Agilkia Island" (Kockelmann 2012)—The metaphysical landscape of the island must be understood in the original location before the high dam construction—the land emerging from the primeval water (flood) and the creation begins as fertility brought to Egypt through stones belt (Antique wood engraved print of the first cataract-Unsigned 1890)

18.1 Introduction to the Ancient Mythology

Philae Complex together with the Abaton "forbidden place" on the neighboring island of Biga/Bigeh, were the most important cultic centers of *Isis* and *Osiris* in Upper Egypt and Nubia. Even though the island housed a rich blend of Egyptian and non-Egyptian cults, *Isis* cult and the Osirian faith were the center of the importance of the islands and their temples, and these Heliopolitan mythology gods are religion symbols within a complex system for the traditional coherent vision of the Egyptian cosmos, *Isis* as the manifestation of the flood, and *Osiris* as the god of vegetation, rebirth, and regeneration. The temples are parts of a metaphysical landscape that is linked to cosmic theological myths which borrowed the historical and geographical facts to give a concrete ontological meanings. The sacred landscape has its significant topography as the point where the river begins and the surge of the sacred Nile is sensible. Here, the cosmic cycle with its static rhyme reappears over and over again, a regeneration of time connected with the essential agricultural cycle of the Egyptians' universe, marking this sacred landscape in the heart of the traditional Egyptian doctrine of creation, and *Isis* Temple by this era was claimed to be the center that existed before everything (see Fig. 18.1).

18.2 The Complex Nature of Philae as a Frontier Zone

Philae is part of the *Dodekaschoenus* (Bierbrier 2011) which is "encounter region" (Rutherford 1998, p. 241), or a frontier zone that had different types of authorities taking over through history. Its original Nubian name is "*Takompso*" as interrupted as "the island of time" or "the island of the past," and the Greek used the Latin

"Philae" derived from Egyptian name "*Pilak*" as friendship (Kockelmann 2012). This explains the importance of this point to different ethnic groups.

The first evidence of religious activity on the island goes back to the seventh century BC by the Ethiopians; the island was not consecrated to *Isis* then, and the Egyptian denomination on the island is attested since the fourth century BC (Kockelmann 2012). The temple of *Isis* dates back to the sixth century BC, that old sanctuary was not an important cult center. While the temple of *Khnum* on Elephantine was the religious center of the First Cataract, the old sanctuary was re-employed with the Ethiopian *Taharqa*'s construction remains as filler of the remaining Ptolemaic temple during the second and third centuries BC (Kockelmann 2012) to become a main Ptolemaic–Roman cult center of *Isis* and *Osiris*. As observed, the central power point of the sacred region of the First Cataract moved over time from Elephantine Island to Philae with the rise of *Isis* cult versus *Khnum*'s. Yet, both reflected the cosmic value of the Nile River—which will be discussed further in the next points.

The complex on Philae included many temples; a birth house "*Mammisi,*" temple of *Arensnuphis* at the southern end of the island, temple of *Imhotep*, all added by the Ptolemies, temple of *Hathor* (to the east of the sanctuary) completed by Augustus, Roman Trajan added his Kiosk, and Hadrian added a monumental gate at the west of the sanctuary as a staging post for the Abaton (Rutherford 1998, p. 233) (see Fig. 18.2).

Bigeh is in the opposite of Philae and related to it ritually; a small rugged granite rocks island and the south most of the islands of the first cataract, it houses the *Abaton*, the burial place of *Osiris-Serapis* (Rutherford 1998, p. 232). What remains of the temple today is just the façade of the outer hall from the Ptolemaic–Roman period (Blackman 1915, p. 1) (see Fig. 18.3).

By the mid of the third century AD, it is not clear who had the most control on the islands, after Diocletian moved the border of the Roman Empire to Syene "Aswan",, there was great Ethiopian and Meroitic influence, and the situation of Philae was complicated until the fifth century (Rutherford 1998, p. 234). With the spread of Christianity and the decline of the traditional temples, *Isis* Temple was the last to be closed (Spencer 2007, p. 83), after it housed annual mixed pilgrimages and festivals until the sixth century AD, it housed a Christian outpost since the fourth century AD (Frankfurtar 1998, p. 64), it was understood that Christians and non-Christians lived side by side indicating that this part of landscape enhanced a sense of common identity, significance and sacredness among all different groups (Rutherford 1998, p. 230).

18.3 Reflections on the Symbolism of Philae Complex

Mythically, the flooding was manifested in the primeval waters "*Nun*" and linked to the underworld or the house of *Osiris* the "*Duat*", and the inundation water was worshiped and feared in the same time. The celestial manifestation of *Isis* is *Sopdet* "Sirius" the Harbinger of Flooding (Belmonte et al. 2010, p. 536); it marked the

A. Gate of Nectanebo/1st Pylon which is the oldest part of the complex

B. Temple of Arensnuphis – a cult association on the island whose worship is rarely attested further north

G. Temple of Hathor – a Ptolemaic temple, later in the mid-first century CE, the cult of the temple focused on Hathor as embodiment of the Sun's Eye that was brought back from Nubia.

M. Great Temple of Isis

R. Gate of Diocletian

I. "Gate of Hadrian": the gate was located opposite the mooring chapel on Biga and served as a departure point for the bark procession of Isis to the tomb of Osiris on the abaton.

O. Temple of Harendotes

Y. Eastern Colonnade

D. Chapel of Mandulis (a Nubian divinity) - K. Kiosk of Trajan

Z. Western Colonnade: (32 columns and 12 openings in the rear wall was decorated under Roman emperors serving as a pronaos of the sanctuaries of the abaton - P. Western Church

F. Temple of Imhotep (dedicated by Ptolemy V)

L. Prostyle Temple Augustus (Roman-style) - Q. Eastern Church

Fig. 18.2 Philae Island before relocating the complex (Kockelmann 2012—originally published by Lyons in 1896)

Fig. 18.3 Philae and Bigeh Islands. Painting in 1838 AD by David Roberts (Heyde 2007), Painting by Louis Haghe after David Roberts, 1849 (Fæ 2014)

Fig. 18.4 *Khnum* creating on potter's wheel (Wilkinson 2003, p. 194)—*Isis-Sothis* standing in the baroque crossing the heavens—Roman period (Wilkinson 2003, p. 168)—double depictions of *Hapi* "spirit of the Nile "representing upper and lower Egypt—The colossal statue of Ramesses II—Luxor Temple (Wilkinson 2003, p. 107)

flooding of the Nile while *Osiris* and the burial rituals were identified with Canopus (Coppens 2004). After the celebration of the birth of *Isis* comes the advent of the appearance of *Sopdet* on the fourth of the Epagomenal days (Apuleius 1975), and this meant that, *Khum,* the creator god will open the gates of his cave releasing the torrent of water fertilizing the fields as *Sopdet* arrives on the first day of *Tekh/Thot*—later *Thout* in Coptic (Canhão 2013) (see Fig. 18.4).

The New Year is celebrated for the regeneration of time and keeping the order, a great cosmic cycle completed and a new one beginning (Holberg 2007, p. 4). The myth of creation is recited in connection with agricultural activities, every practice is part of the sacred and has a mythical meaning (Eliade 1991, p. 22–25), and the beginning of the year is *Akhet* "the season of Inundation", for four months followed by *Peret* "growth" and *Chemu* "harvest" (Wilkinson 2000, p. 98) representing the cosmic cycle of "rebirth", "transformation" and "becoming". Later, the Coptic calendar kept the ancient cosmic one (Gabra 2008, p. 70).

Recognizing the manifestations of the myths of this sacred landscape is reflected in the orientation of the temples, it was presumed that *Isis* Temple was oriented south welcoming visitors from Kush/Nubia (Rutherford 1998, p. 233), but according to the accurate advanced measuring the main axis was oriented to the setting of Canopus, while the perpendicular one for the temple of *Hathor* was with the declination of *Sopdet* "Sirius" (Belmonte et al. 2010) (see Fig. 18.5). As they are the brightest stars of Egyptian sky and linked mythologically to the Nile cycle, this alignment adds a new underlayment, the terrestrial (topographical) and celestial (astronomical) emphasis on the cosmic significant value of *Isis* and *Hathor* Temples.

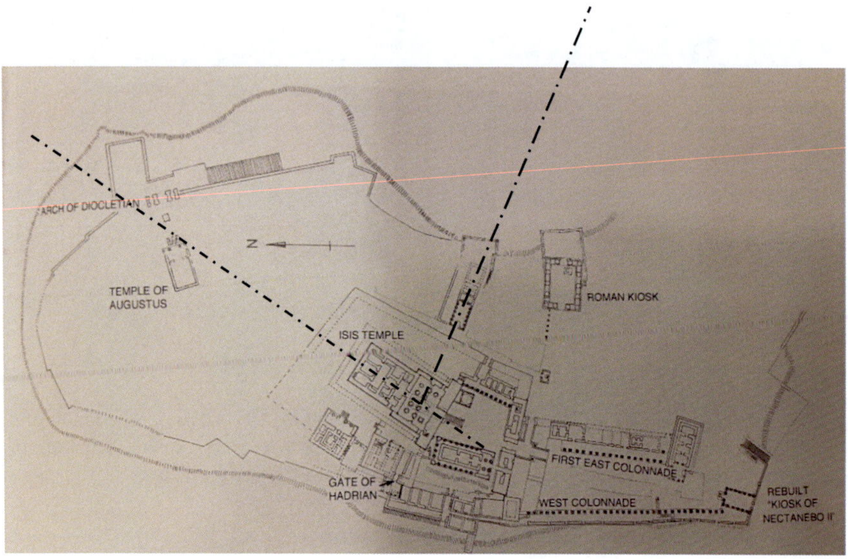

Fig. 18.5 Complex temples during the Roman Era (McKenzie 2007, p. 129)

18.4 Pilgrimages, Festivals, Rituals and Understanding the Metaphysical Position

Archeological evidences indicated the mixed heritage of Philae. The priesthood of Philae in the fourth and fifth centuries AD was a family of mixed Egyptian and Blemmye "Nomadic tribal Nubia third century AD", it acted as an axis for numerous traditional powers, and it held a regional sacredness with pilgrimage all year around. The temple walls are covered in inscriptions highlighting the significance to the Egyptians and the sacred texts spoke of the transportation of *Isis* to the *abaton* (on Biga) during *Osiris* festival (Frankfurter 1998, pp. 105–106). There were plenty of Egyptian graffiti specially in the birth house (Rutherford 1998, p. 235), some by locals and others records of Pilgrimages (Rutherford 1998, p. 240), and this confirms the position of Philae as a center of traditional Egyptian religion. From another side, the Meroitic Pilgrimage to Philae was old, it survived for long time, and the cult of *Isis* was well established in Ethiopia; the demotic inscriptions indicated both Ethiopian Pilgrimages and Meroitic Pilgrimages "from Meroe the capital of Kush". Since the first century AD until the third century AD (Bumbaugh 2009), the celebrating of the "festival of entry" is well documented (Rutherford 1998, pp. 242–243), and the festival included ferrying the shrine of *Isis* across the river to Biga Island (Bumbaugh 2009) (see Fig. 18.6).

Different types of Pilgrimages were the Greek "Ptolemaic" and later in time the Roman ones (Rutherford 1998, pp. 236–240), visitors left *proskunemata* "prayers of praise" showing to which degree the Egyptian priests were involved in

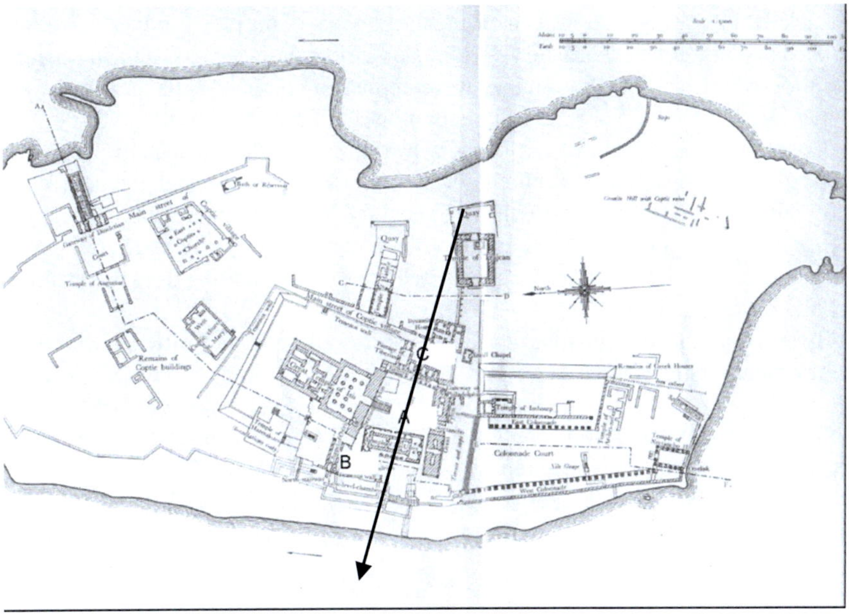

Fig. 18.6 Meroitic worship of *Isis* (Bumbaugh 2009)—(A) the Birth House where most inscriptions in both Demotic and Meroitic are concentrated—(B) the Gate of Hadrian leading from the *Osiris* chapel toward the point of departure for Biga Island—(C) Meroitic chamber

Hellenism surrounding of *Isis,* becoming a *pantokratrix* (a world ruler) or a universal goddess(Keegan 2014, p. 92). Yet, the very nature of *Isis* remains Egyptian in her local context with fertilizing capacity and *Isis*'s position as a Nile cult, this can be understood as there is no evidence that *Isis* here was approached as a healing deity (Rutherford 1998, p. 253) like *Isis* Temple in *Menouthis* (Alexandria)—where *Isis* was replaced by a Christian healing cult (Frankfurtar 1998, p. 47).

The last Demotic inscriptions in Philae—identified as pagan—dated to A.D 452. While the bishop of the island was created in the first half of the fourth century (Moawad 2013, p. 28), the first bishop of Philae *Macedonnius*- "probably part of the monastic community" asked for a church on Philae where the few Christians there can celebrate the Eucharist (Moawad 2013, pp. 29–32), after him the election of Coptic bishops continued until the fifth century, Emperor Justinian ordered the official closer (535–537 AD) followed by the transformation of the temples into a churches (Dijkstra 2011, p. 423) and the campaigns to Christianizing Nubia started (Moawad 2013, pp. 28–30).

After Christianizing the complex, at the *Isis* Temple was reported to have two sanctuaries, dedicated to St. Stephen and the other for the Holy Cross, Pilgrimage flasks were found on Elephantine inscribed by the name of St. Stephen and honoring the Holy Cross, there were groves scraping the walls to obtain holly dust as *eulogion* to be carried mixed away with water carrying the blessing of the place, indicating archeological evidence of the Christian Pilgrimage to the region in the sixth and seventh centuries AD, Christian Pilgrimages were argued to continue until the fourteenth century despite the lack of sources in Papyri discussing it (MacCoull 1990).

This means that the importance of Philae was transformed from a main cult center of *Isis* to an episcopal seat, the churches were described as exceptionally large, probability served as the cathedral until the tenth century when Christian establishments in *Syene* (Aswan) supplanted Philae as the center (Adams 1991), which means the religion center of the First Cataract zone shifted again over time starting from Elephantine to Philae reaching Syene.

By investigating the locations of Pilgrimages inscriptions, they differed according to the groups (south Pylon for Greek, *Mammisi* (the birth house) for demotic, Hathor temple for women, etc.). Later inscriptions appeared in the shrine of *Osiris* on the roof of the temple in the fifth century AD, only after Christianizing; the inscriptions appeared in the Pronaos, the creeping of inscriptions closer to the sanctuary identifying the complex as a "contested space" (Rutherford 1998, p. 253) reaching *Isis* temple as the center.

18.5 Christianizing Philae

Mythically, Dijkstra debated in his thesis that *Isis* Temple was not functioning in the fifth century AD, as it was previously assumed, and the official closer of the temple was only symbolic, he explained the transformation of the complex to churches as being done as a practical solution, for abandoned appreciated monuments and not for ideological reasons (Dijkstra 2011, p. 426). While this research considered that the examination the local mythical and cultural heritage cannot be separated from understanding the meaning of transforming the whole landscape, from traditional to Christian, this landscape that its significance was related to the Nile, which represents the Syncretism of Coptic Christianity (Cannuyer 2001, p. 11). As the Nile was a main feature of the Egyptian cultural landscape, it was Christianized as it rose through the power of Christ or by the intercession of the saints (Kakosy 1982, p. 297).

For the timeline, most churches cannot be dated exactly, before the sixth century they were erected outside the temples (Moawad 2013, p. 33).The oldest churches were constructed in the north of the island behind the complex and far from it, the east church (the cathedral church/episcopal church) and the west church, both of them remained under water with most of the Christian monuments (Dijkstra 2011, p. 425), the historical documents referring to the establishment of "God's holy churches at Philae ", was in the fourth century, the east church (with its large size and systematic four naves) can be dated to the sixth century, while the west to the beginning of the seventh century AD with blocks taken from *Harendots* Temple, with inscription in the street leading to it " the *Topos* (monastery) of the lady of us all the holy *Theotokos* Mary on Philae" dated 752 AD, which was interpreted by Grossmann as

Fig. 18.7 Remains of *Harendots* Temple "the Western Church contained blocks from it" (Kockelmann 2012), *Hathor* Temple from the southwest (Kockelmann 2012)

renovation or rebuilding of an older church dedicated to the Virgin Mary (Moawad 2013, pp. 33–34).

The temples were transformed into eight churches as a complete Christianization. The most important church is the one inside *Isis* Temple, which was dedicated to St. Stephen, the transformation was shortly after the official closer of the temple. Another one in *Hathor* and *Arensnuphi* Temples after it, one in front of Caesar Augustus (seventh or eighth century) and a chapel for individual prayers in *Imhotep* Temple (Moawad 2013, p. 35), and there were also some Coptic inscriptions dating to the sixth century inside Trajan's Kiosk (داود2003, p. 208). Inside the "Abaton" on Biga, there was also a church dating back the sixth century (داود2003, p. 200) (Fig. 18.7).

Grossman's reconstruction of the church inside *Isis* Temple made it clear that church was in the Pronaos—avoiding the Naos "holy of holies" where the ancient god lives,—a three aisled church with new orientation on west–east axis (Dijkstra 2011, p. 427). The transformation of the pharaonic temple's Naos into a church with its altar and recess was only completed in the tenth century, and the Coptic Church was abandoned between the eleventh and the thirteenth century (Volokhine 2014) (Fig. 18.8).

For the passage through the pylon gate, there are crosses that have been carved in place of the goddess *Isis*' face, it can be observed on both sides, there are several types that can be dated according to style, the most ancient are surrounded with a kind of halo or are set inside a circle, and the most recent are carved inside a square (Volokhine 2014). The ones in circles were probably for apotropaic purpose ordered by Bishop Theodore (Pagoulatos 1994) (Fig. 18.9).

Crosses in multiple stylistic criteria have been added to the columns and walls of the temple, some represent exorcism rites and other Coptic rituals for the consecration of churches—church pillars represent the apostles, supporting God's house, and are as such consecrated (Pagoulatos 1994), all giving a new birth to a routed traditional sacred space.

The phases of the Christianization process over time combined with getting closer to the holy of holies, this confirms the perception of the temple as a "contested space," engraving the crosses gradually as a new birth for this point of access to power, a

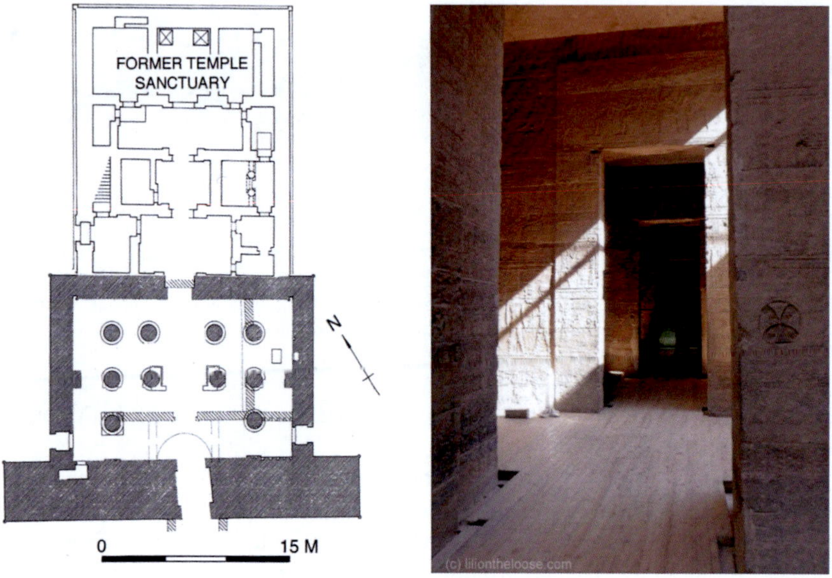

Fig. 18.8 Church of St. Stephen inside the Pronaos of *Isis* Temple (McKenzie 2007)—Passage to Naos (Lili)

Fig. 18.9 Crosses carved in place of the goddess *Isis*' face (Volokhine 2014)—the cross erected in the Naos had won over the demon *Isis*

related power to the cosmic mythology of the holy Nile that *Isis* cult represented it within this whole sacred landscape. Virgin Mary's power replaced *Isis*—which became a "demon"—in controlling the flow of the Nile. For centuries, the Christians had kept the idea that the coming of the Nile was one of her miracles.

18.6 Conclusion

The landscape of the First Cataract was part of the ontological conception of the Egyptian universe or cosmos, the conception was reflected terrestrially (topographically) and celestial (astronomically), demonstrating the meaning of the sacredness of this location. It was at the center of the cosmic order controlling the agricultural cycle with the Nile in the center of it. The interpretation of the transformation of the temples in the First Cataract in general must be understood within that horizon of the traditional spirituality, and the feeling of identity for the Egyptians . The transformation could be recognized as a new birth for an ancient belief to what is sacred, invading the temple which is in the center of it, as part of this phenomenon.

After the temple of *Khnum* on Elephantine, Philae was the center of the world, as an inherited mythical belief, the lines between "magic" and "religion" did not exist within the host culture for Christianity, where the Nile played an essential role in the center of its cycle. Philae carries the symbolism of the center with *Isis* sanctuary as the ultimate zone of sacredness, it was labeled as demonic power by the new faith and her powers became Virgin Mary's. The temple was a profane space that held access to power, and must be Christianized, it must be transformed into a consecrated space, the engraved crosses were a signature of the new birth, started from the courtyards, approaching the holy of holies, announcing the victory for Christianity with its nationalistic impulse, it was Egyptized, and it dominated the traditional cosmological worldview.

By the interpretation of *Isis* Temple transformation, this study puts stress on approaching the value of ancient religious buildings on the base of its cosmical value, instead of adopting the absolute practical vision. The fate of each temple depends on the public ontological understanding. Within Philia Complex, *Harendots* Temple was used for retrieving building materials, while *Isis* Temple took gradual transformation over years, after the transformation, the temple's abandonment was a new phase, after shifting the center of power again to Syrene, and it could be also be interpreted as the final perception shift of the traditional man away from the cosmos or away from nature.

References

Adams WY (1991) Philae. Retrieved June 2019 from the coptic encyclopedia, vol 6: https://ccdl.libraries.claremont.edu/cdm/singleitem/collection/cce/id/1601/rec/1

Apuleius L (1975) The Isis-Book: (Metamorphoses, Book XI)- Edited with an Introduction. Leiden, Brill

Belmonte JA (2010) Keeping ma'at: an astronomical approach to the orientation of the temples in ancient Egypt. Sci Direct 532–539

Bierbrier ML (2011) Dodekaschoenus. Retrieved from Ancient Egypt: http://ancient_egypt.enacademic.com/343/Dodekaschoenus

Blackman AM (1915) The temple of Bigeh. Cairo: Institut français d'archéologie orientale du Caire

Bumbaugh S (2–4 October 2009) Meroitic worship of isis at Philae. Egypt in its African Context. The Manchester Museum,University of Manchester, Archaeopress pp. 66–69
Canhão T (2013) A timeless legacy: the calendars of ancient Egypt. In: Sousa R (ed) Alexandrea ad Aegyptum: The legacy of multiculturalism in antiquity. Coimbra, CITCEM, pp 283–301
Cannuyer C (2001). Coptic Egypt. The christians of the Nile. Thames & Hudson, London
Coppens P (2004) The canopus revelation: the stargate of the gods and the ark of osiris. Adventures Unlimited Press, llinois
Dijkstra JH (2011) The fate of the temples in late antique Egypt. In: Lavan L (ed) The archaeology of late antique 'Paganism'. Leiden, Boston, Brill, pp 389–463
Eliade M (1991) The Myth of the Eternal Return Or, Cosmos and History. Princeton University Press, New Jersey
Fæ (2014) Temple ruins on the island of Biggeh, Egypt. Coloured lithog. Retrieved from View Wikimedia Commons: https://commons.wikimedia.org/wiki/File:Temple_ruins_on_the_island_of_Biggeh,_Egypt._Coloured_lithog_Wellcome_V0049338.jpg
Frankfurtar D (1998) Religion in Roman Egypt, assimilation and resistance. Princeton University Press, New Jersey
Gabra G (2008) The A to Z of the Coptic Church. Scarecrow press, Plymouth
Green K (2010) Bigeh island ruins in the Old Aswan Dam reservoir. Retrieved from wikipedia: https://en.wikipedia.org/wiki/Bigeh#/media/File:Flickr_-_schmuela_-_IMG_6830.jpg
Heyde M (2007) Bigeh (foreground) and Philae with temple, in the first cataract of the Nile River. Retrieved 13 Apr 2019, from wikipedia: https://en.wikipedia.org/wiki/Bigeh#/media/File:David_Roberts_Temple_Island_Philae.jpg
Holberg JB (2007) Sirius: brightest diamond in the night sky. Springer Science & Business Media, Chichester
Keegan P (2014) Graffiti in antiquity. Routledge, Oxon
Kakosy L (1982) The Nile, euthenia, and the nymphs. The Journal of Egyptian 290–298
Kockelmann H (2012) Philae. In: Willeke W (ed), UCLA encyclopedia of egyptology, Los Angeles. http://digital2.library.ucla.edu/viewItem.do?ark=21198/zz002bwv5j
Lili (n.d.) Coptic crosses at philae temple. Retrieved from lilionthеloose.: https://lilionthеloose.com/2016/06/23/coptic-crosses-philae-temple/
MacCoull LS (1990). Christianity at syene/elephantine Philae. Bull Am Soc Papyrol. 27(1/4) 151–162
McKenzie J (2007) The architecture of Alexandria and Egypt. C. 300 B.C. to A.D. 700. New Haven press: Yale university
Moawad S (2013) Christianity on Philae. In: Gabra G (ed) Christianity and monasticism in aswan and nubia. The American university in Cairo press, Cairo, pp 27–38
Pagoulatos G (1994) The destruction and conversion of ancient temples to christian churches during fourth, fifth and sixth centuries. Theologia 65:152–170
Rutherford I (1998) Island of extremity: space, language and poewer in the pilgrimage traditions of Philae. In: Frankfurter D (ed) Pilgrimage and holy space in late anitique Egypt. Brill, Leiden, pp 229–256
Spencer AJ (ed) (2007) The British museeum book of ancient Egypt. The American university in Cairo press, Cairo
Volokhine Y (2014, 27 June) Honouring the gods in the classical mediterranean realm and on its fringes—The "End" of pharaonic religion. Retrieved from Hemed Modules: https://hemed.univ-lemans.fr/cours2012/en/co/module_Hemed_3_10.html
Wilkinson RH (2000) The complete temples of ancient Egypt. Thames & Hudson, New York
نبيه كامل داود. (2003). *تاريخ المسيحية و آثارها في أسوان و النوبة*. القاهرة: مؤسسة القديس مرقس لدراسات التاريخ القبطي
Wilkinson RH (2003) The complete gods and goddesses of ancient Egypt. New York, Thames & Hudson

Chapter 19
Historic Urban Landscape Approach for Historic Coastal Cities Regeneration: The Case of Rosetta, Egypt

Sara Adam, Marwa Khalifa, and Amr Attia

Abstract The concept of Historic Urban Landscape (HUL) was introduced as the latest UNESCO approach considering the ongoing debate around the identification, preservation, and valorization of cultural heritage. The purpose of this paper is to explore the potentials of this approach for the regeneration of Rosetta City in Egypt. This paper attempts to develop a comprehensive framework to finalize and perfect the existing plans. Rosetta City is known for its rich tangible and intangible cultural values and its unique coastal position that acquires an effective innovative approach to urban regeneration to direct future polices for the coastal city. The paper initially begins with brief overview of regeneration policies and precedent heritage conservation theories elucidating their transformation from place to people-oriented approaches. Followed by the rational of potency of (HUL) approach then, it demonstrates the different stances toward (HUL) approach from past best practices case studies in order to learn from their experience. Consequently, the paper attempts to apply and transfer the knowledge gained in the processes of regeneration of the city of Rosetta. Finally, this paper concludes that (HUL) seeks a balance between different dimensions through the maintenance of the human capital, the improvement of the efficiency of urban spaces, and integrating the objectives of conservation of urban heritage with socioeconomic development. Foremost, it explored the operation of HUL in the Egyptian context, promoting the mindset shift and revolutionized method mandatory by the (HUL) approach presenting a constructive initial step in the (HUL) journey.

Keywords Historic urban landscape (HUL) · Historic coastal cities · Regeneration

S. Adam (✉) · M. Khalifa · A. Attia
Urban Planning Department, Ain Shams University Cairo, Cairo, Egypt
e-mail: Sarafawzi1982@gmail.com

Architecture and Urbanism: A Smart Outlook,
https://doi.org/10.1007/978-3-030-52584-2_19

19.1 Introduction

Historic cities in developing countries are mostly located in poor, congested, and degraded areas where underprivileged and marginalized groups reside and earn their livelihoods. Historical centers in the Egyptian cities managed to endure as living cities in spite of physical decline and economic melancholy.

Aim: This paper aims to adapt the HUL approach as a regeneration tool through linking them through their conceptual transformation from place-based to people-oriented approaches, and then applying the adapted approach to the chosen case study of Rosetta, Egypt.

Methodology: This research has been done through the literature review of documented studies regarding the regeneration concept, ideologies, and policies tracing its alteration from place-oriented to people-oriented, and then relating the preference of Historic Urban Landscape through a comparison between recent conservation modes. Following a brief discussion on HUL approach stances in order to turn the theoretical knowledge to applicable tools through the adaptation of HUL approach as a regeneration method. The empirical part was carried out through a field study took place from April 14–17, 2019, interviews with Rosetta Municipality, Rosetta Environmental Management Unit (EMU), Random local residents including youth, women, men, and children through different areas of the city, heritage center, informal areas, fishermen, taxi drivers, ferry boats owners, and boat manufacturers. The aim of the field study was to assess the current situation compared to the literature review.

Findings: This paper applied and transferred the acquired knowledge to the chosen case study by applying the proposed adapted approach to set priorities according to local resident's objectives, comparing those to the latest plans and suggesting consequent actions. This research limitation was the need of further fact finding regarding the legislative tools and funding resources for the proposed actions.

19.2 Rationalizing the HUL Approach

The significance of cultural heritage originates from the fact that it represents the manifest ethos of human experience that holds within the essences of our ancestor's joys and sorrows. Cultural heritage is non-renewable and inimitable. In historic context, the concept of urban regeneration evolved between 1930 and 1960s, firstly, through "The Era of Bulldozer" in USA and UK designated for slum clearance and (Barosio et al. 2016) then, following the WWII when the wrecked European cities had to face the dilemma of choosing an adequate intervention method in their damaged historical city centers, either the comprehensive reconstruction or the conservative renewal (Cifitzi et al. 2010). It was then when the "Integrated protection and regeneration" approach was introduced in England to revive consumed historic sites (Sharghi et al. 2018).

During the 1970s, cities were facing urban decay in the form of rapid urbanization, unstoppable growth, increasing poverty, spatial fragmentation, social exclusion, environmental vulnerability, and the expenditure of resources along with globalization ambitions and aspirations (Ragozino 2016). Historic cities with its sensitive fabric and fragile assets became stressed out (AnnBeswick and Tesnkova 2010). Then, in the late 1990s, the "Integrated Regeneration Approach" called for the importance of public and private partnership to achieve urban quality transforming the concept from "place-oriented" approach to a "people-oriented" one (Barosio et al. 2016). According to Oxford dictionary (2019), urban regeneration is a term that aims to restore profitability and repopulate declined areas. It is deputed to solve urban problems and comprehensively reverses the decline through boosting the economic, social, and environmental aspects of the cultural site. Urban regeneration contributes largely in upgrading historic environments, their socioeconomic quality and is based on governance (DeRosa and DiPalma 2013).

The conservation approaches for cultural heritage were also once about the physical restoration of monuments or buildings. Then after the WWII and the development of the United Nations Educational, Scientific and Culture organization "UNESCO" called for the concept of "Landscape" advocating the change of conservation concepts from physical to comprehensive embracing the multilayered historic landscape (Zeayter and Mansour 2018). The "International Centre for the Study of Preservation and Restoration of cultural Property" ICCROM in 2005 launched the "Living Heritage Regeneration Approach"; it commended that culture in all of its forms matters (Ragozino 2016). Later, the UNESCO launched the framework of the implementation of 2030 Agenda for Sustainable Development, "Culture for Sustainable Urban Development Initiative", to raise awareness on the contribution of culture to sustainable development in cities. The approach urged on the creation of a community-based cultural heritage approach (Table 19.1).

Consequently, on October 2005, the 15th General Assembly of States Parties to the WHC adopted the Historic Urban Landscape approach promoted; it advocates that culture, in its manifold expressions ranging from cultural heritage to cultural and creative industries and cultural tourism, is both an enabler and a driver of the economic, social, and environmental dimensions of sustainable development (UNESCO 2011).

Historic Urban Landscape was introduced by the UNESCO as an inclusive multilayer urban regeneration engine that had yielded from cumulative knowledge and experiences in conservation of cultural heritage achieving the notion of comprehensiveness (HUL Guidebook 2016). Roberts and Skyes (2000) defined urban regeneration as an intervention action which is likely to consider social, economic, environmental, and political frameworks changes to respond to circumstances. Consequently, urban regeneration policies were summarized to incorporate the five layers of place: physical environment, quality of life, social welfare, governance, and economic prospects. In terms of Historic Urban Landscape policy, culture heritage with its tangible and intangible content should be integrated and the policies be regrouped according to relevance (Table 19.2).

Table 19.1 Historic Urban Landscape Milestones and the concept shift from material-based to people-based (Barosio et al. 2016; Cifitzi et al. 2010; DeRosa and DiPalma 2013; Roberts and Skyes 2000; Zeayter and Mansour 2018; Ragozino 2016)

Approach	Timeline	Evolution
Material based approach	Fourth and fifth century	Memorial conservation and Athens conservation of buildings according to demand
	Fifteenth century	Renaissance era and collection of antiques
	Seventeenth century	Regeneration after destruction of the European civil war building replicas
	Eighteenth century	Stylistic restorations with maximum intervention versus conservative preservation
	John Ruskin	Anti-restoration movement opposing intervention during restoration for authenticity
	William Morris	"Society for protection of Ancient Buildings"
	Patrick Geddas	Surgical approach restoration; minimum intervention and materials
	Gustavo Giovani	Scientific restoration; initializing the concept of area conservation
	1931 The Athens Charter	Modern conservation policy influenced by scientific restoration
	Reigel	Extended authenticity definition from building related to urban area concerned
World War II and UNESCO formation		
	1964 Venice Charter	Modern heritage policy keystone
ICOMOS formation		
People-based approach	1976 UNESCO	Recommendation concerning the safeguarding and contemporary role of areas
	1982 sustainability approach	Linking conservation to communicate present and future social, economic, and environmental needs
	1987	ICOMOS General Assembly (Washington Charter)
	1998	Stockholm Intergovernmental Conference on Cultural Policies for Development
	1999 ICOMOS cultural tourism	Creating sustainable setting for tourism "Facadism" improvement
	2001	UNESCO declaration of cultural diversity
	2003	UNESCO Convention for the safeguarding of intangible cultural heritage
	2005 ICCOROM living heritage	Conservation and regeneration in both tangible and intangible

(continued)

Table 19.1 (continued)

Approach	Timeline	Evolution
	2005	15th General Assembly for state parties WHC Historic Urban Landscape Approach Urban heritage is a catalyst for urban regeneration
	2015	UNESCO Habitat 3rd Conference; Culture: Urban future People, Places, Policies

Table 19.2 Adapted comparison between conservation approaches in terms of regeneration polices (Barosio et al. 2016; Cifitzi et al. 2010; DeRosa and DiPalma 2013; Roberts and Skyes 2000; Zeayter and Mansour 2018; Ragozino 2016)

Regeneration policies	Sustainability approach	Cultural tourism	Living heritage	HUL
Place oriented				
Physical improvement	✓	✓	✓	✓
Environmental actions	✓	×	✓	✓
Economic prospects				
Economic development	✓	✓	×	✓
Quality of life	✓	✓	✓	✓
People oriented				
Governance	×	×	×	✓
Social welfare	✓	✓	✓	✓
Community = based	×	×	✓	✓
Participatory	✓	×	✓	✓
Cultural				
Tangible	✓	✓	×	✓
Intangible	×	×	✓	✓

19.2.1 Historic Urban Landscape Action Plan and Case Studies Stances

The HUL institutional action plan consists of four main tools according the HUL Guidebook (2016);

- **Community engagement tool**; publicity, dialog, community empowerment, and cultural mapping
- **Knowledge and planning tool**; planning, policy assessment, and vulnerability assessment
- **Regulatory system tool**; laws and regulations, plans and policies
- **Financial tool**; economics, grants, and private–public partnership.

Previous applications advocated that HUL is not limited to a range of techniques rather than the need to innovate in planning and policy practice; it is more than the execution of the recommended toolkit to the present practice; successful intervention requires the need to create opportunities and governance (Jones et al. 2018). For example, participation and engagement tools were used in Cuenca, Ecuador, to allow the community to value the importance of cultural heritage and the possible challenges that they could help detain. Ballarat, Australia, engaged citizens in a collaborative mapping process of the tangible and intangible components of heritage (HUL Guidebook 2016).

The case of Shanghai, China, was motivated by the need to improve poor facilities in historic buildings, social, and economic recession; therefore, new fabric and spaces were created, economic opportunities and local projects to conserve the heritage and improve the quality of surrounding space by reusing the idle spaces (Jones et al. 2018). The case of Naples, Italy, depended on the incentive to develop local entrepreneurship, revitalization of traditional crafts through an integrative conservation approach with focus on tourism promotion. Also, in the case of Zanzibar, Tanzania, HUL purpose was to document and archive cultural heritage and to enhance the quality of life of the citizens. Finally, the case of Amsterdam, where purpose-built taxonomy-based policy analysis tool was formed in order to assess and integrate urban and heritage polices (HUL Guidebook 2016). From the previous HUL stances, the HUL approach proved to be an inclusive approach that is not bounded by a certain set of tools rather than guidelines. Therefore, different approaches of intervention could be merged or emphasized Bbsed on the HUL recommended tools and the regeneration policies.

19.3 Application: Rosetta Case Study

On the north tip of Al Behiera Governorate along the western bank at the mouth of Rosetta branch of the Nile 65 km away from North–East Alexandria lays Rosetta City with its historic core. The area of the city forms an estuary of 12 kms and is approximately 670.9 acres generated from the Nile sediments. Rosetta was inscribed in the World Heritage site tentative list for criteria (ii),[1] (iv),[2] and (v)[3] (UNESCO 2019).

The history of Rosetta or Bolbitine goes back to the era of Pharaohs, throughout history it acted as a defensive portal of Egypt. When Amr Ibn AlAss opened Egypt, the city was not very popular, and then the Abbasid Caliph ordered a fortress to be built on the site of the Ptolemaic City. The growth of the city started around the fortress (Belgacem 2009). Under the Mamlukes, the city became an important commercial

[1] Exhibit an important interchange of human values, over a span of time or within a cultural area of the world.

[2] An outstanding example of a type of building, architectural, or technological ensemble or landscape.

[3] Outstanding example of a traditional human settlement, land use, or sea use which is representative of a culture (or cultures), or human interaction with the environment especially when it has become vulnerable under the impact of irreversible change.

center, and then during the Ottoman era, houses, mosques, public baths, windmills, and gates were established. During the French occupation on Egypt, Rosetta gained its popularity due to the discovery of the 196 BC Rosetta stone and deciphering the old Hieroglyphic language (Azzam 2013).

The city's growth during the sixteenth century was mainly toward the west, then during the seventeenth century the expansion headed toward east on the sediments provided by the Nile flooding, the southern part of the city was limited by the cemetery that turned in the eighteenth century due to rapid population into rice depots and factories. Rosetta's monumental buildings were built using Rosetta's significant red and black bricks known as Mangur bricks, the manufacturing knowledge of those bricks was lost, red brick factories located in the north of the city produce modern bricks (UN-Habitat RUSPS 2008).

19.3.1 Knowledge, Planning, and Policy Assessment

Rosetta was confirmed as a tourist destination by the law 113/1986. The first phase of preservation was initiated by the Supreme Council of Antiquities in 1985. Rosetta is the second-largest city owing Islamic monuments prior to Islamic Cairo (Belgacem 2009). There are 38 Islamic registered monuments including 22 houses, 14 mosques, and set of wekala, windmill, and public bath. There are 25 monumental buildings lie in Rosetta's historic center while the rest are scattered along the city (Fig. 19.1).

Rosetta's strategic urban plan targeting 2022 vision was to transform it to an open-air museum; it had won the "Islamic Cities and Capitals Planning Projects" award held in Doha, Qatar, in 2010 Gharib and Mostafa (2008), Azzam (2006). The vision was based on four main pillars; the establishment of New Rashid residential area to contain the population increase, the lodging of a craftsmanship area to encourage

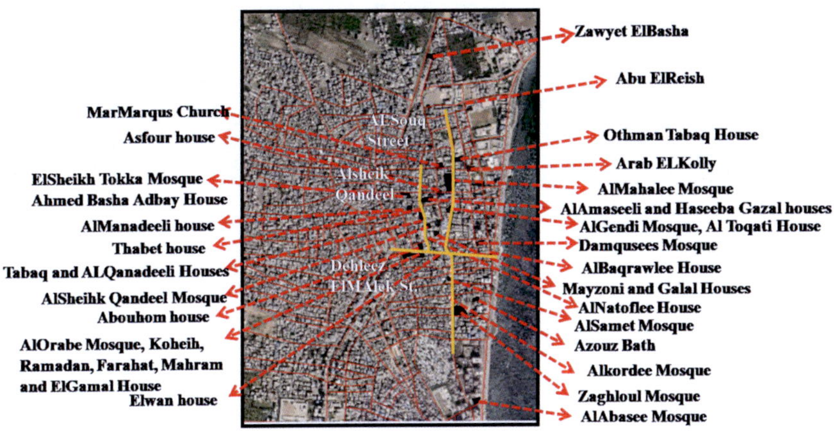

Fig. 19.1 Rosetta main features and cultural heritage areas. The Author (2019)

local crafts, tourism consolidation along Rosetta strait, the coast and the Nile banks and the maintenance of Rosetta cultural heritage. The plans for Rosetta were delayed due to the “Egyptian Spring” in 2011 and the constant lack of financial resources. (Rosetta Municipality representative 2019) cultural heritage areas suffered from great deterioration, then the dilapidated historic assets attracted governmental attention calling for the execution of the strategic plan actions in 2015.

According to Madbouly (2019), President Abdelfattah Elsisi set a three years development plan for Rosetta; the Egyptian Prime Minster announced the establishment of new Rosetta on 3100 acres. The mandate of New Rosetta is to be transmitted to the New Urban Communities Authority that will carry the planned developments on its expenses; yet, the share of AlBeheira Province (20%) is to be maintained. A cooperation protocol was signed on the April 10, 2019 between the Governorate and the Authority that will also develop the old city of Rosetta (Fig. 19.2). The plan targets:

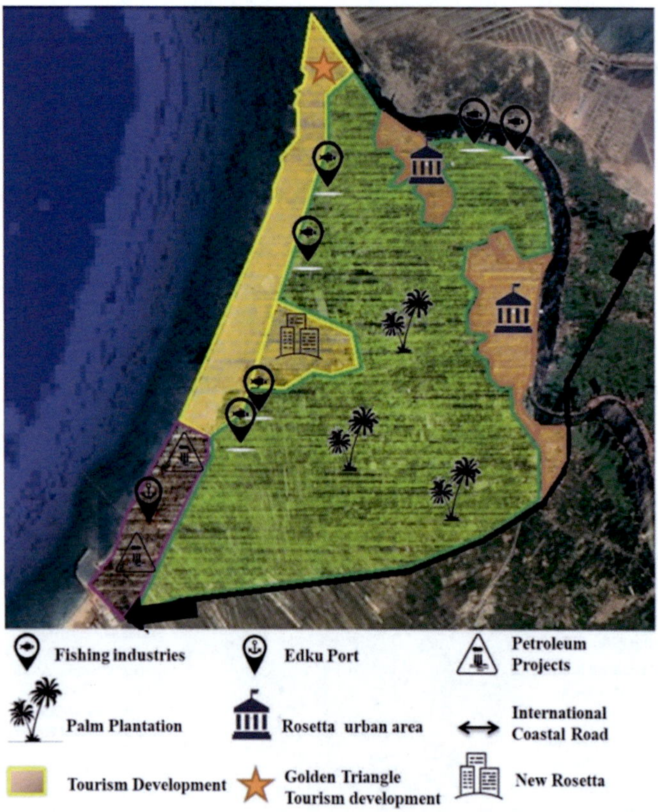

Fig. 19.2 Rosetta strategic plan. The Author adapted from Ministry of Housing, Utilities and Urban communities 2019

Tourism: building resorts, tourist villages, a water sports center, and yachts and rowing clubs. Development of: the Nile banks at cost of L.E. 2 million, the Nile docks and river bus at cost of L.E. 23 million, Abo Mandour's area at a cost of LE 32 million and 43 acres public beach is planned for Rosetta residents after the removal of the existing land violations on the proposed area (Fig. 19.3).

Industrial Investment: New vegetable and fish market at a cost of L.E. 18 million on 59 acres was constructed. Palm products, fishing nets, boats, and yachts factories along with fish and rice paddies packaging and export industries are planned. The establishment of 60% of a new 12 acres fishing port with initial cost of 300 million LE including industrial projects in the field of fisheries with the aim of development of the fishing wealth would serve more than 7500 workers.

Housing and Community: The National Service of the Armed Forces executed four out of five phase project "Bashayer El Kheir" in new Rashid as an alternative housing for informal areas; 204 buildings, 13,971 apartments, and an approximate

Fig. 19.3 Latest development plans for Rosetta. Ministry of Housing, Utilities and Urban communities 2019

cost of 3592 million LE for the four phases. The National Project for Local and Community Development, the governorate, has granted loans to young developers worth 231 million and 879 thousand pounds, providing 21,418 job opportunities for fresh graduates across the governorate through 8268 small and medium investment projects.

Infrastructure and Services Projects: A total of 48 million was targeted for the enhancement of sewage system and another 20 million pounds for the hygiene equipment supplies for the city. President Abdul-Fattah Al-Sisi issued in 2017 during the Youth Forum in Alexandria decree No. 49 concerning the allocation of 24 acres of state-owned land to the Egyptian General Petroleum Authority for the use of petroleum projects, and this would provide jobs and economic revenue. British Petroleum (PB) declared a community participation of LE 20 million to complete development needs. According to Abdo (2017), an agreement regarding the establishment of the University of Rosetta with cost of 250 million pounds was settled on. The Ministry signed a contract regarding a local raw materials and environment-friendly project with Egyptian–Chinese financing aiming the conversion of rice straw to paper.

19.3.2 Assess Current Vulnerabilities: Site Visits Compared to Reports

Cultural heritage: The historic center consists of four streets; Dehleez el Molok, Al Amaseely Street, Al Mahali Street, and Port Said Street. The street widths in Rosetta range between 6 and 16 m wide and in the historic center it is 6 m only leaving no premises buffer zone for maintenance and preservation, there are no building codes. In general, each historic building is dealt with separately and not in its context along with deficiency of recorded data related to historic assets. Resources are scarce; the Supreme Council of Antiquities depends on self-generated resources such as tickets and exhibits (Farid 2019).

The urban landscape around the historic buildings suffers from inadequate scattered daily uses, peddlers, vandalism, wastes, garbage, pollution, meat, and vegetables shops are attached to some historic buildings. The local residents do not have heritage knowledge of the vulnerability of the historic sites, the threats those sites are subjected to, and their capability as a valuable asset. Tourism infrastructure is limited in the historic areas, narrow streets, and improper transportation. The public sanitation network is poor leading to high level of underground water and related problems; corrosion and separation of building materials, change of colors, breakdown of soldering materials, collapse of foundations, cracks, and bacterial caused dispersion of building materials (Fig. 19.4).

Socio-economic Problems: Rosetta's dwellers are mostly modest; they mainly work in fishing, boat manufacture, palm and dates industries, handmade products,

Fig. 19.4 Shows the inadequate uses spread along the Historic Landscape, the spread vandalism on the monuments, The vandalism on the Nile. Author (2019)

and agriculture. According to site visits and interviews, there is high unemployment rate that has led to the increase in the rates of illegal immigration. 22% of the population live below poverty line. 14% is the women contribution to outdoor jobs (CAPMAS 2017). The percentage of residents graduated from schools is 29.5%, and the non-enrolled children in education is 32.46%; hence, there is a high level of illiteracy. For services, 99% are connected to electricity network, and although 92.24% are connected to freshwater; field study has shown that there is a general complaint among them from the quality of water. Informal areas are Elkassara with the area of 37,500 m^2, Ezzbet QatQat with area of 45,000 m^2, Abou el Reish of area of 45,000 m^2, Ezzbet Hassan Ali with area of 10,000 m^2, and finally Behind Transformers with area of 77,000 m^2. Ezzbaet Qatqat and Abou el Reish were upgraded, and Alkassara is considered to be unsafe with massive urban problems. There are also three deteriorated urban areas including 2500 residential unit in which 50% acquires removal (AL Shahid 2010).

Environmental: The intensive human activities along the shoreline accompanied with the reduction of Nile sediments after the construction of the high dam and high sea currents have led to the constant change in the shore shape, water quality, and biodiversity. The General Authority for Beach Protection established a protection head in the strait area in the eastern wing causing the presence of large deposits on the eastern side extending westward to the entrance of the strait reaching the widths of about 40 m with a high level of the bottom, which led to the impotence of fishing vessels to navigate from the Nile to the sea and the concentration of pollution. And also, this problem requires constant cleansing of the strait (EMU 2019).

The rich natural palm groves that enclose the urban area are endangered due to the expansion of the town causing many handcrafts to be subjected to extinction such as palm-based industries and furniture. According to the Environmental Description of Rosetta (EEAA 2007) Rosetta branch has high concentration of soluble salts and ammonia, high alkaline level, and saturated oxygen than the permissible levels. Also, Escherichia coli bacteria, typical colon bacteria, and spherical streptococcus gynogenic bacteria were detected. Another problem is the presence of brick factories that cause pollution; five of them were located along Rosetta branch; those were relocated to the craftsman areas and the others are strictly managed by the Environmental

Management Unit that obliged their owners to improve their performance to comply with environmental regulations.

19.3.3 People-Based HUL Approach

The main stakeholders in Rosetta are the local government including the Governor, Heads of neighborhoods, and the local authority that are clearly popular to residents, and on the other hand, the public council, NGOs, private sector, and experts. There is a clear lack of understanding of urban governance among citizens and the local authority; during interviews with the local authority representatives the information provided lacked credibility and was influenced by governmental reports, interviews with dwellers were more transparent. A partnership with UN-Habitat was carried in 2006 with the aim of achieving sustainable development supported by German Technological Cooperation, World Bank and Governments of Italy, Belgium, and Netherlands (UN-Habitat RUSPS 2008). Starting from 2017, the main engine for development became the intervention of president Abd El-Fatah El SiSi and the Armed Forces which according to local residents was the optimum solution for the city that was neglected for a long time, along with the New Urban Communities Authority intrusion which was also an important milestone for Rosetta's development.

As HUL endorsed that cultural heritage could generate revenues from existing assets for local communities that helps provide new opportunities for poor communities through their cultural knowledge and local productions. It strengthens the ability of local residents to pursue activities that enhance their income and quality of life. Interviews with local residents yielded with their main objectives based on those objectives, a proposed plan is suggested with regards to cultural heritage which according to Historic Urban Landscape is a catalyst for local development that could generate jobs, provide employment, and regenerate communities (Table 19.3).

19.4 Discussion

This paper adopted the Historic Urban Landscape approach as a regeneration method for Rosetta, Egypt. The aim was to investigate the approach potency toward Rosetta's historic center to finalize and perfect the existing plans. The main outcome is that HUL is an adaptive approach not bounded to a set of rules or steps; it demands the retroversion of concepts from physical conservation of separate historic monuments to the belief that improving lives and knowledge is the genesis of preservation. A field study has shown that there is a clear transparency deficiency among governmental bodies and local residents; also another critical point was the lack of local heritage knowledge regarding both the potentials and vulnerability of cultural heritage. The latest plans for Rosetta lacked the societal dimension; it was based on a revolutionary vision of turning it to touristic destination mainly through the development

Table 19.3 HUL adapted to regeneration approach, objectives according to site interviews, priorities, planned, and proposed actions

	Objectives	Priorities	Proposed
Place oriented Physical improvement Environmental actions	Freshwater access Access to sanitation systems Solid wastes collection Slums upgrading Pollution Deteriorated buildings renewal Urban landscape enhancement	Prevent drainage to the Nile Preventing fishing cages scattered along the Rosetta Nile branch Desalination and purification water plant Increase waste management resources Hardscape and landscape supply in historic routes	The adaptive reuse of Historic Urban Landscape including buildings and urban landscape Reducing population pressure on historic areas through Bashayer el kheir, payment facilities and appropriate compensation methods Encourage the reuse of local bricks and previously used materials The necessity of activation of environmental decisions and monitoring system Afforestation and hardscape for historic areas
Economic prospects Economic development Quality of life	Provide adequate jobs and good salaries Easy access to funds for small projects Informal area upgrading	Enhancement of workforce, employability, and entrepreneurial ship skills Promoting projects and access to education Widening the youth range who benefit from the funds Heritage marketing through creating a promoting heritage landscape Establish leadership from local residents in tourism and heritage fields	Enhancement of local knowledge; providing illiteracy programs, supporting unfortunate children and those with learning difficulties Awareness campaigns; on the importance of education and the disadvantages of dropping out of schools Enhancing of youth skills and resources through providing technology, creating various learning programs and flexible training centers Cultural education and consciousness through media, journalism, advertisements, street campaigns, workshops, and civic meetings Encourage local residents to work in tourism field

(continued)

Table 19.3 (continued)

	Objectives	Priorities	Proposed
People Oriented Governance Social welfare Community based Participatory	Public participation Informal settlements upgrading Women empowerment; absence of suitable representatives for the locals	Participation, transparency decentralization, and accountability NGOs funding Upgrading programs for slum dwellers and society development The importance of upgrading EMU equipment's, budgets, and resources	Stakeholders should participate in creating training plans and programs for local residents Develop leadership skills in local council and provide them with adequate tools and information Local residents should be trained on touristic relevant activities Create various training programs for youth to improve their eligibility for various tourism businesses and projects Encouraging the youth to create start-ups Encourage women to attend training workshops
Cultural Tangible Intangible	There is a great lack of heritage knowledge among residents Cultural heritage is out of their objectives	Relocate all non-relevant activities from the historic area Coordination between responsible stakeholders to renew, remodel, and reutilize the surrounding areas of cultural heritage Rehabilitation projects targeting the Historic Urban Landscape areas as a whole New tourism friendly transportation	Formal intercultural education curricula to strengthen cultural identity Creating a supervision committee to follow up and supervise the historic landscape status Transforming heritage streets to pedestrian areas A buffer green zone should be provided around the heritage areas, inadequate uses should be replaced by crafts shops, bazars, touristic facilities, and retail areas Building touristic facilities; transportation, shops and bazars, and guidance offices Encouraging owners in cultural area landscape to the adaptive reuse through transforming those to cultural themed recreational facilities

The Author (2019)

of the Golden Triangle zone along with elite tourism along the shore line. The plan targeted the development of the historic center yet, it has yielded to the division of the center into separate physical conservation areas, which thus become ghettos of historic preservation. Therefore, HUL approach in Rosetta's case was formulated to focus on the enhancement of public awareness, heritage knowledge, and its notion is the significance of augmenting local residents' quality of life. This took place through interviews held during the field study with local residents to understand their hopes and aspirations. Finally, proposed actions were suggested as a supplement for the existing plans. This paper concluded that (HUL) approach is an unconventional method of Rosetta city's regeneration. HUL approach seeks a balance between different dimensions through the maintenance of the human capital, the improvement of the efficiency of urban spaces, integrating the objectives of conservation of urban heritage with socioeconomic development on the basis of a balanced and sustainable relation between the built and the natural environment.

References

Abdou N (2017) Retrieved on 04 15, 2019. from Akhbarak website: https://www.akhbarak.net/news

AnnBeswick C, Tesnkova S (2010) Overview of urban regenration policies. vol. Urban Rgeneration, 9–16

Azzam Y (2006) Preparation project of the general strategic and detailed master plans of the city of Rashid. Al-Beheia Governorate, GOPP Egyptian House for Architecture and Planning

Azzam YA., Ayad HA, Abd El-Wahab MA (2013) Effect of physical development on urban communities using a strategic planning process and its implementation. WIT Trans Ecology Environ 179:211–224. https://doi.org/10.2495/SC130181

Barosio M et al (2016) From urban renewal to urban regeneration: classification criteria for urban interventions. Turin 1995–2015: evolution of planning tools and approaches. J Urban Regeneration and Renew 9:367–380

Belgacem O (2009) Rosetta. A little guide to the city. Egypt, Supreme Council of Antiquities

CAPMAS. (2018). National Census

Ciftci C et al (2010) Urban regeneration strategies in historic city centre of Beysehir-Konya, Turkey. 14th IPHS Conference Urban transforation; controverseries, contrasts and challenges, Turkey

DeRosa F, DiPalma M (2013) Historic urban landscape approach and port cities regeneration: naples between identity and outlook. MDPI. (13/5). 4268–4287

EEAA (2007) Tthe environmental description of El-Beheira governorate. The ministry of environmental affairs issues, Egypt

Farid M (2019) Rosetta mancupility representative. Personal interview

Gharib N, Mostafa S (2008) Upgrading archaeological heritage as an approach for sustainable development in Rosetta: a case study of King street. https://www.researchgate.net/publication/267758004

Jones K et al (2018) Pursuing a historic urban landscape approach to heritage in Edmoton. Faculty of Extension University, Alberta, City Region Studies Centre

Madbouly M (2019) Retrieved 04 Nov 2019, from AlShourok website: https://www.shorouknews.com

Roberts P, Skyes H (2000) The evolution, definition and purpose of urban regeneration. Sage 2000:17–18

Ragozino S (2016) Tools for regenration of the urban landscape social enterprise as a link between people and landscape

Sharghi A et al (2018) A study of evolution and development of urban regeneration with emphasis on the cultural approach. Turk Online J Des Art Commun 18(3):271–284 (March Special Edition)

UN-Habitat (2006) United nations human settlements programme rapid sectorial survey. Max Planck Encyclopedia of Public International Law, Rosetta, Egypt. https://doi.org/10.1093/law:epil/9780199231690/e1715

UNESCO (2011) Recommendation on the historic urban landscape. United Nations Education, Scientific and Cultural Organization, Paris

UNESCO (2016) The HUL Guidebook. The HUL Guide 59. http://historicurbanlandscape.com/themes/196/userfiles/download/2016/6/7/wirey5prpznidqx.pdf

UNESCO. (2019). Historic quarters and monuments of Rosetta_Rachid - UNESCO World Heritage Centre. 2020. Retrieved January 5, 2020, from https://whc.unesco.org/en/tentativelists/1831/

Zeayter H, Mansour AMH (2018) Heritage conservation ideologies analysis—Historic urban Landscape approach for a Mediterranean historic city case study. HBRC J 14(3):345–356. https://doi.org/10.1016/j.hbrcj.2017.06.001

Chapter 20
The Wall, An Increasing Phenomenon in Cairo: Main Reasons and Potential Implications

Yosra G. Elnaggar, Yasser M. Moustafa, and Ayman F. Wanas

Abstract The wall, this stand-alone structure, which materialises in many different ways in the urban context, is the substance of the study. The wall in all its forms has an impact on the built environment and urban space users. Walls divide, segregate, discern, screen, protect, prevent and sometimes guide and redirect. The study investigates the wall phenomenon in Cairo and its increasing manifestations in the urban context; walling public buildings, privatisation of public spaces, street barriers, security walls, fences of residential buildings and finally walls of gated communities. The study critically argues that the decline in the sense of security, the terrorism problem, the changed nature of privacy needs and the neoliberal urban politics are the potential reasons for the increasing phenomenon in Cairo. Using related international urban design literature, the paper identifies the wall potential implications on the urban environment in terms of urban space vitality, right to the city, social cohesion, crime prevention and image of the city.

Keywords Cairo · Urban wall · City of walls · Privatisation of public space · Right to the city

20.1 Introduction

The notion of the wall that this paper addresses refers to the stand-alone structure that divides the urban space to enclose or fortify an area. Walls include fences, parapets, barbed wires, barricades and all sorts of physical artefacts that can form a boundary. Walls are linear interruption of the space preventing continuity and acting as spatial

Y. G. Elnaggar (✉) · Y. M. Moustafa · A. F. Wanas
Department of Architecture, College of Engineering, Arab Academy for Science, Technology and Maritime Transport, Cairo, Egypt
e-mail: y.elnaggar@aucegypt.edu

Y. M. Moustafa
e-mail: yasser_moustafa@aast.edu

A. F. Wanas
e-mail: awanas@aast.edu

Architecture and Urbanism: A Smart Outlook,
https://doi.org/10.1007/978-3-030-52584-2_20

and visual barriers (Lynch 1960). Walls transform the urban landscape, affect patterns of circulation and limit the use of streets and public spaces (Caldeira 2000). Walls transform the spatial, perceptual experience and social integrity of the city (Ghonimi et al. 2010). It appears that the wall is now constituting an increasing phenomenon in Cairo. Cairo streets are commonly limited by physical barriers on both sides, walls or fences; walls of a school, fence of a public park, walls of a military base, walls of a public facility or walls of a residence in a gated community. The paper addresses three main objectives. First, the paper documents the increasing nature of the wall phenomenon in Cairo relying primarily on a field photographic survey. Second, through critical argumentation (Groat and Wang 2013), the paper identifies the main reasons behind the phenomenon in Cairo. Third, through a review of the relevant urban design literature, the paper identifies the potential implications of the phenomenon on the urban environment of the city.

20.2 The Wall, An Increasing Phenomenon in Cairo

20.2.1 Manifestations of the Phenomenon in Cairo

The walling of public buildings and public urban spaces. Most public spaces in Cairo are in fact restricted areas. Generally, urban spaces in Cairo such as plazas and public parks are surrounded by fences greatly limiting public access (Abdelbaseer 2015). The term urban walls or urban gating refers to a series of gated blocks dispersed through the cityscape (Brighenti and Kärrholm 2019). These could be single or multiple buildings, individual plots or blocks forming a sequence of gated urban spaces. This section presents the manifestations of urban walls in the city and the resulting walling out of the public from public buildings and public urban spaces.

Walling parks and statues in squares might be accepted for reasons of maintenance and are not new in Cairo. What is new is the increased height of the walls and their decreased transparency. Most public parks in Cairo are fenced with the recognisable green metal barrier that limits accessibility but allows visual continuity (Tadamun 2014). However, in most cases, vegetation grown along the fences limit visual accessibility and prevent enjoying the interior landscape. This is the case with the fence of El-Azhar park (see Fig. 20.1). Also, El-Hurriyah Park was originally designed to be more accessible; it was bordered by a low brick boundary (see Fig. 20.2) before the installation of a higher steel fence that restricts accessibility (Cairobserver 2011a). Public parks should be more open especially when accessibility to green spaces in a city like Cairo is very limited; Cairo suffers from a very low value of green space area per capita (0.8 sqm/cap) compared to the global average of (4 sqm/cap) (Tadamun 2014).

Culture buildings aim to spread knowledge and promote art awareness. Hence, spaces surrounding culture buildings should be open to attract people and welcome all sorts of gatherings and interactions. However, Cairo Opera House, the cultural

Fig. 20.1 El-Azhar Park fence taken by author (2019)

Fig. 20.2 El-Hurriyah Park fence taken by author (2019)

complex in Zamalek is enclosed by walls and lacks any interaction with the surrounding urban setting. Entry points are gated and controlled by guards while the metro station inside the complex has been closed for security reasons. Recently, opaque sheets have been installed on the premises external walls thus blocking the view to the inside (see Fig. 20.3). The complex has become a culture gated community that offers *culture behind the fence* (Cairobserver 2011b). Similarly, the New

Fig. 20.3 Cairo Opera complex wall taken by author (2019)

Museum of Civilisations in Misr El-Qadima is surrounded by a steel fence that locks its vast entry plaza. While the plaza is created for people, it is deserted and lacks life. Museums should be safeguarded as they display important artefacts; however, people have the right to enjoy their surroundings since they are public buildings that belong to the public.

The administrative building of Mogamaa El-Tahrir has a spacious plaza to accommodate the thousands of daily visitors. Unfortunately, the plaza is restricted with a standard green metal fence (see Fig. 20.4). It is also the case for the Cairo main train station. After its renovation, the large plaza in front of it was fenced off with steel barriers, disturbing the flow of the thousands of travellers (see Fig. 20.5). In other countries, such public buildings are rarely fenced off, but are more typically treated to be active parts of the city and are designed for inviting public activities.

The privatisation of public space. Starting in the 1990s, due to the privatisation and liberalisation of the Egyptian economy, Cairo witnessed an increasing trend towards the privatisation of public space (Abaza 2006; Marafi 2011). The construction of commercial centres boomed in the capital. These centres were widely embraced by the upper socio-economic classes who were looking for a convenient exclusive urban experience (Kuppinger 2004). Secured exclusive private urban spaces in Cairo include private social clubs, malls, resorts, clubhouses and vast green areas inside upscale gated communities. One example of the fortress leisure (Hutchison 2010) is the new extension of Mall of Arabia inaugurated in 2018. The exclusive recreation park attached to the commercial complex contains restaurants, pleasant green spaces, jogging track and kids' playground. The park accessibility is exclusive to members of a virtual group on social media monitored by the park authority who carefully selects its members. Similarly, accessibility to most of the Nile waterfront in Cairo is

Fig. 20.4 Mogamaa El-Tahrir fence taken by author (2019)

Fig. 20.5 Cairo Train Station main plaza taken by author (2019)

controlled by upscale restaurants and hotels. Sometimes, local authorities governing public spaces restrict public access under the pretext of protection and security and ignore the citizens right to the public space (Tadamun 2014).

A new pattern of control affected the use of public spaces. The new private order has been created by building up walls that fragment the city. Politically, the privatisation trend is the withdrawal of the state from its responsibility in providing public services in favour of private institutions that create "… private schools, private recreation, private parks, private roads, even quasi-private governments" (Calthorpe 1993, p. 37).

Street barriers for the control of behaviour. Street barriers usually frame Cairo streets and sidewalks. Their purposes vary between controlling pedestrian behaviour, prohibiting car parking, closing roads, blocking a view or securing entities. Barriers installed on roads and sidewalks control the behaviour of individuals on a daily basis (Grant and Mittelsteadt 2004) which reflects the exercise of power and discipline over the space (Foucault 1977). Barriers can define people's flow, their circulation mode, determine what they can or cannot see, what spaces they can or cannot access and thus influence the possibilities of encounters (Brighenti 2009). Barriers installed in the middle of main roads are intended to control pedestrian's behaviour and prohibit them from crossing the streets but at specific points (see Fig. 20.6). However, when perceived to be erected at the detriment of their convenience, people tend to disrespect these barriers. They may climb over them or even break them to permit access through (see Fig. 207). A massive 2.8 km-long concrete wall lies on the edge of Cairo's main ring road at Dar El-Salam. This wall was installed to hide the slum of Izbet Khayrallah (Tadamun 2013), to minimise car accidents and to secure the road (see Fig. 20.8). There are openings in the wall that allow residents of the slum to directly access the road and use means of transportation. But, for more convenience, residents have broken the wall to create new openings and even installed staircases for easier access.

Security walls. In Cairo, security measures around sensitive state buildings and embassies relied on light temporary features such as signs, chains and barricades. However, after the events of the 2011 revolution and attacks on governmental bodies, security measures have been increased to fortify all governmental and sensitive buildings in the capital: police stations, city halls, churches, ministries, military bases,

Fig. 20.6 Street barriers at El-Azhar St. taken by author (2018)

Fig. 20.7 Crossing street barriers on Sixth October Bridge taken by author (2015)

Fig. 20.8 Concrete barriers at Dar El-Salam taken by author (2019)

embassies, etc.… (see Fig. 20.9). Security authorities used military methods and installed high concrete walls, checkpoints and gates (Mustafa et al. 2017). In some cases, sidewalks, streets or even entire areas were totally closed off depending on the sensitivity of the establishment. For example, a pleasant street in Zamalek district has been greatly affected by the installation of four metre high concrete walls next to the original walls of China's Embassy (see Fig. 20.10). The street became very narrow for traffic and the sidewalks disappeared behind the grey installation which led to an unpleasant pedestrian experience.

Fig. 20.9 Ministry of Exterior at Maspiro taken by author (2019)

Fig. 20.10 Security walls of China Embassy at Zamalek taken by author (2019)

Security walls symbolise misguided policies (Shama 2014). They mark segregation and tension which is clear in overcrowded spaces like Cairo's downtown where more than eight concrete barricades were installed after 2011 (Abaza 2013). The downtown has been divided by concrete security walls into compartments to keep protestors away from governmental buildings. These walls were meant to be temporary but they were left in place for several years (Shama 2014). Walling off the entire area had a paralysing impact on the downtown residents, commercial streets, small cafes and existing businesses (Abaza 2013). Today, some of the concrete walls have been removed by the residents themselves, some by the authorities and some have been replaced with metal gates to be closed when needed (Egypt Independent 2012; Messieh 2012; Shama 2014).

Fences of residential buildings. It appears that over the last decades there has been an increasing need for privacy and security in Cairo. This has materialised in the increase of the height and opacity of fences and walls around residential properties. Fences of residential buildings built up until the early 1970s tended to be much lower than those erected today and did not exceed the height of a normal person (see Fig. 20.11). These low fences or walls permitted a visual continuity between the public space of the street and the adjacent private domain. This openness of residences towards the outside reflects well the more open and less conservative society of the time. The low residential fences also demonstrate that security was not such an important concern at the time as it is today. Indeed, these low fences could easily be climbed over by intruders. They were erected primarily to mark the boundaries of the private property. Today, reflecting changes in privacy needs and increased security concerns, walls around newer residential properties are not only much taller, often exceeding two metres in height, but they are also more opaque. In addition, trees and vegetation are often used to further block visibility from the street and neighbouring buildings (see Fig. 20.12). In many of the older residences, original fences have been made taller and more opaque by the addition of solid plastic sheets. Following the events of 2011, many residential fences have also been made more secure by the use of barbed wires (Abd Elrahman and Mahmoud 2016). In some instances, when residents felt the deficiency of security authorities, they collaborated in constructing informal secured gates to bound their buildings and districts (Mustafa et al. 2017).

Fig. 20.11 Older private residence in Heliopolis with a low transparent fence taken by author (2019)

Fig. 20.12 Recent residential building in New Cairo with high opaque walls taken by author (2019)

Walls of gated communities. The development of gated communities is a global trend, and Cairo has not been isolated from this trend (Singerman 2011). Indeed, a large number of private gated communities have been constructed in the new satellite cities situated on the edges of the capital, such as Sheikh Zaid city to the west and New Cairo, El-Obour and El-Shorouk cities to the east. These gated communities occupy vast areas of lands that was initially intended to be public but had been sold to private investors (Denis 2006). Today, the total area of the gated communities constitutes a large percentage of the urbanised and developed area of Greater Cairo (El-Husseiny 2015; Singerman 2011). These private developments consume a vast part of the city and detach their exclusive facilities and roads from the main public road network (Landman 2006). These developments are typically targeting upper socio-economic levels (Yousry 2010). And, the demand has been high, primarily fuelled by the search for exclusivity and an increased sense of security. Most of these gated communities or compounds are surrounded by high solid walls that prevent any visual or physical interaction between the public space of the street and what happens inside. The walls allow residents of these compounds to enjoy the green spaces, clubhouses and all sorts of exclusive amenities provided in total isolation from the external urban environment and the rest of the inhabitants of the city (see Fig. 20.13).

20.2.2 The Reasons Behind the Wall Phenomenon in Cairo

The decline in the sense of security. Many factors are associated with the need for walling and the decline in the sense of security; chief among them are the increased violent crime and the decreased confidence in public security provided by the state

Fig. 20.13 Walls of a gated community inside a gated city in New Cairo taken by author (2019)

(Lemanski 2010). In Cairo, the declined sense of security is associated with different dynamics and incidents that occurred in the last 20 years. First, the sense of security in Cairo decreased after the terrorist attacks in the 1990s (El-Husseiny 2015). At that time, the media exaggerated the danger of terrorism and Islamists threat, who were hiding inside the city's informal settlements and worked on the construction of fear (Kuppinger 2004). Security then became a necessity and a commodity advertised by the real estate developers. Afterwards, the city's problems started to worsen; the congestion, the traffic, the pollution and the dirt urged the elite to find a refuge and retreat in the walled gated communities (Denis 2006). The absence of the state's role in the organisation of the city's urban planning and the decline in the provision of adequate public spaces led to chaos in Cairo streets (Marafi 2011). Also, the failure of legislation, corruption and the arbitrary actions had worsened the situation and spoiled the remaining good features of the city.

In 2011, after the revolution of 25 January and the failure of the police and security systems in Egypt, the sense of safety in Cairo had declined again. The political violence and the repeated militant attacks were evidence of the failure of the pursued policies (Abdelbaseer 2015) and contributed in the loss of confidence in public security. The media served in broadcasting the discourse of fear by their constant warnings of dangers in the city (El-Masry 2017). Also, social media speeded up the spread of crime news, car accidents, harassments and rumours that were further increasing the sense of fear and uncertainty. Moreover, the escalation of poverty rates in recent years and the growing wealth of a small segment of the population increased polarisation in the society and augmented the fear of the "other" (Shama 2014).

The declined sense of security contributed to the appearance of urban walls. Walling most buildings and urban spaces in Cairo became a norm. Privatisation of public spaces was favoured since citizens preferred to avoid public spaces. Repeated accidents urged the installation of street barriers and drove citizens to rely on individual safety measures and the fortification of their properties. Also, the declined sense of security was one of the main factors behind the increasing demand for gated communities.

The terrorism problem. The problem of terrorism lies in the fact that it is a hidden and unexpected danger. The fear of militant attacks had a role in influencing Cairo's urban spatial dynamics (Kuppinger 2004). The state responded to the problem by taking all the possible security measures for protecting its interests and securing sensitive sites and buildings. Instead of dealing with social and political roots of the problem, the state, with a *fortress mentality* erected higher protective walls, increased numbers of security personnel and built more prisons (Shama 2014). Policy-makers must find innovative political solutions to tackle the issue (Shama 2014). The terrorism problem is one of the main reasons behind the erection of security walls in Cairo streets and walls around most of the public buildings and urban public spaces in Cairo.

The changed nature of privacy needs. There has been a noticeable change in privacy needs within Cairo society. During the first half of the twentieth century, society espoused a westernisation trend, and until the early 1970s, it was more open and less conservative than it is today. For example, women were less conservative regarding the way they dress (Egypt Today 2017). However, the 1980s saw a rise in stricter religious ideologies imported from gulf countries and society became more conservative. This was reflected in the more modest ways women started to dress in the streets of Cairo (Ezzidin 2017). In parallel, with an increased demand for the privacy of domestic life, copying enclosed types of housing typical to gulf countries and walls around residential properties in Cairo became taller and more opaque. The purpose was to shelter domestic life from views from the street and neighbouring buildings. Even private villas within gated communities that are supposed to be protected by the secured walls of the compound are typically surrounded by high opaque fences and tall vegetation to achieve the domestic privacy desired, in a way, creating a system of walls within walls.

Urban politics and neoliberalism. After 1975, as a result of the open-door economic policies, the state responsibilities have been limited to the construction of low-income housing and the government relied on the private sector for the provision of other types of housing (Fahmi and Sutton 2008). The neoliberal policies shifts had been reinforced in the 1990s and the private sector participated in the spatial development of Cairo's new towns (Denis 2006). This contributed to the expansion of the construction of enclosed and demarcated gated communities on the peripheries of Greater Cairo's Metropolitan area by privately owned firms (Kuppinger 2004). The private sector divided the capital into gated communities and the rest of the city causing spatial segregation (Singerman 2011). These policies encouraged segregation based on class differences, leading to the polarisation of Cairo's society (Marafi 2011) and accentuating social injustice. The neoliberal orientation of governments encouraged the trend of urban privatisation. Since then, the private sector dominated in providing various citizens needs and facilities from housing, education, health and recreation which led to the *commodification* of citizens lives and needs. Developers invested in such projects as private schools, private universities, private hospitals, private clubs, private malls; all of them are gated, walled, secured and monitored.

Neoliberal policies contributed in the manifestation of the wall phenomenon in terms of the privatisation of public space and the appearance of gated communities. "Gated communities represent an extreme form of citizen retreat in the global era of neoliberal state withdrawal […] the private sector has emerged as the dominant service provider, alongside the growing privatisation of space in contemporary cities" (Lemanski 2010, p. 4). The issue is not merely the social fragmentation or the physical isolation by walls, it is the withdrawal of citizens from participating in public life, exercising their civil rights, shaping their city and their choice to retreat into secluded walled neighbourhoods (Lemanski 2010).

20.3 The Wall Phenomenon and Potential Implications

20.3.1 The Wall and the Vitality of Urban Public Spaces

Much of the urban design literature discusses the necessity of designing open urban spaces that encourage the occurrence of activities and stresses how urban spaces divided by walls lack vitality. Lynch (1981) emphasised the importance of a continuous fabric that permits accessibility and helps create liveable environments. He described walls as *edges* that prevent the continuity of the space and work as physical barriers (Lynch 1960). Similarly, Jacobs (1961) indicated that successful neighbourhoods should be continuous, intersecting and linked together. They should be diverse, mixed use and bringing people together (Wendt 2009). She argued that for safer environments; buildings must be oriented towards the streets to *keep eyes on the street* and to encourage street activities on sidewalks which, in turn, increases the sense of security in the city. The unplanned daily encounters on sidewalks are essential for a vital public life in the city (Jacobs 1961). For Gehl (1980), lively streets depend on the occurrence of longer activities like "stay and do" and "stay and play" rather than short activities like "come and go". Lively streets are influenced by the street design and the occurrence of activities which tend to be concentrated near the street's *edge* adjacent to buildings (Gehl 1980). Hence, the edge design is very important as it directly affects the liveliness of the public space. When the edge does not work; the street acts as a "come and go" traffic corridor and when the edge works the street becomes a lively place having higher potential for interaction activities (Gehl 1980). "The life of the public space forms naturally around the edge. If the edge fails, then the space never becomes lively" (Alexander et al. 1977, p. 600).

Permeability affects the entire urban sensory experience. The liveliest activities occur beside the most permeable edges. Blank opaque walls that one cannot see or pass through are considered as dead spaces (Mehta 2006). Since opaque walls restrict visual and physical permeability, they affect the urban experience and urban vitality.

Accordingly, the edge design affects the liveliness of the street, and when the edge is a blank wall, the street becomes a *dead corridor* that lacks any liveliness. Hence, urban walls closing or framing public urban spaces in Cairo are affecting their vitality. The wall phenomenon is with all its manifestations whether urban walls, street barriers, security walls, walls of residential or gated communities have negative implications on the vitality of streets and urban spaces in Cairo.

20.3.2 The Wall and the Right to the City

According to Lefebvre (2003), Lefebvre and Nicholson-Smith (1991), the right to public space is essential for achieving the right to the city. It is the right to access, occupy and use the space. It is the right to belong to and not to "be alienated from the spaces of everyday life" (Mitchell and Villanueva 2010, p. 2), the right to claim social and economic services as housing, culture and work, the right to all citizens without exclusion to a space in the city (Mitchell and Villanueva 2010).

Harvey (2008) claimed that the right to the city is one of the critical neglected human rights. It is not only limited to urban accessibility; it is the citizens' right to change and shape their own city (Harvey 2008). Citizenship and sense of belonging have been threatened by the neoliberal principles. The privatisation of public space has transformed the urban space into a *commodity*, promoting fragmented *urban fascism*. A new lifestyle has been introduced; an urban political economy is based on consumerism, tourism and industries. In developing countries, cities have been encroached by private microstates and fortified fragments (Balbo 1993). Private neighbourhoods offer for their wealthy residents all kind of exclusive services such as private schools, golf courses, sports clubs and private police while informal settlements are deprived of all the means of a proper human life. In the last decades, the right to the city has been restricted to a minor group of elite having the political and economic power to shape the city according to their own private interests (Harvey 2008).

Consequently, urban walls dividing the city are diminishing the public right of moving freely, accessing and using all parts of the city especially for gated public buildings and urban spaces. Also, the privatisation of public space excludes the right to use private facilities to a minor group of the society who have the means. Street barriers imposed on pedestrians control their behaviour and exclude them from the reshaping or participating in the design of their city. Finally, gated communities, with their enclosed amenities and road networks, subtract a vast area of land from the accessible map of the city.

20.3.3 The Wall and Urban Social Cohesion—Segregation

Lynch (1981) stressed the importance of mixing urban uses in "a fine mosaic of public and private places" (1981, p. 272) at the neighbourhood level as this increases social cohesion. Placing urban uses and people of different cultures in proximity to one another is essential for achieving a healthier community. Private realms should be small and should exist within a matrix of a democratic public space. The opposite could be dangerous for the community (Lynch 1981). Unfortunately, the increased rate of violent crime drives the ones who have the means to enclose themselves behind the walls, to retreat from public life and to protect themselves from the dangerous "other" (Carolyn 2011). A city of walls prevents social interactions in public spaces and creates spaces of fear, discrimination and suspicion (Caldeira 2000). When people isolate themselves and interact less, their sense of community and security decline, and their sense of fear increases (Calthorpe 1993). Spaces that used to be public are turned into private walled spaces for wealthier groups and are restricted for the majority. The remainder public space is left for the vulnerable "other", and the wall protects the insider "us" from the external "them" (Lemanski 2010). Urban segregation trends are reinforced through the privatisation of public spaces and restricted services which produce a "fortress city" (Brighenti and Kärrholm 2019; Low 2003). In urban design, the separation of adjacent different social classes by walls is the most blatant pattern of segregation as it confirms rules of inclusion and exclusion. This pattern establishes a new culture that accentuates class division (Caldeira 2000).

Thus, the increasing wall phenomenon in Cairo, the privatisation of public spaces, walls of exclusive gated communities and the fortification of residential buildings have a great implication on destroying social cohesion and increasing polarisation and segregation.

20.3.4 The Wall and Crime Prevention—Vandalism

In order to solve the problem of rising crime rates, some urban planners led by Newman (1972) and Coleman (1985) promoted the notion of "defensible space". They claimed that controlling access and improving surveillance can contribute to building safer environments. These ideas merged in a way or another with the neoliberal policies and the appearance of privatised urban spaces and gated communities. "Newman and Coleman's work fed into the neoliberal urban agenda that was on the ascendant by the 1980s" (Carolyn 2011, p. 666). Poyner (1983) encouraged the total separation between houses of the middle class and poorer residents. According to him, houses should not only be separated by roads but preferably by walls or edges (Poyner 1983). Consequently, he gradually justified the development of urban walls and gated communities. Policies implemented based on these theories promoted the design of exclusive public spaces and have been criticised for increasing the socio-economic polarisation in cities (Carolyn 2011). By the admission of planners who were promoting defensible architecture, lack of surveillance over the street and low lighting make the place vulnerable to vandalism (Newman 1996). But, the use of physical barriers prevents *keeping eyes on the street* (Jacobs 1961). The wall is destined to fail in providing the expected protection and is vulnerable to vandalism. While walls and gates are a tool for securing internal edifices against vandalism and theft (Grant and Mittelsteadt 2004), and their external side is exposed to vandalism and damage; the wall itself is a fragile entity that needs to be defended (Young 2019). Graffiti drawings over blank walls are considered as a type of antisocial behaviour that leads to a decline in urban quality (Iveson 2010).

While the main purpose of the wall is to define a property line, to limit access and to provide security and control, the physical barrier provokes vandalism, prevents surveillance over surrounding streets and the wall and, in turn, needs to be defended.

20.3.5 The Wall and Image of the City

Walls that surround vast areas of the urban environment have a huge impact on forming the image of the city. Visual accessibility is an essential factor that influences spatial orientation and wayfinding (Gärling and Golledge 1989). Visual permeability is the possibility of viewing different parts of the environment from different points. The wall acts as a visual barrier in the urban environment, and its opacity affects the degree of visual permeability. Walling buildings and closing streets create a restricted environment which affects the experience of walking and observing. The restricted movement prevents the observer from forming a mental map of the city (Caldeira 2000). Urban walls are a design tool that shapes the urban experience by influencing urban mobility. Urban mobility influences the image of the city (Young 2019). Politics of visibility determine politics of possibilities and *appropriacies* like who "can" or "cannot" see and what "can" or "cannot" be seen. Hence, the wall offers politics of verticality (Brighenti 2009).

Concealing large spaces of lands in Cairo using walls weakens the mental image of the city and diminishes the visual connection between the city's different parts. In that sense, the wall prevents the formation of a complete image of the city and complicates wayfinding.

20.4 Conclusion

The paper proposed the main reasons behind the increasing wall phenomenon in Cairo. First, the decline in the sense of security was due to the escalation of the city urban political problems, the spread of the discourse of fear and the increase of the socio–economic gap. Second, the terrorism problem and militant attacks prompted the state in taking defensive measures and erecting security walls that tend to spread a feeling of fear. Third, the changed nature of privacy needs and the drift towards a more conservative lifestyle are the reasons behind the increased height and opacity of residential walls. Finally, the neoliberal politics of the state supported the privatisation of urban spaces and the manifestation of private gated communities which turned urban life into a commodity.

It is clear that there are many potential problems related to the wall phenomenon. The paper identified the potential implications based on the related literature. Mainly, the wall affects urban space vitality as it prevents space continuity and the occurrence of vital activities. Unfortunately, Cairo streets lack vitality. Urban spaces are not inviting for possible interactions as spaces are either private, closed, fenced or deteriorated. Also, the walls dividing the city are against the right to the city defined as the right to access and use the city. Cairenes do not have the full right to access all parts of the city due to the presence of urban walls, exclusive amenities inside private communities and privatisation of public spaces. Furthermore, their behaviour is controlled through barriers and they do not have the right in participating and reshaping their city. The wall has a negative impact on social cohesion as it accentuates class differences. Cairo citizens live in separate isolated spheres. Those who have the means reside, shop, entertain, educate and live in a total exclusive sphere while they fear the poor "other". The wall is also used to show status and power. The poor live in the rest of the city. They are banned from using or accessing facilities inside walled urban spaces and buildings. Actually, the wall is a tool for urban segregation. While the wall is a tool for crime prevention, some authors argue that it is not successful and has negative impacts on urban space. Signs of vandalism and graffiti are spread along most of public and security walls in Cairo. Incidents of climbing and overpassing the walls are very common particularly walls around schools, clubs and public gardens. Lastly, urban walls and street barriers restrict the urban environment, and as a visual screening, they prevent the observer from forming a complete image of the city. Cairo is losing its identity behind the walls that hide most of its urban spaces and buildings.

To conclude, the paper suggests that the wall has an extended impact on the urban environment from the physical, social and visual aspects that need to be highlighted. Furthermore, the wall's initial purpose of securing, banning, isolating or controlling has failed in many cases. This should drive decision-makers to review the matter and to find alternative solutions with the least negative implications on the city.

References

Abaza M (2006) The changing consumer cultures of modern Egypt: Cairo's urban reshaping. In: Eickelman DF (ed), vol 101. Brill, Boston

Abaza M (2013) Walls, segregating downtown cairo and the mohammed mahmud street graffiti. Theor Cult Soc 30(1):122–139. https://doi.org/10.1177/0263276412460062

Abd Elrahman AS, Mahmoud RA (2016) The post-revolutionary effect on the urban harmony of Cairo's built environment in relation to the collective memory of the population. Ain Shams Eng J 7(4):1099–1106

Abdelbaseer AM (2015) Send in the wall: a tale of a forbidden city. Retrieved 27 Feb 2019, from This Big City website: https://thisbigcity.net/
Alexander C, Ishikawa S, Silverstein M (1977) S. A pattern language. Oxford University Press, New York
Balbo M (1993) Urban planning and the fragmented city of developing countries. Third World Plann Rev 15(1):23
Brighenti AM (2009) Walled urbs to urban walls–and return? On the social life of walls. The Wall and the City. Professional Dreamers, Toretno, Italy. pp 63–71
Brighenti AM, Kärrholm M (eds) (2019) Urban walls: political and cultural meanings of vertical structures and surfaces, 1st edn. Routledge, New York
Cairobserver. (2011a). Andalus Park, pt 1 (El Tahrir Garden). Retrieved from Cairobserver website: https://cairobserver.com/post/8571646315/
Cairobserver (2011b) Culture behind the fence. Retrieved from Cairobserver website: https://cairobserver.com/post/7913846646/
Caldeira TPR (2000) City of walls: crime, segregation, and citizenship in São Paulo. University of California Press, Berkeley
Calthorpe P (1993) The next american metropolis: ecology, community, and the american dream. Princeton Architectural Press, New York
Carolyn W (2011) Secure cities. In: Companion to urban design. Routledge
Coleman A (1985) Utopia on trial: vision and reality in planned housing. Hilary Shipman, London
Denis E (2006) Cairo as neoliberal Capital? Cairo cosmopolitan. The American University in Cairo Press, Cairo, New York, pp 47–71
Egypt Independent (2012) MP demands the reopening of downtown streets [News]. Retrieved 2 June 2019, from Egypt Independent website: https://ww.egyptindependent.com/
Egypt Today (2017) Difference between Egyptian fashion in 1950s vs. Now. Retrieved from EgyptToday website: https://www.egypttoday.com/
El-Husseiny MBM (2015) Compounds of modernity: national order and the other in Egypt (1940–present). University of California, Berkeley
El-Masry E (2017) From gated city to gated communities: evolution of Cairo's walls. Retrieved 11 Jan 2019, from Egyptian Streets website: https://egyphutttianstreets.com/
Ezzidin T (2017) Depiction of hijab in Egyptian Media: why does it often backfire? Retrieved 23 June 2019, from Egyptian Streets website: https://egyptianstreets.com/
Fahmi W, Sutton K (2008) Greater Cairo's housing crisis: contested spaces from inner city areas to new communities. Cities 25(5):277–297
Foucault M (1977) Discipline and punish: the birth of the prison (trans: Sheridan A). A Divion of Random House Inc. Vintage Books, New York
Gärling T, Golledge RG (1989) Environmental perception and cognition. In: Zube EH, Moore GT (eds) Advance in environment, behavior and design, vol 2. Springer, Boston, MA, pp 203–236
Gehl J (1980) The residential street environment. Built Environ 6(1):51–61
Ghonimi I, Al Zamly H, Khairy M, Soilman M (2010) Against the great divide between theory and practice: gated communities versus urban liveability. Presented at the REAL CORP 2010: Liveable, prosper, healthy CITIES for everyone, Vienna
Grant J, Mittelsteadt L (2004) Types of gated communities. Environ Plan B: Plan Des 31(6):913–930. https://doi.org/10.1068/b3165
Groat LN, Wang D (2013) Architectural research methods, 2nd edn. Wiley, Hoboken, NJ
Harvey D (2008) The right to the city. New Left Rev 53
Hutchison R (ed) (2010) Cairo, Egypt. In: Encyclopedia of urban studies. https://doi.org/10.4135/9781412971973.n37
Iveson K (2010) Graffiti. In: Encyclopedia of urban studies, pp 327–329. https://doi.org/10.4135/9781412971973
Jacobs J (1961) The death and life of great american cities. A Divion of Random House inc. Vintage Books, New York

Kuppinger P (2004) Exclusive greenery: new gated communities in Cairo. City & Society 16(2):35–61. https://doi.org/10.1525/city.2004.16.2.35
Landman K (2006) Privatising public space in post-apartheid South African cities through neighbourhood enclosures. GeoJournal 66(1):133–146. https://doi.org/10.1007/s10708-006-9020-5
Lefebvre H (2003) The urban revolution. University of Minnesota Press, Minneapolis, London
Lefebvre H, Nicholson-Smith D (1991) The production of space (Nachdr.), vol 142. Oxford Blackwell, Malden, MA, USA
Lemanski C (2010) Gated community. In: Hutchison R (ed) Encyclopedia of urban studies, pp. 290–292. https://doi.org/10.4135/9781412971973.n109
Low S (2003) Behind the gates: life security and the pursuit of happiness in fortress America. Routledge, New York
Lynch K (1960) The image of the city, vol 11. MIT press, MA
Lynch K (1981) A theory of good city form. MIT press, Cambrige, MA
Marafi S (2011) The neoliberal dream of segregation rethinking gated communities in greater Cairo a case study Al-Rehab City. The American University in Cairo, Cairo
Mehta V (2006) Lively streets: exploring the relationship between built environment and social behavior. University of Maryland, USA
Messieh N (2012) Taking down walls is the new tahrir [Blog]. Retrieved 20 June 2019, from Atlantic Council website: https://www.atlanticcouncil.org/blogs/menasource/
Mitchell D, Villanueva J (2010) Right to the city. In: Hutchison R (ed) Encyclopedia of urban studies. https://doi.org/10.4135/9781412971973.n238
Mustafa N, Khalifa M, Abd Elrahman A (2017) Safety and security of cities and their impact on the city life. SSRN Electron J. https://doi.org/10.2139/ssrn.3164155
Newman O (1972) Defensible space. Macmillan, New York
Newman O (1996) Creating defensible space. Diane Publishing, New York
Poyner B (1983) Design against crime: beyond defensible space. Butterworths, London, Boston
Shama NM (2014) City of walls. Le Monde Diplomatique. Retrieved from https://mondediplo.com/2014/09/13cairo
Singerman D (2011) Cairo contested: governance, urban space, and global modernity. The American University in Cairo Press, Cairo, New York
Tadamun (2013) Izbit Khayrallah. Retrieved 18 June 2019, from Tadamun website: https://www.tadamun.co/
Tadamun (2014) The right to public space in the Egyptian constitution. Retrieved 18 June 2019, from Tadamun website: https://www.tadamun.co/the-right-to-public-space-in-the-egyptian-constitution/
Wendt M (2009) The importance of death and life of great American cities (1961) by Jane Jacobs to the profession of urban planning. New Visions for Public Aff 1:1–24
Young A (2019) On walls in the open city. In Mubi Brighenti A, Kärrholm M (eds) Urban walls: political and cultural meanings of vertical structures and surfaces, 1st edn. Routledge, New York, pp. 29–46
Yousry A (2010) The privatization of urban development in Cairo. Cairo University, Cairo, Faculty of urban and regional planning

Chapter 21
Adaptive Reuse as a Revitalization Tool: The Case of Downtown Cairo Post Relocation of Governmental Buildings to the New Administrative Capital

Nada Hassan, Marwa A. Khalifa, and Johannes Hamhaber

Abstract Although Egypt has several experiences in building new satellite cites, the case of the New Administrative Capital is different in having a direct effect on Cairo as it is relocating a major function, its administrative center. Accordingly, this new development should be planned and implemented with a holistic approach including how to revitalize Cairo and re-establish it as the cultural and historic capital of Egypt. As announced by the government, the first phase will include the relocation of the Egyptian Parliament and all ministries from Cairo and Giza to their new headquarters at the administrative capital. This research explores the effects of adaptive reuse of abandoned buildings in Downtown due to the relocation to the New Administrative Capital on their neighborhoods and collectively on the city. It focuses on the governmental buildings through categorizing government strategies, mapping buildings with planned reuse if available and assessing planned government reuse through conducting interviews with involved government officials. After exploring the current framework for the relocation and reuse processes, a generic categorization for the abandoned governmental buildings was developed.

Keywords Adaptive reuse · Urban revitalization · Purpose-built capitals · Downtown Cairo · New Administrative Capital

---

N. Hassan (✉)
Urban Planner and Researcher, Cairo, Egypt
e-mail: nada.hossam.eldin@hotmail.com

M. A. Khalifa
Ain Shams University, Cairo, Egypt
e-mail: marwa_khalifa@eng.asu.edu.eg

J. Hamhaber
TH Koeln, Köln, Germany
e-mail: johannes.hamhaber@th-koeln.de

Architecture and Urbanism: A Smart Outlook,
https://doi.org/10.1007/978-3-030-52584-2_21

21.1 Introduction

The research focuses on the buildings that will be abandoned in Downtown Cairo due to the relocation to the New Administrative Capital through categorizing government strategies, mapping buildings with planned reuse if available and assessing planned government reuse based on the information extracted from the conducted interviews with officials from Administrative Capital for Urban Development (ACUD), Ministry of Housing, Utilities and Urban Communities, General Organization for Physical Planning (GOPP) and through reviewing related government announcements and news articles to confirm which buildings will be totally or partially abandoned, ownership, specifications of the building and if there are any plans regarding the reuse or intervention for the building. Aiming to explore the potential of these abandoned buildings as a tool for urban revitalization for Cairo's Downtown neighborhood, interviews were conducted with officials from GOPP and UN-Habitat to learn more about the urban challenges and opportunities facing this neighborhood.

Due to the exploratory nature of the research, a qualitative approach was adopted during the data collection phases in order to help explore the different stakeholders, entities and frameworks involved in the reuse process in Egypt.

21.1.1 Data Collection

Conducting interviews is one of the most effective data gathering methods in Egypt and many developing countries where many important information is not published or accessible. As interviews are an effective tool for obtaining data that cannot be directly observed (Patton 1990), hence, it is suitable for the case with the future plan for reuse of buildings.

The primary date collected through the interviews was revised and backed-up by secondary sources from information and documents published by government entities, statements and interviews of public officials and official Web sites of involved agencies. The interview guide was adopted for this research as it is the commonly used approach for qualitative research where the interviewer prepares the outline of topics or questions to be covered but remains free to change the order and/or the wording during the interview as seems fit (Patton 1990).

21.2 Adaptive Reuse as a Revitalization Tool for Capital Cities

During the last century for various reasons, several countries have changed their capital cities through either transferring it to another already established city or establishing a newly purpose-built capital city such as the examples of Brasilia,

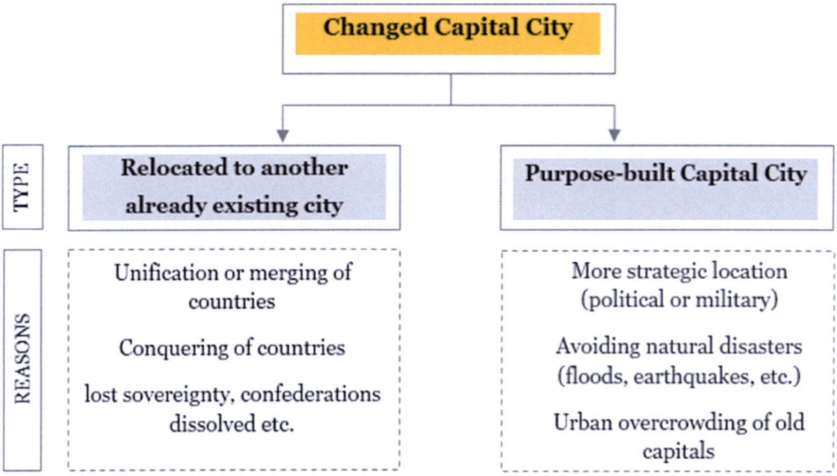

Fig. 21.1 Reasons behind changed capitals. *Source* Authors

Washington and Putrajaya. According to Vale (2006) during the 2000s, about three quarters of the 1900s capital cities were not serving as capital cities anymore. Joining the list of countries recently developing its purpose-built capitals is not only Egypt but also South Korea. Relocating a capital city is not a simple decision to be taken and should rather be planned and studied well, as Gottmann (1985) argued capital cities are not only cities where the seat of the central government is located, as it also acts as an anchor between different regions of the country hence why the rearrangement or change of these complex structures and relations is complicated. Through reviewing different examples of changed capital cities, it was clear that this development decision is usually planned in order to address challenges faced by the old capital such as increased urbanization and densities. Figure 21.1 illustrates the main reasons behind changing capitals whether to a newly purpose-built capital or to another already established city.

Figure 21.2 illustrated how relocation of capital cities could result in functional changes that would have spatial implications such as change in traffic patterns and unused spaces and buildings. Although this could be perceived as a negative impact, but if planned well could act as a major opportunity. For instance, in compact neighborhoods such Downtown Cairo, the sudden availability of vacant spaces and buildings could be efficiently utilized.

21.2.1 Defining Adaptive Reuse

Adaptive reuse is considered an effective strategy when it comes to sustainability as it enhances the optimization of built assets through changing the function the

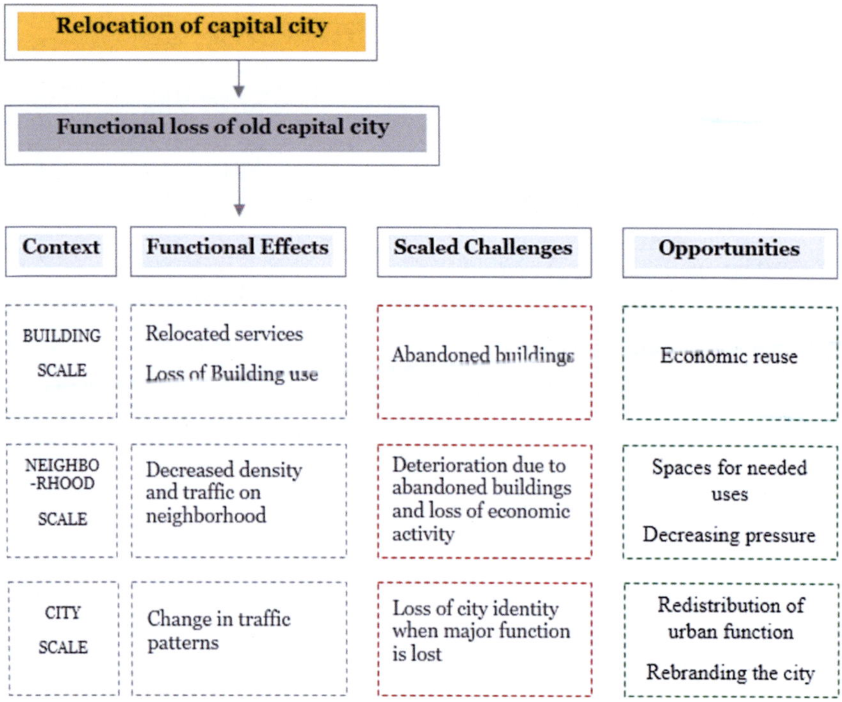

Fig. 21.2 Functional Changes, challenges and opportunities imposed by relocation of capital cities

building was originally designed and built for to a new use (Bullen and Love 2011a, b). Some might define adaptive reuse as the aesthetic process of adapting buildings for new functions while maintaining their old features Joachim (2002); argues that a more accurate definition is that adaptive reuse is the process of retaining most or all of a building's structural system while also trying to retain as much as possible of the other elements such as cladding and interior partitions in order to extend the building's cradle-to-grave period.

21.2.2 Adaptive Reuse on the Neighborhood Scale

According to Bullen and Love (2011a), architectural conservation and reuse have changed from only preservation to being an effective part of urban regeneration strategies and sustainable development. At the macro level, the dynamic factors of urbanization and economic growth could act as triggers for urban renewal and revitalization strategies, where building adaption acts as a significant part of these revitalization schemes (Douglas 2002). Furthermore, SDG goal 11, Sect. 11.4 focuses on the significance of heritage and building conservation to the sustainable development agenda (ICOMOS 2015). The term of urban revitalization according to Cowan (2005) could be used in referring to actions or procedures of enhancing a place and mentioned that finding new uses for old buildings and promoting cultural activities are two examples of urban revitalization tools. Furthermore, according to Bullen and Love (2010) and Ball (1999), the conservation of buildings especially those with valuable heritage has become an important tool in regeneration, explaining that conserving parts of the city's history through the adaptive reuse of buildings for a variety of new functions is driven by growing calls of urban regeneration.

21.2.3 Adaptive Reuse Decision-Making Process

Although the adaptive reuse requires many qualifying factors to be executed and assures continuity such as responding to the user requirements and satisfaction (Günçe and Mısırlısoy 2019), yet it is still more efficient in terms of energy consumption and reducing emissions (Bullen 2007). According to Bullen and Love (2011a, b), the adaptive reuse decision-making process is influenced by three underlying factors which are capital investment, asset condition and regulation. Günçe and Mısırlısoy (2019) argued that in many cases of adaptive reuse, the physical aspects were focused on the most whereas the sociocultural and economic aspects were neglected through the decision-making process resulting in unsuccessful adaption. In that sense, determination of the most suitable future use for the buildings could be done by decision-makers only after the comprehensive understanding of a complex set of considerations including location, heritage, architectural and structural assets and market trends (Bullen and Love 2011a). Jansz (2012) argues that in cases where the market is uncertain and difficult to be predicted, where the lifespan of new investments is expected to be relatively short, reuse will probably be more sustainable than demolition and new build.

21.3 Abandoned Governmental Buildings in Downtown Cairo

21.3.1 Relocation of the Administrative Center

The New Administrative Cairo is not Egypt's first attempt to relocate the administrative center of Cairo as there have been several unsuccessful attempts under different regimes (Awatta 2015). The first attempt was during the late 1950s during President Abdelnaser's regime, when Naser City, a new urban suburb east of Cairo, was constructed in 1958 with a plan to relocate governments buildings there. However, the plan did not succeed with only few ministries now located in the residential

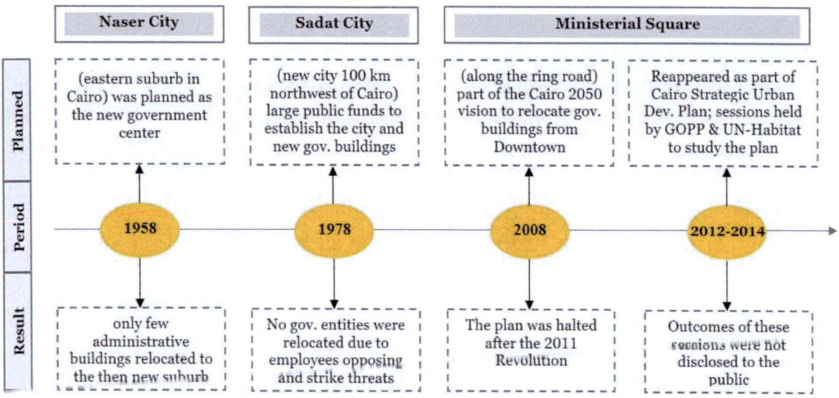

Fig. 21.3 Previous Attempts to relocate Cairo's administrative center. *Source* Authors

commercial neighborhood of Naser City. Another attempt was under President Sadat by establishing the Sadat City, a new city 100 km northwest of Cairo. Large funds were allocated to establish the city and construct the new government buildings there; however the plan was opposed fiercely by government employees, threatening to strike if the relocation plan was put to action. The last attempt was part of the Cairo 2050 plan under Mubarak Regime in 2008, where the proposal was to move the ministries and government buildings from Downtown to another location along the Ring Road; nevertheless the plan was halted because of the 2011 revolution where Mubarak Regime was overthrown. This proposal reappeared as part of Cairo Strategic Urban Development Plan where GOPP and UN-Habitat held what was referred to as community sessions to study the proposal, nonetheless the outcomes of these sessions were never disclosed to the public. Figure 21.3 illustrates the previous different attempts of the relocation of the administrative center.

New Administrative Capital was first revealed at Egypt's Economic Development Conference in March 2015 and is being currently developed by The New Administrative Capital for Urban Development (ACUD), a holding company with a capital estimated by 6 billion pounds. The company consists of the New Urban Communities Authority (NUCA), National Service Products Organization (NSPO) and Armed Forces National Lands Projects Agency. Some of the major contracting and development companies working on this project include Orascom, Hassan Allam, Petrojet, Concord, Talaat Mustafa, Arab Contractors and Wadi El Nile (State Information Service).

For the first phase, the target population is about 0.5 million inhabitants and around 50,000 governmental employees. The first phase will include the relocation of the Egyptian Parliament and all ministries from Cairo and Giza to their new headquarters at the administrative capital's government district. The planned capacity to be reached by phase three is 100,000 governmental employees with the expected relocation of further governmental entities (New Urban Communities Authority n.d.). According to Fateen (2019), phase one of the relocation acts as an attraction pole to initiate

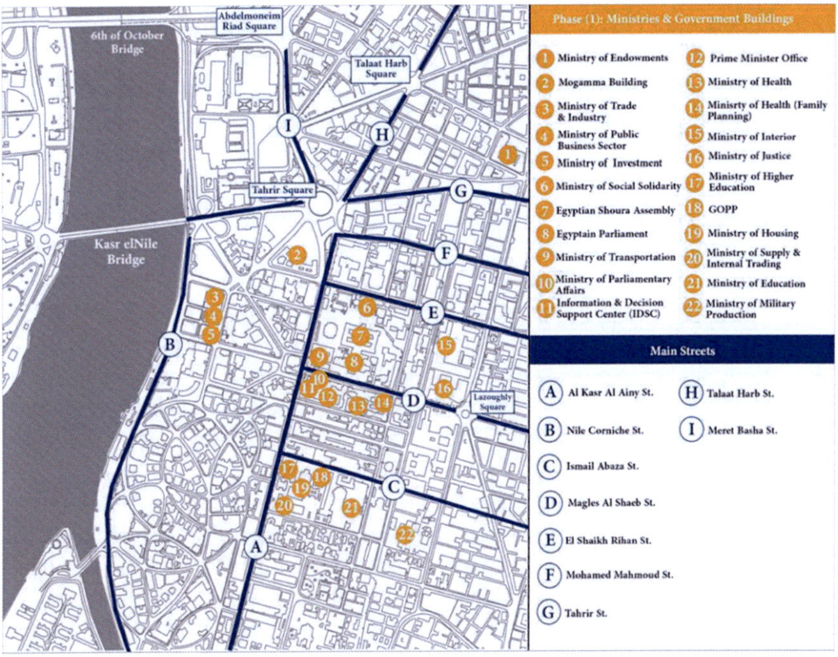

Fig. 21.4 Mapping of governmental buildings relocating during the first phase. *Source* Authors

movement of other governmental entities and private business which will eventually also act as an attraction for more inhabitants and visitors. According to the same source, the list of other governmental entities that will be relocated after phase one is not yet finalized but there is an initial plan of moving embassies and consulates to the diplomatic zone at the New Administrative Capital's government district. The below map illustrates the locations of the buildings planned to relocate during the first phase (Fig. 21.4).

21.3.2 Framework

The relocation of government buildings is generally composed of two main tasks: The first is the relocation of manpower which is carried on by the Ministry of Planning and Administrative Reform where manpower of each government building and ministry is studied in terms of number of employees and their grades as it varies in the required office space for each (ElShaarawy 2019). The second task is the design and construction of the new government buildings in the New Administrative Capital which is supervised by the Ministry of Housing, Utilities and Urban Communities, as it is considered the government's arm when it comes to issues related to the built environment (Abdelwahab 2019). As some ministries have larger manpower to be accommodated in their new buildings at the New Administrative Capital, some departments will remain at Downtown Cairo (Abdelwahab 2019).

According to the conducted interviews and government announcements and news articles, the government will deal with each abandoned building individually according to different factors such as building condition, location, area, possibility of new use and most importantly the architectural or historic value. The interventions whether it is demolition or reuse will not be decided unless the whole building becomes vacant; however there is a proposal that some of the high-rise buildings could collectively host departments from various ministries that have not relocated to the New Administrative Capital instead of partially occupying several buildings and hindering their reuse.

Regarding the reuse process, two different frameworks are available: The first is the establishment of "Sandook Misr" in August 2018. The Egyptian Official Gazette published the decision of President Abdel Fattah al-Sisi to issue Law No. 177 of 2018 establishing a new entity called Egypt Fund "Sandook Misr" (ALYoum el Sabee 2018). This law states that the President may, after counseling the Prime Minister and the Minister of Planning, transfer the ownership of any of the unutilized assets owned by the state or any of its entities or subsidiaries to the established "Egyptian Fund" or any of the funds or entities it establishes and owns. The ownership of functioning assets may as well be transferred upon the agreement between the Prime Minister, the Minister of Planning and the Minister of Finance in coordination with the concerned Minister.

The other reuse framework was established to deal with listed or protected buildings that are of heritage or architectural value. According to Abdelwahab (2019), during the ministerial committee organized between 2016 and 2017, under the auspices of the Prime Minister, compromised of representatives from different ministries such as Ministry of Planning and Administrative Reform, Ministry of Finance, Ministry of Housing where GOPP was mainly responsible for the technical aspects. The committee worked on developing a reuse framework for the soon to be abandoned buildings, by first reviewing the work done by GOPP and a private consultancy firm called Mckinsey during 2007–2009 for a proposal of reuse of the ministries district when it was planned to be relocated during the Mubarak Regime as part of the Cairo 2050 plan. According to the conducted interviews and literature review, Fig. 21.5 illustrates the relocation and reuse framework of the governmental buildings.

According to the conducted interviews with officials from ACUD, Ministry of Housing and GOPP, Table 21.1 explains how the above framework in Fig. 21.5 will be applied on some of the mapped buildings and also explains the planned reuse or intervention if already announced.

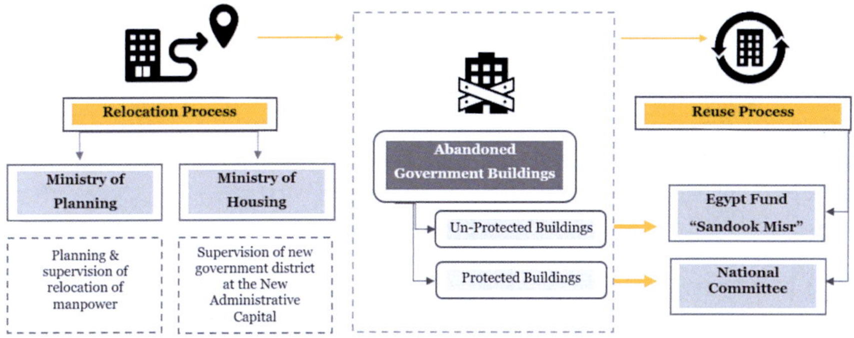

Fig. 21.5 Relocation and reuse framework of governmental buildings. *Source* Authors

Table 21.1 Examples of how the reuse framework will be applied on some buildings

Building	Current status	Government proposed intervention
Mogamaa Building	Partially occupied, only the passports immigration and nationality administration are still in the building but will be soon moved to the Police Academy	Will mostly be reused not demolished due to its significant value into an office building/ hotel/or both combined
Parliament Building	N.A	Will mostly be reused not demolished due to its significant value into an events/conferences venue and not as a museum as some forecasted
Ministry of Interior Building	The building is already vacant as the ministry relocated to its new premises along the Rind Road. According to an interview with GOPP, the assets were transferred to the Ministry of Housing to manage the reuse as this building is listed under protected buildings	N.A
Ministry of Housing & GOPP	Still occupied	There is a proposal to join both small compounds, demolish the high-rise building of Ministry of Housing and reuse the two palaces of GOPP and Ministry of Education as boutique hotels, museums or cultural venues
Ministry of Education	N.A	

Source Authors

21.3.3 *Categorization*

Through the mapping and observation during site visits, the below generic categorization was developed although there should be a more comprehensive categorization that would also take into consideration the protected buildings categories developed by the National Organization for Urban Harmony (NOUH).

21.3.3.1 Ownership

Most of the buildings hosting government entities within the study area are state-owned, which in turn will ease the framework of managing its reuse after the relocation with the assistance of involved ministries. In some other cases such as the case of the Ministry of Environment building in Maadi, the building is rented from a landlord and therefore will be released without having a role in deciding or managing the new function of the building (Fateen 2019).

Misr Real Estate Asset Management Company announced that according to the ruling of the Constitutional Court issued in 2018, all governmental or private companies that rent historic or unique buildings from Misr Real Estate Asset Management Company are obliged to handover the buildings to the company after moving to New Administrative Capital (Ahram 2019). Misr Real Estate Asset Management, a subsidiary of Misr Insurance Holding Company, owns 351 historic buildings in the governorates of Egypt, 75 of which are in Downtown (with a total of 140 buildings in Downtown).

21.3.3.2 Architectural Style

Based on the observation, the below categories in Table 21.2 were derived.

Table 21.2 Categorization according to architectural style

Category	Category (A) low-rise buildings with architectural value (originally villas or palaces)	Category (B) medium–high-rise buildings with architectural value	Category (C) medium–high-rise buildings (relatively modern or eclectic style)
Example	Ministry of education	Ministry of social solidarity	Ministry of supply and internal trade
Photo			

Source Authors

21.3.3.3 Buildings Grouping

This aspect explains the grouping of related or non-related adjacent buildings or buildings that lie on the same plot of land. It is very crucial to understand the relation of each building with its surrounding when considering reuse, as this aspect might act as an opportunity or risk. For instance, buildings that are connected or lie within a fence could have supplementary functions or have common facilities. In other situations, some functions may hinder the functions of other adjacent buildings if placed near each other. After few site visits to the study area, the following typologies were observed: Category (1) Single Buildings: The typology includes government buildings that are allocated in land plots with other privately owned or educational buildings that will most likely remain occupied. Category (2) Buildings sharing the same land plot are the most common typology within the study area, as most of the government buildings are allocated on common land plots but each having their own entrances. Category (3) Buildings grouped together by a fence, this category is a very interesting typology where buildings that are mainly affiliated to the same entity or have common mandates are located inside a small compound surrounded by a fence. This could be very resourceful when deciding a new use as they most probably have shared outdoor areas that could be useful for certain activities.

21.4 Conclusion

As established earlier, the relocation of the capital city especially to a newly purpose capital is usually a development direction that is somehow linked to addressing issues imposed by the old capital. Accordingly, extensive planning should be done to take this as an opportunity to revitalize the old capital; however, there is very scarce literature regarding this issue. In the case of Cairo and the New Administrative Capital, the government announced the plan of relocating the administrative center in order to decrease the pressure and traffic on Cairo; however, the plan of the phasing out and how this would affect Cairo remained unclear and unannounced to the public.

Due to the ambiguity surrounding the development of and relocation of governmental buildings to the New Administrative Capital, the main aim was to explore the current government's framework in dealing with the relocation as a first step and then be able to assess the adaptive reuse potential of the abandoned buildings in Downtown Cairo. Although there is still no complete reuse framework available as mentioned earlier, but at least the main stakeholders included in the decision-making process were identified as follows: The Prime Minister, Ministry of Planning and Administrative Reform, Ministry of Finance, and Ministry of Housing, Utilities and Urban Communities, GOOP. Other stakeholders that are related/involved are ACUD, Ministry of Investment, Misr for Real Estate Asset Management Company and National Organization for Urban Harmony (NOUH).

Adaptive reuse should not only be addressed as an intervention on the building scale due to its effect on the surrounding neighborhood and on the city scale as a means of heritage and identity conservation. The main recommendation regarding the reuse process is conducting a comprehensive need and potential assessment for Downtown Cairo to allocate uses accordingly. It is important to focus on Downtown Cairo's strengths or booming activities such as arts, culture and entrepreneurship.

References

Abdelwahab M (2019) Housing minister assistant for Urban planning—GOPP. (N. Hassan, Interviewer)

Ahram (2019) Retrieved from Ahram online: http://www.ahram.org.eg/News

ALYoum elSabee (2018) Retrieved from https://www.youm7.com/story/2018/8/21/%D9%86%D9

Awatta H (2015) Whose downtown is it anyway? the urban transformation of downtown Cairo between state and non-state stakeholders. The American University in Cairo

Ball R (1999) Developers, regeneration and sustainability issues in the reuse of vacant industrial buildings. Build Res Inf 27(3):140–148

Bullen PA (2007) Adaptive reuse and sustainability of commercial buildings. Facilities 25(1/2):20–31

Bullen PA, Love PE (2010) The rhetoric of adaptive reuse or reality of demolition: views from the field. Cities 27(4):215–224

Bullen PA, Love PE (2011a) Adaptive reuse of heritage buildings, vol 29. No. 5. Structural Survey, pp 411–421

Bullen P, Love P (2011b) A new future for the past: a model for adaptive reuse decision-making. Built Environ Project Asset Manage 1(1):32–44

Cowan R (2005) The dictionary of urbanism

Douglas J (2002) Building adaptation. Butterworth-Heinemann Publishing, London

ElShaarawy B (2019) Technical advisor to the minister of housing, utilities and urban communities (N. Hassan, Interviewer)

Fateen A (2019) Administrative capital for urban development (ACUD). (N. Hassan, Interviewer)

Gottmann J (1985) The study of former capitals. Ekistics 52(314/315):541–546

Günçe K, Mısırlısoy D (2019) Assessment of adaptive reuse practices through user experiences: traditional houses in the Walled City of Nicosia. Sustainability
ICOMOS (2015) Cultural heritage, the un sustainable development goals, and the New Urban Agenda
ICOMOS Concept Note for the United Nations Post (2015) Agenda and habitat III. International council on monuments and sites (ICOMOS).
Jansz S (2012) The effect of the estimated service life on the sustainability of vacancy strategies MSc, TU Delft
Joachim M (2002) Adaptive reuse, massachusetts institute of technology, Cambridge, Massachusetts. Retrieved from https://www.archinode.com/lcaadapt.html
NUCA (n.d.) Retrieved from New Urban Communities Authority: https://www.newcities.gov.eg/know_cities/NewCapital/default.aspx
Patton M (1990) Qualitative interviewing and research methods second edition. London
State Information Service (n.d.) Retrieved from state information service official website: https://www.sis.gov.eg/section/352/5238?lang=en-us
Vale LJ (2006). The urban design of 20th century capitals. In: Planning 20th century capital cities

Chapter 22
Aspects of Regenerative Cities

Lina Fayed, Abeer Elshater, and Rowaida Rashed

Abstract A regenerative vision is mainly about ensuring that future generations inherit a robust and intact world, in which individuals can fully realize their human potential. Regenerative cities provide all residents with opportunities to improve their quality of life. Modern cities continuously offer their citizens the possibility of a better quality of life, as well as reaching their full potentials. Scholars from various disciplines mention that cities can recognize and assure that the earth's ecosystem stays healthy. The aim of the current study is to provide concepts that go beyond the ambiguous notion of sustainability toward regenerative urban development. The objective is to improve the currently degraded condition of our ecosystems. This contemporary concept provides a better comprehension of how modern cities function. The results explain the possibilities that a city have to shift from extraction to regeneration. The concluding remarks show that the regenerative cities allow their citizens to consume among its local as well as its global ecosystem limits.

Keywords Sustainability · Ecological footprint · Ecopolis

22.1 Introduction

Across history, cities were able to grow, as a result of agricultural surplus, which allowed some citizens to engage in other activities. As cities grew bigger and richer, urban citizens sourced their food from further afield. It became easier to consume beyond the city's means, and faster than ecosystems can regenerate (Schurig and Roumet 2014). Cities are increasingly relying on resources from all around the world; while lacking the knowledge and prioritization of ensuring long-term availability of these resources.

L. Fayed (✉)
German University in Cairo (GUC), New Cairo, Egypt
e-mail: arch_linafayed@live.com

A. Elshater · R. Rashed
Ain Shams University, Cairo, Egypt

Architecture and Urbanism: A Smart Outlook,
https://doi.org/10.1007/978-3-030-52584-2_22

All around the world, people move to urban areas with the dream that the city can offer a better life for them, providing a space for them to realize their full human potential (Schurig and Roumet 2014). As the world continues to urbanize rapidly, human needs are also evolving. Needs such as energy generation, transportation, industrial production and urbanization became the main factors of environmental degradation, climate change and the depletion of biodiversity leading to natural resource consumption and global warming (UN-Habitat 2002).

A new urban agenda is inevitable to ensure that cities not only become resource-efficient but go beyond to enhance their ecosystems, thus providing them with their needs. The solution, as a result, lies in thinking beyond the notion of sustainability. It is argued that paradigms of sustainability reach their limitations because of their support of the modernization project that prevents effective participation in a dynamic world (Plessis 2012).

The research tackles today's challenges, as cities can no longer *just* be sustainable, but they *also* need to be regenerative. Cities should not just become resource-efficient, but positively enhance the ecosystem rather than undermining it (Girardet 2014a, b). The concept of regenerative cities aims to efficiently restore the resources they consume, in addition to recycling and reusing waste. These cities support local biodiversity and create public areas that enhance the quality of life (UN-Habitat 2016).

This research aims to provide a holistic approach on various levels of city planning and design. The research starts with a literature review that includes definitions, e.g. regenerative cities, ecological cities, Ecopolis, urban ecology and circular urban metabolism. It then analyses the policies and strategies of regenerative cities to comprehend how this could alter the ecological and environmental aspects of modern cities.

22.2 Regenerative Cities

The notion of regenerative cities was guided and influenced by Jane Jacobs' philosophy of "urban vitalism", which stated that cities needed to develop their internal vitality to "find the seeds of their own regeneration". A regenerative city is an urban development based on an environmentally enhancing, restorative relationship with the natural systems that sustain the city. The term "regenerative" encloses the actions that lead to the continuous renewal and restoration of the ecosystem, on which the existence of the city's urban system depends. A regenerative city sustains a mutually beneficial relation with its hinterland, not only by minimizing its environmental impact, but also by actively enhancing the city's ecosystems (Girardet et al. 2013).

Regenerative cities seek to address the continuous degradation of cities' ecosystems by allowing the city to function efficiently, thus benefiting both the human and the environment. In addition to the previously mentioned characteristics, regenerative cities surpass modern cities by adopting renewable energy systems and by accommodating citizens' participation, through responsible lifestyle choices, in the transformation process (Girardet 2014a, b).

The core of a regenerative city concept lies in the notion of going beyond the restrictive definition of sustainability and embracing an even broader model of urban development. A regenerative city takes sustainability a step further, as it seeks to be more than just "sustainable", especially that "sustaining" implies that no improvement is being done to the current degraded state.

This shift from a built environment that is degenerating its ecosystems, to a city that restores local environments and regenerates the capacity of ecosystems, should not only be a gradual process of improvements but should also incorporate a fundamental rethinking of architectural design and urban planning approaches. This new paradigm has the capability to shape cities' future where damaged ecosystems can be restored, focusing on the emerging problems of sustainability (Plessis and Brandon 2015).

22.2.1 The Regenerative City as an Ecopolis

Regenerative cities need to be built inside out (Girardet et al. 2013). Citizens need green spaces free from both pollution and waste pills. They should ensure that their daily practices are contributing to the regeneration of the ecosystems beyond the city's boundaries. Ecopolis proposes a new integration of the human habitats within its local environment, where regional needs are developed using local, renewable energy supplies (Girardet 2014a, b).

Ecopolis is defined as a regenerative urban system that is inspired by how the Earth's ecosystems function in an essentially circular manner (Fig. 22.1). In such a system, all waste emanating from nature would have the ability to continuously regenerate its ecosystems, enrich soils and facilitate plant growth (Girardet et al. 2013). The Ecopolis presents a clear framework of strategies that consist of three main

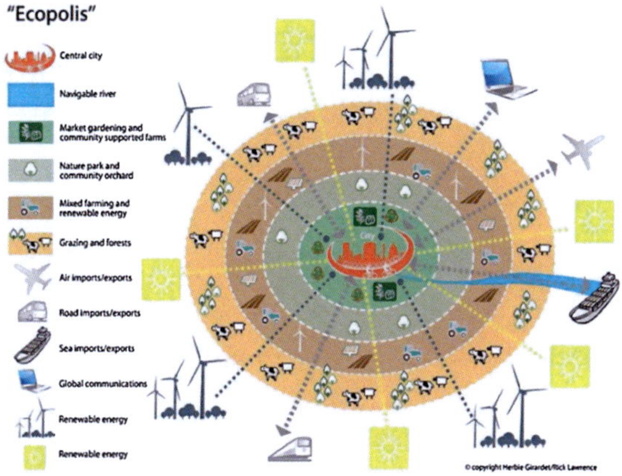

Fig. 22.1 Ecopolis reconnects to its local hinterland, using renewable energy, soil-restoring and food production systems. *Source* Girardet (2014a, b)

themes (Tjallingii 1996). The first is t**he responsible city**.**In this**, the city doesn't pass on its problems to the next generations. Second, the **living city** integrates local ecological potential. The third is **the participating city** that emphasizes citizens' participation.

Ecopolis has goals for long-term viability. It does this by redefining the existing urban ecology and mimicking nature's circular metabolic systems. However, one challenge facing the Ecopolis is the lack of policies to improve cities and ecological conditions. Mainstream architectural and urban planning practices don't focus on integrating aspects of eco-cities, as well as ecological engineering within the planning and design of buildings and cities (Bohemen 2009).

22.2.2 Redefining Urban Ecology

Urban ecology is the study of all living organisms and their relation to each other, and their relationship to their surrounding urban context (Wahl 2016). As urbanization has a farther-reaching impact on biodiversity and climate, the influence of urban ecology expands to include all the territories involved in sustaining the urban systems and clarifying how regenerative urban planning can be implemented (Fig. 22.2).

This approach moves towards enhancing the city's urban ecology, which is essential due to problems related to resource consumption and exceeding the capacity of these resources, allowing them to regenerate themselves annually. Not only do regenerative cities strive to limit the consumption of natural resources to rates that the ecosystems can recover, but they also work on reversing the trend by actively improving the regenerative capacity of their ecosystems.

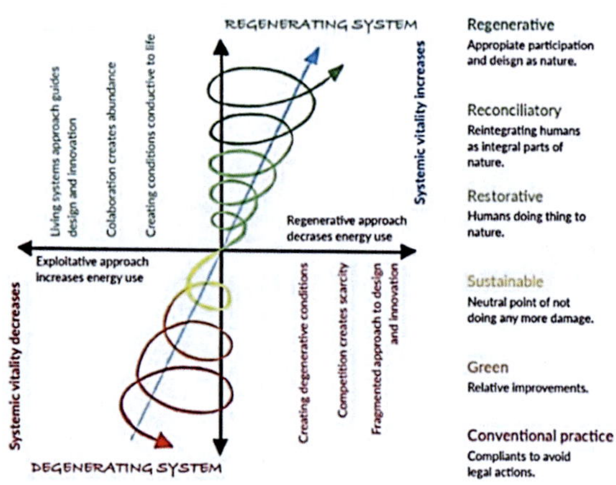

Fig. 22.2 Moving from a degenerating system to a regenerating system. *Source* Wahl (2016)

22.2.3 *Changing the City's Metabolism*

Cities, as "eco-technical organisms", have their own metabolism, where the transformation of resources into vital functions occurs. Nature has a zero-waste metabolism, while cities have natural resources flowing through their urban systems with no concern about their origin. The metabolism of cities examines interactions between nature and humans in the urban context.

The current urban metabolism of cities operates in an inefficient and wasteful manner with a linear input–output system. This linear metabolism allows resources to flow through the urban system and waste to flow out, which prevents the fulfilment of regenerative cities' principles. This input–output system urgently needs to be substituted with a resource-efficient circular system (Woo et al. 2014). In a predominantly urban world, it is inevitable that cities adopt circular systems to assure their viability and their rural environments', which they depend on (Fig. 22.3).

Urban metabolism is a representation of "the total of the technical and socio-economic processes that take place in cities resulting in growth and production of energy, while circular metabolism seeks elimination of waste also" (Kennedy et al. 2008).

22.3 Ecological Footprint of Regenerative Cities

Today's human activities consume resources and produce waste. As cities' populations grow, global consumption increases. Therefore, cities must measure their nature's capacity to meet their demands. The ecological footprint concept has emerged as one of the world's leading measures for human impact on nature. The ecological footprint of regenerative cities examines whether cities' natural resources are sufficient enough to keep up with citizens' demands.

Meanwhile, cities rely on resources and ecological services from distant ecosystems. The prosperity of cities' citizens is influenced by the conditions and availability of ecosystems. The Earth's capability to maintain its biodiversity, which is vital for the

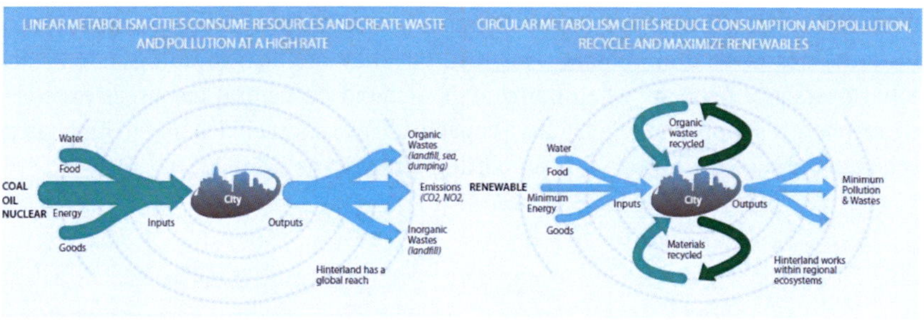

Fig. 22.3 Visualizing the regenerative city input–output system. *Source* (Girardet et al. 2013)

well-being and stability of the biosphere, depends directly on the ability of humans to support other species. The ecological footprint of a regenerative city is an indication for its citizens' impact on global biological resources. Ecological footprint is used as an indicator of environmental performance and product sustainability (Wiedmann and Barrett 2010).

22.3.1 Ecological Footprint

It is defined as the amount of productive area necessary to produce the renewable resources humans are consuming and to absorb its waste. The ecological footprint explains the relation between cities' population and the area of their biologically productive land and sea required to produce the needed renewable resources while reusing the waste they generate. And so, ecological footprints document the extent to which cities stay within the regenerative capacity of their biosphere (Wackernagel et al. 2006).

The ecological footprint is a measure of human impact on both countries and cities ecosystems. By using the concept of the ecological footprint, cities can estimate how many Earths it would take to support current citizens' lifestyles (Global Footprint Network 2018). The ecological footprint can also be measured in the number of planets, where one planet represents the biological capacity of the Earth in a given year (Wiedmann and Barrett 2010).

Several concepts explain the dimensions associated with the ecological footprints of cities, e.g. biocapacity. This term is defined as the area of productive land available to produce resources or absorb carbon dioxide waste. It demonstrates the capacity of cities' ecosystems to regenerate what citizens' demand from those surfaces. Ecological Deficit is another concept that explains these dimensions. An ecological deficit happens when the ecological footprint of a city's population exceeds the biocapacity of the area available to that population. "When national ecological deficit occurs, it means that cities are importing biocapacity through trade, liquidating national ecological resources" (Anon 2019), (Fig. 22.4).

Cities may exceed global biocapacity entering overshoot, where natural resource consumption rates are faster than regeneration rates. Within this current demand on natural resources, overshoot is becoming a global phenomenon. Nowadays, not only are cities consuming nature's interest, but they are also invading its capital. Global overshoot happens when humanity's demand on natural resources exceeds the biosphere's supply and regenerative capacity. This is quickly leading to high rates of consumption of the Earth's life-supporting ecosystems (Fig. 22.5).

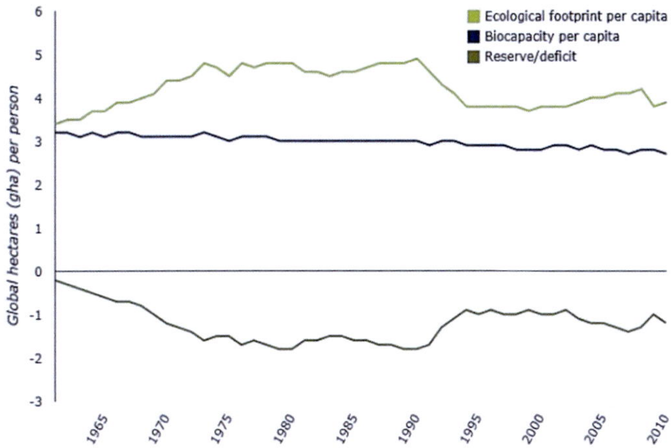

Fig. 22.4 Diagram illustrating the ecological reverse/deficit. *Source* National Footprint Accounts 2014 Edition

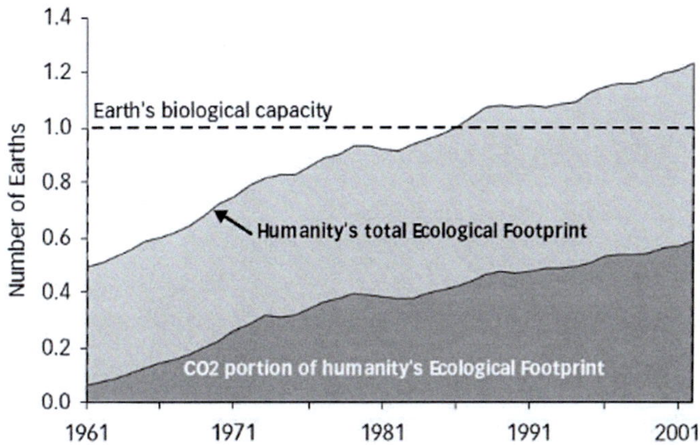

Fig. 22.5 Ecological overshoot from 1961 to 2002. *Source* Global Footprint Network National Accounts (2018)

22.3.2 *Cities' Current Ecological Situation*

Human consumption, unlike animal consumption, is not only limited to the satisfaction of biological needs, but it is also influenced by the consumer's culture, which encourages the pursuit of non-essential luxury goods.

This high demand for natural resources will inevitably create substantial changes in the resource base, which- *in turn*- will undermine both economic performance and human welfare. Most of the countries are dangerously undergoing ecological deficits, by demanding more than what their ecosystems can regenerate. Despite the increased global biocapacity, it can't compensate for the global ecological footprint. In 2020, Earth's ecosystems will surpass what the natural capital can compensate by 75%. (Moore et al. 2012), (Fig. 22.6).

Many countries are in red, when it comes to consuming more natural resources than their ecosystems can regenerate. In this case, their biocapacity is running on an "ecological deficit". Cities can run ecological deficits through liquidating their natural capital, such as overfishing, importing resources transmitting more carbon dioxide in the air, more than what their ecosystems can absorb. Meanwhile, this global deficit represented by the overshoot cannot be compensated, as there is only one planet available. Since the 1970s, humanity has been in the state of ecological overshoot with the annual demand on natural resources exceeding what Earth can regenerate each year (Fig. 22.7).

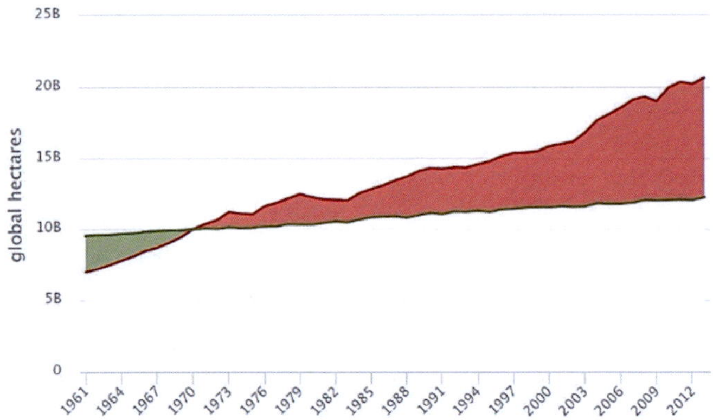

Fig. 22.6 Ecological deficit shows the gap between total human demands on nature and its capacity to meet them. *Source* Global Footprint Network National Accounts (2018)

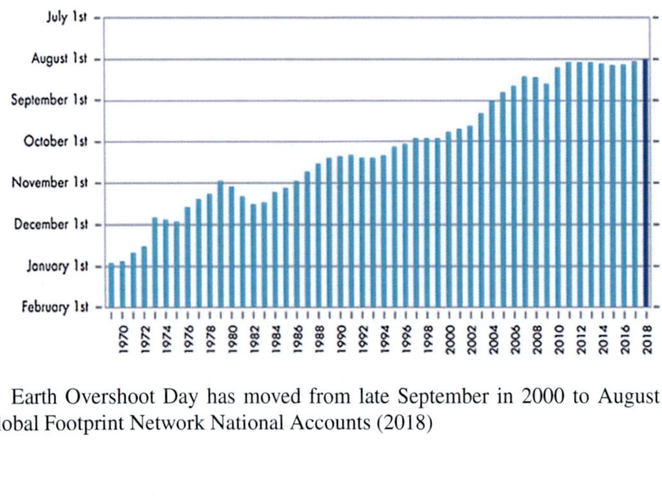

Fig. 22.7 Earth Overshoot Day has moved from late September in 2000 to August 1 in 2018. *Source* Global Footprint Network National Accounts (2018)

Fig. 22.8 How many Earths are needed if the world's population lived like these countries citizens. *Source* Global Footprint Network National Accounts (2018)

Earth Overshoot Day marks the date when humanity's annual demand on nature exceeds the natural resources that Earth can regenerate for the same year. Earth's Overshoot Day has moved from late September in 1997 to reach the 2nd of August in 2018. "Today humanity consumes the equivalent of 1.7 Earths to provide the resources cities use and absorb the waste they produce" (Anon 2019) (Fig. 22.8).

Taking into consideration that city inhabitants require a productive ecosystem outside the borders of the city itself to produce the food, water and renewable resources that are consumed within the city, they also depend on ecological systems to provide clean air and to process waste. However, modern city planning tends to take the ecosystem for granted (Folke et al. 1997). The ecological costs of this global excessive consumption of natural resources are evident such as deforestation, drought, freshwater scarcity, soil erosion, biodiversity loss, overgrazing, extinction of species, as well as increased carbon concentration in the atmosphere. Variations in the ecological footprint of cities are a result of mainly socio-economic influences, like disposable revenue, infrastructure, social and citizens' behaviour (Anon 2019).

This increase in global overshoot leads to liquefying natural capital and loss of biodiversity (Toderoiu 2010). Ecological overshoot is only possible for a limited period before the cities' ecosystems begin to deteriorate and possibly collapse. Additionally, almost 70% of humans live in countries with an ecological deficit and below-average income. Hence, they are unlikely to buy themselves out of the resource crunch. Accordingly, if these current trends of consumption continue, by 2050, countries will need the equivalent of almost three planets like the Earth to support the consumption rates of resources and lifestyles of their citizens (Fig. 22.9).

In an increasingly urbanizing world, ecological issues are crucial to the development of cities. The proper management of the city's ecosystem and effectiveness of the urban scheme, such as transportation and energy consumption can be taken as

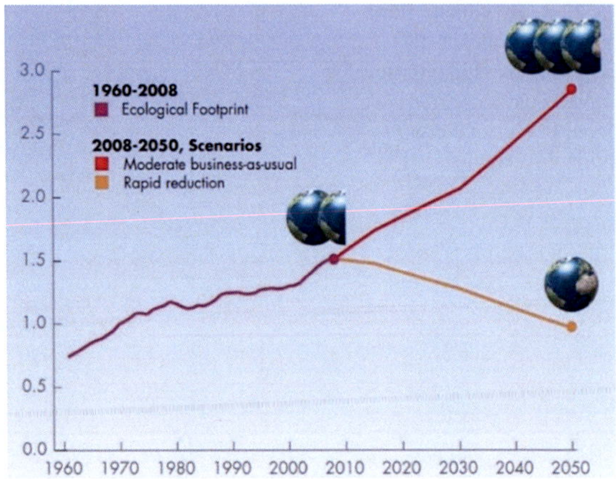

Fig. 22.9 Cities ecological footprint scenarios in 2050 with and without the paradigm shift to Regenerative Cities. *Source* https://www.everythingconnects.org

part of the urban surrounding. To achieve the above, cities need to acquire certain characteristics, as well as adopting eco-friendly urban policies that have the notion of reaching a restorative relation with the city's ecological systems rooted through the decision-making and implementation phases.

22.4 Characteristics of Regenerative Cities

Regenerative cities' characteristics support a paradigm shift from the modern cities of today, demonstrating a method of urban development that seeks a restorative relationship with the city's ecosystems (Hand et al. 2017). Regenerative cities need to have seven main characteristics achieved in the city's urban strategies, such as Livability, Equity, Ecology, Mobility and Accessibility, Resilience, Energy and Natural Resource Efficiency, and finally Smartness.

Livability, as a regenerative city, uses every chance to develop, though using the growth of cities to improve quality of life. By recognizing the influence of built environments on people's behaviours and their sense of belonging, a regenerative city can dramatically contribute to the happiness of its citizens. **Equity** in regenerative city is crucial, as cities are becoming economic engines that drive innovation and productivity. This economic dynamism provides both the social and economic mobility needed to promote good mental health, through supporting living wages and empowering workforce, leading to stability (Dahiya and Bigio 2004).

Ecology is the main key, and a regenerative city is *Ecologically-friendly* where it has a symbiotic relationship with nature. Population growth tests citizens' abilities to effectively develop cities in an ecological way (Bartlett 2016). Degradation of natural systems and citizens' irresponsible consumption damage the ecosystem; cities need to live within their natural constraints. A regenerative city is *Waste-free*, with a circular metabolism. Transforming the cities' waste into a resource is a challenge for cities. It requires a shift in both the perception and approach, along with effective service delivery and citizen satisfaction. It has *Food self-dependent*, as consumer's preference for local foods is growing. Cities need to meet this market demand by incorporating food production and development, by preserving agricultural lands and retrofitting cities for urban farming (Bartlett 2016).

Mobility and Accessibility, a regenerative city, have to be open and accessible. In the last century, cities increased reliance on automobiles, which made them become proportionally more dependent on fossil fuel, consuming more land and diminishing public health. Thus, regenerative cities are shifting back to people's movement instead of cars to rebalance mobility options. As for **Resilience**, naturally a regenerative city is highly responsive, where cities should be flexible enough to address multiple issues immediately. More than just being ready, cities need to be proactive to produce positive progress and meaningful change.

Energy and Natural Resource Efficiency are the main focus. Regarding *Energy*, a regenerative city produces more energy than it consumes. Cities must diversify their power systems supply and depend on cleaner renewable sources, thereby equally distributing risk and production. Cities should stop consuming natural resources at a rate faster than they can regenerate. As for *Water*, being a vital natural recourse. A regenerative city cleans more water than it consumes, as clean water is the plant's most valuable resource. Water is critically being undervalued as an asset and is often over-engineered to the detriment of its natural functionality. Reducing impermeable surfaces, increasing water recharge, besides reusing stormwater as a source is crucial to engage in water ecology.

Smartness, typically the regenerative city, adopts smart solutions. Smart cities are innovative cities that use information and communication technologies (ICTs) to enhance citizens' life, urban activities and facilities, and attractiveness (Allam and Newman 2018). Achieving those city's characteristics lead to transit comfort, small business generation, enhancing air quality, waste diversion rates, decreasing stormwater run-off and energy demand. This, *in turn*, promotes both public and individual health and increases citizens' satisfaction, leading to the creation of a regenerative city (Fig. 22.10).

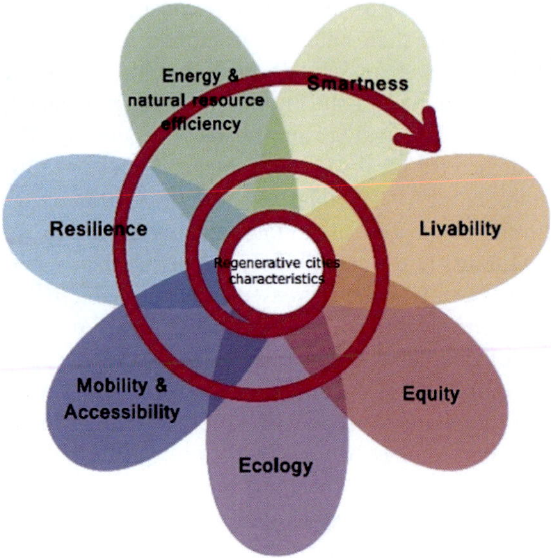

Fig. 22.10 Regenerative cities characteristics. *Source* The authors

22.5 Polices for Regenerative Cities

Nowadays, cities need to become regenerative to overcome the environmental and ecological challenges. Adopting regenerative urbanism strategies is compelling due to local ecological degradation, such as (Girardet 2014).

Aspects	Implementation policies	
The solar city model	– Endorsing the concept of solar city as national policy – Prioritizing the feed-in bill for renewable energy systems, allowing citizens to sell electricity – Supporting the shift to renewable energy as a key strategy for the industrial functions – Creating empowering policies mean for developing renewable energy in the city's hinterland	(Girardet 2017a)
Energy sufficiency	– Using strategies such as the 2000 W Society concept as an operative policy principle – Modifying building codes to make resource-efficient building practice the norm	(Girardet 2014)

(continued)

(continued)

Aspects	Implementation policies	
Nature reintegration	– Encouraging replanting trees to regulate soil erosion and enhance city's biodiversity – Making carbon appropriation the main goal of tree planting initiatives – Developing strategies that support restoring forests and wetlands in remoter areas	(Darinka Czischke 2015)
A culture of restorative urbanization	– Utilizing both global networks and local expertise to develop restorative urbanization – Assuring the vision is introduced in the education system and events and ensure citizens participation – Endorsing the creative reports on urban restoration measures through the media – Producing regular reports on the implementation of eco-restoration plans or their applications	(Girardet 2014)
Regional food production	– Encouraging local urban food production for local markets – Endorsing any community-supported agriculture as well as farmers' markets – Ensuring usage of composted bio-waste in urban farming	(Juliana Lutz 2016)
Water security	– Balancing the urban, agricultural and commercial water usage with their social, economic and environmental benefits – Encouraging "waterproof" cities by promoting water efficiency practices and rainwater collection – Making wastewater recycling and stormwater reuse the core of water policy	(Xin (Cissy) Ma 2015)

(continued)

(continued)

Aspects	Implementation policies	
Sustainable transport	– Creating new pedestrian zones and a comprehensive network of dedicated cycling lanes across cities – Encouraging public transport by improving attractiveness, frequency and flexibility – Stimulating the development of new electric and fuel vehicle technology – Encouraging carpooling as the main form of urban transport	(The United Nations Economic Commission for Europe 2015)
Waste free	– Developing new enterprises to process organic wastes into soil enhancing materials – Making sewage reprocessing and nutrient capture the main focus of waste management – Implementing policies that promote cost-effective reprocessing of all technical wastes – Applying a zero-waste policy to create new green businesses and jobs	(Girardet 2017b)
Green investments	– Encouraging the establishment of green business through usage of city's procurement – Endorsing resource-efficiency through businesses – Creating "green business incubators" all over, as a norm – Making environmental resilience a base for all newly generated businesses	(Girardet et al. 2013)

These strategies are achieved through the framework of urban policies. These urban policies include five aspects. The first is the national policy where a framework that enables the regenerative urban development is executed. The second is the urban planning policy that is integrated regenerative urban planning and becomes the key organizing norm. The third comes for the green savings, where cities adopt reducing waste, recycling materials and cutting-costs strategies. The fourth is green economy, where cities' economy reduces ecological scarcities, with new businesses and jobs. Finally, the green talent of cities encourages investing in technical and entrepreneurial skills.

Fig. 22.11 Regenerative cities policies. *Source* The authors

Through these urban policies cities can reach a strong restorative relationship with their natural resources, which, in return, enhances its ecological footprint. Cities must start to make ecological strategies that focus on the use of human ingenuity to find new ways to consume, within the Earth's resources (Fig. 22.11).

22.6 Conclusion

The paper concludes by arguing that for humans in the twenty-first century, it is clear that cities in one part of the planet cannot continue to consume at an unsustainable rate, while other populations survive on very little. Modern cities consume resources and produce wastes that are incommensurate with their populations, which have negative influences on the environment and the ecological footprint. Scanning relevant literature showed that regenerative cities offer an alternative path forward, beyond the trajectory that modern cities are currently on, as it enables human systems and habitats to positively contribute to the ecological system.

Regenerative cities adopt this shift from merely addressing issues, such as climate change, green buildings, biodiversity, city's ecosystem, urban development, in isolation to working with them simultaneously. These regenerative cities represent a drastic change from perceiving the world as a mechanical system, where citizens are isolated from their ecosystems, to perceiving the city as an interconnected, living and flexible, socio-ecological system that is continuously changing. The results here showed that actions should be taken to deal with this overshoot gap within regenerative strategies; this directed investment into the technology and infrastructure which enable cities to function efficiently. It also required focusing on both individual action, and citizens' demand on investments, and policymakers to participate in creating regenerative cities.

A regenerative city efficiently restores the resources it consumes while minimizing waste. It encourages the process of ecosystem restoration as well as depending on its hinterland for food, together with urban and community-based agriculture. The present findings confirm that regenerative cities enable urban wealth generated from natural capital in the region and allow value to remain in the area with local people, where cities are not only self-sufficient by focusing on local assets, but they also improve the ability of ecosystems to regenerate resources. The core of the regenerative city is to establish a symbiotic relation between cities' urban areas and their hinterlands.

The concluding remarks confirm that regenerative cities can regain some power in ensuring basic service provision, where citizens are capable of sustaining their quality of life by finding natural resources closer to home and regenerating them. People living in regenerative cities could have a huge influence on the resources cities consume, when cities' strategies focus on the local stock. Unlike cities that consume their hinterland, without returning anything except waste, a regenerative city has a mutually beneficial relationship with its peri-urban territory.

Regenerative cities emerged as a solution to ecological challenges arising from rapid urbanization. Such cities create a new urban system, where applying sustainability synergies is a must in this unprecedented era of urbanization. Future direction of research should seek to engage people worldwide to focus on integrated systems' approaches for creating neighbourhoods and urban areas through a "net positive" or "regenerative" lens. This new concept is still evolving and not mainstreamed due to technological, economic and policy barriers. The changes, required for cities to become regenerative, need comprehensive strategies as well as longstanding policies, not temporary compromises like the majority of political decision-making schemes worldwide.

References

Allam Z, Newman P (2018) Redefining the smart city: culture. Metaband Governance. 1(1)

Anon (2019) Global fooyprint. Retrieved from Ecological Footprint: www.footprintnetwork.org

Bartlett A (2016) Reflections on sustainability, population growth, and the environment.

Bohemen HV (2009) Ecopolis: architecture and cities for a changing climate, P.F.Downton. 35(5)

Dahiya B, Bigio AG (2004) Urban environment and infrastructure: toward livable cities. World Bank, Washington DC

Darinka Czischke CM (2015) Sustainable regeneration in urban areas. URBACT

Folke C, Jansson A, Larsson J, Costanza R (1997) Ecosystem appropriation by cities. Roy Swed Acad Sci 26(3):167–172

Girardet H (2014a) Creating regenerative cities. Routledge, New York: NY

Girardet H (2014b) Creating regenerative cities. Routledge, New York

Girardet H (2017a). Regenerative cities. In: Economics SI (eds) Green economy reader. Springer, Cham, pp 183–204

Girardet H (2017b) Regenerative cities—making cities work for people and planet. Cooperative Research Centre for Low Carbon Living

Girardet H, Schurig S, Woo F (2013) Towards the regenerative city. World Future Council, Hamburg

Global footprint network (2018) Footprint network. Retrieved from global footprint network: www.footprintnetwork.org

Hand GH, Weber R, Bluestone N (2017) Regenerative cities: moving beyond sustainability. Urban Design and Mental Health

Juliana Lutz BS (2016) Farmer cooperation as a means for creating local food systems—potentials and challenges. Sustainability

Kennedy C, Cuddihy J, Engel-Yan J (2008) The changing metabolism of cities. J Industerial Ecol 11(2):43–59

Moore D, Cranston G, Reed A, Galli A (2012) Projecting future human demand on the Earth's regenerative capacity. Ecol Ind 16:3–10

Plessis CD (2012) Towards a regenerative paradigm for the built environment

Plessis CD, Brandon P (2015) An ecological worldview as basis for a regenerative sustainability paradigm for the built environment. 109.

Schurig S, Roumet C (2014) Imagine a reganerative city. Munich: the world future council and energy cities.

The United Nations Economic Commission for Europe (2015) Sustainable urban mobility and public transport in UNECE capitals. United Nations

Tjallingii SP (1996) Ecopolis: strategies for ecologically sound urban development 36(3)

Toderoiu F (2010) Ecological footprint and biocapacity—methodology and regional and national dimensions

UN-Habitat (2002) Sustainable cities programme 1990–2000. UN-Habitat

UN-Habitat (2016) The city we need 2.0 towards a New Urban Paradigm. United Nations Human Settlements Programme, Nairobi

Wackernagel M, Kitzes J, Moran DSG, Thomas M (2006) The ecological footprint of cities and regions: comparing resource availability with resource demand. Environ Urbanization 18(1):103–112

Wahl DC (2016) Designing regenerative cultures. Triarchy Press Ltd

Wiedmann T, Barrett J (2010) A review of the ecological footprint indicator: perceptions and methods. Sustainability 1645–1693

Woo F, Wortmann J, Schurig S, Leidreiter A (2014) Regenerative urban development : a roadmap to the city we need. World Future Council, Hamburg

Xin (Cissy) Ma XX-M (2015) Sustainable water systems for the city of tomorrow—a conceptual framework. Sustainability

Chapter 23
Behavioural Pattern of Pedestrian Negotiation in Public Spaces

Khushbu Soni, Ritesh Agrawal, and Sheetal Pikle

Abstract Public spaces are a crucial intersection within the lifestyle in an urban scenario. Human behaviour and response to public space are unique and can be found in all cities. With similar narratives of consistent demands on the infrastructure, rising densities and increasing concerns in traffic conditions, cities in developing countries face comparab-le issues with respect to the usage of their public spaces. However, very little is still known about the behavioural pattern of users and pedestrians in the ever-changing public spaces in these cities and still leaves a lot to be explored. This paper attempts to bridge this gap in information and endeavours to bring out some common derivatives of behavioural patterns of pedestrian negotiations in a public space. The scope of the paper concentrates on studying the requisites in an urban context albeit within a set of defined conditions. These conditions are inclusive of, but not restricted to, consideration of cities in developing countries with a **historical business and financial district.** The study emphasises on **important and existing busy nodes** of such situational manner to derive on parameters of ever-shifting user behaviour patterns. The paper culminates to conclude by drawing parallels for user response; keeping in account the variables of personal, situational and environmental factors; to public realm ideals in developing cities using an observational study.

Keywords Public realm · Placemaking · Behavioural pattern · Urban design, pedestrian life · Pedestrian life

K. Soni (✉) · R. Agrawal · S. Pikle
VIVA School of Architecture, Mumbai University, Mumbai, India
e-mail: khushbu.viva@gmail.com

R. Agrawal
e-mail: ritesh.viva@gmail.com

S. Pikle
e-mail: psheetal.viva@gmail.com

Architecture and Urbanism: A Smart Outlook,
https://doi.org/10.1007/978-3-030-52584-2_23

23.1 Introduction

Human response to public spaces is an interesting study in the sociology of urban spaces. A public space can be defined as a space of built or unbuilt environments which is accessible to all people equally, whether planned (organised) or unplanned (organic). Sarah W. Goldhagen in her book *Welcome to your World- How the built environment Shapes our Lives* has beautifully brought about various concepts of the interface between people and how they perceive their surroundings. '*What's more, the built environment shapes our lives and the choices we make in all these other arts do- combined*' (Goldhagen 2017). It has already been seen that architects have been able to create profound effects on individuals with their planned and designed built environments.

Consecutively, the unbuilt spaces or the undefined public realms also start playing an essential role which affect the individuals in various ways. The users of these spaces introduce social characteristics and elements such as culture, ethnicity, gender, age, and these elements together with the physical and non-physical features of the public space are capable of having a profound effect on the way that the people behave, experience and interact with the public spaces (Madanipour 2003). Public spaces capture the vitality, tension and excitement of the city, acting as social and spatial integrators, not separators (Hanafi et al. 2013).

Developing countries in particular have a unique situation, wherein many a times, the effects due to the rate of globalisation, and therefore, the changes in the corresponding city lifestyle are much faster than the actual physical development of the city. Changes within major 'city functions' and 'illustrated activities' are related to a society shift towards leisure and consumer-oriented lifestyles; however, life in public spaces has not changed materially (Hanafi et al. 2013). This can lead to a void and an imbalance in the way users respond to public spaces. The introduction of vehicular networks and worldwide dependency on everyday transport has also largely contributed to tremendous transformations in the scenario of usage of public spaces all over the world.

This paper is the result of an observational study which focused on the variations in user behaviour as a pedestrian navigating the available public spaces for their usage in the contemporary times. This is done taking into consideration a case study of a region with pre-defined conditions pertaining to the Indian scenario as an example to create a database for outreaches with similar scenarios and parallels.

23.2 Story of Urban Spaces in India

23.2.1 Worldwide

A public space is as community-driven as its process (Peinhardt and Storring 2019). The relation between people and their public spaces has always been cohesive and has observed a gradual development through the ages. Over the years, the activities observed in public spaces have shifted, added or subtracted multiple times to the existing circumstances. Urban spaces of public domain have become slightly more complex in their purpose and utility, less for relaxing and more of multidimensional practicalities. Over about fifty years ago, the primary focus in public activities included shopping; however, in the recent scenario recreational activities like cultural events, parades and exhibitions are often hosted in public spaces. The introduction and heavy usage of the vehicular commodity have also led to a huge shift in public realm leading to the introduction of newer forms of public spaces within the morphology of the urban scenario.

According to Gehl and Gemzoe, the main 'three' vital functions of a city space are: **meeting place**—Social Information Exchange, **market place**—Goods and Service Exchange and **connection space**—Access to city functions (Gehl and Lars 2001). In *Life between buildings: using Public Spaces,* Jan Gehl further defines outdoor activities into three categories (Table 23.1):

Table 23.1 Author created, with reference from Jan Gehl (Life between buildings: using Public Spaces, 1971/1987)

Sr. No	Category	Description	Example
1	Necessary activities	• These are almost compulsory activities and give no choice • They are perennial in nature • Will happen irrespective of conditions of Infrastructure, space or climate	• Walking to school or work • Shopping etc.
2	Optional activities	• These are choice related activities • The participants have a desire to participate • Will happen if users have time and place to do • Can depend upon the climate, weather and other physical conditions	• Taking a walk or jogging outside • Playing outdoors • Picnic etc.
3	Social activities	• Occur when people congregate and socialise • Often spontaneous in nature • Frequently evolve from activities happening in the other two categories Therefore, also called resultant activities	• Children playing • Conversations etc.

23.2.2 India

'*The street was the outside, the space for which one did not have responsibility, or which was not one's own, and it therefore lacked any association with obligation, because it did not symbolise any significant principle, did not express any values. It was merely a conceptually insignificant negative of the inside, which was prized and invested with affectionate decoration. Thus, the outside—the streets, squares, bathing ghats, and other facilities used by large numbers—were crowded, but they did not constitute a different kind of valued space, a civic space with norms and rules of use of its own, different from the domestic values of bourgeois privacy*'—(Kaviraj 1997: 98).

Human behaviour and relative social interactions in public spaces depend to a large extent on the different features of the spaces. These features could be physical, social, cultural or even sensory and have the power to affect people's behaviour in the public realm. Cynthia Nikitin, senior vice-president at non-profit organisation: *Project for Public Spaces*, has noted that high-density cities across Asia, long working hours and smaller homes mean people often entertain at night and in the outdoors. Perhaps the best evidence for the vital importance of public urban spaces can be found in our histories. Even from ancient times, there is evidence of a basic human impulse to govern streets and open spaces, to make them more useful in the necessary and desired activities of the old cities and for the citizens of the community (Hanafi et al. 2013).

Public spaces in pre-colonial Indian cities were also important cultural and ceremonial spaces that temporarily connected different communities and private spaces within the city. Bound in official rules and norms of conduct, the city's bazaars, streets and squares functioned as the commons whose location and use had specific meanings for different users. Everyday spaces like the bazaar (market) and street were as much spaces of economic and social interactions as they were spaces of unexpected encounters (Chakrabarty 1991). Various historical public space usage scenarios of the Indian context can be categorised into Gehl's definitions (Table 23.2).

This historical evidence of public spaces is even to this day, a large presence in the urban scenario. This has to do, a lot, not only with the acceptive climate conditions but also the social structure of the society. Even within the confines of one's home, there is a strong sense of connecting with the nature and living in the open. This belief system can be strongly emphasised in the courtyard arrangement of most residential layouts of old homes. Interestingly, this inherent concept about openness spills into the public domain of people's lives.

Table 23.2 Author created matrix to define historical public space usage scenarios in India, Reference from Table 23.1

Sr. No	Category	Location	Description
1	Necessary activities	• Near a water source like river, lake, stream well • At the market place • At the village square (*Chowk*) or ground	*(Female-Dominated)* • Daily chores like washing, bathing • Compulsory for drinking water facility • To buy everyday necessities • To conduct meetings for village-related jurisdictions. *(Male-Dominated)*
2	Optional Activities	• At the village square • Temple and its grounds	• In the evenings for a chat, after the daily chores and jobs are over. *(Male-Dominated)* • During Prayer times. *(Generic)*
3	Social Activities	Community Celebrations: • Squares and Grounds (*Chowks* and *Maidans*) • Streets • Temple/Prayer Grounds • Community Halls	• These were generally during festive seasons. *(Generic)* • Marriage • Birth or death • Announcements • And any other celebrations where large crowds could gather

23.3 Case Study: Key Considerations for Observational Study Conducted

The ability of public spaces to educate and offer knowledge is a significant aspect, particularly when those spaces play an important role in the history of the city and the social life of its citizen (Madanipour 2003). If we look at world history, many countries in the Asian and African continents at some point of time or the other were either under the Western rule or influence. Often these influences left behind some very typical and strong traits of design, planning and architecture. The Western conceptualisation of public space took root in the Indian cities and Indian imaginations through British colonialism and modern master planning. In some cases, these have been so well executed, that they still remain interesting examples of architecture and planning in the region, sometimes even becoming or serving as the urban epicentres or business districts of the city.

Several key factors are taken into consideration while selecting the site for study.

23.3.1 City in a Developing Country

India is a part of the BRICS and is a developing country. Mumbai is the capital of Maharashtra State in India and is a thriving city in the Indian developing economy. It is the largest city in India (by population) and is the financial and commercial capital of India. Mumbai has also been ranked as the seventh in the list of 'top 10 cities for billionaires' by Forbes magazine in April 2008.

The context of Mumbai is very interesting due to its peculiar physical and geographical profile (Fig. 23.1). Today, with an area covering more than 600 sq kms, Mumbai Metropolitan region extends into the districts of Thane and Palghar and its satellite cities like Navi Mumbai (Fig. 23.3 showing railway connections to all regions). Mumbai sits as a 'V' with its south end tapering into the Arabian Sea and the north landmass spreading upwards (Fig. 23.2). However, this was not the story earlier.

23.3.2 Historical Relevance

Colonial and post-colonial planning in India mirror the vocabulary of the boulevards of Haussmann's Paris, the Garden Cities of Europe and North America and the city beautiful movement. (Vanka 2014). The development of Mumbai and its region as they seem now can be traced back to the colonial times into the 1800s when it was recognised as a land with massive potential due to its location. The history of the

Fig. 23.1 Current day Mumbai, (Google Earth 2019)

Fig. 23.2 V-shape tapering of Mumbai

formation of Mumbai, inevitably, is a saga for romance. Initially a cluster of only 7 small islands, this city has been majorly formed by reclaiming the land between the islands and joining them to the landmass up north (Figs. 23.4, 23.5 and 23.6).

Development of the **southern region of Mumbai** by the British colonies took over 150 years to join as a landmass. The seven islands were initially lush green and had a range of 22 hills. The original island of Mumbai was only 24 km long and 4 km wide from Dongri to Malabar Hill. The other islands were Colaba, Old Woman's Island, Mahim, Parel, Worli and Mazgaon (Fig. 23.4). The first phase of reclamation started in full form only in the 1700s. The Mumbai that we see today is the result of various reclamation projects by several authorities post the plague epidemic scenario which hit Bombay in 1896–97.

23.3.3 Business and Financial District

For the purpose of study, a region at this **historical southern tip** of Mumbai, Churchgate and its vicinity, has been identified. The stretch under consideration is about 1.5 kms in length and connects from Churchgate Railway station to the *Kala Ghoda*

Fig. 23.3 Railway line connections through Mumbai, (m-indicator, 2019)

(Black Horse) Art District (Fig. 23.8). It is a historic quarter and has developed with an eclectic mix of architectural styles peppered along the way (Fig. 23.7). The path from the station to the district is spotted with innumerable heritage buildings from various eras and stories which mark the development of Bombay. This is a gateway to not only a strong cultural milieu but also houses several business and commercial sectors.

23.3.3.1 Churchgate Station

Churchgate is the southernmost train station or terminus in the western line of local suburban railway connections running north to south in the city (Fig. 23.3). It was initially one of the three gates of the fortified island city of Mumbai. In 1870, it was constructed and started to operate as a railway station (Fig. 23.9).

Churchgate (Fig. 23.9) is home to several prominent banking sector, business organisations and prestigious institutes. Several civic and official headquarters are

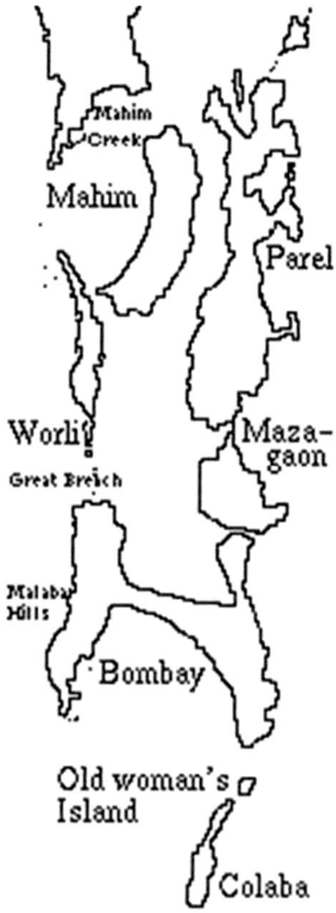

Fig. 23.4 Seven Islands of Bombay

also located in this region. Due to its early development in the growth of Mumbai, buildings of varied architectural styles right from Gothic, Victorian, Indo-Saracenic to contemporary architectural styles can be witnessed in the area. In fact, Mumbai has the second largest collection of Art Deco buildings after Miami, and Churchgate houses a huge stretch of this.

23.3.3.2 Kala Ghoda District

Kala Ghoda is an important district in the commercial or business centre of south Mumbai. It has an interesting mix of buildings functionalities including art galleries, museums, synagogue, libraries, educational institutes and some very good restaurants and cafes. It has been developed into an **art district** and hosts a very popular annual art-fest called '*Kala Ghoda* Festival' after the Black Horse statue which landmarks its precinct (Fig. 23.10).

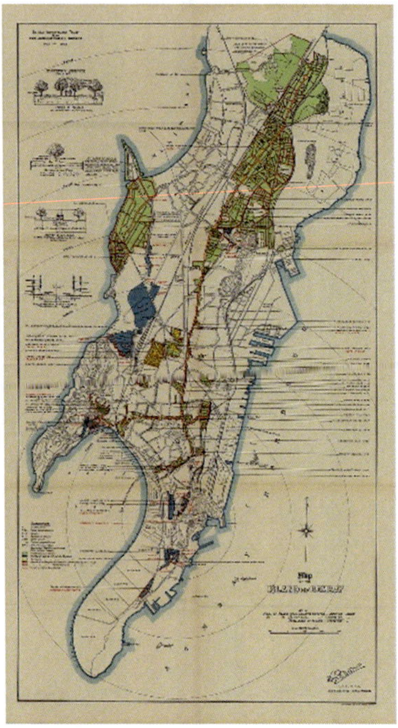

Fig. 23.5 Historical map of Bombay

23.3.4 Important and Existing Busy Node

The district is centrally located to **Mumbai Port Docklands** in the east, Regal Cinema and **Gateway of India** in the south, **Flora Fountain** in the north and **Oval *maidan*** to the west (Fig. 23.11). All of these are important junctions leading to various business, commercial or cultural and recreational places in the region.

This exceptional location with its proximity to various structures of historical prominence, its connection to various areas of business and commercial importance and its own narrative of historicity with a strong identity and sense of place makes *Kala Ghoda* an apt precinct to study pedestrian negotiation with comparatives of change through time.

The built environment conveys a deceptive sense of permanence. In truth, statistics show that it is constantly under renovation, renewal and expansion' (Goldhagen 2017). The study tries to understand and map how human behaviour pattern adapts to this age-old design which by itself has also undergone changes over the course of the years.

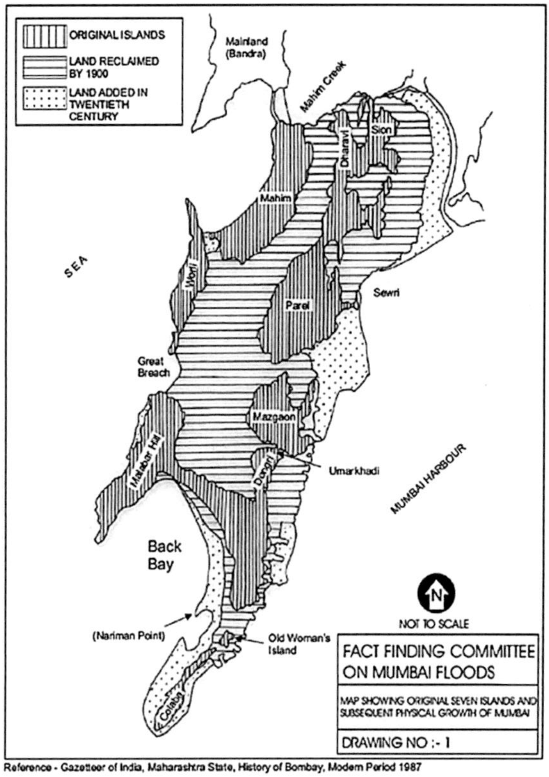

Fig. 23.6 Reclaimed growth of Mumbai

23.3.5 *Defining* Kala Ghoda *Fest as a Live Intervention*

An interesting element that was an added advantage to the study was in the form of an annual week-long festival which occurs in the precinct during the month of February. The festival is a celebration of art, culture and architecture. Visual arts, dance, music, drama, cinema, literature, workshops, heritage awareness programs and even food become a part of the event. Several abstract installations, artworks and related paraphrenia showcasing social themes are often created for public awareness towards these sensitive issues and have attracted people from all age, genres and walks of life. It is very interesting to understand how the pedestrians negotiate this alteration and form their own public realm during this period.

23.4 Methodology and Task

'Observation is entrapping' (Whyte 1988). One of the guiding features for this study was inspired by the methods adopted by William Whyte in his research 'The Street

Life Project', described in the book *The Social Life of Small Urban Spaces* (Whyte 1980). He uses photography, time-lapse sequences, interviews and direct observation of users. A similar approach has been used to conduct this study to map the behaviour pattern of user at various locations in the identified stretch. ***The method of case study has been done using time-lapse techniques, photography, mapping the streetscape and direct observation to record the various patterns of behaviour along the route.***

The study aims to derive upon information statistics related to multi-level parameters that come into play in the urban scenario. Objectives of study include but are not restricted to understanding the occurrence and behaviour of pause points and their importance in pedestrian life and to understand the social behaviour of the public at these various inter-references. It also looks at the transformation of public spaces and the reciprocative behaviour of the public in presence at the duration.

23.4.1 Methodology

The selection of region for study and the identification of the locations as per the parameters have been detailed in Sect. 23.3. The methods for observational study

Fig. 23.7 **a** Map showing several heritage buildings in the vicinity, **b** Arial view towards flora fountain (UDRI files)

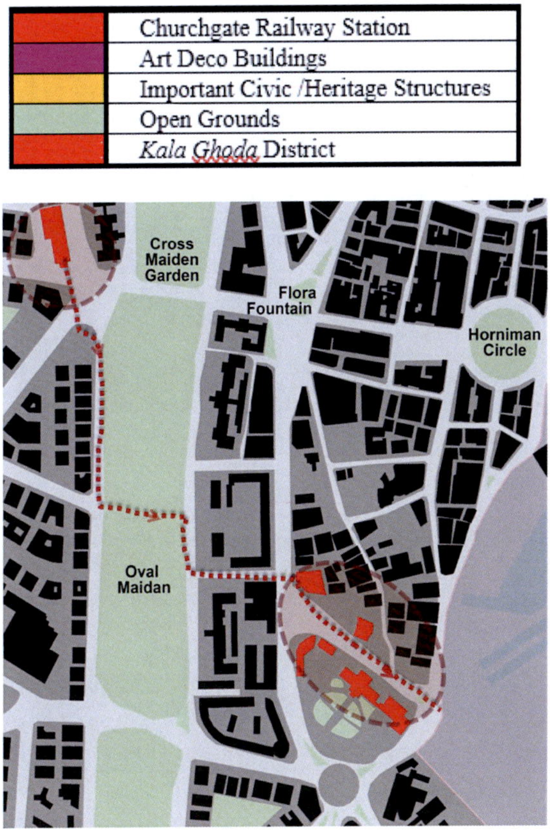

Fig. 23.8 Route taken for study **C**- Churchgate station, **K**- Kala Ghoda District

Fig. 23.9 Old Churchgate station, 1900s; from Rail archives

have been defined in the introduction of Sect. 23.4. Section 23.4.1 will define the various steps which will lead to the findings of the case study.

Fig. 23.10 Old Statue of Kala Ghoda (Black Horse) initially installed at Elphinston College

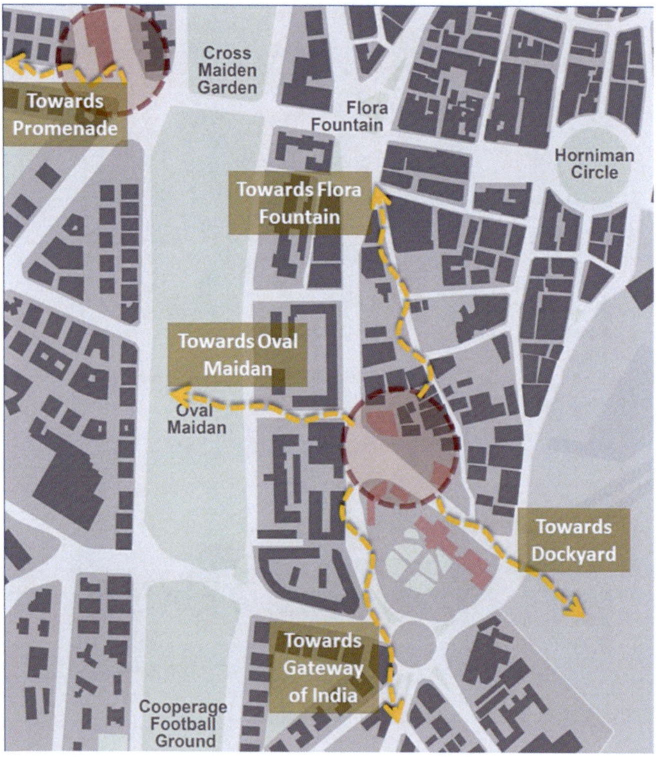

Fig. 23.11 Proximity chart for area of study

Observation and Mapping in 3 time zones at the mapped junctions mentioned in Figs. 23.11 and 23.13 of the following patterns as explained. The **study of movement pattern and circulation** of both vehicular traffic and pedestrian user in the identified locations, **mapping of pause points** due to pedestrian activities and also the variation in the activities created by pedestrian movement to be noted. Parallel study and understanding with the support of observational study and analysis about reflections due to **interventions** in the identified locations to be noted and recorded. The data collected will be followed by analysis of findings and conclusion for further use (Fig. 23.12).

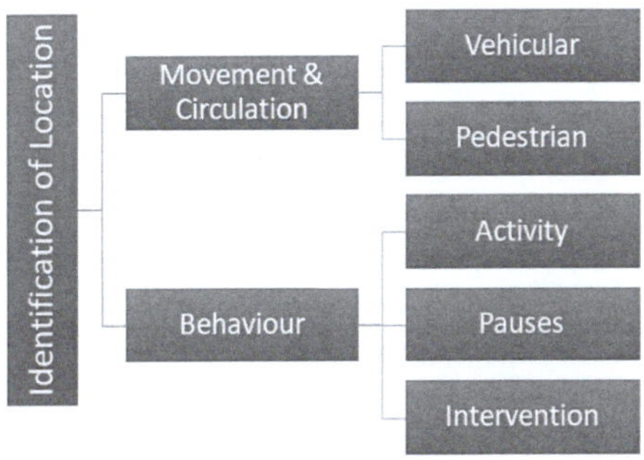

Fig. 23.12 Methodology diagram (Author, 2019)

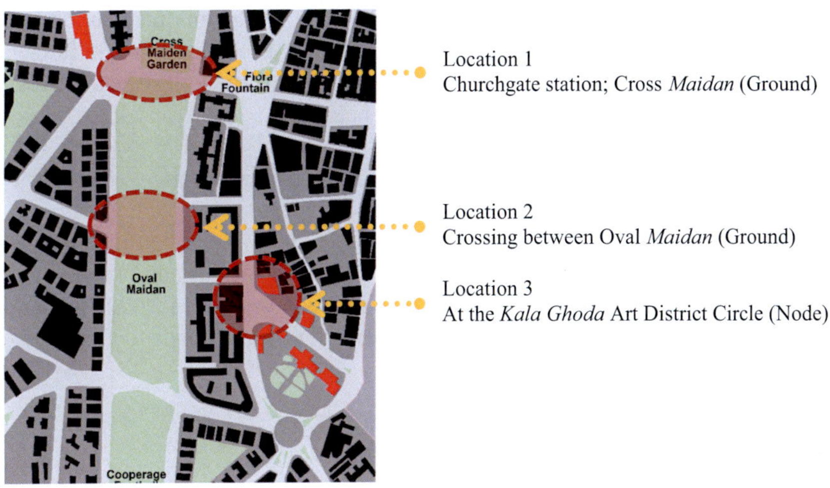

Fig. 23.13 Identified locations of study

23.4.2 Identifying the Location of Study

Once the area of study was identified, an observational study was undertaken along the route thus explained earlier. Three major intersections of activity zones were fixed along the route for further inspection of human behaviour in the specific scenarios (Fig. 23.13).

23.5 Findings

To represent the observations and findings of the study, a tabulation has been created with various heads including the **locational mapping** with and without the pedestrian movements at the location. A list of **interventions** observed in the location has been recorded. Another important observational graphics explain the percentages of **typology of people** observed in the area. The listed pedestrian activities observed on site were later bifurcated as **necessary, optional and social** activities in reform with the matrix derived from Tables 23.1 and 23.2.

The most important finding was with reference to the **interrelationship between the user, their activity and the intervention in place**. The interventions listed in the table strongly influence and inform the bifurcation of the various pedestrian activities observed during the study at the locations. An activity that could be social or optional at one location could become necessary at another due to the introduction of an alien element in terms of an intervention on the site. Again, this intervention and activity could be of a permanent or temporary nature. This condition would further influence the totality of the space that is being studied. Therefore, the categorisation of activities cannot be done by its nature alone, but it depends to a very large extent upon the external prevailing conditions at the time of study.

To simplify this complicated relationship of the three interdependent entities (the user, the activity and the intervention), it became mandatory to create a matrix to determine the segregation or categorisation of the activities as per their importance. The matrix defines, with reference to, the various standing local conditions the categorisation of the pedestrian activities leading to some conclusive concepts of human-urban associations and behaviour which in turn lead to the recommendations provided at the conclusion (Table 23.3).

23.6 Conclusion and Recommendations

The initiation of *Kala Ghoda* festival brings about an interesting change in the vitality of the region. Although the district has a very strong identity, it undergoes a transformation during this time which is not stark but blends within the fabric of the district creating a new identity for the timeline of the festival. Several vehicular

Table 23.3. Author created matrix to establish the findings during study w, r, f, Gehl's process in Table 23.1

Location	Mapping	Intervention	Pedestrian activity observed		
			Necessary	Optional	Social
Location 1: Churchgate station; Cross *Maidan* (Ground) 	1. Figure ground 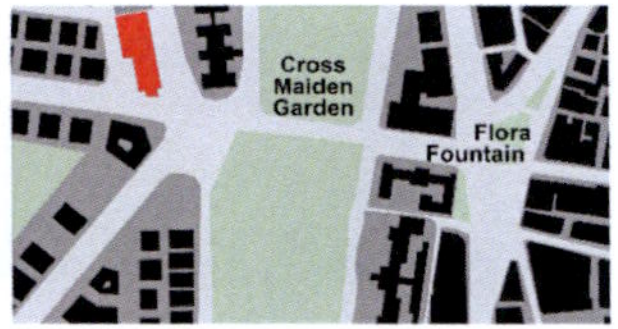	*Permanent* • Bus stop • Entry to Cross Maidan *Temporary* • Hawkers • KGF	• Purposeful walking towards place of work or study • Catching bus/taxi • To or From Railway station • '*Dabbawala*'—The lunch bearers	• Eating/refreshments at the Hawker stalls • Talking on phone • Smoking • Resting/sitting before moving ahead	• Selfie enthusiasts • Gathering point for KGF • Visitors
Typology of People 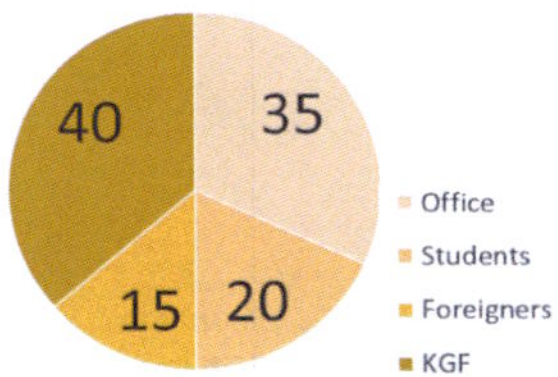	2. Pedestrian movement 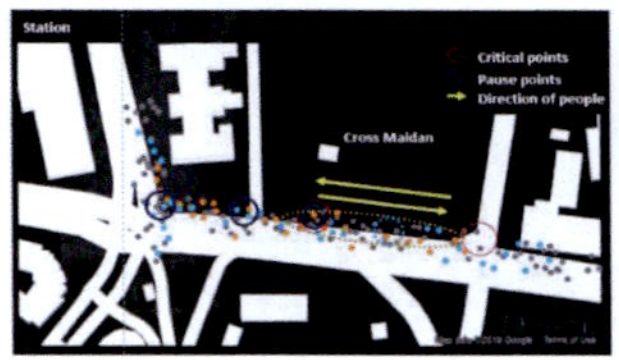				
Location 2: Crossing between Oval *Maidan* (Ground) 	1. Figure ground 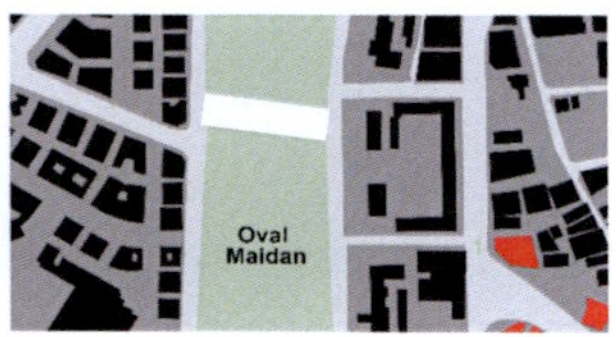	Permanent • Benches at the peripheral locations • Tall palm trees Temporary • Scheduled cricket matches	• Athlete/trainees • Short cut rute connecting to Civic Centre • Students • Taxi/bus queue in the peripheral edge	• Joggers • Walking Heritage tours • Photography enthusiasts due to wide scope of panaroma • Yoga club • Dog Owners	• Match spetators • Resting on the benches

(continued)

Table 23.3. (continued)

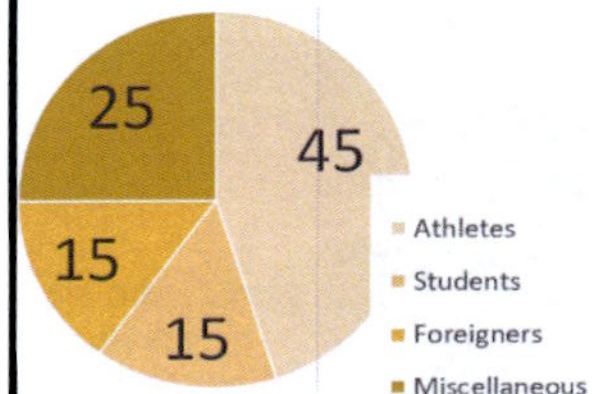	2. Pedestrian movement 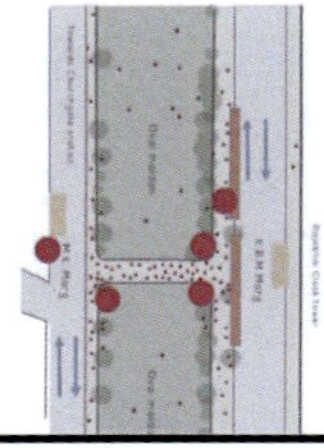	  			
Location 3: At the *Kala Ghoda* Art District Circle (Node)	1. Figure ground 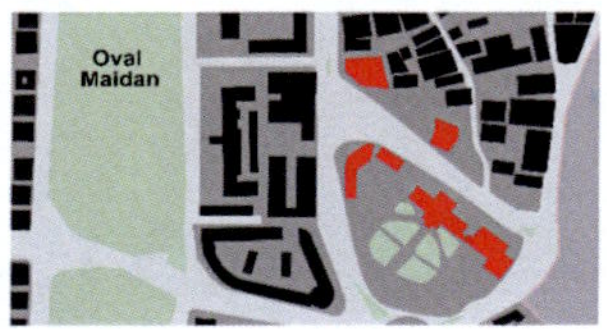	*Permanent* • Parking • Spill over from cafes etc. *Temporary* • KGF	• Purposeful walking towards place of work or study • Volunteers for the festival, cleaning, maintenance etc.	• Café/restaurant customers • Shoppers • Attending the KGF (stalls, installation, workshop etc.)	• Meeting with friends • Shopping • Meal at the cafes or restaurants
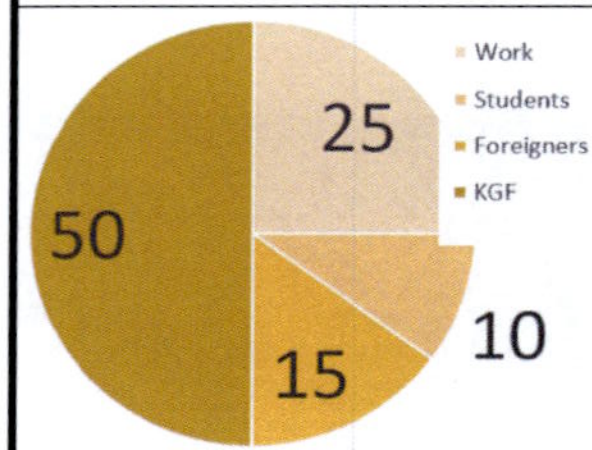	2. Pedestrian movement 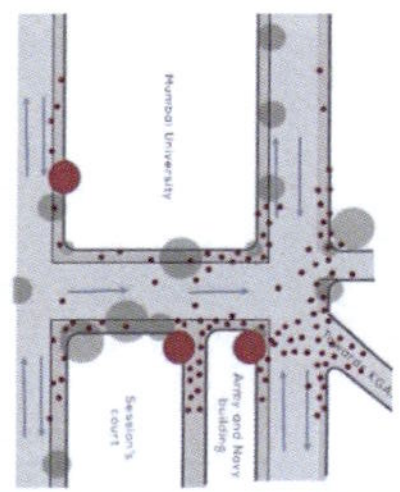	  			

Note KGF- Kala Ghoda Festival | Any non-English words, used locally have been put in italics for ease of understanding and explained
All information in the matrix is sourced during the research process

roads are converted into pedestrian-only zones leading to vehicular route diversions, which affect the fabric within the region. There is a marked increase in the flow of pedestrians around the area. A unique trade-off or understanding is seen between the pedestrians and visitors who seem to form their own interaction and pause points apart from the physical existing infrastructure.

Although the path towards the festival location and points is not designed for pause, several temporary pause points were observed in places like bus stops, entrances or transition areas for exhibitions, wider footpaths and near refreshment stalls. Even on regular days, fences and walls seem to become the support elements for people wanting to take a pause before moving towards their destination. Hawkers seem to be the largest places for pauses for pedestrians both during and beyond KGF. Another important observation is the adaptation of the individual with the phone to create a space around themselves for the activity. Often, group and individual selfie, along with photography are the most seen activities throughout the study.

Throughout the study, two most important things come out very clearly. **One** is that the vitality of a place although visually seems to be depending on the magnanimity and intricacy of the space designed, it is the small intangible and temporary interventions that connect the user to the space. The **second** factor is very interesting, because this comes as a by-product of the inherent nature of human being, which is to creatively adapt to spaces as per their inclinations. This means that the rigidity of space often led to it being overruled for its function and the more flexible the space, the more it was utilised.

As a race, humans consistently strive to keep their historic quarters of the city intact and preserved, but what we forget is the ever-evolving nature of humans and their fast-paced approach to technology which is overpowering most aspects of their life. It is very important to dive into these small peculiarities while we try to rethink our spaces for the need of the new public user. As observed in the study, the solution lies, not in creating completely new dynamics for the people, but merely in tackling the finer and smaller aspects of urban public experiences. Here the term '*People -friendly dynamics*' can be introduced, wherein the user is the authority to decide how they can mould a space for their use. The choice to make decision of participation, interactive tools and voluntary approaches make a public space more vital and approachable. With the kind of massively accelerating populations of cities of developing countries, the task for urban designers of the new age is to achieve this fine balance of *system-defined* and *user-defined* functionalities to create more vibrant and swarming public spaces. The key would be to take lessons from our history and give importance to the flexibility of spaces rather than the more formal principle of design, albeit while keeping in considerations the advancing effect of entry of technology on individual lifestyle.

References

Bandra Info (1600's) The 7 Islands of Bombay: with Bandra and Salsette [Map]. Retrieved from https://bandra.info/history-of-bandra-the-queen-of-bombay-suburbs/sevenislandsbombay/
British Library (2009) Old Map of Mumbai,1933 [Map]. Retrieved from https://www.oldmapsonline.org/map/britishlibrary/IOR_24_2812_1933
Chakrabarty D (1991) Open spaces/ public spaces: garbage, modernity and India. S Asia J S Asian Stud 14(1):5–31
Gazetteer of India, Maharashtra State (1987) Seven Islands of Mumbai Reclaimed [Map]. Retrieved from https://thatandthisinmumbai.files.wordpress.com/2014/02/the-seven-islands.jpg
Gehl J (1987) Life between buildings: using public space. Van Nostrand Reinhold, New York
Gehl J, Lars G (2001) New city spaces. The Danish Architectural Press, Copenhagen
Goldhagen S (2017) Welcome to your world- how the built environment shapes our lives. Harper Collins, New York
Hanafi I et al (2013) Human social behaviour in public urban space: towards higher quality cities. Spaces Flows: Int J Urban Extr Urban Stud 3(2):23–33
Kaviraj S (1997) Filth and the public sphere: concepts and practices about space in Calcutta. Publ Culture 10(1):83–113
Madanipour A (2003) Public and private spaces of the city. Routledge, London
Old Churchgate Station (1920) [Photograph]. Retrieved from https://www.rvcj.com/these-22-rare-images-of-old-mumbai-local-are-breathtaking/old-churchgate-station-in-1920/
Peinhardt K, Storring N (2019) A playbook for Inclusive placemaking: community process. Retrieved from https://www.pps.org/article/a-playbook-for-inclusive-placemaking-community-process
UDRI (2014) *Churchgate Revival—Organising Public Space at Churchgate Station* [Photograph]. Retrieved from http://www.udri.org/projects/churchgate-revival-organising-public-space-churchgate-station/
Vanka PS (2014) Public space and life in an Indian city: the politics of space in Bangalore (Doctoral Dissertation). Retrieved from https://deepblue.lib.umich.edu/bitstream/handle/2027.42/108825/salila_1.pdf
Whyte W (1980) The social life of small urban spaces. Washington D.C.: The conservation foundation
Whyte W (1988) City: rediscovering the centre. Doubleday, New York

Chapter 24
Quality of Life and the Architecture of Low-income Neighborhoods

Aliaa Ahmad, Yasser Mansour, Muhammad Gabr, and Marwa H. Khalil

Abstract The issue of quality of life has been discussed by many theorists and philosophers since ancient times relating it to many aspects of life, but the scientific research on quality of life started being evident from the 1970s, especially with the first publications of the social indicators research journal. Yet, there is still no concrete and agreed-upon definition of quality of life, what constitutes it, nor the parameters affecting it. More specifically, there is limited research that focuses on the relationship between quality of life and architecture and it is starker with low-income communities. Only the past couple of years witnessed an increase in the number of researches conducted in this field. Hence, this paper aimed to review and discuss the research carried out, precisely in the new millennium, investigating the relationship between quality of life and the architecture of low-income neighborhoods focusing on its main components. This review included a variety of researches addressing different issues such as measuring residential satisfaction, investigating the effects of the built environment on quality of life and conducting post-occupancy evaluations. Based on this comprehensive analytical review, the paper defined three main domains that constitute residential satisfaction, namely "housing unit, neighborhood, and social environment" and also the attributes of each domain. In addition, the paper highlights the shortcomings of the surveys technique and the missing attributes compared to studies conducted on other types of housing projects. In its conclusion, the paper discusses the implications and suggestions for future research in this field.

Keywords Quality of life · Residential satisfaction · Low-income neighborhoods · Built environment

---

A. Ahmad (✉) · Y. Mansour · M. Gabr
Ain Shams University, Cairo, Egypt
e-mail: arch.aliaa.fayek@gmail.com

M. H. Khalil
Arab Academy for Science Technology & Maritime Transport, Cairo, Egypt

Architecture and Urbanism: A Smart Outlook,
https://doi.org/10.1007/978-3-030-52584-2_24

24.1 Introduction

Discussing quality of life (QOL) is not a new stance, and it has been discussed by people and philosophers since ancient times and they still have not agreed on an exact definition for it. Aristotle defines the quality of life as, "*the quality of a person's life is associated with the community he/she lives among and the existing conditions in this community*" (Michalos 2015). In modern times, we find that the WHOQOL group (World Health Organization measuring Quality Of Life group) defines QOL as: "*An individual's perception of his/her position in the context of culture and value systems in which they live in and in relation to their goals, expectations, standards and concerns. It is a broad-ranging concept incorporating, in a complex way, the person's physical health, psychological state, level of independence, social relationships, personal beliefs and relationships to salient features of the environment*" (Mohit 2013) and the Ontario social development council and social planning network of Ontario, Canada (Massam 2002) has defined QOL as "*The product of the inter-play among the social, health, economic & environmental conditions which affect human & social development.*". The quality of life model (Research unit, University of Toronto 2018) defines it as "*The degree to which a person enjoys the important possibilities of his or her life.*" It still does not have an exact definition, and researchers claim that trying to find a fixed definition for "Quality of Life" is not an easy thing. It has to be drawn from multiple disciplines, combining human needs with subjective well-being or happiness and relating it to the opportunities provided for the human being (Costanza et al. 2007).

Also, there have been many attempts to define the indicators constituting quality of life in order to measure people's satisfaction with these indicators and hence stand upon their satisfaction with their quality of life. In 2013, the EU started the 8 + 1 indicators measuring technique which broke the indicators to material living conditions, productive or main activity (covering employment), health, education, leisure and social interactions, economic and physical safety, governance and basic rights, natural and living environment, and overall life satisfaction. Hughes (1990) divided the indicators into eight indicators: subjective satisfaction, physical environmental factors, social environmental factors, socioeconomic factors, cultural factors, health status factors, personality factors, and personal autonomy factors. Different empirical work had different indicators to assess but they all shared the same hierarchy, each indicator is divided into domains which in turn are divided into attributes constituting them (Mercy et al. 2015), and this is generically explained in (Fig. 24.1) for a full understanding of how the hierarchy flows.

24.1.1 Quality of Life and the Built Environment

Whether it is called "natural and living environment" or "physical environmental"; the built environment is considered one of the main domains constituting quality

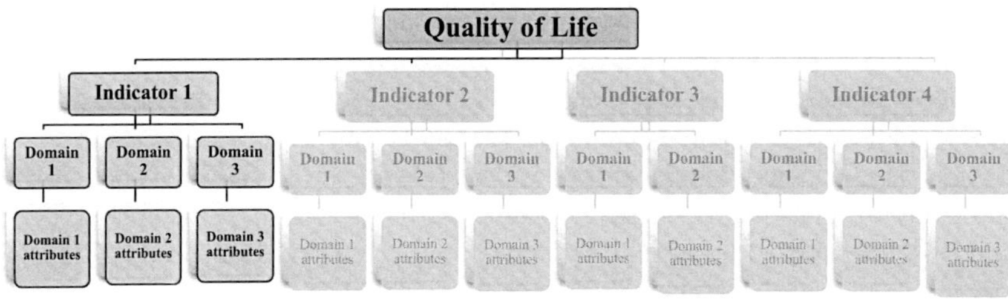

Fig. 24.1 A generic diagram explaining the hierarchy of quality of life, indicators, domains, and attributes

of life, investigating the built environment in neighborhoods and assessing it is an indicator of residents' overall quality of life. From here came the term "residential environment." A residential environment comprises of three components: neighborhood, house, and neighbors and the mutual interaction between these three components is what needs to be studied. As people live in the same place for a long time, they develop a sense of "dwelling." They become psychologically fused with their residential environment as they build up a cognitive awareness of this environment due to their habitual routines. (Lynch 1960).

The book "The quality of American life" stands as one of the first and most important trials of assessing the quality of life in relation to five basic indicators: The Residential Environment, The Country as a Domain of Experience, The Experience of Work, Marriage and Family Life, and Personal Resources and Personal Competence (Campbell et al. 1976). The chapter on "The residential environment" was co-authored and built upon previous studies of Robert Marans (Marans and Rodgers 1975). It paved the way for the studies linking quality of life and the built/residential environment. In this chapter, it was the first registered time to state that studying a residential environment is divided into four main domains: the nation (the region), the community, the neighborhood, and the dwelling (the residence), while studying the region is left for nationwide studies rather than one concerning certain neighborhoods.

24.1.2 Low-income Neighborhoods

Governments around the world are continuously building new neighborhoods and housing projects. Measuring the residents' satisfaction with these projects and understanding the attributes that make residents more satisfied and those that make them less satisfied will help in the design process of the future projects in order to enhance residents' satisfaction with their built environment and hence enhance their overall life satisfaction. In the USA, the department of housing and urban development has recognized the importance of good design in the development of public housing and how providing people solely with a roof above their heads was not enough anymore

(HUD 2000). This lead to an increase in the research performed to measure the quality of life in housing projects provided by governments and the private sector for different income groups, especially low-income groups, as reviewed; a majority of these studies were conducted in the past decade.

Understanding the perceptions of residents' living in a relatively poor urban community of their quality of life can guide decisions and inform actions to help build stronger and more resilient communities (Ciorici and Dantzler 2018). The empirical research is based upon the idea that residential satisfaction is a measure of the residents' existing conditions and their aspirations for their residential environment (Mohit et al. 2010). Accordingly, evaluators established objective and subjective attributes and introduced them for residents' subjective assessment, and collectively, they serve as a basis for assessing the domains of residential satisfaction in low-income neighborhoods. Those residents represent a relatively disadvantaged group and lack the ability to choose their houses or solve their housing problems by themselves (Huang and Du 2015).

This paper is a part of a larger study that focuses on studying the domains and attributes of the residential environment of a low-income neighborhood in New Cairo. Yet this paper focuses on the following objectives:

1. To critically review the attributes assessed and categorize them.
2. To point out the missing attributes.
3. To propose future approaches for studying QOL in low-income neighborhoods.

24.2 Methodology

Literature from previous research was selected and reviewed. This literature included research that dealt with measuring and studying quality of life specifically in low-income neighborhoods (provided by both public and private sectors). The research was done using database available online, mainly through Egyptian Knowledge Bank (Elsevier, Wiley, JSTOR, Taylor & Francis online, Sage and Springer). The keywords used to direct the research were: quality of life, assessment of/measuring residential satisfaction, neighborhood characteristics, and low-income neighborhoods/housing. The research was restricted to the body of research done in the twenty-first century. The purpose of the analysis is to extract the domains and attributes assessed before and highlight the apparent gaps that need to be researched, while analyzing the methods used with their advantages and disadvantages (Wee and Banister 2015).

24.3 The Residential Environment of Low-income Neighborhoods: Domains and Attributes

The focus of the research was on the literature studying quality of life and the residential environment of low-income housing, basically research concerned with measuring/assessing it. Although the contents had great similarities, the titles and aims have some differences. There were studies that aimed at measuring QOL in residential environment (Ismail et al. 2014; Gou et al. 2018; Kowaltowski 2006) and studies that aimed at measuring residential satisfaction with its domains housing/community/ neighborhood (Alnsour and Hyasat 2016; Jiboye 2010; Huang and Du 2015; Addo 2016; Mohit et al. 2010; Mohit and Azim 2012; Mohit and Nazyddah 2011; Riazi and Emami 2018; Ibem et al. 2018; Goh 2011). Some of the studies focused only on one of the domains, either housing quality (Emmanuel 2011) or neighborhood satisfaction (Cioric and Dantzler 2018) and some research measured the satisfaction with the residential environment and its relation to a certain aspect or feature, like housing performance (Hashim et al. 2012) or safety performance (Hussein et al. 2014).

All these studies used a survey questionnaire with a Likert scale to measure the residents' satisfaction with the designated attributes, and in some occasions, it was accompanied by more qualitative work like field visits or semi-structured interviews to support, confirm, and explain the results (Addo 2015; Riazi and Emami 2018; Hashi et al. 2012; Husin et al. 2015). Yet, the results of the questionnaires were statistically analyzed to reach the residents' level of satisfaction with their residential environment and correlate it to their overall life satisfaction. The base model provided by Campbell et al. (1976) suggested three domains for measuring residential satisfaction "Housing," "Neighborhood," and "Community." The literature found on low-income housing used the term "Social Environment" instead of "Community" with some differences in the attributes assigned for it (Mohit et al. 2010; Mohit and Azim 2012; Mohit and Nazyddah 2011; Ibem et al. 2018; Addo 2016). Although not all research found used the same terminologies, others even divided them into more domains. Yet all attributes assessed where found to fall under the three designated domains: "*Housing, Neighborhood and Social Environment.*" These attributes were collected, categorized, and analyzed while mentioning the sources they were extracted from.

24.3.1 Housing Unit

Measuring housing satisfaction is a common domain in almost all of the studies assessing residential satisfaction and some studies even only assess housing satisfaction (Aigbavboa and Thwala 2012; Emmanuel 2012; Ismail et al. 2015; Ibem et al. 2013). Housing satisfaction assessment was mostly moderate, with the exception of (Jiboy 2010; Huang and Du 2015; Ibem and Aduwo 2013), where it was reported to be high. In those cases, the attributes correlated with high satisfaction were unit design, unit area, number of rooms, finishing material, sanitary fittings, buildings materials, structure, and privacy. Low satisfaction was perceived in Ghana (Addo 2015) in a multi-habited housing project, which was correlated with number and size of the room, the units had a shared bathroom and kitchen and each unit compromised of one small room and a veranda which nearly all the residents turned into an extra room.

After collecting the attributes assessed and classifying them, the housing domain was found to be constituted of two main categories: ***internal qualities*** (Physical qualities, spatial qualities, and services) and ***contextual qualities*** (Typology and Accessibility). In (Table 24.1), the attributes are classified and stated with the number of times they appeared and the literature they were extracted from. They were classified collectively per category for space limitations.

24.3.1.1 Internal Qualities

They are present in all research work with some differences in the attributes assessed; room, kitchen and toilet areas, ventilation, natural lighting, and privacy are the most repeated and the most correlated as well with housing satisfaction. Small room along with sharing the kitchen and toilet with neighbors for both male and female (Addo 2015) leads to a low assessment, as well as having bad ventilation and air circulation in the unit that leaves it cold in winter and hot in summer (Aigbavboa and Thwala 2012). Large rooms and enough kitchen space can help make the residents slightly satisfied with their housing, although they were unsatisfied with the toilet and dry area (Ibem 2018).

The difference in the conditions and luxuries between low-income projects in different parts of the world can be seen in Hashim (2012) assessing water proofing, rain penetration, and cracking, which are basic attributes suffering major defects, where on the other hand Alnsour and Hyasat (2016) were assessing the residents' satisfaction with their homes' private gardens and balconies, which the residents were quite satisfied with.

24.3.1.2 Contextual Qualities

Contextual attributes are present fewer times than internal attributes and in lesser studies. Although all case studies were multistory apartment buildings "floor level" attribute appeared only once (Huang and Du 2015), "lift" as well was assessed few times and residents were very unsatisfied with its service (Ismail et al. 2015; Husin et al. 2015). These are basic elements that help the elders and the handicapped and are missing, which again shows how the conditions and luxuries of low-income projects are different in different parts of the world.

24.3.2 Neighborhood

The neighborhood is the second most assessed domain after the housing domain. Its attributes where divided into two main categories: *physical features* and *services* and are shown here in (Table 24.2), with the literature they were extracted from.

Table 24.1 Housing unit domain with its attributes, number of times assessed, and literature extracted from

Domain	Attribute			Literature extracted from
Housing Unit	Internal qualities	Physical qualities	Rooms areas (18)—Kitchen area (15))—Toilet/Bathroom area (13))—Unit layout/Design (11))—Number of rooms (7))—Sanitary fitting (7))—Finishing materials (6))—Building materials (5))—Size of unit (5))—Windows/Doors (5))—Balcony/Yard (4))—Structure (3))—Dry area (3))—Garden (1)	Alnsour and Hyasat (2016); Jiboye (2010); Huang and Du (2015); Addo (2015); Mohit et al. (2010); Mohit and Azim (2012); Mohit and Nazyddah (2011); Riazi and Emami (2018); Ibem et al. (2018); Goh and Yahya (2011); Hashim et al.(2012); Ismail et al.(2015); Husin et al.(2015); Aigbavboa and Thwala (2012); Ibem and Aduwo (2013); Mohit and Iyanda (2016); Karim, (2012); Iyanda and Mohit (2016); Salleh (2008); Ibem et al.(2013); Byun and Ha (2016)
		Spatial qualities	Ventilation (14) Privacy (11) Safety (8) Noise (7) Natural lighting (6)	Jiboye (2010); Huang and Du (2015); Addo (2015); Mohit et al. (2010); Mohit and Azim (2012); Mohit and Nazyddah (2011); Riazi and Emami (2018); Ibem et al. (2018); Goh and Yahya (2011); Hashim et al.(2012); Ismail et al.(2015); Husin et al.(2015); Aigbavboa and Thwala (2012); Ibem and Aduwo (2013); Mohit and Iyanda (2016); Karim, (2012); Iyanda and Mohit (2016); Salleh (2008); Ibem et al.(2013); Byun and Ha (2016)

(continued)

Table 24.1 (continued)

Domain	Attribute			Literature extracted from
		Services	Electric wiring (13))—Garbage collection/cleanliness (10)—Water supply (7)—Maintenance (plumbing, electrical, drainage) (6)—Firefighting (5)—Waterproofing (3)—Rain penetration (1)—Cracking (1)—Telecommunication (1)	Alnsour and Hyasat (2016); Jiboye (2010); Mohit et al. (2010); Mohit and Azim (2012); Mohit and Nazyddah (2011), Ibem et al. (2018), Goh and Yahya (2011), Hashim et al.(2012), Ismail et al.(2015), Husin et al.(2015), Ibem and Aduwo (2013), Mohit and Iyanda (2016), Salleh (2008), Ibem et al.(2013)
	Contextual qualities	Typology	Location (6))—Appearance of dwelling (6))—Number of units in each building (2)—Outdoor views (2) Floor level (1)—Residential setback (1)	Alnsour and Hyasat (2016); Jiboye (2010); Huang and Du (2015); Mohit and Nazyddah (2011); Riazi and Emami (2018); Ibem et al. (2018); Goh and Yahya (2011); Hashim et al.(2012); Ismail et al.(2015); Aigbavboa and Thwala (2012); Ibem and Aduwo (2013); Ibem et al.(2013)
		Accessibility	Corridor (5) Staircase (5) Lift (5)	Mohit et al. (2010); Mohit and Azim (2012); Mohit and Nazyddah (2011); Riazi and Emami (2018); Husin et al.(2015)

Table 24.2 Neighborhood domain with its attributes, number of times assessed, and literature extracted from

Domain	Attributes		Literature extracted from
Neighborhood	Physical features	Education facilities (15)—Health services (14)—Shopping area (11) Public transport (10)—Parking (9) Market (8)—Children playground (8)—Workplaces (7)—Open green areas (7)—Place of worship (6) Recreational facilities (6) Multipurpose/community hall (5) Overall appearance of housing project (5)—Town center (3)—Public library (2)—Sport facility (2) Outdoor furniture (1)—Public toilet (1)	Alnsour and Hyasat (2016); Jiboye (2010); Huang and Du (2015); Addo (2015); Mohit et al. (2010); Mohit and Azim (2012); Mohit and Nazyddah (2011); Riazi and Emami (2018); Ibem et al. (2018); Goh and Yahya (2011); Hashim et al.(2012); Ismail et al.(2015); Husin et al.(2015); Aigbavboa and Thwala (2012); Ibem and Aduwo (2013); Mohit and Iyanda (2016); Karim, (2012); Iyanda and Mohit (2016); Salleh (2008); Ibem et al.(2013); Byun and Ha (2016)
	Services	Street lighting (7)—Garbage collection/cleanliness (7)—Water supply (6)—Police (6)—Firefighting/station (6)—Roads (6)—Pedestrian walkways (4)—Sewage system (3) Public phone (3)—Drainage system (1)	Alnsour and Hyasat (2016); Huang and Du (2015); Addo (2015); Mohit et al. (2010); Mohit and Azim (2012); Mohit and Nazyddah (2011); Riazi and Emami (2018); Goh and Yahya (2011); Ciorici and Dantzler (2018); Aigbavboa and Thwala (2012); Mohit and Iyanda (2016); Karim (2012); Iyanda and Mohit (2016); Salleh (2008)

Some studies focused solely on assessing neighborhood satisfaction (Ciorici and Dantzler 2018; Kowaltowski et al. 2006). Many studies found it highly correlated with overall residential satisfaction (Alnsour and Hyasat 2016; Huang and Du 2015; Salleh 2008; Byun and Ha 2016). Unlike housing unit, neighborhood satisfaction was low in some studies, because of cleanliness, street light, pedestrian walkways (Mohit and Nazyddah 2011), education and health facilities (Ibem and Aduwo 2013; Mohit and Iyanda 2016; Iyanda and Mohit 2016) and public transport and proximity to workplace (Ibem et al. 2018). High satisfaction was found in (Huang and Du 2015; Karim 2012) because of green areas, playgrounds quietness, and security.

Health and education services are the most assessed attributes. In most studies, residents stated that it is the main concern for their families where these facilities are located, their proximity to their housing unit as this saves families a lot of transportation time and has a great effect on their neighborhood satisfaction and overall life satisfaction.

24.3.3 Social Environment

Fewer studies focused on studying and assessing the social environment like (Huang and Du 2015; Addo 2015; Mohit 2010, 2011, 2012; Riazi and Emami 2018; Gou 2018), as most studies were focusing on housing and neighborhood domains. Social environment was sometimes highly correlated with overall life satisfaction (Riazi and Emami 2018; Mohit 2012; Addo 2015; Ciorici and Dantzler 2018). In some studies that namely assessed housing unit and/or neighborhood satisfaction, some of the attributes that belonged to the social environment were assessed within the neighborhood features and so in the analysis they were separated and added herein (Table 24.3) for the social environment. The attributes were categorized into *Community Interaction* and *Community Conditions* and like before are shown alongside the literature they were extracted from.

Social environment was the least domain studied and as follows the attributes assessed were the least in number and diversity. Social relations and community safety were the most assessed, and they were associated with each other. Communities where there are evident relations among the residents, and they share some activity can help reduce crime rates in their neighborhood and help enhance their quality of life (Ciorici and Dantzler 2018).

To sum it up, as per explained in Fig. 24.1 that showed how the hierarchy of the indicators, domains and attributes flow, the residential environment indicator is explained by the same hierarchy in Fig. 24.2. With its three domains: housing unit, neighborhood, and social environment, with each of their respective main categories for attributes.

24.4 Discussion

After all the attributes became apparent, with the loads of their presence in the previous case studies and literature, it is crucial to understand the meaning of these studies and their results. They can help governments and the private sector that build for the low-income group, understand the residents' perception of the residential environment they live in and hence be able to build future housing projects that will enhance people's quality of life and promote social inclusion (Huang and Du 2015).

Regarding data collection tactics, most of them are considered subjective although being quantitatively analyzed, as the results are driven from the residents' answers and not from official records. This research would like to point out some main shortcomings of using survey questionnaires:

Table 24.3 Social environment domain with its attributes, number of times assessed, and literature extracted from

Domain	Attributes		Literature extracted from
Social environment	Community Interaction	Social isolation/relation (10) Community support (5) Community stability (2) Neighbors conflict (1)	Huang and Du (2015); Addo (2015); Mohit et al. (2010); Mohit and Azim (2012); Mohit and Nazyddah (2011); Riazi and Emami (2018); Goh and Yahya (2011); Ciorici and Dantzler (2018); Karim, (2012); Iyanda and Mohit (2016)
	Community Conditions	Community safety/crime rates (12) Noise level (9)	Addo (2015); Mohit et al. (2010); Mohit and Azim (2012); Mohit and Nazyddah (2011); Riazi and Emami (2018); Ibem et al. (2018); Goh and Yahya (2011); Ciorici and Dantzler (2018); Aigbavboa and Thwala (2012); Ibem and Aduwo (2013); Karim, (2012); Iyanda and Mohit (2016); Byun and Ha (2016)

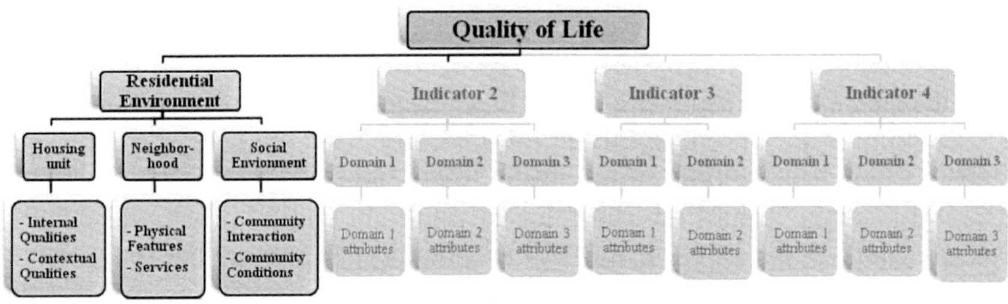

Fig. 24.2 Domains and attributes of the residential environment indicator in low-income neighborhoods

First: measuring satisfaction with an approach based on people's own assessment is determined to a great extent upon their own aspirations, which will consequently measure how well they have adapted to their situation, whether the situation is adequate or not. As Maslow (1943) explains it, there is a frustration tolerance, when one is deprived of a need for so long that they get accustomed to its unsatisfaction and are able to function with its extreme deprivation without much noticing it. People may adjust their needs according to the situation they find themselves living in but one should note that a person may learn to live with the conditions present in his/her life while smiling and acting happy but that does not mean they do not deserve to be compensated for the bad conditions that they live in. The background of the residents also plays a role in their assessment. Those who had a higher standard of living and moving into a certain project was a downgrade for them might show lower level of satisfaction than those who lived in menial conditions before and the move represented an upgrade for them (Nussbaum and Sen 1993).

Second: measuring residential satisfaction using only a Likert scale does not allow the residents to give reasons for their answers, and hence, the research loses the ability to solve the problem correctly. People may be dissatisfied with the same attribute but for two totally different reasons and their lacking of a proper way to express themselves might not help the researchers find the correct solution. However, researchers claim that a questionnaire is a preferable option when dealing with low-income and disadvantaged groups (Aigbavboa and Thwala 2012), a repeated opinion which no reason was given for.

Third: although it is considered to be a bottom-up approach, as people's own assessment is the guide of most study results, yet in no case was it mentioned that the attributes used were extracted from the residents themselves, but they were rather almost always enforced by the researchers. Whether built upon previous research (Huang and Du 2015) or based upon their exploration of the existing conditions of the communities they are studying (Riazi and Emami 2018). This again leaves no space for the residents to voice their needs, opinions, and preferences which could be an added value for the research to help it reach an accurate assessment strategy.

Fourth: it was noted that very few studies focused on the social environment, even so the attributes measured lack ones found in other research that is dealing with other groups than the low-income group like sense of belonging, attachment, and ownership (Easthope 2004; Fleury-Bahi et al. 2008; Adriaanse 2007) or homemaking, self-actualization and identity (Canter 1977; Cuba and Hummon 1993; Mee 2007), and other similar social and psychological attributes. This could either suggest that the low-income housing projects provide only the basic needs of Maslow's (1943) hierarchy of needs, namely the physiological and safety needs, or it could suggest that the researchers working with low-income groups focus only on these basic needs above else. This stands on contrary to what is now understood by many, that it is not enough anymore to merely provide people with a roof upon their heads in order for them to have a good quality of life (HUD 2000). The psychological needs of this group need to be studied and assessed as well for the researcher to be able to state if these neighborhoods provide a good quality of life or not.

24.5 Conclusion

Quality of life has many definitions and is constituted of many indicators. This research has focused on the built environment as it has been found to be constantly one of these indicators. Researchers who attempted to study and assess quality of life and the built environment measured residential satisfaction and correlated it to the overall life satisfaction. The research focusing on low-income neighborhoods is still in its infancy stages; yet it has shown diversity in choosing the attributes being assessed. The paper's main objective was to review the domains and attributes found in this body of research and categorize them, which resulted in three main domains for the residential environment which are "housing unit, neighborhood, and social environment." The categories that constitute the attributes of each of these domains, respectively, are internal qualities and contextual qualities, physical features and services, and community interaction and community conditions. This could serve as a basic guide for future research focusing on assessing residential satisfaction in low-income neighborhoods. However, the body of research studied depended mainly on survey questionnaires while assessing residential satisfaction and the shortcomings of using survey questionnaires were highlighted. They might give conflicting messages according to the residents' adaptation to their living conditions and the different backgrounds they come from. They lack the chance of people voicing their opinions on why they gave such assessments. They lack the residents' contribution in deciding the attributes that need to be measured in their neighborhoods. Finally, they lack the ability to cater for the psychological needs of the residents and focus only on their physiological and safety needs.

In its next phase, the research will continue to study the residential environment of a low-income neighborhood in New Cairo and will use an ethnographic approach relying on marginal participant observations, site visits, and semi-structured interviews, guided by the attributes concluded in this stage. This might help in defining the attributes based on the residents' narratives, opinions, and perceptions.

References

Addo IA (2015) Assessing residential satisfaction among low income households in multi-habited dwellings in selected low income communities in Accra. Urban Stud 53(4):631–650

Adriaanse CC (2007) Measuring residential satisfaction: A residential environmental satisfaction scale (RESS). J Hous Built Environ 22(3):287–304

Aigbavboa CO, Thwala WD (2012) An appraisal of housing satisfaction in South Africa low income housing scheme. Int J Constr Manage 12(1):1–21

Alnsour JA, Hyasat AS (2016) Residential satisfaction with low income housing in Jordan: Salt city as a case study. Jordan J Econ Sci 3(1):31–44

Byun G, Ha M (2016) The factors influencing residential satisfaction by public rental housing type. J Asian Archit Build Eng 15(3):535–542

Campbell A et al (1976) The quality of American life: perceptions, evaluations, and satisfactions. Russell Sage Foundation

Canter D (1977) The psychology of place. St. Martin's Press, New York
Ciorici P, Dantzler P (2018) Neighborhood satisfaction: a study of a low-income urban community. Urban Aff Rev 1–29
Costanza R et al (2007) Quality of life: an approach integrating opportunities, human needs, and subjective well-being. Ecol Econ 61:2–3
Cuba L, Hummon D (1993) A place to call home: identification with dwelling, community, and region. Soc Q 111–131
Easthope H (2004) A place called home. Hous Theory Soc 21(3):128–138
Emmanuel JB (2012) "Housing quality" to the low income housing producers in Ogbere, Ibadan, Nigeria. Procedia Soc Behav Sci 35:483–494
Fleury-Bahi G, Félonneau M-L, Marchand D (2008) Processes of place identification and residential satisfaction. Environ Behav 40(5):669–682
Goh AT, Yahya A (2011) Public low-cost housing in Malysia: case studies on PPR low-cost flats in Kuala Lumpur. J Des Built Environ 8(1)
Gou Z, Xie X, Lu Y, Khoshbakht M (2018) Quality of life (QOL) survey in Hong Kong: understanding the importance of housing environment and needs of residents from different housing sectors. Int J Environ Res Public Health 15(2):219
Hashim AE, Samikon SA, Nasir NM, Ismail N (2012) Assessing factors influencing performance of Malaysian low-cost public housing in sustainable environment. Procedia Soc Behav Sci 50:920–927
Huang Z, Du X (2015) Assessment and determinants of residential satisfaction with public housing in Hangzhou, China. Habitat Int 47:218–230
HUD (2000) Principles for inner city neighborhood design: Hope VI and the new urbanism. U.S. Dept. of Housing and Urban Development, Washington
Hughes B (1990) Quality of life. In: Peace SM (ed) Researching social gerontology. Sage Publications, London
Husin HN, Nawawi AH, Ismail F, Khalil N (2015) Correlation analysis of occupants' satisfaction and safety performance level in low cost housing. Procedia Soc Behav Sci 168:238–248
Ibem EO, Aduwo EB (2013) Assessment of residential satisfaction in public housing in Ogun State, Nigeria. Habitat Int 40:163–175
Ibem EO, Ayo-Vaughan EA, Oluwunmi AO, Alagbe OA (2018) Residential satisfaction among low-income earners in government-subsidized housing estates in Ogun State. Nigeria. Urban Forum 30(1):75–96
Ibem EO, Opoko AP, Adeboye AB, Amole D (2013) Performance evaluation of residential buildings in public housing estates in Ogun State, Nigeria: users satisfaction perspective. Frontiers Architectural Res 2(2):178–190
Ismail F, Jabar IL, Janipha NA, Razali R (2015) Measuring the quality of life in low cost residential environment. Procedia Soc Behav Sci 168:270–279
Iyanda SA, Mohit MA (2016) Measuring the dimensions and attributes of liveability of low-income housing communities in Nigeria. Plann Malays J 14(4)
Jiboye AD (2010) The correlates of public housing satisfaction in Lagos, Nigeria. J Geogr Reg Plann 3(2):017–028
Karim HA (2012) Low cost housing environment: compromising quality of life? Procedia Soc Behav Sci 35:44–53
Kowaltowski DC, Silva VG, Pina SA, Labaki LC, Ruschel RC, Moreira DD (2006) Quality of life and sustainability issues as seen by the population of low-income housing in the region of Campinas Brazil. Habitat Int 30(4):1100–1114
Marans RW, Rodgers W (1975) Toward an understanding of community satisfaction. In Amos HH, Vincent PR (eds) Metropolitan America in contemporary perspective, Sage Publications
Maslow AH (1943) A theory of human motivation. Psychol Rev 50(4):370–396
Massam BH (2002) Quality of life: public planning and private living. Prog Plann 58(3)
Mee K (2007) I ain't been to heaven yet? Living here, this is heaven to me": Public housing and the making of home in inner Newcastle. [Article]. Housing, Theory Soc 24:207–228

Mercy JL et al. (2015) Quality of life: facts and views. Publications Office of the European Union
Michalos AC (2015) Ancient views on the quality of life. Springer, Berlin
Mohit MA (2013) Quality of life in natural and built environment—an introductory analysis. Procedia Soc Behav Sci 101:33–43
Mohit MA, Azim M (2012) Assessment of residential satisfaction with public housing in hulhumale', maldives. Procedia Soc Behav Sci 50:756–770
Mohit MA, Iyanda SA (2016) Liveability and low-income housing in Nigeria. Procedia Soc Behav Sci 222:863–871
Mohit MA, Nazyddah N (2011) Social housing programme of Selangor Zakat Board of Malaysia and housing satisfaction. J Hous Built Environ 26(2):143–164
Mohit MA, Ibrahim M, Rashid YR (2010) Assessment of residential satisfaction in newly designed public low-cost housing in Kuala Lumpur, Malaysia. Habitat Int 34(1):18–27
Nussbaum M, Sen A (eds) (1993) The quality of life. Clarendon Press, Oxford
Riazi M, Emami A (2018) Residential satisfaction in affordable housing: a mixed method study. Cities 82:1–9
Salleh AG (2008) Neighborhood factors in private low-cost housing in Malaysia. Habitat Int 32(4):485–493
The Quality of Life Model (n.d) Retrieved August 16, 2018, from quality of life research unit: Home Page https://sites.utoronto.ca/qol/qol_model.htm
Wee BV, Banister D (2015) How to write a literature review paper? Transp Rev 36(2)

Chapter 25
The Last Train Traces in Beirut: Towards an Ecological Urbanism Perspective

Nadine Hindi

Abstract The last train journeys tying Damascus to Beirut stopped operating gradually during the Lebanese Civil war. While many parts of the railway were subjected to vandalism by militiamen, the last section in Beirut still keeps tangible fragments. It is with a renewed perspective that this last railway segment adjacent to the Mar Mikhayel train station is studied. The train track passing across a rather underprivileged residential Armenian settlement in Badawi neighbourhood, represents a *terrain vague* (Solà-Morales Rubió in Terrain vagues. MIT Press, Cambridge, pp 118–123, 1995), where informal greenery wildly established itself along the railway, over decades of abandonment. This neglected space undergoes a different urban metabolism, in tandem between the dual informalities of both the social and natural appropriations. The paper problematic outlook doesn't consider the natural recovery phenomenon as a threat, but rather as an opportunity for thinking paradigms for urban regeneration. While this is not a new idea in rethinking from the perspective of ecological urbanism, it presents a real situation for understanding new flows between human existence and the natural recovery phenomenon in urban settings. In this case study, Google Earth maps serve as the main tool in mapping the growth of greenery over time and the study of the social practices informality reveals a realistic first-hand depiction of this urban landscape. Though this paper doesn't claim to offer an urban solution for similar landscapes, it however sheds the light on identifying opportunities out of the natural recovery phenomenon and paving the way for a new form of designing in more ecological urban environments.

Keywords *Terrain vague* · Urban · Ecology · Train · Ecosystem

25.1 Definition of a Problematic: A Terrain Vague Quality

Across the Armenian neighbourhood known as *Badawi*, the railway traces of the last train segment appear clearly all along with the nineteenth-century metal bridge

N. Hindi (✉)
Department of Architecture, RC_FAAD, Notre Dame University NDU, Zouk Mosbeh, Lebanon
e-mail: nhindi@ndu.edu.lb; nadine.hindi@gmail.com

Architecture and Urbanism: A Smart Outlook,
https://doi.org/10.1007/978-3-030-52584-2_25

that connects the rail to the *Mar Mikhayel* main train station. This train track which connected Beirut to Damascus since the late nineteenth century,[1] stopped gradually during the Lebanese Civil War (1975–1989). While most of the parts of the railway were subjected to vandalism by fighting militiamen and successive occupying armed forces, this last section still keeps tangible fragments of its railway.[2] This rather underprivileged neglected urban slice where informal greenery wildly established itself along the abandoned railway, share the characteristics of undefined dysfunctional spaces coined by Solà-Morales (1995) as *'terrains vagues'*.

From the outset, this paper doesn't consider the natural recovery phenomenon as a threat, but rather an opportunity to contemplate another paradigm for urban regeneration (Mostafavi and Doherty 2010). Based on an urban case study, this paper highlights the opportunities of regenerating a space out of understanding the flows generated from people behaviour and the natural recovery process. Google Earth maps and field observations served as main tools in mapping the dual informalities of the patterns of social practices and greenery growth. Consequently, the paper is structured in an approach to understand the balance between nature, inhabitants and space, through observing the natural phenomenon recovery initiated along the train track and spreading across the area, over the last two decades. It also highlights the current inhabitants territorialization, and moments of appropriation and adaptation of the train track space during their daily life practice.

This paragraph frames some different definitions of the natural and human ecosystems in urban contexts from a trans-disciplinary perspective, related to urban ecology and ecological landscape. It attempts as well to frame a definition of the condition of the urban context. Then it will cross-relate them with the case study area, taking into consideration its physical and social specificities. On the one hand, Beck and Forman focus on understanding a fluid balance unfolded in space and time, based on the reading of the fluid expansion of natural ecosystems (Beck 2013) and the urban patterns flows and changes over time (Forman 2014). On the other hand, Corner (2006) frames a condition of *'terra fluxus'*, informally and fluidly shaped by human beings and the natural ecosystems in urban contexts. This framework implies the fluid participation of different stakeholders in the same space, over time. From the perspective of the ecological landscape realm, Beck observes the transformation and persistence of natural systems within the urban system itself. Alternately, Forman observes the built environment as affected by the infiltration of the natural ecosystem from the perspective of urban ecology. By considering the city as a *'human ecosystem, and urban design is one of our tools for adaptation'* (Palazzo and Frederick 2011, 265), the authors highlight the responsibility of urban design in considering the impact

[1]Inaugurated in 1895, the railway linking Beirut to Damascus formed part of a larger train network implemented during the late Ottoman period to connect the Empire territory. This inauguration placed it along the larger Muslim pilgrimage *Al-Hijaz* railroad network of the Muslim pilgrimage.

[2]Following the fall of the Ottoman Empires and their instatement into different States, the train network became confined within the countries borders. There is a scarcity of research that addresses the destiny of the train railway system and the actual condition of the abandoned tracks across what used to be the Ottoman territory.

on the environment, and the potentials brought forth by the natural ecosystem to architecture and cities.

25.2 Contextualizing the Last Train Segment

This paper highlights opportunities out of the natural recovery phenomenon, while taking into consideration that similar abandoned urban spaces often turn into an implicit threat of different forms of appropriations in underprivileged social circumstances. In the context of Beirut, abandoned postwar landscapes became spaces of economic contestations. Abandonment and neglect of heritage buildings became an undeclared demolition objective for developers, under the alibi of their physically unsafe and irreversible condition.

The case study area is the last train section in Beirut and the surrounding underprivileged neighbourhood. The last train remnants in Beirut are not the Mar Mikhayel train station itself, which became a trendy night destination in the past few years along Armenia Street. Actually, they are the track traces which tangibly appear in the last train segment, crossing the predominantly Armenian residential neighbourhood known as Badawi. This context is an urban pocket detached from the main attractions and daily activities. This train track as we know it today is characterized by abandonment and lack of maintenance, just a few metres from a vibrant surrounding infrastructural system. To the surprise of the stroller in the city, it reveals itself like an isolated pocket inside the city. At its northern edge, the train used to reach the station crossing a steel bridge over Armenia Street, today a vibrant road aligned with night pubs. While at the northern part the old bridge still exists in its dysfunctional condition, at the southern part Yerevan Bridge constructed after war across the old train path forms a physical infrastructural barrier. The context is entangled between this threshold to the abandoned train station and the edge created by the city developing infrastructure. Social marginality is added to the physical abandonment of this neighbourhood. This context suffered from degradation not only because of the halting of the train during the civil war, but also from the post-war lack of consistent and constant maintenance. It is currently socially suffering from the flow of Syrian refugees, who benefit from a low rent and further squat the neighbourhood by trespassing over the train track legal limits.[3] In tandem between the dual informalities of the social and natural ecology appropriations, I attempt to understand in this paper a different urban metabolism considering the synergy of three main actors in this area, *people, train track* and *nature*.

[3] As per Decree Law No. 148, dated 16/9/1993, the setback from the train track lot property (Chemins de Fers Libanais) is a minimum of 2 m from the lot limit, from each side, unless another law determines a higher setback.

25.3 An Informal Natural Recovery

Abandonment and neglect from the governing agents paved the way for the natural ecosystems to infiltrate along the train track remnants, halted in use since decades, and to grow across the surrounding residential neighbourhood. Based on Beck's ecological landscape and Forman's urban ecology, this chapter highlights two perspectives of the natural recovery process in this urban context. First it frames the spontaneous natural recovery process and transformation in this urban context area, by establishing its own systems of *flows, mosaics* and *patterns* within the urban environment (Beck 2013). The first perspective tackles the hard and soft landscape centred along the existing traces of the track, while the second highlights the natural transformation across the built and unbuilt urban environment (Forman 2014).

25.3.1 The Hard and Soft Landscapes

The growth of greenery happened at times of abandonment and lack of track traces maintenance, basically from the part of the municipality. Furthermore, the neighbourhood itself consists of an urban pocket with dead-end streets, entangled by Yerevan flyover that forms a barrier at its southern edge. This urban isolation further fostered the contagious growth of plantation by creating a natural reservoir (Mostafavi and Doherty 2010) inside the city. Google Earth maps reveal an informal growth of unexpected greenery that gradually took over the abandoned train traces over the last decades, particularly during the last decade (Fig. 25.1). The remaining steel train rails and the steel and wood bridge over Armenia Street represent the last evidence in

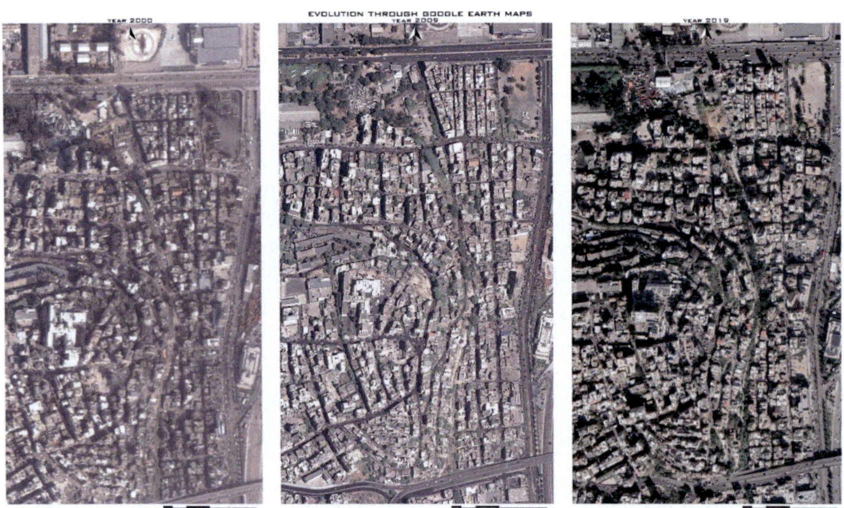

Fig. 25.1 Google Earth maps show development from 2000 to 2018, consulted February 2019

Beirut of the late 19th century Imperial infrastructural expansion. Despite the short length of this span, it presents different ecological conditions. Drawing on Beck's discourse of ecological landscape as *'patches'* and *'patterns'* (Beck 2013), we can trace the linear growth pattern guided by the remaining train steel rails. Several types of trees shrubs and plants grew wildly along the track but also across it invading the non-constructed empty sites and abandoned houses and buildings. The development of this landscape created a pattern at different levels, starting from the train steel rails, the unused empty lots, the abandoned built environment, the infrastructural edge (Yerevan flyover) and the flora itself. The contagious growth of different types of plants across the unbuilt lots established a kind of nested hierarchy (Beck 2013, 212). Likewise on a different level, the growth of trees established a vertical growth network, while different types of shrubs and rampant plants established horizontal growth (Beck 2013, 213), evolving dynamically in time and space in response to the Mediterranean seasonal changes. The most preserved segment is the bridge above Armenia Street, privileged by a private local bank initiative coupled with the neighbourhood community efforts. They restlessly keep it maintained by placing gravel on the track to prevent unwanted shrubs from growing, plant adequate trees along it and constantly remove garbage (Fig. 25.2). The situation gets gradually worse southward towards the wall created by Yerevan flyover and where the sense of community

Fig. 25.2 Left photograph shows a well maintained segment of the train, closer to Armenia street—Photograph by author (2018)

loses ground to the presence of different ethnic groups and Syrian refugees. In the absence of any maintenance of the train rail by the public or private initiative, plants grow wilder, expand faster and extend into the non-built empty lots and abandoned or poorly maintained houses. Poor garbage disposal adds to the informality of the landscape. Some more informal moments along the train track range from the placement of electrical generators on the rail at couple of locations, to pouring concrete patches at other locations, and sand levelling at some parts in order to make it more pedestrian-friendly (Figs. 25.3 and 25.4). This linear pattern along the train track presents a combination of hard and soft landscape, defined by the steel train railway

Fig. 25.3 Upper right photograph shows an altered use of the train track—Photograph by author (2018)

Fig. 25.4 Lower left photograph shows the location of an electrical generator along the train track—Photograph by author (2018)

as hard landscape and the different plantations ecosystems as soft landscape, like *'fragments of natural ecosystems that persist in the mosaic of our built environment'* (Beck 2013, 219).

25.3.2 *Transformation of the Built and Unbuilt Landscape*

Creating a soft and hard landscape starting from the train track, the landscape ecology process followed a linear pattern, different forms of transformations developed by establishing a nested flow across the physical nature of the environment. These vary from flowing and expanding across the built environment, but also forming networks across the unbuilt landscape. During the last decade, the ecological succession manifested itself in spontaneously finding its ground across the mass and the voids of the urban context. Due to the low maintenance from the part of the municipality, nature colonized itself spontaneously as per Forman's description of what he calls *'An ecology of urban mosaic'* as follows: "*Also all cases are dominated by 'spontaneously' colonized (unplanted) plants, mostly herbaceous. These early successions are on small and large patches in diverse locations, as well as on many types of corridors*" (Forman 2014, 9). The nesting system spread contagiously following the prevailing wind and the site natural topography, across the brown sites and unused plots creating inner green pockets in the area, except for the maintained open spaces, mostly used as parking lots (Fig. 25.5). This dynamic natural infiltration found persistently its way inside the abandoned houses, growing inside unreachable gardens and blossoming out of unused architectural elements like walls and balconies. The different man-made moments along the rails created diverse ecological situations and *'patches'*. A broader perspective of this area over time reveals that the spontaneous infiltration of nature within the urban context, developed into a phase of adaptation into urban patterns. The contagious growth paved the ground within private yards and gardens for growing plants and vegetables and fostered the inhabitants to maintain the plants growth in their backyards (Fig. 25.6). The curvilinear character of nature ecosystems growth was adapted to the urban grid patterns and created over time a balance between natural and urban ecosystems. While this paper doesn't trace the pattern growth and transformation between ecological landscape and urbanism, it nevertheless highlights this resilience over time.

25.4 An Informal Social Engagement

If instability and marginalization are characteristics of *'terrains vagues'* (Solà-Morales Rubió 1995), Badawi neighbourhood evokes socially these characteristics. This socio-natural ambiguous entanglement depicted dually with the train track landscape and the natural ecology of the context, echoes the *'terra fluxus'* condition (Corner 2006, 30). It reveals a diversity of social practices and needs that

Fig. 25.5 Left map shows the spread of network of greenery—Prepared within ARP 556 studio, Spring 2019

were expressed over time in different types of appropriations and practices. This informal engagement by the local inhabitants anticipated physical configurations and re-configurations to the cityscape. The quality of the changes is not judged against aesthetical design values, but rather highlighting the social engagement practices as tools for understanding and negotiating physical changes over time in cityscapes. The insalubrious negative aspect of some local practices and the physical violation over the train public property remain though an alarming risk in this study. Apart from the train passage and the French army barracks presence not far from it, this area was

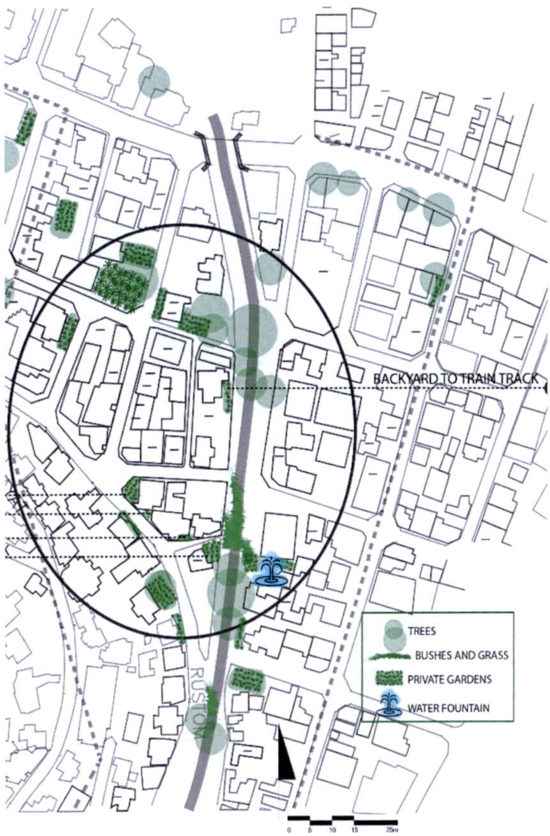

Fig. 25.6 Map shows the contagious growth of greenery along train track and private backyards—Prepared within ARP 556 studio, Spring 2019

an empty terrain known as Ghabi, before the arrival of Armenian refugee and their gradual distribution in Badawi Street and its neighbourhood. The area social history is shaped by waves of refugees, starting with the Armenian settlements which still constitute the majority of this neighbourhood. Demographic changes marked with the gradual settlement of Asian and African migrant workers during the last few years, culminated in the arrival of forcibly displaced Syrian refugees in the past few years, attracted by low rents in the area. Currently, they live in the less favourable part of the neighbourhood, located on a steep topography and entangled by a dead-end pocket street and the Yerevan flyover wall. The lack of public surveillance in this vulnerable context made it possible for the local inhabitants to establish informal relations along the train track, turning it into a pedestrian passageway by covering the track with sand. Almost in complete abandonment, this track section became recently a space of illegal constructions along its track (Fig. 25.7). Social marginalization evoked different types of functional and daily practices along the train track. They vary from the creation of pedestrian shortcuts from one neighbourhood to the other by informally laying few steps across the rail, to using it as a car passage between two different roads by pouring some concrete car ramp patches. Reciprocally to the

Fig. 25.7 Illegal informal construction along the train track limits, and addition of sand along the train track—Photograph by author (2019)

changes instigated by the natural phenomenon recovery, local inhabitants territorialized the train track and redefined new negotiable geographies, typical of *'terrain vague'* characterized by *'imprecise limits and vague definition'* (Solà-Morales Rubió 1995). While temporality rules the construction infringements and the connection shortcuts, this informal regenerative process turns this context into both a negotiable and reversible space at once. Despite the illegal and threatening urban condition, this self-regeneration process between nature recovery process and people appropriation of the public train track could be seen as an opportunity for setting guidelines for real needs and their possible solutions.

25.5 Conclusion

The last train is situated along a dead-end pocket in Beirut. Furthermore, its location amidst the increasingly socially marginalized Badawi neighbourhood increases its inaccessibility. Not to underestimate the complexities of this area, this paper tackled only one aspect of the area highlighting the dual social and natural informalities processes. The area underlies far more complex layers that form part of urban complexities, including the economic and political ones. The primary focus aimed at understanding the types of physical, social and nature informalities unfolding along the train track over the last decade and the interrelations between them. This brings me to the primary conclusion of how to cautiously intervene as urban designers in similar *'terrain vague'*, where seemingly they appear to operate by themselves within a certain balance. Despite the *'terrain vague'* vulnerable condition for all what it presents currently, in relation to ecological, social and construction trespasses, it is equally important to approach similar spaces as opportunities in their broader context, rather than ecological threats. The importance of this context is related to the broader ecological logic of the city or can become part of this ecological network, yet to be thought of. As small as any area could be in size, the potential lies in setting a larger ecological network out of small-scale ones, particularly in the dense context of Beirut. By looking at the potentials of the case study, the paper aims at introducing other parameters for consideration in negotiating a cityscape. These parameters bypass considering the degraded urban landscape as a major threat, into contemplating ecology as the main potential for urban regeneration paradigms. While considering the delicate balance between the built environment, people and ecology, more resilient solutions to abandoned spaces can be considered, tapping on the preservation of heritage aspects, the train track in this case, and avoiding social gentrification results.

References

Arnaud JL (Dir.) (1997) Beyrouth, grand-beyrouth. Les Cahiers du Cermoc No. 16s

Beck T (2013) Principles of ecological landscape design. Island Press

Corner J (2006) 'Terra Fluxus' in Waldheim. Princeton Architectural Press, Charles. Landscape Urbanism Reader, pp 21–33

De Solà-Morales Rubió I (1995) Terrain vagues. Anyplace, MIT Press, Cambridge M, pp 118–123

Fawaz M, Peillen I (2003) Understanding slums: case studies report for the global report, DPU and UN-Habitat

Forman R (2014) Urban ecology: science of cities. Cambridge University Press

Mapping Beirut through its Tenants' Stories (2016) Retrieved 10, 30, 2019, from Public Works Website: https://publicworksstudio.com/en/projects/mapping-beirut-through-its-tenants-stories

Mariani M, Barron P (eds) (2014) Terrain vague. Interstices at the Edge of the Pale, Routledge London and New York

Mostafavi M, Doherty G (eds) (2010) Ecological urbanism. Zürich Lars Müller Publishers

Palazzo D, Frederick RS (2011) Urban ecological design: a process for regenerative places. Island Press

World Cities Report (2016) Urbanization and development—Emerging futures. Retrieved April 2019, from UN-Habitat Website: https://unhabitat.org/books/world-cities-report/

Part IV
Smart Public Spaces

Chapter 26
Cittaslow and Applicability in the Egyptian Context: The Case of El-Fayoum

Ahmed S. Abd Elrahman

Abstract Cittaslow is a movement that became organization, which concentrate on economic welfares for local societies, meaning of cultural capitals, host community contribution, small-scale green tourism. It tries also to refine the quality of life in small cities by reducing down its general pace, particularly, the usage of spaces and activities inside the city and their mode of transportation and traffic. The Cittaslow is fragment of a cultural tendency known as the slow movement. This movement introduced its concepts by Paolo Saturnini through the different ways of development, based on the improving the quality of life. It can be implicit as an alternate, accountable procedure of sensitive tourism rehearsal that discards modernist obsessions with hustle and entrepreneurial principles. "Small cities and towns around the world start developing numerous developments intended to endorse and strengthen the Cittaslow values. Recently, cities from Africa started to join the movement like Sedgefield in South Africa. In Egypt, it is vital to discuss to what extent the Cittaslow principles can be applicable to be applied on selective Egyptian cities, as Egypt already contains small cities and towns that have the potentials and characteristics to join the Cittaslow movement. Consequently, it is important to identify these potentials, and to what extent the charms of sustainable tourism provoke their growth. Tunis town in El-Fayoum City can be one of the Egyptian towns that comprise respectable potentials to link with the Cittaslow Charter, and the researcher tends to verify the Cittaslow principles on the town to identify the applicability using experts' evaluation. This paper is an attempt to understand the principles of Cittaslow movement, its applicability in the Egyptian context through verifying the Cittaslow Charter goals on Tunis village in El-Fayoum.

Keywords El-Fayoum · Cittaslow · Slow city · Slow movement · Sustainable development

---

A. S. Abd Elrahman (✉)
Department of Urban Design and Planning, Faculty of Engineering,
Ain Shams University, Cairo, Egypt
e-mail: ahmed.sami@eng.asu.edu.eg

Architecture and Urbanism: A Smart Outlook,
https://doi.org/10.1007/978-3-030-52584-2_26

26.1 Introduction

Nowadays, the connection between city tourism and the natural setting becomes very essential. Hence, to make a city pleasant to most of stakeholders like residents and travellers and to feast welfares of urban tourism to its environments, consequently strengthening its influence and handling overcrowding (Maćkiewicz and Konecka-Szydłowska 2017). These principles are being supported by international initiatives or movements, Some of these movements as the Cittaslow concentrate on economic welfares for local societies, meaning of cultural capitals, host community contribution, small-scale green tourism. Cittaslow accordingly followers for arrival to a pre-globalization governance and party of indigenous societies that re-invigorate inner-city public space, attach citizens to indigenous, maintainable fabrication and trades, and emphasize a method of pleasing living together with ingesting issues, which, in particular, discreet and leisurely (Osbaldiston 2013).

The Cittaslow established in 1999, however, it did not have a wide publicity, though, its membership increased swiftly to comprise 147 towns by 2011. Honoré (2004), and this number had been reached 252 cities nowadays within 30 countries around the world (Cittaslow 2019), and recently, it comprised an African town which is Sedgefield in South Africa. Consequently, there is a great potential to comprise more towns from Egypt, as, Egypt already contains small cities and towns that have the potentials and characteristics to join the Cittaslow movement.

Certainly, if the Egyptian towns that have the potential started seriously to work on joining the movement, they would have a great chance to refining their exclusive potentials they can hostage any of these contests through enticing guests from extended places, thus revolving the uncountable latitudinal interface potentials of the current world to great economic advantages. (Ball 2015) Hence, this paper discusses the different principles of Cittaslow movement, and its applicability in the Egyptian context through verifying the Cittaslow Charter goals on the probable towns as Tunis village in El-Fayoum.

26.2 The Cittaslow

During the past decades, global awareness increased regarding the green tourism, the world trend seeks the imaginable route between the tourism development, sustainability and natural environment (Maćkiewicz and Konecka-Szydłowska 2017). Many civic establishments have turn out to be influential associates and promoters for generating liveable places (http://www.cittaslow.org.2019); consequently, many efforts done in order to keep that balance, where they can be recognized as initiatives or movements (Maćkiewicz and Konecka-Szydłowska 2017). These movements are rapidly multiplying border-crossing establishments; they are occupying a significant part in many aspects in many countries; and some of them interested in urban aspects like Active Living by Design. (www.activelivingbydesign.org), Public Architecture (www.publicarchitecture.org), Walkable Communities (www.walkable.org), Shaping Footprints (www.sfpinc.org), or interested in other dimensions like Public Interest Design. (www.publicinterestdesign.org), Design for Health (www.designforhealth.net), SEED Network and Design Corps (seednetwork.org), and the Cittaslow (www.cittaslow.net) which promotes the main values of ecotourism, countryside preservation, tutoring, financial assistances for indigenous societies, significance of traditional possessions, smallest ecological influence and extreme ecological sustainability, multitude public contribution, and small-scale tourism turn out well with the philosophies of the Cittaslow principles.

Cittaslow made an obvious shift in the preparation and exercise of specialized urbanists, where some institutes are not with the "crit" (critique). That indicates shortfalls in scholars' work, to "conversations", where labours and educators involved in a cooperative knowledge procedure categorized by "appreciative inquiry" (Ellin 2013); it frameworks the responsibilities for a Cittaslow concerning ecological security and sustainability, along with quality of life, cordiality and generosity (Mayer and Knox 2009).

26.2.1 The Establishment and Spreading of Slow City Concept

In Paris in 1989, a manifesto signed to set ground rules for slow made movement, like Slow Economy, Slow Urbanism, Slow Fashion and the Cittaslow or slow city, and others. Slow Made is constructing an original financial model that avoids extraordinary amount and small rate in favour of excellence scheme, resources and manufacturing, consumer and manufacturer bond (Pinkhasov and Nair 2014).

In 1999, the Cittaslow movement itself began (Cittaslow 2019) by Paolo Saturnini in Italy, the mayor of the small hill city in Tuscan (Maćkiewicz and Konecka-Szydłowska 2017); a meeting was organized by other mayors of other metropolises (Orvieto, Bra, and Positano) to put a definition for the characteristics of a citt'a lente—slow city (Cittaslow 2019). Through this establishment meeting, the mayors agreed to set a succession of ideologies that comprise functioning to more calm and less contaminated urban surroundings, in addition, preserving indigenous appealing ethnicities, and development native crafts, products and gastronomy (Cittaslow 2019). In 2001 in Chianti in Italy, an association named "Cittaslow—Rete Internazionale delle città del buon vivere": inshort, "Cittaslow", officially established in. The association is a not for profit entity, and it has goals to indorse the culture of good living. The identity, reminiscence, environmental fortification, integrity and communal presence are amongst the morals that Cittaslow endorses. The essential philosophies for this association defined in the Manifesto of Cittaslow, principles the conception of robust microeconomics, this argument of respectable performs through the network, and indorses residency movements in indigenous societies, through the towns and its terrains, in all existing or futuristic aspects for next generations (Maćkiewicz and Konecka-Szydłowska 2017); they also pledged to use technology to sort inhabitants' conscious of the importance of more restful measures to life, and staking their knowledge in pursuing directorial explanations for healthier existing (Cittaslow 2019).

In 2015, the Cittaslow comprises 209 small cities in 30 countries at different continents; in addition to European countries, other towns in Australia, North America, South America, Asia, in addition to South Africa join the movement. Yet, the Italian cities prevail: as they considered almost one-third of the network (Maćkiewicz and Konecka-Szydłowska 2017); although the association has augmented throughout the past decade, the movement residues insignificant (Ball 2015). In 2016, this movement embraces more than 200 towns in other countries in Europe, Asia, America and Africa (Maćkiewicz and Konecka-Szydłowska 2017); this slow movement prioritizes sustainable living, and the excellence of life in towns within the limits of 50,000 capita (Rumford 2014).

Furthermore, to evolving geographic extension of the movement's network, Cittaslow has as well in progress to banquet its impact more up the urban hierarchy. While the population limit to have the membership is limited to fifty thousand capita, the mayor of Paris presented in 2002 through the yearly reward in the non-Cittaslow commissioners segment. In 2005, a Cittaslow sponsor grouping established for small cities and towns with a number of inhabitants that exceed fifty thousand capita, affirming that this network has protracted from its prior limited attention (Ball 2015). In addition, many local efforts had been exerted to implement the model of Cittàslow in many countries, where a novel ethical add-on to home-grown foodstuffs (Dietzsch 2014).

Because of securing the movement membership, cities start developing numerous plans intended to endorse and strengthen the philosophy of Cittaslow (Sirgy et al. 2009). Furthermore, the members had been encouraged by the movement to imitate with international assessment criteria recognized outside Cittaslow (Ball 2015). Cittaslow indicator system stimulates home-grown financial, ecological, and community sustainability through inspiring several arrangements subsidized by various representatives and interventions (Mayer and Knox 2009).

More cities expected to join the network as centres number is presently working over the self-assessment implementation mandatory for Cittaslow affiliation (Ball 2015). The first twenty-eight slow towns qualified in 2001, where all these members were Italian, the majority situated in northern Italy, mainly in Umbria and Tuscany. By 2006, number of certified cities increased to seventy-seven cities (Cittaslow 2019). Then, more cities joined the Charter but still the widely held from Italy, but towns in Germany (Hersbruck, Waldkirch, Schwarzenbruck), Britain (Aylsham, Ludlow, Diss), in addition to Norway (Levanger and Sokndal) are now qualified as Slow Cities (Sirgy et al. 2009). Nowadays, more than 300 other cities from other parts of the world have requested to joining certified Slow Cities in order to enjoy a stable local and international visitors' stream attentive in the movement (Mayer and Knox 2009).

26.3 The Philosophy

The "slow tourism" concept mounting interest in the "slow movement" (Dickinson and Lumsdon 2010); the contemporary concept tight to the appearance of the "slow food" occurrence. Originated through Carlo Petrini's confrontation contrary to the foundational of a restaurant of McDonald in Rome nearby the famous Spanish Steps located in piazza di spagna (Honoré 2005). Appointing into a refusal of "fast", slow food has then headed to a series of other "slows" containing slow cities (Cittaslow), slow design, and slow economy and slow living (Fullagar et al. 2012). The slow food movement is now big and international with more than hundred thousand associates (Slow Tourism 2013). The evolution outline in contemporary culture requests to reoccurrence to the antiquated evolution; consequently, the Cittaslow network can be a substitute to the maintainable growing decision and to the tragedy of international ecosystems and tainted civilization (Sohn et al. 2015).

Slow beliefs castoffs acceleration and its multiple consequences for communal and environmental relationships, where it is accompanying through the beliefs of anti-consumerism and anti-globalization (Honoré 2005). Imbedded inside slow is a number of principles which recount to ecological reliability and sustainability, societal fairness, residence and values, and it is not basically anti-fast provision but revels indigenous cuisines and accepts viable applies in nutrition production and spreading.

The Slow tourism can be implicit as an alternate, accountable method of sensitive tourism practice that discards modernist obsessions with promptness and entrepreneurial philosophies (Dickinson and Lumsdon 2010). Using this sense of sensibility, slow can be considered a nearby counterpart of luxury (Pinkhasov and Nair 2014); this luxury remains an influential narrative alongside the tide of fast-paced globalization that frequently interrupts and relocates native societies (Osbaldiston 2013). Luxury's scarcity cannot be considered the modest deficiency of supply. On the contrary, luxury affords to one's daily practice. Appreciative luxury also means appreciative people's lives. You need to recognize how society incorporates this through their believed progressions, and it is important to realize the community's dynamics (Pinkhasov and Nair 2014); this new emphasis is not basically constructed on confined uniqueness; it is providing an alternate to international powers, not disregarding them (Honoré 2004).

The Cittaslow is actually a trademark for cities who respect the nature significance, ethnic antiquity, sustainability and generosity; in order to gain this qualification, the selected municipalities must devote in the excellence of their visitors and inhabitants living (Hospers and Reverda 2015).

There are commitments starting with conserving the distinguishing urban character and initiating to generate additional open spaces. In addition to intensification of cycling and "pedestrianized" routes, reduce noise and prohibition of car alarms reduce all kind of pollution, adoptive the use of diverse energy sources, public transport developing and encourage ecological architecture within urban extensions

(Mayer and Knox 2009). However, not all inhabitants are passionate about the "slow" profile of their municipality (Hospers and Reverda 2015).

Short-term occurrences—cost-effective sequences, shopping customs and customer tendencies–are symbols of higher powers, like people's insights of safety, admittance to opportunities, the wellbeing, the upcoming predictions, and their principles. These forces are significant to recognize the prediction and respond to speedily developing market conditions successfully (Pinkhasov and Nair 2014).

Slowing down the travel's stride with bikes, canoes or walking offers occasions for tourists to increase closer happenstance with the indigenous environment. Not only is a vacationer, this slow tourist amending his/her temporality to the stayed sites. Slow tourism avoids the typical mass tourism through prominence on extensive, globalized, standardized and commodified practices (Fullagar et al. 2012).

Towns and cities need to mature "place-sustaining" strategies to develop the exclusivity of urban spaces, through the extensive context of contemporary powers plus the standardizing powers of globalization. Cittaslow or Slow City movement is considered as a new method of "re-territorialization" in a globalization setting (Parkins and Craig 2006). This new emphasis may be realized as a restricted revitalization philosophies of French geographers during the early twentieth century, like Paul Vidal de la Blache, who described the uniqueness of the indigenous "pays" which had diverse atmospheres for history and exceptional "terroir" that formed such distinct and reliable "genres de vie" (ways of life) (Sirgy et al. 2009).

Historic towns or cities, agonized from two problems. The first one is the threat of bigger urban ties to surge their purposeful significance, which minimized the amount of roles in these minor centres. The other threat is coming from the forces to accept the apparently unstoppable adjustment of modernist lifestyles that make these centres appear old-fashioned and lacking chances or enthusiasm (Ball 2015).

Although the Cittaslow has the similar ethnic origins as the Slow Food movement, the goals of both movements are not the same but corresponding. Both organizations are supporting home-grown, traditional ethos, leisurely contemplation and pleasantness. Every movement opposes enormous trade and globalization. Genuineness is slow; normalization is fast. Individualism is slow; franchise is fast. Quietness is slow; noise is fast. Trees are slow while concrete is fast (Mayer and Knox 2009).

A domestic analysis conceded by an organization in 2002 presented that inhabitants worth their city's uniqueness also it assured that their urgencies support thoroughly the Cittaslow agreement and the indicators of quality of life. They need additional relaxation spaces in the centre, different amenities, enhanced labelling, for old places and buildings fortification. With securing the movement's membership, the city started to develop some projects that promoting and supporting the Cittaslow philosophy (Sirgy et al. 2009).

26.4 The Characteristics of Cittaslow and Goals of the Charter

Villages, towns or small cities realized the usefulness to follow and register in the Cittaslow network (Maćkiewicz and Konecka-Szydłowska 2017). Cities and towns admitted to membership simply after preparation of preliminary report on the city's obligation to Cittaslow ideologies; this report (Sirgy et al. 2009) trailed by a comprehensive assessment report that expending a communal quality of life indicator system and cover six main aspects (Mayer and Knox 2009). These aspects starting with the ecological strategies "regulations" and organization; infrastructure usage; technology incorporation; the advancement of home-grown produce and life habits; the generosity and life pace; and sense of place (Sirgy et al. 2009); each town or city requisite improve a group of indicators grounded on the Charter (Mayer and Knox 2009). When a city or a town has been nominated and certified as a Slow City, a re-certification should be done every four years. Characteristically, the certification is done after preparation a comprehensive assessment of the city agendas and guidelines (Sirgy et al. 2009).

The Cittaslow Charter classifies the essential concepts of the movement that are defined as the "requests for excellence" that are mentioned in Table 26.1, which comprise 43 policy objectives collected to seven sets, dealing with the ecological aspects, infrastructure, facilities and technologies, Preservation of autochthonous production, hospitality, alertness in addition to other extraordinary categories.

The Cittaslow Charter considered these concepts as fundamentals for Cittaslow. It necessity be emphasized that nearly half of these policies are considered obligatory features for cities wishing to become members (Ball 2015). Furthermore, Susan Ball added additional elements to these policies in 2015, and these mentioned polices had been summarized and rephrased the ideas of the Charter.

26.5 The Egyptian Case

In Egypt, it is vital to discuss to what extent can the Cittaslow principles can be applicable to on selective Egyptian cities, and to what extent the temptations of sustainable tourism provoke their progress. In addition, the necessity for the preferment of small cities as indigenous development centres expending their local possibilities. Many towns comprise the essential elements that Cittaslow requires as a high quality of its natural environment in addition to the high tourist potential.

El-Fayoum City is one of the Egyptian cities that comprise small cities and towns with respectable potentials; it is located in the southwest Cairo. It considered the green valley that bounded by desert, apart from for southeast direction (EEAA 2008). El-Fayoum City comprises continuing past (Khalifa Marwa and El-Khateeb Samah 2011) that symbolizes in occurrence of pre-historic civilizations (EEAA 2008) additionally to the striking protectorates, one of them is El-Hitan valley that was selected

Table 26.1 Cittaslow Charter goals; author depending on results developed by (Ball 2015)

Ecological Policies
• Existence of municipal strategy for renewable sources and energy saving • Prohibition the usage of genetically adjusted organisms • Existence of light pollution control system • Implementation of environmental management systems • Air, soil and water quality verification with law • Spreading guidance on separated waste collection • Presence of urban sewage distillation plant and usage of waste containers with regular evacuating timing • Existence guideline of profitmaking art and transportation symbols • Existence of noise pollution
Infrastructure Policies
• Maximizing use of public transportation, incorporate traffic amongst public modes of transportation and walking routes • Accessibility to public places for disable people • Program promotion to expedite including (family life/linkage between school and community/Providing social centres/Municipality services/public toilets) • Existence of urban reforming and upgrading programs • Integration of community information roles with information windows of Cittaslow • Improving and retrieving historic centres and mechanism of historical/ethnic significance • Existence of safe mobility/traffic policies • Existence of cycle pathways to attach educational buildings and public areas • Existence of a medical support centre • Existence of good and consistent green areas and facilities infrastructures • Existence of natural products distribution centres • Creation of "friendly shops" agreement with shopkeepers for assistance people in trouble
Facilities and Technologies for Urban Quality
• Existence of bio-architecture/training people/training methods • Plantation of environmentally suitable and local plants • Support use of the Internet for municipalities and citizens/optical fibres/wireless systems/monitoring electromagnetic fields • Review colour schemes
Methods of Preservation Autochthonous Production
• Using and developing organic farming • Certification artisan products/artistic crafts • Existence of plans to protect craftsperson and crafts products facing extermination • Protection of traditional work techniques and vocations at threat of extermination • Creation of nutrition/taste tutoring programs in collaboration with Slow Food concepts • Developing a region's products census and supporting commercialization and publicizing • Preservation and promotion of native cultural occasions
Hospitality Features
• Providing training courses to convey excellence hospitality practices • Creation of guided journeys and usage of global signage in historic centres • Preparation of "slow" city guides, by websites and flyers

(continued)

Table 26.1 (continued)

Ecological Policies
Alertness Policies
• Development official information campaign for city residents about movement's objectives and procedures of joining the movement • Initiation of programs supporting Cittaslow values and schemes, comprising informative green areas, libraries and improvement seed and species bank • Provision the Slow City and Food Presidia, that indorses a variety of financial, social, ecological and ethnic intentions
Extraordinary Requirements
• Promotion the intentions of the Stow City movement through displaying the Cittaslow trademark (Letterhead/Website …etc.) • Establishment of local slow food convivial/Enhance adult educational programs for tasting and nutrition • Establishment of school vegetable gardens, with the assistance of Slow Food members especially species at risk of extinction • Using nutritional traditions for food collection food service • Implementation the concept of the "Mercati della Terra" (home-grown and produced products' market) in collaboration with Slow Food associates

as a UNISCO World Heritage Site because of exceptional and important fossils that are present (Khalifa Marwa and El-Khateeb Samah 2011). Many of tourism attraction fundamentals are in the city that supporting ecotourism evolution. Likewise, agricultural, coastal and desert environments are present in the city's location. El-Fayoum also holds diverse fundamentals, which can be amid scientific researches benefits (EEAA 2008). The city has constantly tormented from government inattention from over the previous decades.

Tourists are inspired to make scientific outings to the world heritage site of Wadi El-Hitan, north Lake Qarun forest, fossils of Gabal Qatrani, as well as the bird-watching events; besides, several desert safari activities can be practised there like horse and camel riding, desert hiking, trails and mountain climbing. Also, fauna and flora observation, entertaining and interaction with Bedouin farmers' societies' lifestyles, water sports like sailboats, feluccas, kayaking, single-line fishing, swimming as well as firmly regulating duck hunting in the fishponds, inspections and tours to the earliest archaeological sites. Ecotourists will like to stay in the environmental residences and eco-lodges like Auberge, New Panorama hotel, Zad El Mosafer lodge (Khalifa Marwa and El-Khateeb Samah 2011) where the individual memories of the ecotourists integrate together with the inhabitants to form the collective memory of this community (Abd Elrahman and Mahmoud 2015).

Many urban and tourism development plans were put for tourism and urban development in El-Fayoum City. Frequently, by the Tourism Development Authority (TDA), like "Indicative tourism development plan for Lake Qarun and Wadi El Rayan" which was set in 1991" and "Ecotourism for sustainable development in El-Fayoum Oasis in 2000" and "Tourism Development Plan for the northern Coast of Lake Qarun in 2004". The Royal Dutch Embassy in Egypt sponsored "the preparatory phase for ecotourism in El-Fayoum project” in 2004, which proposed to assess the ecotourism accomplishments, plus, performing training developments for different resident target groups. Additionally, the TDA planned to make create investment occasions in Lake Qarun to realize the development potentials there through growing the carrying capacity of tourist housing and hotels units and poverty extenuation through tourism industry (Khalifa Marwa and El-Khateeb Samah 2011).

26.5.1 Tunis Village and Cittaslow Potentials

Tunis or (‘izbat Tunis) is well thought out one of the most distinguished villages in El-Fayoum because of the unique location, which is bounded by fertile lands and bosses the waterfront of lake.

Qarun on a high hill, while on the other side, there is endless desert. Two eminent Egyptians poets exposed Tunis village; currently, there are several pottery studios in Tunis village where many skilled citizens are working (Fig. 26.1).

Numerous people from Egypt and other countries particularly Europe used to come there to enjoy both the vernacular architecture and indigenous traditions of the village, many guesthouses and homestays, and a diversity of activities are existing in the village and small-sized museums like Caricature and Egyptian photography museums (Design with Nature 2016).

People build the houses from local resources, such as earth, limestone rocks, clay, dried straw and trees’ starches. Traditional practices for roofs structures used like domes, vaults and wooden ceiling. Trials to produce energy through renewable sources are occur as biomass, pre-packaged photovoltaic systems, solar hot water heart; furthermore, the vernacular methods stand frequently used to accommodate in the hot climate like arcades, cooling towers, inner courts, wind catchers. Besides,

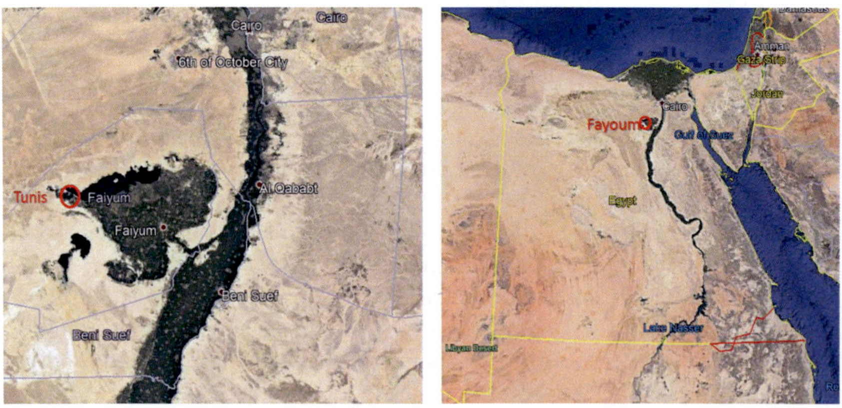

Fig. 26.1. El-Fayom and Tunis village location (Googleearth 2019)

environmental methods for filtering and storing water, as sand soil strainers and bio-filters. There are some initiatives that incline to spread and feast the philosophy of Cittaslow, even without existing of real application or awareness for the goals of the movement, one of these initiatives is "design with nature" tends to spread concepts about construction by means of old vernacular practices with new enhancements through contribution of villagers. This initiative inclines to spread the knowledge and tutelage about the organic agriculture, native arts, preventive healthcare, integrity and education on small rule to empower the villagers.

26.5.2 Application of the Goals of Cittaslow Charter

In order to examine the applicability of verifying Cittaslow goals, several experts in urban planning and tourism development had evaluated the goals of the Cittaslow Charter, in order to verify the applicability of these goals on Tunis village. There are three grades for applicability level, strong, weak or inconveniency of application, and then the average is taken for all the evaluations (Table 26.2).

Experts' evaluation stated that 34 goals out of the total policies that comprise 43 goals can be accomplished, where nine of the goals are weakly applicable, however, that does not mean the way to join the Cittaslow Charter movement is effortlessly appropriate. On the other side, the ambiguity towards the applicability of any of the mentioned goals relates to different municipal provisions.

26.6 Conclusion

The research highlights the emergence of Cittaslow movement and the opportunities that cities gained from joining the movement; it also identified the importance of Cittaslow, as follows:

- The regeneration of architectural assets or further methods of upgrading the spaces' eminence could produce to be inadequate. There are many challenges regarding many issues that have to consider starting from renewable sources and energy-saving strategy, control of light pollution, environmental management systems, air and soil quality and water quality.
- There are futuristic concerns concerning the integration of public information roles with Slow City information windows, usage of the Internet for municipalities and citizens.
- Regarding the Egyptian case, the achievement of the application policy of the Cittaslow principally relies on to what extent the municipalities and interested entities can promote this idea amongst their local communities.
- The challenge that Cittaslow membership for Egyptian towns or villages face is to mature a legislation, which can draw on a road map to maximize the benefits, and

Table 26.2. Application of Cittaslow Charter goals on Tunis village

Ecological polices	Strongly applicable	Weakly applicable	Can't be applied
• Renewable sources and energy-saving strategy		•	
• Prohibition of genetically adjusted organisms	•		
• Light pollution control system		•	
• Environmental management systems		•	
• Air, soil and water quality		•	
• Waste collection separation and industrial domestic composting	•		
• Distillation of urban sewage and regular evacuation	•		
• Commercial art and traffic signs guidelines	•		
• Control of noise pollution and electromagnetic sources	•		
Infrastructure Policies			
• Public transportation, integration with public transport and pedestrian areas	•		
• Accessibility for disable people	•		
• Program promotion	•		
• Urban reforming and upgrading programs	•		
• Integration of community information roles with information windows of Cittaslow		•	
• Improving and retrieving historical centres	•		
• Safe mobility/traffic plans	•		
• Bicycle pathways	•		
• Medical support centre	•		
• Good and consistent green areas and facilities infrastructures	•		
• Natural products distribution centres	•		
• "Friendly shops" agreement	•		
	•		

(continued)

Table 26.2. (continued)

Ecological polices	Strongly applicable	Weakly applicable	Can't be applied
Facilities and Technologies for Urban Quality			
• Bio-architecture/training people/training methods	•		
• Environmentally suitable and local plants	•		
• Usage of the Internet for municipalities and citizens/optical fibres …etc.		•	
• Colour schemes	•		
Methods of Preservation Autochthonous Production			
• Organic farming	•		
• Certification artisan products/artistic crafts	•		
• Plans to protect craftsperson and crafts products facing extermination	•		
• Traditional work techniques and vocations at threat of extermination	•		
• Nutrition/taste tutoring programs		•	
• Census of the region's products		•	
• Local cultural occasions and events	•		
Hospitality Features			
• Hospitality practices courses	•		
• Guided journeys and global signage	•		
• 'Slow' guides to the city	•		

(continued)

Table 26.2. (continued)

Ecological polices	Strongly applicable	Weakly applicable	Can't be applied
Awareness Policies			
• Official information campaign	•		
• Programs supporting slow city philosophy and its projects	•		
• Slow city and food presidia		•	
Extraordinary Requirements			
• Promotion the intentions of the slow city movement	•		
• Educational programs for tasting and nutrition	•		
• School vegetable gardens	•		
• Nutritional traditions for food collection food service	•		
• Concept of the "Mercati della Terra"	•		

assets that candidate towns or villages like Tunis village have. Moreover, trying to make the challenges astonished through a comprehensive plan including the municipalities side by side with NGOs and interested volunteering institutions.

- There is a great potential in Tunis village to join the movement; however, a considerable efforts regarding renewable sources and energy-saving strategy, pollution, environmental management systems, integrity of public information, nutrition programs, census of the products and Slow City Presidia that shall be exerted.

Accordingly, the Cittaslow movement will act as a reaction counter to the pressures that face Tunis village or any other nominated Egyptian town. By emphasizing the importance of replacements to the traditional trends, and by stressing the role of these minor momentous cities in generating more satisfying, pleasant, calming and even sensual outlines of life in city places that are linked to the Cittaslow traditions.

References

Abd Elrahman AS, Mahmoud R (2015) The post-revolutionary effect on the urban harmony of the Egyptian built environment in relation to the collective memory of the population: Urban context of the "after revolution" between contravention and elaboration. Ain shams Eng J Sci Direct. www.elsevier.com/locate/procedia. http://dx.doi.org/10.1016/j.asej.2015.10.007

Ball S (2015) Slow cities. Theme cities. In: Solutions for Urban problems. GeoJournal Library, vol. 112

Cittaslow (2019) http://www.cittaslow.org/content/association

Dickinson L, Lumsdon D (2010) Slow travel and tourism. Earthscan, London

Dietzsch I (2014) Life-worlds of deceleration: reflections on the 'new mobilities paradigm' through ethnographic research in post-socialist Germany. In: Mobilities in socialist and post-socialist states, societies on the move. Palgrave Macmillan, p 255

Design with Nature (2016) http://designwithnature.eu/design-with-nature/tunis-village. Accessed on Sept 2016

Ellin N (2013) From good to great urbanism: beyond sustainability to prosperity. Good urbanism: six steps to creating prosperous places, metropolitan planning + design, p 106

Egyptian Environmental Affairs Agency (2008) Environmental action plan Fayoum governorate, Cairo, Egypt, pp 17–18

Fullagar S, Markwell K, Wilson E (eds) (2012) Slow tourism: experiences and mobilities. Channel View, Bristol

Googleearth (2019) https://www.earth.google.com/web/@29.38440812,30.73500632,10.4800551a,67033.52156871d,35y,0h,0t,0r

Hospers G-J, Reverda N (2015) Managing population decline in Europe's Urban and rural areas. Springer Briefs in Population Studies, p 45

Honoré C (2005) In praise of slow: how a worldwide movement is challenging the cult of speed. Orion, London

Honoré C (2004) In praise of slow: how a worldwide movement is challenging the cult of speed. Harper, San Francisco

Khalifa Marwa A, El-Khateeb Samah M (2011) Fayoum oasis between problems and potentials: towards enhancing ecotourism in Egypt. In: 4th International Congress on Environmental Planning and Management Green Cities: A Path to Sustainability. December 10–13, 2011. Cairo and El-gouna. Egypt. pp 4–8

Maćkiewicz B, Konecka-Szydłowska B (2017) Green tourism: attractions and Initiatives of polish Cittaslow cities. In: Tourism in the city: towards an integrative agenda on urban tourism. Springer International Publishing, Switzerland, pp 297–309. ISBN 978-3-319-26876-7. https://doi.org/10.1007/978-3-319-26877-4

Mayer H, Knox P (2009) Pace of life and quality of life: the slow city Charter. In: Community quality-of-life indicators: best cases III, vol. 1, pp 24–37

Osbaldiston N (2013) Consumption and public life culture of the slow social deceleration in an accelerated world. Palgrave Macmillan, pp 24–73

Parkins W, Craig G (2006) Slow living. Berg, Oxford

Pinkhasov M, Nair RJ (2014) Real luxury: how luxury brands can create value for the long term. Palgrave Macmillan

Rumford C (2014) Cosmopolitan borders. Mobility & politics series. Palgrave Macmillan, p 31

Sirgy MJ, Phillips R, Rahtz DR (2009) Community quality-of-life indicators: best cases III. Best cases series, vol. 1

Slow Tourism (2013) About us: slow food international. Retrieved 18 May 2019. http://www.slowfood.com

Sohn D, Jang H, Jung T (2015) Go slow and curvy: understanding the philosophy of the Cittaslow slow city phenomenon. Springer, Berlin, p 27

Chapter 27
City Branding Through Porter's Competitive Models: Investigating El Shorouk City Market Position

Mahmoud Elnably, Yasser Mansour, Ayman Assem, and Aya Elsisy

Abstract Globalization has affected the transformation and growth of our cities, where some became highly progressive and others are degenerating. The forces of globalization are driving cities to be more competitive in order to attract foreign direct investments and enhance its local economic development while also attracting population to sustain its real-estate capital. City Branding is one of the tools that cities use to market their position and their competitive advantage. Improving cities performance is linked to understanding cities as competitive corporates that follow that same rules of marketing strategies along with understanding the human perception of city and placemaking. The aim of this paper is to analyse the competitive advantage for one of the New Cities in Egypt, El Shorouk City, to be able to formulate a City Branding strategy based on the competitive position of this city within New Cities in Egypt. The methodology for this analysis is depending on Porter's Competitive Diamond for Nations and his Five Forces Model of Competition. The first model is used for the internal analysis for city's potential while the second model is used for the external analysis. The paper is using both models in analysing the market position of El Shorouk City and how to better allocate its competitive potentials to formulate a City Branding Strategy that attracts both population and investments. This will in return impact the strategies for placemaking in New Cities which could be directed to capture the economic potentials of these cities.

Keywords City branding · Porter's competitive models · New cities · El shorouk city

M. Elnably (✉) · Y. Mansour · A. Assem · A. Elsisy
Faculty of Engineering, Ain Shams University, Cairo 11517, Egypt
e-mail: mahmoud_nably@eng.asu.edu.eg

Y. Mansour
e-mail: yasser_mansour@eng.asu.edu.eg

A. Assem
e-mail: ayman.assem@eng.asu.edu.eg

A. Elsisy
e-mail: a.elsisy@eng.asu.edu.eg

Architecture and Urbanism: A Smart Outlook,
https://doi.org/10.1007/978-3-030-52584-2_27

27.1 Introduction

The New Urban Agenda has stated that cities are becoming more important in leading the economic development of nations, as they contribute with more than 80% of the global GDP (UN-Habitat 2015). They are also the major consumption hubs for resources and the main location for population rise around the world. Thus, cities are considered to be the building block for any sustainable development strategy and there is a strong competition among cities to attract both investments and population to sustain themselves (Andersson 2015). Accordingly, Egypt has started a New Cities Program in the 70s to absorb the increasing population and to deliver an adequate quality of life for its urban citizens. However, New Cities in Egypt are facing slow growth patterns causing low occupancy rates and poor local investments (Sims 2015). Thus, cities are degenerating as they are not achieving the targeted economic or urban development. Despite that increasing urban agglomerations are occurring simultaneously in other areas like in peri-urban areas and in informal settlements (Sims 2011). The New Cities aren't meeting the habitation needs for most population, and some of them are not conducting successful market strategy. Branding is of a paramount importance when it comes to laying down the marketing strategy (Kasapi and Cela 2017). One of the issues of this poor marketing strategy is weak City Branding policies that don't position New Cities within the market properly for attracting people, resources and business. City branding is considered the main strategic step to increase city's competitiveness (Papp-Váry 2018).

Throughout different literature, cities are encouraged to be better managed as corporates in order for these cities to function effectively and efficiently (Kotler and Gertner 2002; Zhang and Zhao 2009). This is different from understanding the cities as a place for urban interactions activities, which could be analysed by city systems or urban metabolism models (Anton and Lawrence 2014; Batty 2007; Riza et al 2012). Accordingly, this research is using two models for analysing cities' competitive advantage which is traced to the studies done for strategic management of corporates and cluster industries within a nation, and not commonly used for analysing cities' performance. The first model is the Diamond Model of Nations developed by Michael Porter, which have four determinants of national competitive advantage and aims to understand the cities' competitive advantage in supporting national industries and formation of clusters, thus improving the local economic development and the welfare of its citizens (Porter 2011). The second model is the Five Forces of Competition developed also by Porter, but for corporates strategic management and how they define their market and rivalry position among other corporates, so they can improve it (Porter 2008).

The aim of this research is to use the factors introduced by both models to be able to analyse the city as a corporate. In the first section of the research, the first model—the Diamond model—is used for the internal analysis of the city, while the second model is used for the external analysis. Both models have factors that need to be operationalized in terms of understanding the city as a corporate. The second section of this research takes the concluded factors to analyse one of the New Cities in Egypt, El Shorouk City. The analysis aims to understand the competitive position of El Shorouk City among its rivals, and the market dynamics and conditions that could enhance the performance of the city, regarding its occupancy rates and rates of investments, or weaken it.

27.2 Competitive Advantage of Cities

As cities are perceived as a corporate, the first thing to illustrate is what type of commodity is the city offering for its citizens and for real-estate investors. For citizens, the city is offering a liveable place where they can interact, learn, work and fulfil their basic living needs (UN-Habitat 2009). Thus, the social value proposition plays an important role in the city brand where it needs to offer safety, security, integration, inclusion and resilience for its citizens. This is not only achieved by creating a sense of place within the public spaces and common services within the city, but also by having viable urban densities that support the commercial services and other amenities (Kavaratzis 2009). On the other hand, for real-estate investors, the city is offering an encouraging context for urban and economic development, which is translated into market value proposition. Other than being an attractive destination for occupants to buy and live, the city should brand its supporting social and urban infrastructure that real-estate developers could benefit from (Parkerson and Saunders 2005). Also, city taxes and governance systems play an important role in positioning the city within the market.

Accordingly, the competitive advantage of cities is measured by their success in branding their commodity to both occupants and investors. The following section is divided into two subsections. The first subsection is the internal analysis for city's potential in reflection to Porter' Diamond Model of Nations which is concerned with understanding the success factors of cities that could enhance its competitive advantage and its local economic development. The second subsection is the external analysis for city's exterior forces that could affect its competitive advantage in reflection to Porter's Five Forces Model for Competition among corporates. Both models are tailored to represent the city as a corporate rather than being used as commonly done by other studies.

27.2.1 *Internal Analysis—Porter's Diamond Model of Nations*

The Diamond Model of Nations has been introduced by Michael Porter to develop an understanding for the context that could host potentials for economic development based on offering a hospital environment to industries and businesses to grow, compete and develop (Porter 1990). Thus, the model is initially meant to encourage cities to develop a competitive strategy that attract direct investments which enhance in return its local economy. The model is based upon four determinants that Porter found essential for these cities to develop which are Factor Conditions, Demand Conditions, Relating & Supporting Industries, and Strategy, Structure and Rivalry among different corporations within the city. As we are discussing New Cities in Egypt, which are not necessarily intending to attract industries and businesses but rather occupants and investors, the four determinants would have different definitions from what is meant by Porter in order to reflect the objectives of this research.

The Four Determinants of the Diamond Model The first determinant is the Factor Conditions which are the resources of the city (natural and artificial) that enhances the city compatibility for the flourishing business development (Porter 2000). Within our analysis, Factor Conditions for occupants could include climatic conditions, natural scenery and recreational areas, existing infrastructure, and proximity to services and favourable destinations. On the other hand, the Factor Conditions for investors could differ a little as they focus on market value proposition such as existing running capital within the city, land availability, the existing of competent and skilled labour (for running local services and administrative units), and the welfare of citizens which affect their purchasing power. These Factor Conditions combined represent the city brand uniqueness, as they are hardly replicated within different cities, which make them the base for any city marketing strategy (Ashworth 2009; Govers 2013).

Secondly, Demand Conditions represent the market dynamics within the city that corresponds to the commodity offered by the corporates (Porter 1990). As illustrated earlier, the city here is the corporate and its commodity is offering a liveable place for its citizens; thus, the focus would be on are the housing demand within the city. This demand is much related with targeted market segment within occupants as it determines the quality level of housing as well as the hosting environment. In order to be highly competitive, the city should provide a diversity in housing provision systems representing different target groups with different needs. The Demand Conditions are determined with the real-estate market's ability of providing adequate and affordable housing. This reflects upon the rivalry condition among real-estate companies (whether local or foreign), where if it is strong it indicates a higher satisfaction for occupants needs and a higher profitability for these companies.

Thirdly, the Relating and Supporting Industries in Porter's Model indicate all the complementary industries that increase the competitive advantage of home industries and facilitate the formation of business clusters (Porter 2000). In our modified model, they are all the activities supporting the city brand as a liveable place and improve its quality of life like education, commercial services, workshops, healthcare and retails. These activities are what add up to the sense of ownership, pride and belonging to a community, and this sense of place is what defines the home for citizens (Anton and Lawrence 2014). Thus, the competitive advantage for the city is depending not only on the quality of houses it provides but also with all the other amenities included in the living experience. This is different from Factor Conditions because they are not inherited characteristics with the location of the city but rather the development around these Factor Conditions.

Finally, Strategy, Structure and Rivalry is directly connected with management systems of strong local businesses and what are the favoured corporate governance systems that reflect also on the rivalry conditions and related strategies (Porter 1990). Reflecting on our city model, the Strategy, Structure and Rivalry is defined by the city governance systems and its market dynamics. It is the legislative and regulative framework that organizes all operations and policies conducted by the city. The efficiency of this framework is measured by its capability of hosting different processes in governing the city, a strong monitory over its resources, and flexible innovative solutions in dealing with upcoming challenges. Poor city management structure could be a repelling agent for both occupants and investors.

Internal Analysis Diamond Model for Cities According to the previous discussion, Porter's Diamond Model is updated to reflect our study on New Cities competitive advantage and their market position between its rivals. The same as his model, the interconnection between the different factors is essential for the city to perform efficiently and effectively. However, unlike his model, the chance and government don't affect these factors but rather the market and branding strategies. The marketing strategy is closely connected with the housing demand and the supporting activities that together formulate the commodity the city's offer from its target groups, i.e. a home. On the other hand, the branding strategy is closely connected with city resources and its developed potentials around these resources (the supporting activities), where city branding focuses on the uniqueness and authenticity of the city to develop a higher brand loyalty for its citizens. Both strategies should fit directly to city governance systems that in return manage the existing resources, the developed urban activities and the demand conditions of the housing market.

In order to develop a comprehensive internal analysis for cities' competitive advantage, each factor in the model must be measured both separately and in relation to the other factors. The best-case scenario is when the city natural and artificial resources encourage its urban development, the housing demand is high and is met with the supply provided by the city, the supporting activities are well-functioning through viable densities, and city governance is supportive and consistent in managing all the factors. However, sometimes the issues are not in the existence of the factor of competitive advantage, but they are in the linkage between these factors and how the whole system operates. An illustration for the developed model is shown in Fig. 27.1, where it presents the four factors of cities competitive advantage and their relationship.

27.2.2 *External Analysis—Porter's Five Forces Model of Competition*

Porter's Five Forces Model of competition was designed to perform an external analysis for company's competitive advantage that could enhance their strategic management performance. The model defines five main forces that affect the market

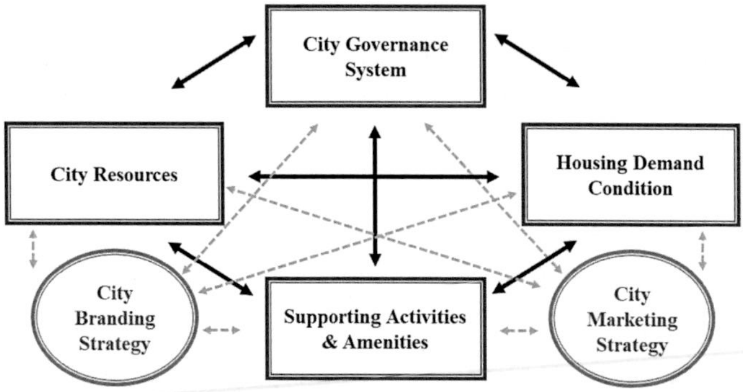

Fig. 27.1 Diamond model of new cities internal analysis by authors

position of a corporate and its ability to sustain its profitability with a competitive market. These five forces are the Existing Rivalry, Buyers Bargaining Power, Suppliers Bargaining Power, New Entrants and Substitute Industries. When any of these forces is strong enough to jeopardize the market position of a corporate, the management strategy should focus on enhancing the corporate's potential in reversing their impact (Porter 2008). Each of these forces is defined by several factors as illustrated by (Bruin et al. 2016). However, both the Five Forces and their defining factors need to be translated in the reflection of this research study substituting commonly known corporates with New Cities.

As for this research, New Cities are perceived as competitive corporates, and accordingly, each force in the model is defined differently than the original model. The clearest ones are the Existing Rivalry, New Entrants and the Buyers Forces, where the first is the existing New Cities that offer the same commodity for occupants and investors, who in return represent the Buyers, while the New Entrants are the New Cities to be developed in the future. On the other hand, the Suppliers are those who deliver the basic commodity component that the New City offers, which are buildings, infrastructure and services. Accordingly, the Suppliers for New Cities are building contractors, infrastructure providers (whether they are public or private companies), and those who provide public services (like education and healthcare) or private services (like owners of retails and workshops). In some cases when the city is contracting its urban development to a private developer, this developer becomes the supplier for the design drawings and planning standards. Finally, the Substitute Industries are areas that could provide a substitute commodity for occupants and investors, but not with the same quality features, social or market value propositions. Accordingly, existing old redeveloped districts within the older city and informal areas are considered substitute for living in New Cities (Ammar 2014), if the later failed to satisfy its users' needs.

There are few attempts to operationalize the factors defining the five forces, yet one of the most promising studies was conducted by (Lee et al. 2012). This research is using their definitions for these factors as a base for translating an upgraded definition to be used for the New Cities perceived as corporates. Table 27.1 summarizes

Table 27.1 Updated description of five forces model factors by authors based on (Lee et al. 2012)

Competitive force	Defining factor (sub-force)	Description
Existing rivalry	Industry structure	The number of competing new cities in the country
	Industry demand and capacity	*"The difference between capacity and demand"*
	Differentiation among cities	The degree of differentiation between city brands
Bargaining power of buyers	Buyer portfolio	The purchasing power of occupants and investors, their number, and the degree of concentration within the area
	Dependence on buyer	The degree that city economic development is dependent on occupancy and investments rates (the proportion of city taxes and trading land value from the city revenues)
	Buyers switch costs	The amount of time, energy, and money for occupants and investors to move out of the city
	Product uniqueness	The degree that this city is differentiated from others
	Importance to buyers	The degree that commodity offered by city is crucial
	Backward integration	The degree of threat that buyers integrate backward to fulfil their demand without the city (in this case, informal housing and informal real-estate investments)
Bargaining power of suppliers	Supplier portfolio	The degree that suppliers are concentrated, well-functioning, and performing in large scale
	Dependence on supplier industry	The degree that city development is dependent on supplier services
	Supplier uniqueness	*"The degree to which product/service provided by supplier is differentiated"*
	Importance to supplier	The degree that the city development is important for the supplier to gain profitability

(continued)

Table 27.1 (continued)

Competitive force	Defining factor (sub-force)	Description
Threats of new entrants	Economies of scale	The degree that cities are dependent on large scale economy to be able to succeed (the need of high occupancy rate and high investments rates)
	Government regulations	"*The degree to which government prohibits New Entrants from entering the market*"
	Brand loyalty	The degree that citizens feel belonging to their city
	Cost advantage	The level of experience needed to generate new well-functioning cities
	Initial capital requirement	The initial needed capital to generate a new city
	Customer switching costs	The amount of time, energy, and money for occupants and investors to move out of the city
Threats of substitutes	Number of substitutes	The number of redeveloped areas and informal areas that compete for attracting residents
	Relative price performance	The degree that social and market value propositions offered by substitutes is matching users' needs
	Switching costs	The amount of time, energy, and money for occupants and investors to move out of the city

these factors. Certainly, there are some factors that aren't at all consistent with the understanding of a city as a corporate like Exist Barrier for Existing Rivalry (as cities can't exit the market) and Forward Integration of Suppliers (as they couldn't upgrade their business to generate a city); thus, they are removed from the analysis.

27.3 El Shorouk City—Egypt (Case Study)

El Shorouk City is one of the New Cities established by the New Urban Communities Authority (NUCA) in 1995, 37 km away from centre Cairo. The city is allocated between two main highways, Ismailia Desert Road from the North and El-Suez Desert Road from the South. Its population hit 120,000 residents in 2015 regardless of having a targeted population of 500,000 in 2000; along these lines, having a 24% occupancy rate. EL Shorouk was chosen as the case for investigation since its accessibility is high with fixed boundaries, with no conceivable further flat extension, as it is likewise limited from the East by New Heliopolis city and from the West by military private venture. Despite not achieving its targeted population, there is no official branding strategy for the city. Thus, the following two sections discuss the competitive advantage of the city (using the previous upgraded models) which in return could be a base for a successful branding strategy.

Both the internal and external analyses of El Shorouk City have been conducted through a qualitative research methodology, where the first part was based on online survey illustrated in a previous research (Elnaby et al. 2019). The second part was based on an in-depth interview designed on the basic results of the online survey, conducted on fifteen interviewers. The respondents of the interview were divided into five categories; residents living in El Shorouk for more than five years occupancy, residents living in El Shorouk for less than five years, recent residents living in El Shorouk for less than one year, dwelling owners who haven't relocated yet and residents who reallocated from El Shorouk after living there for a while. Each interview was divided into three sections.

Section one covered individual and household information like age, daily activities, work location, commuting behaviour, family members, number of car ownership and previous dwelling location. Section two covered residents' perception including their expectations and what they faced, starting with how they heard about El Shorouk City and who recommended it for them, and why they choose specifically to live there. Finally, the last section covered residents demands and needs, going through the available services and missing ones that they considered vital and important, stressing on what they consider the main advantage that shape El Shorouk City and what they consider as its main flaws. The interviews covered different spectrum of people with various age range, marital status, income status, and job status to have a wider perspective on the community perception.

27.3.1 El Shorouk Diamond Model—Internal Analysis

El Shorouk City is central to the new urban areas situated along with the Eastern augmentation from Cairo, with a favourable position of closeness to the New Capital where Egypt's power is moving all legislative central stations. Also, El Shorouk hosts within its limits a diverse community. As illustrated earlier, the location of the city is one of the main city resources that have a high social and market value proposition giving it an edge in the competition. From the main points highlighted by the online questionnaire was El Shorouk proximity to vital locations, considering the availability of transportation and the need of private car ownership in order to reach daily destinations. Furthermore, the in-depth interview validated that most residents who reallocated to El Shorouk City as they suffered from high traffic and crowded streets, despite that they all declared that it wasn't always like that and things got worse year after year. Some interview respondents stated that despite the good location of the city it is not connected to other cities with accessible transition nodes, which makes it very difficult for daily trips outside the city, the most common for all residents who work in other places. Despite the city isn't walkable and doesn't have natural scenery or attracting economy, the location of the city is what attracted most occupants towards it, clarifying that this is the most important factor condition for the city to work on to enhance its competitive advantage by increasing its accessibility.

On the other hand, the housing demand conditions are also one of the strong factors that helped the competitive advantage of El Shorouk City. Respondents declared that the city offers the best value for money, lower rental options compared to other cities with better living conditions. When they were asked why they decided to leave their home and relocate to El Shorouk City. Some stated that they were seeking a place where all their family members living within the same house, for him and his children. Others seeking a better community, higher standard of living, quieter, less congested place with better air quality. The previous features are what represented the definition of Home to those residents, the features that the city should use for its marketing and branding strategy.

However, related to commuting options to workplace, shopping and entertainment, they stated that on one hand, at their old residence areas, they used to walk to nearby services or entertainment zones as all what they need is within walkable distance. On the other hand, now in El Shorouk City they use their private car or uber as most services or entertainment areas are not nearby, with no walkable routes. They mentioned that most daily basic needs and services are available and sufficient, with the availability of few insufficient entertainment zones, and many prefer to go to Madinaty (a nearby gated community) for shopping or entertainment. They mentioned that they never tried to use public transportation at El Shorouk City and knowing from social media that it is not sufficient nor comfortable. Thus, the relating and supporting activities main weakness is internal transportation.

The city governance systems weren't highly perceived by El Shorouk occupants. All of them mentioned that they can sense a progress regarding city infrastructure like roads paving and lighting, yet it is very slow and inadequate compared to private compounds like Madinaty or El Rehab city (a nearby New City). Other than that, the lack of governance and poor city management is what perceived by most occupants, where unjustified decisions (like removing all the planting in front of houses to allow for a sidewalk space), or not properly responding to occupants' complaints are the main issues within the city. Adding to that the lack of a city marketing or a city branding strategy to enhance the performance of the city could be indicators that governance systems are considered weaknesses for El Shorouk City.

27.3.2 El Shorouk Five Forces Model—External Analysis

In order to determine the strategic position of El Shorouk City, the research develops the Five Forces analysis affecting the city competitive advantage. This analysis is

Fig. 27.2 El Shorouk five forces model of competition by authors

demonstrated in Fig. 27.2 that shows that El Shorouk City has high rivals' power with diverse nearby cities that could be compared to it and with high suppliers and buyers' power, yet with minimal New Entrants or substitute threat.

As for the definition of forces, the Rivals are other New Cities competing within the area of El Shorouk City offering the same social and market value propositions. Accordingly, cities like New Cairo, Badr, 6th October, Sheikh Zayied, and others fill into that category. There are more than 20 cities within this category that differentiates itself between being closer to Cairo, bigger, have more job opportunities, more luxurious, cheaper, or have the biggest shopping mall in the country. Most of these cities have lower occupancy rates than targeted making them compete strongly to attract occupants and investors. However, unlike most of these cities, El Shorouk has an unintended differentiation strategy as it doesn't market itself as a shopping hub or a luxurious zone, but as a safe, friendly, quiet, and comfortable neighbourhood that could host different income strata and a strong connected community. This proposition is perceived in by occupants as validated by the online survey and in-depth interviews, which makes El Shorouk better in attracting occupants rather than investors, unlike its rivals. Other cities have sold most of their housing units and expanding their territories, while having high vacancy rates (IDSC 2009; Sims 2015).

Both the suppliers bargaining power and the buyers bargaining power are strong forces for the city to maintain its competitive advantage. As from one side, the city has a fully developed infrastructure but not for internal transportation, which as illustrated earlier is affecting the city's market position. The service providers and contractors also play a vital role in keeping the costs of living within the city at the market value it competes with. If they raised their prices, the city won't be able to compete on the basis of low prices as it is already doing, and that would affect its strategy severely. On the other hand, the buyers bargaining power is high because they formulate the basic resource for the city, but not as high as the suppliers' power. The city could progress with minimum number of occupants—like other New Cities—as its local economy isn't based on collecting taxes from occupants. Also, the switching costs are very high regarding reallocation of residents, not just economically but also socially. The survey and interviews validated that Egyptians don't favour reallocating constantly and prefer settling within a certain area for the rest of their lives (Elnably et al. 2019).

On the other hand, threats of substitutes and new entrants are quite low. For the first, the value proposition introduced by El Shrouok can't be replicated by other redeveloped areas within the city (as they are not as quiet and comfortable as declared by interview respondents), and the income strata that target El Shorouk are unlikely to target informal areas. However, private gated communities do have a competitive edge over El Shorouk City as respondents stated they have better service, management and accessibility—but they don't compete on the same price level (they are much higher). On the other hand, the threat of new entrants is weak because the entry barriers are high, and it isn't easy to start developing a New City which would attract occupants towards it—the main target group for El Shorouk. Despite that, New Cities are being developed constantly but to attract investments rather than occupants, as in the case of the New Capital.

27.4 Conclusion

As the competition between cities is fierce, each city needs to be distinguished through a unique brand identity in order to recognize and differentiate themselves among other cities to attract both residents and investors. The qualitative analysis of El Shorouk City for its internal and external competitive advantage potentials has developed an understanding for the areas that the city could work upon in enhancing its branding among other cities. The strongest potentials of the city are its prime location—which needs to be supported by accessible and convenient internal and external transportation—, its price value proposition, and its social value proposition as an intimate community place. The weaknesses are the city governance systems, poor management and poor branding strategies. If the strongest potentials were properly and clearly branded with the purpose of attracting both existing and potential target groups, the occupancy rates could increase to have viable densities to support existing services and encourage future investments. However, further empirical investigation for each factor is needed to develop a comprehensive city branding strategy for the city, especially regarding attracting investors, not just occupants.

References

Ammar H (2014) Informal areas & new cities: the link & impact. Paper presented at the 6th international conference on responsive urbanism in informal areas: towards a regional agenda for Habitat III, Cairo, Egypt

Andersson M (2015) Metropolitan regions as governance systems: unpacking metropolitan governance for sustainable development. GIZ-UN-Habitat, Eschborn

Anton CE, Lawrence C (2014) Home is where the heart is: the effect of place of residence on place attachment and community participation. J Environ Psychol 40:451–461

Ashworth G (2009) The instruments of place branding: how is it done? European Spatial Res Policy 16(1):9–22

Batty M (2007) Complexity in city systems: understanding, evolution, and design. MIT Press, Cambridge, MA, USA

Bruin Ld, Marques M, Kanani P, Roekel Mv (2016) Porter's five forces. industry analysis. Retrieved from https://www.business-to-you.com/porters-five-forces/

Elnably M, Mansour Y, Assem A, Elsisy A (2019) the impact of city branding and community perception through social media on housing occupancy rates: the case of El-Shorouk CİIty, Egypt. Paper presented at the SPACE international conference 2019 on sustainable architecture, planning and urban design, London, United Kingdom

Govers R (2013) Why place branding is not about logos and slogans. Springer, Berlin

IDSC (2009) Learning from experience: potentials and preconditions for new cities in desert areas. Retrieved from Egypt

Kasapi I, Cela A (2017) Destination branding: a review of the city branding literature. Mediterranean J Soc Sci 8(4):129–142

Kavaratzis M (2009) Cities and their brands: lessons from corporate branding. Place Brand Public Diplomacy 5(1):26–37

Kotler P, Gertner D (2002) Country as brand, product, and beyond: a place marketing and brand management perspective. J Brand Manage 9(4):249–261

Lee H, Kim M-S, Park Y (2012) An analytic network process approach to operationalization of five forces model. Appl Math Model 36(4):1783–1795

Papp-Váry ÁF (2018) Country branding: what branding? Relevant terminologies and their possible interpretations in the case of countries. Econ Reg Stud/Studia Ekonomiczne i Regionalne 11(4):7–26

Parkerson B, Saunders J (2005) City branding: can goods and services branding models be used to brand cities? Place Brand 1(3):242–264

Porter ME (1990) The competitive advonioge of notions. Harvard Bus Rev

Porter ME (2000) Location, competition, and economic development: local clusters in a global economy. Econ Dev Q 14(1):15–34

Porter ME (2008) The five competitive forces that shape strategy. Harvard Bus Rev 86(1):25–40

Porter ME (2011) Competitive advantage of nations: creating and sustaining superior performance, Simon and Schuster

Riza M, Doratli N, Fasli M (2012) City branding and identity. Proc Soc Behav Sci 35:293–300

Sims D (2011) Understanding Cairo: the logic of a city out of control, American University in Cairo Press

Sims D (2015) Egypt's desert dreams: development or disaster? American University in Cairo Press

UN-Habitat (2009) Planning sustainable cities. Earthscan: for UN-Habitat, Nairobi

UN-Habitat (2015) HABITAT III issue papers 22—informal settlements. UN-Habitat, OHCHR and UNOPS, New York

Zhang L, Zhao SX (2009) City branding and the Olympic effect: a case study of Beijing. Cities 26(5):245–254

Chapter 28
Investigating the Urban Structure of Newly Planned Cities in Egypt: The Case Study of New Cairo City

Walid S. Abdeldayem and Tamir El-Khouly

Abstract Egypt is characterized by an abundance of cities built along the narrow valley of the Nile River and into the delta since ancient times, accounting for nearly 90% of the total population. This has led to a need to find alternatives to urban development and expansion beyond the borders of the narrow valley. One of these alternatives is the establishment of new urban communities extending deep into the Sahara, characterized by a wide horizontal extension, highways and arterial roads, entirely vehicle-oriented as a means for movement throughout the city because of a lack of public transportation. This paper evaluates this trend by analysing one of the newly established cities outside Greater Cairo, known as New Cairo City, which requires changes to its physical structure to increase the pedestrian crossing routes because of the high number of accidents involving pedestrians crossing main roads. Using space syntax as an informative theory, the study was able to test the designed structure of the road network in new desert cities in Egypt. Furthermore, relate it to the pedestrian perspective in the city using local-scale analysis. The study concluded that there is a connectivity issue on the local/pedestrian scale. Due to the design of the road network structure, roads are acting more like barriers than connectors between the city boundaries. Furthermore, the study proposes a methodology that enables us to locate the exact spots needed to provide pedestrian accessibility over the road network. The study provides further value by discussing the impacts at the pedestrian level and the diversity of land uses in the city.

Keywords Space syntax · Sustainable development · Spatial configuration · Planned cities

W. S. Abdeldayem (✉)
Cairo University, Giza, Egypt
e-mail: walid.201910308@eng-st.cu.edu.eg

T. El-Khouly
Department of Architecture, Ain Shames University, Cairo, Egypt
e-mail: t.el-khouly@eng.asu.edu.eg; t.el-khouly@aucegypt.edu

Department of Architecture, The American University in Cairo, New Cairo, Egypt

Architecture and Urbanism: A Smart Outlook,
https://doi.org/10.1007/978-3-030-52584-2_28

28.1 Background

In the early 1970s, Egypt's policy was orientated to construction, aiming to transfer the country to a new era. Development began by establishing three new cities along the Suez Canal as a first step towards returning life to the region after the migration caused by the regional wars that had lasted for seven years. The construction of new cities was an urban strategy for the comprehensive development of the country.

Since then, the vision has been to redesign the urban system of Egypt and to redistribute the population that was concentrated along the narrow valley of the Nile River and across the Nile Delta to review Egypt's potentials that Egypt holds in order to achieve the desired development objectives. This was followed by a series of changes to the policies determining development priorities. The establishment of new cities and the distribution of communities have been a major part of the development process ever since. Yet the total area of land remains unfairly exploited in Egypt.

Exponential growth in population and continuous migration from the countryside to cities seeking employment opportunities has caused an imbalance in the distribution of population and urbanization. Egypt's urban population increased from 17 to 42% in the early twentieth century (Abdelkader and Ettouney 2009), which negatively affected economic and social levels, signified by:

1. A high density in populated areas reaching around 500 people per hectare, with a particularly high concentration in Greater Cairo (the capital) and Alexandria as the largest urban centres.
2. An infringement of the agricultural land due to the urban expansion of cities located on the delta with serious consequences for agricultural production, national food security, and per capita green areas.
3. An urbanization, nearly 90% of the total population, is concentrated at the heart of the delta and along the valley of the Nile river while more than 90% of the country's land is still deserted.
4. Rising urban problems such as the decline of services, the erosion of infrastructure, and the appearance of informal settlements in major urban cities.

28.2 The Case Study of New Cairo City

The greater Cairo region was the one holding the largest number of new satellite cities. By 1982, the population in this region reaches 9 million inhabitants. The main role of these new cities was to stop the urban sprawl over the agriculture land as well as provide alternative solution in the surrounding desert.

Located in the desert east of Greater Cairo, New Cairo's population is expected to reach one million inhabitants most of them escaping the overpopulated centre of Old Cairo. It is intended to achieve a balance in the distribution of urban communities, social services, and transportation networks. The master plan of the city includes

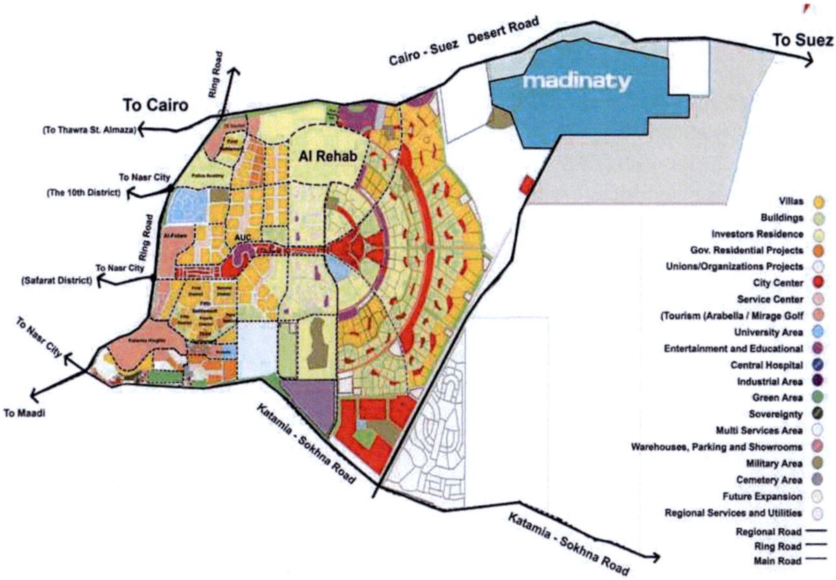

Fig. 28.1 New Cairo master plan. *Source* https://www.madinaty.com/Ar/project.aspx

three settlements—the 1st, 3rd, and 5th—which are separated by two running flood torrents (Farid and El Shafie 2002). Each settlement has been planned to include at least one major service core to provide various activities for four to five districts, each of which is to include a sub-central core for four to five residential neighbourhoods; see Fig. 28.1.

As a result of the existing situation of the new city, the upper economic class with the highest income (constituting 5% of Egyptian society) makes up most of the occupancy rates. This class is more motivated to leave the old city with its environmental and administrative problems to move and contribute to a newly planned built environment. As a consequence, land prices have significantly increased to levels that had been reached before in the Egyptian real estate market. In spite of the motivation to migrate to new cities, their overall occupancy is still low (about 40% occupancy in New Cairo) (Abdelkader and Ettouney 2009).

In New Cairo City, although the master plan contains residential, commercial, and industrial zones, it lacks institutional or governmental presence. Settlements are suffering from the controversy arising from the low growth in occupancy despite the planned objective of targeting a high population. In consequence, this has created a barrier in shaping the economic growth of the new city. Thus, the entire real estate development in New Cairo City is characterized by luxury housing and expensive land plots mainly being built by the high economic-social classes leading to the establishment of gated communities.

28.3 Space Syntax Theory

Planners are facing a challenging issue regarding the continuous need for design modification to meet the users' needs and that is due to the lack of accordance between the design and the users. As so, the relationship between the built-up environment and the user is a significant relationship in the planning process.

Contrary to different planning theories that concern about urban morphology, space syntax is one of the most efficient ones in representing the relationship between the human dimension and the spatial configuration. The theory adopts the notion that urban structure has a considerable influence on human social activities (Hillier and Hanson 1984).

The city in general consists of a number of integrated components; the "physical city" comprises physical components that consist of all the material artefacts—the buildings and the spaces around those buildings—in the city; while the "functional city" describes how humans interact with the surrounding physical city (Hillier 1996). The physical and functional cities are interrelated. The physical city comes into being out of the functional behaviours and activities of the people in it and has emerged, and it, in turn, reshapes people behaviour and activities. As a result of this back and forth relationship, cities' appearance stays under the state change. In addition, external representations in the physical city may be at variance with internal representations that exist as thoughts, ideas, intentions, and memories in individuals and as history, tradition, and culture in institutional agents. With its networks of spaces, forms, and functions, it could be concluded that both the physical city and its people and institutions are complex (Portugali 2016).

The theory articulates the network between urban spaces in a city using their physical boundaries. In the network's analysis, the relationship between urban spaces varies according to how they are enclosed, divided, overlap, and intersect (Al-Sayed 2014).

28.3.1 *The Configuration of Spaces*

Space syntax aims to represent the network that holds different urban spaces in a spatial sequence. These networks aim to describe the relationships between similar individual spatial components as the traditional layouts represent space networks as a way to related spatial units to each other. The theory argues that the configuration of space could be used to reflect different social phenomena as well as other means of analysis (Rashid 2019).

> All these ways of looking at space can be seen as layers of spatial structuring, co-existing within the same plan, each with its own contribution to intelligibility and function. A spatial layout can thus be seen as offering different functional potentials. What is it like to move around in it? Does it have potential to generate interaction? Can strangers understand it? And so on. All these questions are about the relationship of space as formal potentials to different aspects of function. (Hillier 1996)

The theory investigates these configurations using advanced mathematical techniques with the help of modern computer technologies. The use of these techniques enables the labelling, analysis, and visualization of various configurations of spaces in the city's physical form at various scales (Rashid et al. 2009), as well as to articulate the relationships that might appear between different configurations.

28.3.2 Axial and Segment Analysis

Space syntax identifies simple two-dimensional spaces in an urban environment. To draw a connection between spaces in a presented layout, the theory uses visibility as an indicator of a possible movement. The visibility that leads to a movement between spaces is represented by an axial line. The indicated space here is called a convex space where each corner in that space is visible to the others. Consequently, for more accurate analysis and results, the use of segments is conducted, and it is displayed as a part of the axial line. The segments represent an interaction between axial lines. Depicting configuration through lines or segments is relevant because people tend to move directly with minimal changes of direction (Rashid 2019).

Urban space could be identified by the space that is shaped using the outline of blocks. Street model could be presented as urban spaces in three types which are axial lines, axial segments, and streets (Peponis 1997). Axial lines could be generated in a connected street model by determining the longest possible visibility line. Furthermore, the generated axial lines could be converted into segment lines by cutting the long line into small segments at intersections. All of the axial and segments use a colour scheme to verify its type (Hiller 1996).

Axial lines represent the possible long line in a space representing visibility. An axial line also considered as the largest visible distance that could be perceived from the human perspective. Axial lines can represent a large urban space referring to the least number of spaces (Jiang and Claramunt 2000). By analysing the way these spaces are linked to each other, it is possible to understand the structure of the larger space and predict social behaviour within the space (Schumacher 1986).

The axial and segment maps enable us to conduct a comparative analytical study among different areas in the urban environment. The structure of spaces needs to be distinguished from the structure of solids. It is also defined as a connected pattern of physical forms, from the structure of solids, where physical forms presented as a mass on the layout (Rashid 2016; Ellis1986).

28.4 Spatial Analysis for New Cairo City

This research uses space syntax theory to analyse the city's urban structure. DepthMap software is used to generate the analysis. It starts by drawing closed polygons for the building plots throughout the city. Instead of drawing axial lines representing roads and open spaces manually, these lines are generated automatically and then reduce the generated number of lines to the minimum. Consequently, the generated map is converted into a segment map for more accurate results.

This research anticipates using the advantage of syntactic analysis to determine the unrevealed potentials and possible opportunities in the city's spatial structure. As mentioned in Sect. 28.3, the essential step towards conducting the required analysis for the spatial configuration is achieved by creating an axial line map to represent the model network and the urban pattern (Hillier 2009). Every axial line has its own value based on its location and connectivity with respect to the whole urban system (Hillier 2002). The analysis is run on a segment analysis map, using variant scale approach to analyse the map on local and global scales. The radii vary between local (e.g. R_1200 m) to global measure (Rn). The analysis is conducted on the weight of segment length for the wide range of street lengths.

New Cairo is a newly planned city, and the gradual hierarchy of roads represents the connectivity for vehicles and pedestrians throughout the city. To analyse the whole urban structure, choice-to-depth was conducted at the radius of (Rn), which, represents the global scale analysis (Azimzadeh and Bjur 2005). Choice represents the street segments that are highly chosen to be passed through as the shortest path. The generated map is produced with respect to the weight of angular connectivity for a critical angular path choice. Angular segment analysis obtains (integration and choice) measurements of multiple radii ranges. The variety in measurements shows outcomes at different urban and spatial scales (Hiller et al. 2012). Figure 28.2 presents a map of the research methodology, Fig. 28.3 shows angular segment analysis_(T1024) choice measure of a global radius Rn (global measure), and Fig. 28.4 shows angular segment analysis_(T1024) choice measure of a radius_1200 m (local measure).

Fig. 28.2 Research methodology

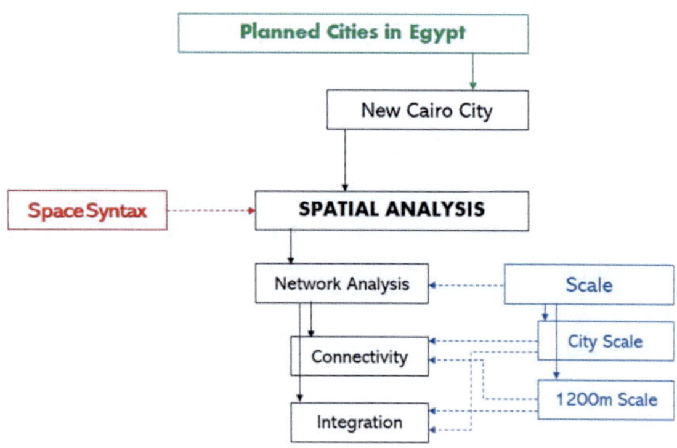

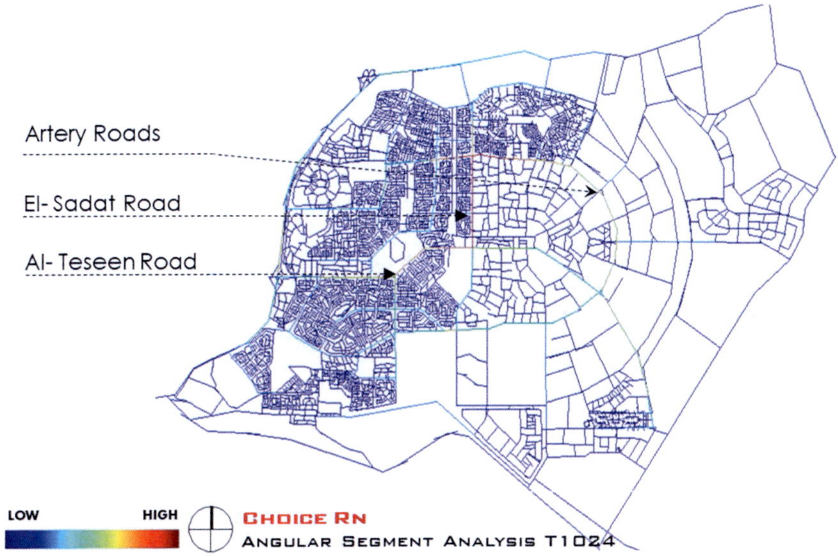

Fig. 28.3 Angular segment analysis_T1024 Choice_Rn (Global). *Source* Authors, using depth map

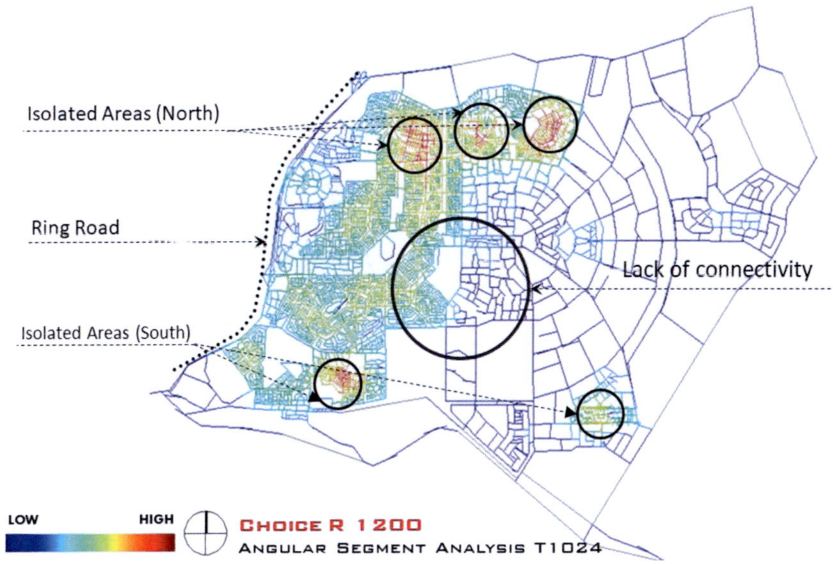

Fig. 28.4 Angular segment analysis_T1024 Choice_R1200. *Source* Authors, using depth map

28.5 Results

As seen in Fig. 28.3, the angular analysis of choice measure shows that on a city-wide global scale, Al-Teseen and El-Sadat roads have the highest value of choice for vehicles, which means that these roads are most likely to be chosen for a vehicle to move from a to b across the whole city. It can also be seen that there is no gradual hierarchy in the value given for the roads. As shown, there are just a few roads with a high degree of choice, while all the rest are almost neglected, which indicates a high traffic flow in these routes and only these will be chosen as a path.

On the other hand, the lack of accessible pedestrian routes across these roads has created a very high level of hazard, and these roads are presented as separators or barriers rather than as connectors. These results justify the recent decision of the central authority of New Cairo City to construct three new pedestrian bridges across the main artery road Al-Teseen (90 m wide) in order to improve connectivity and reduce the number of pedestrian accidents.

Figure 28.4 shows angular segment analysis with a different radius (1200 m) to represent the scope of pedestrian movement. It shows that there are certain areas in the north and the south with a high degree of choice, which means that on a district and neighbourhood scale, there is a good range of connectivity for pedestrians. However, there is a lack of connectivity between some other areas in the north and the south, which indicates that Al-Teseen and El-Sadat roads are unlikely to be chosen as an accessible route to connect the northern areas with areas in the south. In fact, every median or low range of choice in these maps indicates a road with high traffic levels that is acting as a barrier between different districts and neighbourhoods against pedestrian accessibility.

Integration analysis on a city-wide scale (n) in Fig. 28.5, shows that the arterial roads with the highest choice are the same as those with the highest integration values. There are some areas in the south isolated (neglected) from the whole city's network, which means that these areas have less connectivity and integration to the whole system; as there are not many connectors to the main arterial roads, they are represented as less accessible areas. This points to the need for an arterial road to run through the southern part of the city to connect areas in the south and assure better connectivity.

28.6 Conclusion

Old Cairo has been suffering from congestion problems because of the extreme density of its street networks. A network was proposed for the newly built cities in Egypt characterized by wide streets. However, the problem of congestion, as well as accessibility for pedestrians, still remains. This has become a significant factor in the master planning of the upcoming generation of new cities in Egypt. The analyses conducted in this research show variations at the local and global scales. The

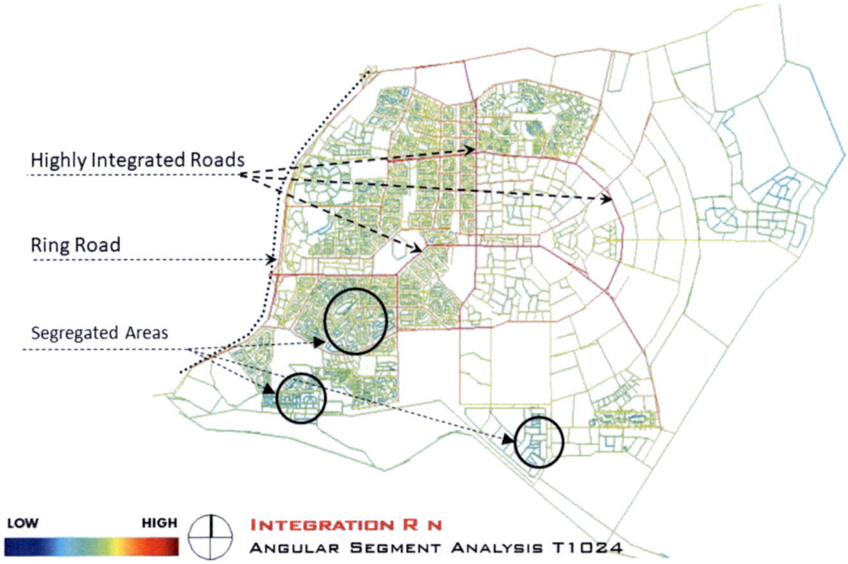

Fig. 28.5 Angular segment analysis_T1024_Integration_Rn (Global). *Source* Authors, using depth map

connectivity analysis on a local scale across the whole city is very much interrupted by the main arterial roads that divide the city into isolated islands and makes it even more dangerous to commute. This is justifying the current adjustments and changes occurring on these arterial roads, as a remedy to reduce the incidents and increase connectivity over these serious barriers.

It is concluded that connectivity and congestion problems are not related as much to the width of the roads as to the structural properties of the network and the way the network is formed in the spatial dimension. It is also concluded that new satellite cities in Egypt are not taking connectivity into sufficient consideration, particularly in early planning phases. To solve the problem of connectivity in New Cairo City, the authority has already intervened by adding new bridges, tunnels, and pedestrian crossovers to the spatial structure. This research approach can be deployed in understanding the aspects of walkability and other social behaviours in the city and warrants further investigation in future studies.

References

Abdelkader N, Ettouney S (2009) The Egyptian new communities, between objectives and realization—a critical discourse, three decades later, research paper, ARCHCAIRO

Al-Sayed K et al (2014) Space syntax methodology, 4th edn, Bartlett School of Architecture, UCL, London, UK

Azimzadeh M, Bjur H (2005) Transforming cities: the role of the configuration of the network of public spaces in urban life (2005). In: Proceedings of the 5th international space syntax symposium. TU Delft Faculty of Architecture, Delft, NL

Ellis WC (1986) The spatial structure of streets. In: Anderson S (ed) On streets. MIT Press, Cambridge

Farid M, El Shafie H (2002) An assessment of the New Cairo city. J Eng Appl Sci-Cairo

Hillier B (1996) Space is the machine: a configurational theory of architecture. Space Syntax, London, UK

Hillier B (2002) A theory of the city as object: or, how spatial laws mediate the social construction of urban space. Urban Des Int

Hillier B (2009) spatial sustainability in cities: organic patterns and sustainable forms. In: Proceedings of the 7th international space syntax, symposium, Stockholm

Hillier B, Hanson J (1984) The social logic of space. Cambridge University Press, Cambridge, UK

Hillier B, Yang T, Turner A (2012) Normalising least angle choice in depthmap-and how it opens up new perspectives on the global and local analysis of city space. J Space Syntax

Jiang B, Claramunt C (2000) Extending space syntax towards an alternative model of space within GIS. In: 3rd European Agile conference on geographic information science. Helsinki, Finland

Portugali J (2016) What makes cities complex? In: Complexity, cognition, urban planning and design: post-proceedings of the 2nd delft international conference

Peponis J, Ross C, Rashid M (1997) The structure of urban space, movement and co-presence: the case of Atlanta. Geoforum

Rashid M (2016) The geometry of urban layouts: a global comparative study. Springer International Publishing.

Rashid M (2019) Space syntax: a network-based configurational approach to studying urban morphology. In: D'Acci L (ed) The mathematics of urban morphology: modeling and simulation in science, engineering and technology. Birkhäuser, Cham

Rashid M, Wineman J, Zimring C (2009) Space, behavior, and environmental perception in open plan offices: a prospective study. Plan Des Environ Plan B

Schumacher T (1986) Buildings and streets: notes on configuration and use, In: Anderson S (ed) On streets, pp 133–148. MIT Press

Chapter 29
GIS-Based Spatio-temporal Analysis for Social Events in Urban Public Spaces

Mohamed Ezzeldin and Ayman Assem

Abstract Urban public spaces have been considered as an essential part of cities throughout history and one of the most significant indicators of the quality of urban life. Over the span of urban life, public spaces such as squares, streets, and plazas have continuously reflected the complexities of their cities' cultural, social, and economic contexts. The social production of a space through different activities that take place when the people are present in the space such as watching, listening, and experiencing and active and passive participation in the environment that make the space live and dynamic, and transforming the physical space, into a place that is full of emotions and senses. The concept of an event is widely ranged in types and activities that happen at a given place and time. Social events are mainly connected to the urban public spaces and can take place day or night, limited in time or continuous. This paper introduces a computational approach that utilizes GIS tool as a technology to represent and visualize the spatio-temporal analysis for the different social events carried out in public urban spaces. Using GIS in creating the spatio-temporal analytical maps will clear the definition of public urban spaces and determine their social characteristics which should be considered when constructing the visions of upgrading and the development of urban public spaces.

Keywords Spatio-temporal · Places · Public spaces · Social events · GIS-based

29.1 Introduction

Urban public spaces represent the relation between space and place, and how the perceived physical space is being transformed into a conceived meaningful place, during the lived-in different social activities that are carried out during social public

M. Ezzeldin (✉) · A. Assem
Department of Architecture, Faculty of Engineering, Ain Shams University, Cairo, Egypt
e-mail: mohamed.ezzeldin@eng.asu.edu.eg

A. Assem
e-mail: ayman.assem@eng.asu.edu.eg

Architecture and Urbanism: A Smart Outlook,
https://doi.org/10.1007/978-3-030-52584-2_29

events. The paper introduces the concepts of space and place and the different dimensions of place (spatial, temporal and social). It introduces briefly Lefebvre (1991) theory of the production of space, showing the different characteristics of public spaces, and introducing the different types of social activities and events that are carried out in these urban public spaces. Using the spatio-temporal maps generated using a GIS-based analysis for social activities that took place during different types of events for several samples in urban public spaces and highlighting the significance of spatio-temporal maps as a computational method of analysis to represent the different types of active and passive engagement of people, during the different public social activities. The paper concludes that using GIS in creating spatio-temporal analytical maps; can help to clear the definition of public urban spaces; and determine their social characteristics to be considered when constructing the visions of upgrading and development of urban public spaces. The next part introduces the methodology used to create that kind of emotional maps and spatio-temporal studies, in order to in-depth study the interaction of people in urban public spaces, and how the place is being produced in a process of a social production of space.

29.1.1 Methodology

This paper introduces a computational approach; utilizes GIS to represent; and visualizes spatio-temporal analysis for the different social events carried out in public urban spaces. Using GIS in creating the spatio-temporal analytical maps could help in clarifying the definition of public urban spaces and determine their social characteristics which should be considered when constructing the visions of upgrading and development of urban public spaces. ArcInfo is the desktop GIS software application selected for the study. It includes all the desired functionality of spatial analysis and extensive data manipulation. This program is usually used for geographical and land spatial analysis. However, it can also be used to perform the required analytical studies for urban public spaces. To undertake the spatio-temporal analysis, it is required to use a database tool that allows connecting the temporal analysis of the social events, to the grid cells of the urban public spaces based on the way social events and activities that are performed according to their types and characteristics. Figure 29.1 shows the spatio-temporal analysis process using the GIS model builder; this analysis gives another physical dimension, in understanding how the space can be transformed into a place through a social production process.

29.2 Producing 'Places' as a Social 'Spaces'

As introduced in the introduction, urban public spaces have been considered as an essential part of cities throughout history and one of the most significant indicators of the quality of urban life. Over the span of urban life, public spaces such

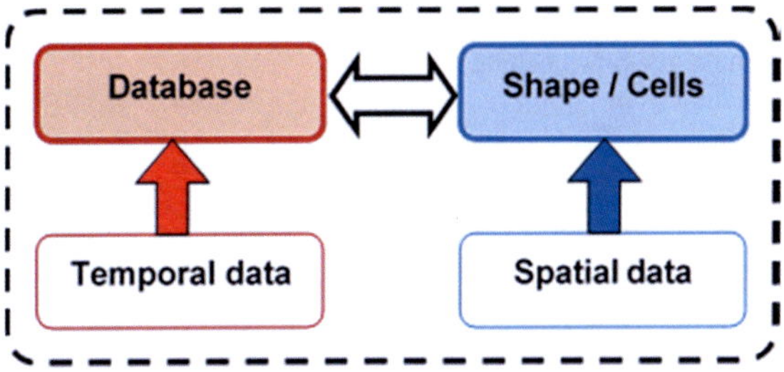

Fig. 29.1 Spatio-temporal analysis process using the GIS model builder

as squares, streets, and plazas have continuously reflected the complexities of their cities' cultural, social, and economic contexts. Public spaces play a vital role in the life of urban areas, whereas memorable, accessible, or meaningful places. In order to make a spatio-temporal analysis for the different social events carried out in public urban spaces, the next part will briefly introduce the basic concepts of 'space' and 'place', and how a physical space is being transformed into a 'place' through a social production process in order to understand the need of spatio-temporal analysis for the different social events that are carried out in public urban spaces.

29.2.1 Space and Place

The ideology of a place can be figured out clearly in any public space. Tuan (2001) defined space as abstract, while place has a meaning and gives senses and emotions. Space and place are not the same in meaning and significance. Space represents the physical, open, abstract, dynamic, threat, free, movement side of any place, while place represents the social, humanist, enclosed, concrete, mental, stable, secure, local, and static part of any space. Places give meaning to spaces, while spaces give the context and familiarity to places. The identity of any place is structured through the social activities of people in that space. According to Tuan, any space can be transformed to a place through different human social interaction (Kafafy 2012).

29.2.2 Dimensions' of Place

According to Edward Relph (Seamon and Sowers 2008), any place has three main elements or dimensions that clarify its identity: the spatial 'physical dimension', the temporal 'historical dimension', and the Social 'cognitive or mental dimension' as shown in Fig. 29.2. In the 'spatial' dimension, the physical elements of the space are being represented on maps, where urban and architectural design decisions could take place (Cighi 2008). The temporal dimension of place takes the importance from the sense of place that is being created during the time people spend in that space. People create their own cognition of the place during the time they spent in events

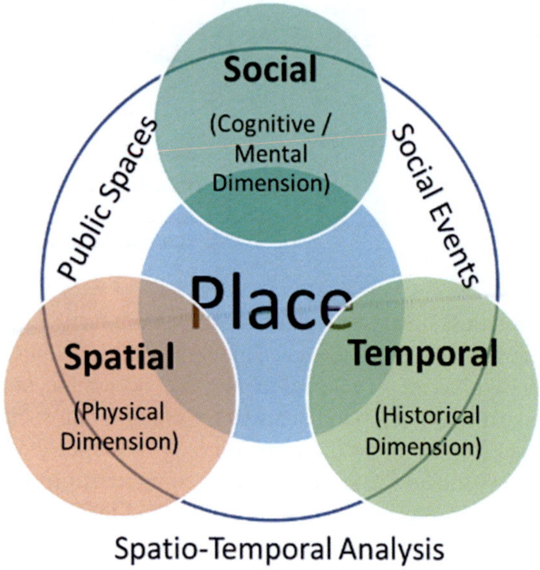

Fig. 29.2 Edward Relph's (Seamon and Sowers 2008) three dimensions of place: spatial, temporal and social

or public activities, establishing their own meanings, memories, and emotions. This involvement of the temporal or historical level develops the sense of place (Harrison and Dourish 1996). The social dimension is related to the behaviour and activity of people; their impressions, feelings, meanings, emotions, perception, and cognition either as individuals or in groups. Here the sense of place is being established through the interaction of humans in their different social activities and events. It relates both concepts of 'space' and 'place' creating a significant part of the identity of place (Harrison and Dourish 1996). The significance of the spatio-temporal analysis, that it gives the ability of the integration between the three dimensions of place, spatial; physical dimensions, with the temporal; historical dimension, in order to interpret and understand in-depth, the social mental dimension, where the different social events are carried out in public urban spaces; giving the place a part of its identity.

29.2.3 Social Production of Place

In order to make a spatio-temporal analysis of public spaces during different social events, it is important to introduce first how the space is socially produced. Lefebvre (1991), in his theory of 'the social production of space' introduced the relation between space and social actions and activities, producing a meaningful space, which could be defined as 'place' to him, space is considered as a social product. The social phenomenon of space, as introduced in Lefebvre theory, produces another dimension of place. In his theory, the space is being transformed to a place during the events and social activities and actions whether it social, cultural, or political (Parkinson 2012).

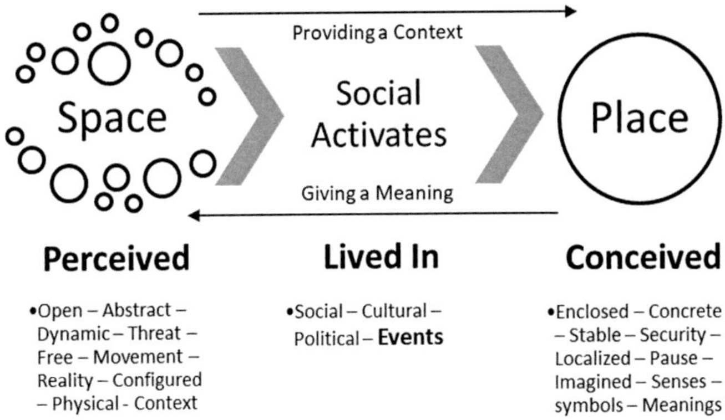

Fig. 29.3 Social production of space—Lefebvre (1991) theory. In this process, the space is being transformed into a place

Lefebvre did not use the expression of 'place', he used instead of the term 'concrete space' (Agnew 2011). 'The production of space' is an endless, and a continuous process consists of three main levels. Figure 29.3 illustrates the theory of social production of space by Lefebvre (1991). In this theory, the first level is the perceived space where the terms of open, abstract, dynamic, threat, free, movement, configured, physical, context representing the 'reality space'. The second level is the lived-in space where the social, cultural, and political events and activities are carried out. The last phase is the conceived space, where the terms of enclosed, concrete, stable, security, localized, pause, imagination, senses, symbols, meanings representing the definition of 'place' (Perdikogianni 2007).

The different types of analysis methods of space syntax such as spatial morphology, axial lines, convex spaces and isovists are used to study the physical perceived space (Perdikogianni 2007). These types of analysis mainly focus on the physical settings, forms, and functions of the perceived space (Salama 2013). The spatio-temporal analysis using a GIS-based integrates the physical aspects with the temporal social and living space that carried out the events or public social activities and performances, producing the meaning of conceived place.

29.3 Public Spaces: Ideology of Place

As introduced, places are produced through a social production of space, creating a place that has meanings, emotions, senses and memories to people who performed or carried out different social and public events in that place. Such experiences can be clearly figured out in social public spaces where visual memories and emotions through events are created and attached to these places (Carmona et al. 2008). Public spaces represent a kind of integration between the physical perceived space, and the lived-in social activities and events (cultural, social and political), where place is socially produced, and the identity of the place is initiated through that process (Rasouli 2013). In social spaces, both characteristics of spatial and temporal (spatio-temporal) are being presented, where integrations and interrelations between reality and imagination, past and future, physical and nature, culture and context (Lefebvre 1991). The physical space hosts the social activities and actions, identifying the area of these activities and giving the basic parameters of these social human interactions. The question of who, how and what for these spaces are going to be used is important in understanding the spatio-temporal relation between the perceived physical space and lived-in social activities. Public spaces are important part of any city for the essential social public interactions that are carried out at that kind of places. Streets, squares, civic spaces, piazzas or plazas, and parks are samples of the different kinds of open public spaces that could be found in any city fabric (Abidin et al.). Urban public spaces have always played an important role in the history of cities, reflecting the diversities of contextual, cultural, social and economical sides of these cities and creating memories and meanings for these places (Rasouli 2013). The physical space shall fulfil the needs of the social public activities that it will host or carry out. The quality of physical elements of the public space is an essential aspect to prepare these spaces to socially carry out kinds of gathering and hosting different types of activities and events although generating a social event itself is more important than these physical elements. The sense of place attachment, and the feeling of belonging, and more other meanings are the most important as a result of the interaction and participation of people in social public activities (Vahid et al. 2013). The number of people, how long these events or activities are going to take place and what are the different types of these activities and events. All these different parameters are essential in the spatio-temporal analysis for social events that are carried out in urban public spaces.

29.3.1 Characteristics of Public Spaces

Any public space shall do its essential part and function in any city fabric, and it shall have basic spatial and social characteristics: first to be defined as a public space, second to be able to host different social and public events and activities. People need to feel safe, comfort and vitality in public spaces. People need safe places to engage and comfort spaces to spend time. The duration people spend in places makes an attraction to others to imitate the same action which makes these spaces more interesting and popular. Both physical and social beauties of public spaces make them essential and vital part of any city (Finlayson et al. 2011).

29.3.2 Social Activities in Public Spaces

Events and activities are the social dynamic lived-in part of any public space; they transform the physical perceived space into a conceived life place. There are different types of people's engagement to any urban public space to meet their needs. This kind of occupation or engagement could be active or passive. In active people's interaction with public spaces should be in a direct connection with the environment, while in passive this connection might be optional. People also might occupy the space only for a kind of discovery, others for relaxation or enjoying the visual beautiful features, watching or listening. All these kinds of different active or passive interaction and participation make the perceived space, live and dynamic in a social production process, as previously introduced. The concept of event includes different types of active and passive interactions; but in this case, time is an important factor. The social public events could take place day or night, the duration and time also could be limited or continuous, depending on the type and purpose of these social activities (Finlayson et al. 2011). There are different types of users; daily users who have urgent activities, visitors or shoppers who have daily entertainment activities, entertainment visitors or visitors who will attend an event and will share in organized activities. These different types of groups can be according to their age or their role in these different social activities (Vahid et al. 2013).

The different types of social activities could be classified as necessary, optional and social activities. Necessary activities could include working, waiting for a bus, running, going to work or school. The quality of the physical space does not take an important influence on them; they are obligatory and repetitive. They are carried out daily or yearly, participants have no choice to pass through interact with the space to do such activities. The happening of optional activities depends on the conditions and quality the space (safe, comfort and vitality) or the surrounding environment such as climate. It may occur individually, including walking for a fresh air, standing or sitting for leisure, eating or playing. It takes place when the space and conditions are attractive and inviting (Vahid et al. 2013). The last category is the social activities, and here the experience might have active or passive participation. It occurs in groups (two at least), including different types of social activities like talking, playing, performing, listening and watching. Such types of events interact and occupy the public space making the space full of life and action. In such types of activates, the space is being transformed into a place through a social production process. Improving the quality of the physical space is important to encourage people to share and participate in these social public events (Finlayson et al. 2011). Figure 29.4 shows the different active and passive engagements during the different types of social public activities and events.

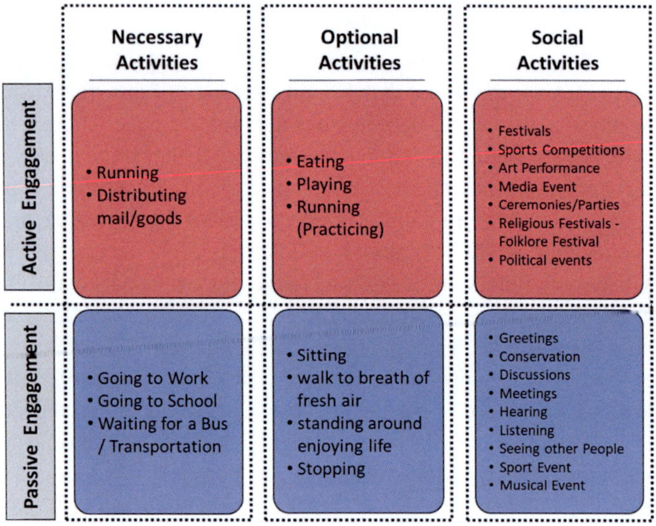

Fig. 29.4 Active and passive engagement during the different types of social public activities and events

29.3.3 Events and Public Spaces

The main concepts of events and activities: (Activity, Happening, Event and Festival). These types of social activities occur in public spaces with different types of engagement. There are different types of gathering: Ceremonies like marriage or funereal, competitions: like sports competitions, conventions or meetings, happening: like an art performance, festival; like musical events, media events, parties, protesting; like revolutions or demonstrating or occupying a place. For example, happening as a kind of an event is more related to art or a performance, people are in a direct interaction with the artwork like music festivals, open art exhibitions, or any kind of other art performances. Time, space and performance are the basic elements of that happening as an event (Şenyol 2010). Events can host different types or activities. They are usually connected to the heart of cities or urban public spaces. Types of events could be classified according to their purpose and objective. For example, there are leisure events like sport, music or recreation; cultural events like ceremonial, religious, art, heritage, folklore; organizational events like commercial, political, charitable, sales or products launching; personal events like weddings, birthdays or anniversaries (Şenyol 2010).

All these types could take place in streets, squares or any public space in the city fabric. During these different types of events and social activities, the physical space is transformed to a social place; according to the interpretive theory of space, 'The Production of Space', of Lefebvre (1991), the main concepts and characteristics of public spaces show how people interact through different social, cultural and political activities and events in the physical space. Activates and functions of urban public spaces change related to the events it hosts and reflects the social, cultural and political transformations of a country or a nation as can be observed.

29.4 Spatio-temporal Maps

This paper introduces analysis of the spatial and temporal (events) that take place in the public urban spaces using a GIS-based, creating a kind of emotional maps that represent the interaction of people during these events, as a way of interpreting the social production of public spaces, and their transformation into places. Spatio-temporal maps and emotional maps in general are kind of analysis that allows creating a dialogue-based map either in the current or future state of public spaces; exploring the different social experiences of these places. This spatial tool can give the chance to study the different social engagement of public spaces creating a feeling and a sense of place and belonging, that could be used to develop the relation between people as (citizens) architect, urban planners, in different communities and active participations (Panek 2018). Creating a place through the social production of space; as introduced before in the previous section, generates a space that full of emotions and feelings, creating a spatio-temporal or an emotional map give the chance to include the experience of people in public spaces (Panek 2018). The use of GIS in spatio-temporal maps gives the chance to sketch, visualize and analyse a mental map that expresses the different experiences of people in these public spaces; not only spatially but also during a duration time of an event, and describing the sense and identity of these places.

29.4.1 Determine Pulses of Place Based on Spatio-temporal Analysis

The spatio-temporal analysis gives us the chance to determine the pulses of spaces, representing the feelings and emotions of people that participate in different types of events. As introduced before the different types of activities in public spaces (necessary, optional and social), during different types of events, people can have an active or passive engagement during them. The spatio-temporal analysis gives the emotional pulses of place according to active or passive engagement of people during different types of events. The next graphs, Figs. 29.5 and 29.6, show a generic sample illustrating the pulses of a place based on spatio-temporal analysis.

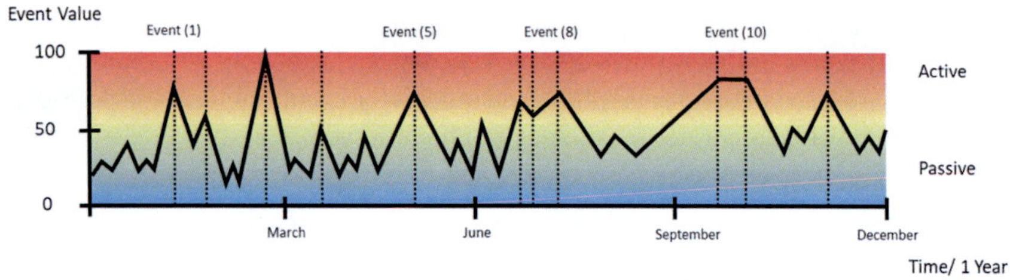

Fig. 29.5 Pulses of a place based on spatio-temporal analysis

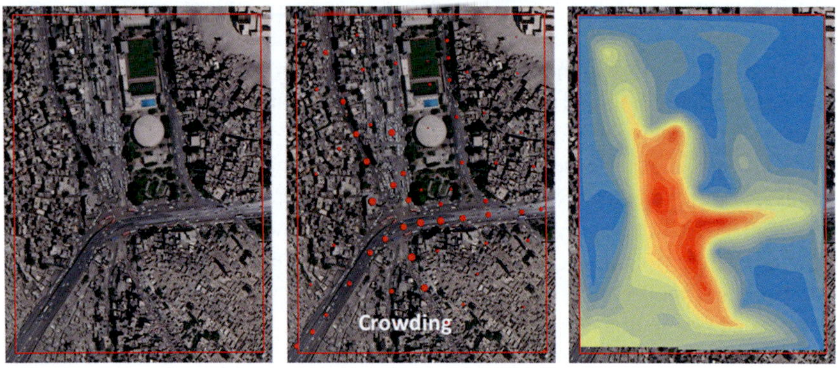

Fig. 29.6 Sample of spatio-temporal analysis of crowding, Sida-Aesha Square

29.5 The Process of the Spatio-temporal Analysis

The proposed spatio-temporal analysis process using GIS is divided into four main steps as shown in Fig. 29.7. The first step defines the effective boundary of the urban space where different social activities are performed regularly or eventually. Additionally, the defined boundary determines the area where the activities will be

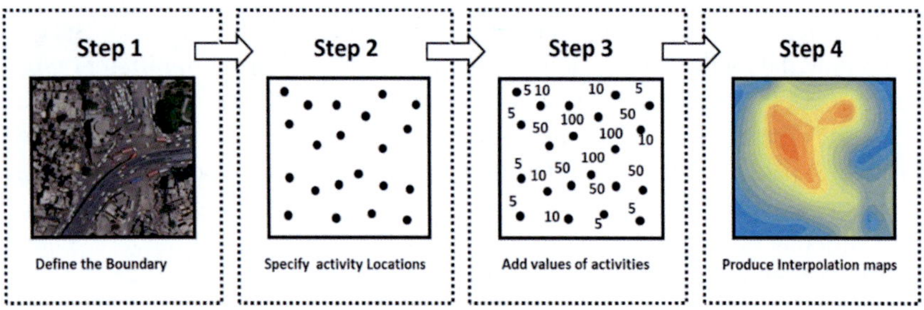

Fig. 29.7 Process of the spatio-temporal analysis, and how the interpolation maps are being produced

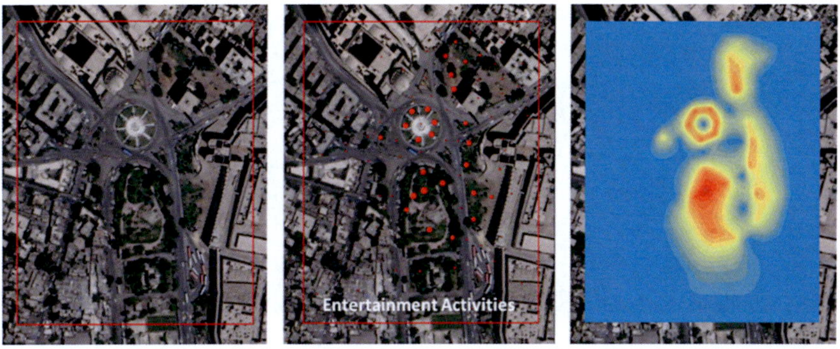

Fig. 29.8 Spatio-temporal analysis for the citadel square entertainment social activities during the event of Eid, as a religious festival, showing the active and passive engagement of the place

observed and recorded. During the second step, the location of the event is specified by defining the zones of the performed activities then calculating the centre of these zones. Adding values is the main function of third step, the calculated values which added to the event's locations are calculated by recording the number of the active people who participate and involved in the performance of the determined activity in the recording duration based on the characteristics of the event. However, a remapping operation is carried-out to scale the calculated values to a predefined range of values (0:100). The remapping of values is necessary to produce homogenous interpolation maps which clear the differences between the spatio-temporal maps for the distinct social activities. In step four, the analytical spatio-temporal maps are produced using the interpolation process. In GIS, there are many types of data interpolation based on the number of data points and the main purpose of the map.

29.5.1 Data Interpolation Using GIS

The natural neighbour (NATURAL NEIGHBOUR INVERSE DISTANCE WEIGHTED (NNIDW) interpolation method has been used to predict the values for remaining cells of the raster image for an urban space, expressing a performed activity based on enough sample data points. NNIDW works accurately with clustered scatter points, and the cell values are estimated by calculating the average of the values of sample data points in the neighbourhood of each processing cell. The closer a point is to the centre of the cell being estimated the more influence; it has in the averaging process. When using NNIDW method, the weight of each sample point is an inverse proportion to the distance, and for the non-sampled locations, an estimation of the values is carried-out based on the distance and values of the nearby known points, that means a smoother surface results from a larger number of sample points (ArcGIS Pro 2019).

Fig. 29.9 Spatio-temporal analysis for Tahrir Square political–social activities during the event of 2011 demonstrations, showing the active and passive engagement of the place

29.5.2 Samples of Spatio-temporal Maps for Different Social Events

The next part of the paper will introduce different samples of spatio-temporal maps during different types of events showing the active and passive engagement of the place. The first sample shown in Fig. 29.8 is a spatio-temporal analysis of entertainment social activities that take place on Citadel Square during the event of Eid, as a religious festival, showing the active and passive engagement of the place. Another example shown in Fig. 29.9 is the spatio-temporal analysis of Tahrir Square during one of the political activities (demonstration) that took place during 2011 events. The last example shown in Fig. 29.10 is the spatio-temporal analysis of Sida-Aesha Square during different types of social activities and events: pedestrian gathering, commercial activities, Sida-Aesha Mooled (Birthday festival), and Friday prayers.

29.6 Conclusion

Public spaces represent the relation between space and place, and how the perceived physical space is being transformed into a conceived meaningful place, during the lived-in; social activities, that are carried out during the different types of social public events. The several examples of these social activities that took place during events represented that the spatio-temporal maps using a GIS-based analysis, are a significance approach and computational method of analysis to represent the different types of active and passive engagement of people, during the different public social activities. Using GIS in creating the spatio-temporal analytical maps can help to clear the definition of public urban spaces and determining their social characteristics which should be considered when constructing the visions of upgrading and

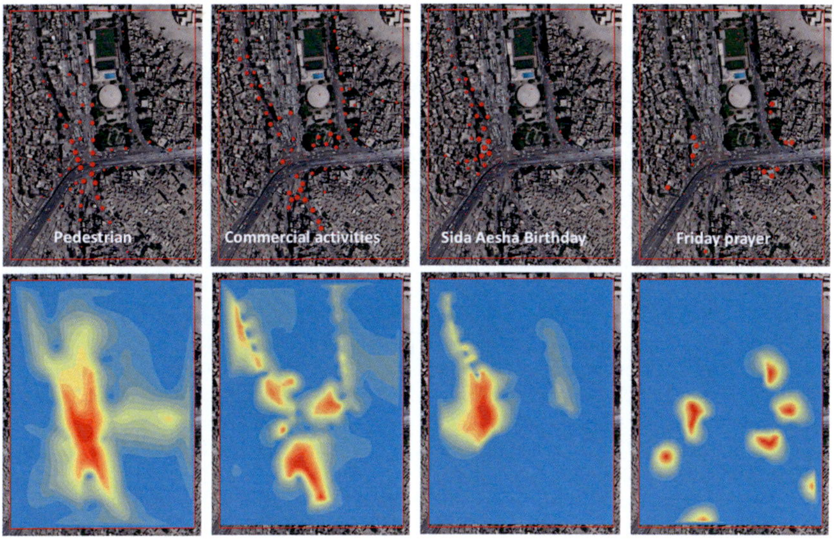

Fig. 29.10 Spatio-temporal analysis for Sida-Aesha Square during different social public events, showing the active and passive engagement of the place

development of urban public spaces. This paper highlighted the importance of using these kinds of emotional maps and spatio-temporal studies, in order to in-depth study the interaction of people in urban public spaces, and how the place is being produced in a process of a social production of space.

References

Agnew J (2011) Space and place. In: Livingstone JA (ed) Handbook of geographical knowledge. (Fourthcoming ed.). Sage, London

ArcGIS Pro (2019) Retrieved from Comparing interpolation methods: https://pro.arcgis.com/en/pro-app/tool-reference/3d-analyst/comparing-interpolation-methods.htm

ArcInfo. (n.d.) Retrieved 25 Feb 2011, from https://www.esri.com/software/arcgis/arcinfo/index.html

Abidin IZ, Usman IM, Tahir MM, Yap YC (n.d.) Characteristic of attractive square as public Space: Putra Square, Putrajaya. Department of Architecture, National University of Malaysia, Malaysia

Bigdeli Rad V, Bin Ngah I (2013) Public spaces and effective factors on social interactions. Int J Current Eng Technol 184–188. Retrieved from https://inpressco.com/category/ijcet

Cighi CI (2008) Senses of place. Graduate School of the University of Massachusetts Amherst. Retrieved Feb 2014, from https://scholarworks.umass.edu/theses/116

Carmona M, de Magalhães C, Hammond L (2008) Public space the management dimension. Routledge, London

Harrison S, Dourish P (1996). Re-place-ing space: the roles of place and space in collaborative systems. In: CSCW'96 proceedings of the 1996 ACM conference on computer supported cooperative work. ACM, New York, NY, USA, pp. 67–76

Kafafy N (2012) The dynamic influence of symbolism and meaning in urban design, the case of Tahrir Square, Cairo, Egypt. In: 48th ISOCARP congress 2012. ISOCARP, Perm, Russia, pp 1–13
Lefebvre H (1991) The production of space. Blackwell Publishing, UK
Panek J (2018) Emotional maps: participatory crowdsourcing of citizens perceptions citizens perceptions of their urban environment. Cartographic Perspect 90(2)
Parkinson JR (2012) Democracy and public space. Oxford University Press, New York
Perdikogianni I (2007) From space to "Place": the role of space and experience in the construction of "Place". In: 6th international space syntax symposium. İstanbul, pp 52:1–14
Rasouli M (2013) Analysis of activity patterns and design features relationships in urban public spaces using direct field observations, activity maps and GIS analysis: Mel Lastman Square in Toronto as a case study. University of Waterloo, Waterloo, Ontario, Canada
Salama HH (2013) Tahrir Square: a narrative of a public space. Int J Architect Res 7(1):128–138
Şentol Ö (2010) Exploring temporary usage opportunities of urban public spaces for event organizations. Graduate School of Engineering and Sciences of İzmir Institute of Technology, ZMİR
Seamon D, Sowers J (2008) Place and placelessness, Edward Relph. In: Hubbard RKP (ed) Key texts in human geography. Sage, London, pp 43–51
Sweeny Sterling Finlayson & Co Architects (2011) Public spaces, activity and urban form. Strategic framework—city centre plan phase 1, City of Saskatoon, Community Services Department.
Tuan Y-F (2001) Space and place: the perspective of experience. University of Minnesota Press, Minneapolis, London

Chapter 30
Mapping Arts and Urban Development in Cairo's Downtown

Elaf Raslan, Marwa Khalifa, and Abeer Elshater

Abstract Recent years have witnessed a widespread cultural-led urban and regional policies in the urban rehabilitation process. Worldwide, urban developers use culture-based economy and industries as a driver for social, physical and economic regeneration through attracting a diverse population. This research **aims** to map arts and culture activities in Downtown Cairo. With the current arts and culture emerging scene in the case, the study investigates state of the art using empirical research tools. The **purpose** is to develop an evidence-based strategic plan for Cairo's Downtown using arts and culture activities. The current research design relies on a literature review, interviews with stakeholders and online surveys. The **results** showed that stakeholders have a similar vision for developing Downtown into an international destination using different methods. Locally translated, our findings indicate that the private sector is fostering arts and cultural activities, while the local authorities with their limited capacities are investing in urban and public infrastructure. The **findings** highlighted the importance of art and culture in the ongoing development approaches in Downtown Cairo. The paper figured out in the main **conclusion** that strategic plans are needed taking into consideration arts and culture events as a tool for urban development.

Keywords Gentrification · Cultural economy · Rehabilitation

E. Raslan (✉) · M. Khalifa · A. Elshater
Ain Shams University, Cairo, Egypt
e-mail: elaf.raslan@gmail.com

M. Khalifa
e-mail: marwa_khalifa@eng.asu.edu.eg

A. Elshater
e-mail: abeer.elshater@eng.asu.edu.eg

Architecture and Urbanism: A Smart Outlook,
https://doi.org/10.1007/978-3-030-52584-2_30

30.1 Introduction

Between 1950 and 2000, the global urbanisation level has increased to 58%, and is expected to reach 70% by 2030, with 65% of the population being youth (UN-Habitat 2016). With such an increase, new development strategies are important to cope up with the world economy and with this urban population increase. Historically, after World War II, the declination and destruction of the old city centres forced the Global North to embark on developing new forms of economic and social development moving away from industrialization and began restructuring the urban realm following this form of economy. Urban policymakers adopted a cultural economy approach to revitalize the city centres using public funding. To encourage capital flow, job production and overall growth, governments developed private–public partnerships allowing the latter to invest in cultural industries like museums, events, theatres and 'creative industries' attracting what Florida termed as 'creative class'. This social class have caused—in some countries—demographic and cultural changes and gentrification. Moreover, cities started to compete internally and globally to raise funds and to attract more tourism to their cities making them more prone to inequality for the sake of investments which does not necessarily trickle down to other neighbourhoods needing further development (Binns 2004).

The literature or secondary resources helped in developing the main research framework that was discussed in the paper, from which emerged certain key highlights that explain the process of gentrification, arts and culture in urban realm and finally how to determine the human needs in certain context. The paper relied on a physical interview with a representative from Cairo Governorate, a phone interview with a representative from Ismaelia's Real Estate and four physical interviews with artists: a previous photographer for Downtown contemporary art Festival (DCAF) and street photographer. Two illustrators participated in art residency in Cairo's Downtown and finally an artist who has taken part in arts events organized by Ministry of Culture. For the artists, the selection of the interviewees was a challenge. However, to overcome this, the researchers reached out on social media to connect with the artists and conduct the interviews based on the availability and willingness to conduct the interview. Similarly, the field survey faced limitations as people were reluctant to answer any of the questions and refused participation.

Looking into Egypt, according to Eldaidamony et al. (2019) when investigating Fatimid Cairo stated that the gentrification process in Egypt takes on a form of cultural tourism and commercial activities. Such a process induces the exclusive entertainment sector under cover of tourism. When local elites and authorities focus on tourism, gentrification takes place. Since 2012, Cairo's Downtown has been witnessing several arts and culture events and workshops led by the private sector while the local authorities are investing in the infrastructure. The current process has yet to be identified or agreed upon collectively between the stakeholders. Through interviews with local governorate, the private sector and the artists, the vision for Downtown is similar between the stakeholders. However, each stakeholder is achieving it in a different manner. Both the private sector and local authorities view it as a district with a distinctive history that needs to be promoted nationally and internationally while the artists look at it as a hub for arts and culture in Cairo. With that background, the paper aims at determining the role of arts and culture events on Downtown on the current development process along with determining the ongoing urban process and its methodology, focusing on arts and culture given they are new tools for urban development. Most of the current literature is produced either through one of the stakeholders and newspaper articles, or conference documentation which only takes into consideration the events and tools (arts and culture), without investigating the probable consequences.

30.2 Research Background

30.2.1 Urban Models, Economic and Social Concepts

In the 1980s, urban regeneration concept emerged with development projects that aimed at increasing the economic competitiveness of cities. Following that during the twentieth century, Downtown were transformed into central business districts containing housing and retail units. Following deindustrialization of the cities and the inclination towards investing in the suburbs, urban revitalization or rehabilitation emerged to attract shoppers and businesses back to the city centre while maintaining the urban and cultural heritage of the centre (Padilla and Eastlick 2009) where some got gentrified. In both urban models, an important aspect that drove development is to compete internally and internationally with other countries and cities to attract investment and to attract cultural tourists. Cities worked on maintaining and marketing their cultural heritage and attracting more arts and culture activities. As the European Union developed the capitals of culture, more cities are racing to get this status through the arts and culture activities (Aboamer and Abdelfattah 2019).

The first mode of response to the destruction after the war was the construction of high-rise buildings, reconstructing the urban fabric while preserving the urban identity and culture heritage of the city with focus on motor vehicles being the main mode of mobility. According to Bigio and Licciardi (2010) as stated by Roberts and Sykes (2000), urban rehabilitation includes preserving the buildings, reusing building creatively, to consider the district as a whole unit instead of individual buildings and to consider the social and economic aspects of the community. In developing countries, these quarters represent the housing stock for poor immigrants and the low-income residents of the city. This is the result of the deterioration of the buildings and lack of maintenance from the owners or renters, whereby the former receives low amount of rent that is not sufficient to maintain the buildings, and hence, the building value and in return the rent decreases. The idea of rehabilitation continued to grow revive the communities while resolving social issues through developing strategies that combat social exclusion (von Hoffman 2008; Steinberg 1996).

As for urban gentrification, it has been rooted in the social and economic aspect of planning while also depending on culture-led (arts and culture) development. This form of development causes the metamorphosis of the neighbourhood's social and economic class (Lees et al. 2008; Franz 2011). Lees and Slater argued that this metamorphosis attracts a population of labelled as the creative class. This class engages in creating meaningful forms and engages in creative problem solving; it includes scientists, writers, editors, analysts, nonfiction writers, actors, designers, poets, professors, actors. For the Department of Culture, Media and Sport (DCMS), a major entity that assisted in coning the term in UK defined the creative industries like media, film, publishing, recorded music, design, new media, visual art, crafts theatre/cinema, concerts, performance, literature, museums and galleries to revive economies and attract cultural tourism. These issues are later developed to include advertising, architecture, arts and antiques market, craft, design, designer fashion, film and video, interactive leisure software (video games), performing arts (theatre, ballet, musicals and opera) publishing, software and computer services and television and radio. Later to theses, the issues include non-state actors interventions like TV, radio, film, music, books, adverts and concerts. Economically, in cases of gentrification, a common aspect is the rent gap that causes gentrification when the buildings are refurbished and released to the housing market. This also develops an attractive environment for the creative class and the industries (CLUSTER & British Council 2010; Lees et al. 2008, British Council 2010; Florida et al. 2008).

Another important aspect in urban development is human needs into five hierarchal categories—in which earlier needs to be satisfied to move on the next (Abusaada and Elshater 2019). The first is physiological which includes, food, air, water and shelter as a basic needs. Following that safety needs (secure working environment free from threats), then social needs like the need to be loved. Other researchers conclude by arguing that the esteem needs are to be represented in self-respect and approval of others and finally self-actualization which refers to the need to become with other (Ahmed et al. 2019). In this, one is capable of being to develop one's fullest potential. When reflecting upon the needs of users, arts and literature find itself in the fifth through achieving self-actualization that releases the artistic and aesthetic needs in the person. This matter is crucial in defining the needs of the community (Lester et al. 1983).

30.2.2 Case Study Background

Egypt has been a tourist destination through its heritage sites and coastal tourism. Tourism is the leading form of economy in Egypt, where revenues increased to 11.40 USD Billion in 2018 (Tradingeconomies.com 2019). Following the cultural economy development approach, cities within Egypt like Gouna have already developed their own yearly festivals, attracting local and international actors. Cairo's Downtown is known for its distinctive architecture, cultural importance, political importance and its commercial activities in the past (couture, international designer houses), etc.

Cairo's Downtown fits the profile of a neighbourhood that would be rehabilitated, and there have been different efforts to do so by the local authorities. When reflecting on Downtown, there is a stream of events that has great potential to attract further international attention.

In Egypt's 2050 development plan (later Egypt 2052), Downtown was envisioned as a world-class destination of elegance, activity, history, culture and liveable neighbourhoods (Aboamer and Abdelfattah 2019). With the revolution in 2011, the plans were halted and Tahrir square became very politicalized. Later, Cairo Governorate in collaboration with different national, academic and private stakeholders started the façade furbishing efforts and also moving sidewalk vendors to Torgoman, which was met with approval from the residents. Currently, the governorate is working with international entities to develop a bike lane in Downtown. This initiative has enhanced the urban realm through traffic lights, turning some of the passages into pedestrian passages, investing in developing public spaces and have started an initiative promoting the sidewalk as a public space (Rabie 2015; Afify 2015; Shukrallah 2015).

Currently, there is a growing arts and culture initiatives and galleries, as seen in Fig. 30.1 which are fostered by Al Ismealia Real Estate, a key private stakeholder. Al Isamelia Real Estate now owns 33 buildings in Downtown like Tamara building, La Viennoise and others, which is estimated as 5% of the real estate in Downtown (Ismailia 2013). The firm invested in renovating buildings like Cinema Radio in Taalat Harb Street. This firm also have invested multiple cultural festivals like Downtown Contemporary Art Festival (DCAF), leased out multiple spaces for art galleries, used renovated buildings as photography studios (Dakhli West ElBalad) and worked on publishing papers like Mantiqti (my neighbourhood) which is available only in Downtown and only depicts events. Their initiatives have invited other forms of cultural activities like Zawya and Cimatheque (alternative cinemas) (Gibaly 2015; Kingsley and Berger 2014; Shukrallah 2015).

It is also stated that by 2018, Al-Ismaeila will be launching megaprojects combining retails concepts, co-working spaces and an entertainment hub, based on restoring the iconic buildings to their previous grandeur while providing a uniquely modern experience. As an art and culture hub, it contains some of the oldest commercial galleries like Townhouse that has a huge influence on the current art scene. A boom in arts and cultural scene could be noticed through the art competitions that have occurred in these institutions like Rozmana, 100 Posters competition, Cairo Comics fair and others. Downtown has created an online presence through social media, presenting art-related events, tours and promoting images from different photographers of from Downtown ran by orient productions. Orient production is a film and theatre production company whose main missions are to develop cultural projects in the fields of visual and performing arts with a focus on Egyptian and Arab artists. The company also encompasses studio Emad ElD in foundation and manages the program for American University in Cairo's theatre located in Downtown and DCAF.

Fig. 30.1 Arts and culture spaces in Cairo's Downtown. *Source* cuip.com and updated by author

30.3 Methods

The research applies a mixed approach of qualitative and quantitative methods (Hart and Hart 1989). Qualitatively, through literature and semi-structured open-ended interviews as a primary resource and quantitatively through the field and online surveys. The paper discussed the following concepts: urban rehabilitation and then urban gentrification to later understand of the current urban development approach in downtown. Following that, the research discussed the cultural economy or industries to showcase the relationship between current global urban development and development in Cairo's Downtown and its link to arts and culture. The research creative class. These concepts are concluded with Maslow's human needs pyramids to know the real needs of the population in Downtown Cairo.

In terms of interviews, several meetings were arranged with the followings target groups. The authors design open-ended questions to investigate the effect of art in stockholders' development planes. **In Cairo Governorate**, the interview questions was scheduled to determine the governorate's vision for Cairo's Downtown, the current and previous efforts of Cairo Governorate in Downtown, the challenges they face in developing their projects, their relationship with Al-Ismaelia Real-Estate work in Downtown, the possibility of replicating Al-Ismaeilia Real Estate work and collaborating with them in the development of Downtown.

In **Ismailia's Real Estate**, the interview was scheduled to understanding the firm's vision for Cairo's Downtown, the methods they are using to achieve this vision, the reason they choose of arts and culture activities to revive Cairo's Downtown and their different interventions. The open-ended questions of the interview allowed the interviewee to discuss the programmes on which the firm builds its events and achieve their vision.

The interview was scheduled with **artists** to understanding their role in the current events and their observations. Also, their vision of arts and culture and then their participation in Downtown arts events, their views of the events, the stakeholders they interacted with, the changes they have witnessed in Downtown and in the art scene in Downtown, their observations of the events attendees and finally their own vision on Cairo's Downtown.

Online survey and on-disc survey were also conducted to investigate how the users perceive the correlation between art and development. Both surveys aimed at determining the reasons for visiting Cairo's Downtown, their needs from the district, if they have attended any of the arts and culture events and if they have noticed any change in Cairo's Downtown. The survey was conducted on both mediums to sample a large population. For the online survey, 40 people responded as for the field survey; it was conducted on three separate days with a total 15 of visitors and five workers. The survey questions are:

1. Have you witnessed any changes in Downtown Cairo? (between 2010 and 2018), if yes, please elaborate on what kind of changes you have observed.
2. How would you rate these changes based on your experience? (1 is the lowest, and 5 is the highest)
3. Have you attended any of the D-tours organized by Ismailia Real Estate Company or Downtown Contemporary art festival or events organized in the Falaki theatre? If yes, did you enjoy it? (1 is the lowest rank, and 5 is the highest)
4. What are the usual activities that you do in Cairo's Downtown?
5. What do you want to experience in Cairo's Downtown?

30.4 Results and Discussion

Through the interview conducted with Cairo Governorate representative in May 2019, the Governorate aims at allowing people to enjoy Cairo's Downtown and becoming an international attraction following the Cairo 2052 plans. Through the interview held with Ismailia Real Estate representative in June 2019, the real estate firm aims at achieving 'Downtown for all', to revive the Egyptian identity, bring back the glory days of Downtown or Downtown's spirit and to become—as a firm—one of the catalyst for change in Egypt and the Middle East. As for artists interviewed in July 2018, they viewed. Cairo's Downtown as an inclusive space for contemporary arts, culture and other daily life activities. For the surveyed population, it was reported that they can see the changes that Cairo Governorate has implemented in Cairo's Downtown like relocating street vendors, adding traffic lights and enhancing the urban realm.

To sum up the role of arts and culture in Downtown, Cairo Governorate focused on urban infrastructure and public amenities. This was done through renovating the facades in Downtown with the assistance of external funding, developing bike lanes, increasing traffic lights for vehicles and pedestrians and relocating street vendors to allow for better walkability and mobility through Downtown. Cairo Governorate has been attempting to transform passages like El-Borsa, Belhar and El Sherifeen passages into pedestrian passages and aims at developing arts and cultural festivals attracting people and enhancing the economic situation (promoting local products). However, due to limited capacities, this has not been achieved yet. Cairo Governorate developed El Shawarbi passage into a pedestrian area, leaving the same commercial activities within the area. Through the interview with Cairo Governorate, they expressed that the passage still needs further development given its incoherence and the sound pollution that is caused by the behaviour of the commercial activities in the passage.

As for Al Ismaelia Real Estate, they aim at reviving the life in Downtown through injecting arts and culture as they view this model, as a common ground for different stakeholders. They are also investing in refurbishing significant historic buildings; inviting arts and culture activities to take place in these buildings and collaborate with other arts and culture entities. The firm has existing programmes that focus on arts and culture like supporting and financing arts initiatives through cash or subsidizing locations for artists and are creating programmes for arts and culture like SCHNIT (short film festival in Cairo) and supports Arts *d'Egypte* (conducted night at a museum, art city walks, etc.), Cairo Photo Week in collaboration with Photopia (photography hub in Heliopolis) with international participation from different photographers and artists, and finally supporting sports by conducting the first Downtown run with Cairo Runners. Also, working on public spaces like Kodak passage and presenting artefacts in Kodak passage raise awareness on public spaces in Downtown. Finally, reintroducing heritage comes through a contemporary lens through Downtown Contemporary Arts festival (DCAF). According to Cairo Lab For Urban Studies, Training And Environmental Research (CLUSTER) and British Council (2018) as shown in Fig. 30.2, the current activities are causing economic slipover within its vicinity. However, through field surveys, this is unnoticeable.

Based on artists interviewed in July 2018, the event attracts certain social class given the nature of the events which focus on contemporary art vs popular art that is more common that could attract a wider population. DCAF remains very eclectic. The artists agree that Downtown arts and cultural scene was very vibrant before 2011 and returned to its prime between 2014 and 2015. After 2015, the arts and culture competitions and residency decreased in number, and hence, there are limited ways for exposure. One artist stated, 'We want more art events from the international sector

Fig. 30.2 Possible economic spill over from the cultural initiatives and businesses. *Source* (CLUSTER & British Council 2018), edited by authors

like before and more art residencies'. One of the artists stated that 'art has become very exclusive to artists or certain class'. The example is given to a man who works with copper or leather from El-Ghurriya and would not have his work featured in Town House and would not be considered art though his crafts as a very artistic issue. The final artist, when asked about the art scene in Cairo and in particular downtown, he stated that Downtown is the centre for art, given its geographical accessibility to Cairenes and people coming from other governorates. He disagrees when it comes to the decline in arts scene. He views it as growing, especially the comic art scene. A comic three days event is held every year, which is one of the consistent events for the past six years. According to the artist, it attracts people from different social classes. However based on observation, it attracts a small certain social class, and this was confirmed by some of the interviewed artists.

As for the online and field survey, respondents of the two methods reported that they had not attended any of the arts and culture activities. Minor remarks were

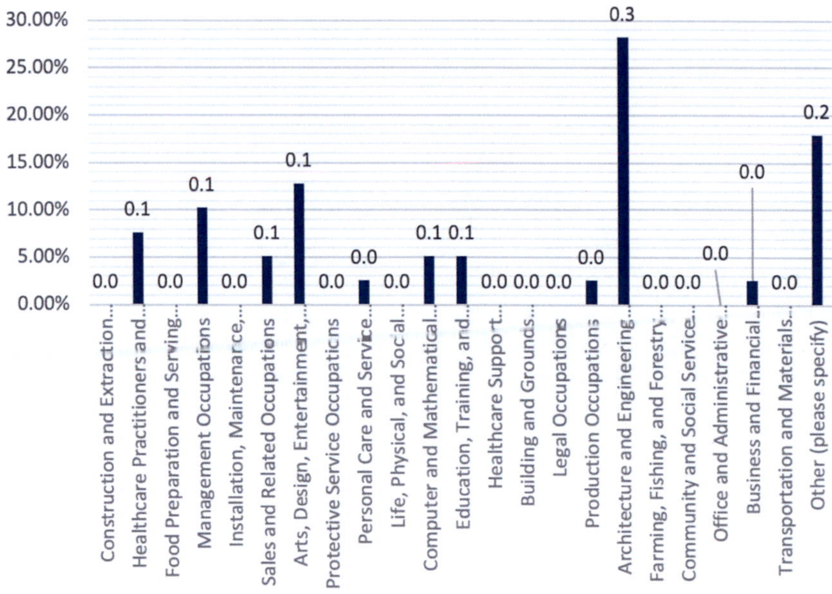

Fig. 30.3 Responses to the question 'Do you witnessed any changes in Cairo's Downtown?'

noticed by the authors regarding the response to the changes happened in Downtown. Figure 30.3 shows that architecture style is main actors in changes happen in Downtown. From the online survey, 2 out of 40 have visited galleries. When asked about their needs for Downtown, they requested more places spaces and more security against harassment. Building on this, these results contradict the spill over effect that CLUSTER and the British Council would suggest. Though the interventions have caused small art pockets, the economic influence is still not measurable and is not recognized by the field-surveyed population.

The authors figure out the response to the question of what the usual activities are in Downtown using Pareto chart (Wilkinson 2006). The events that the users participate in the site were innumerable; some users select the given another choice, but others prefer to add explicit discretion of other activities such as visiting friends or walking around (Fig. 30.4). The results of questioning how users can rate their experience in site showed that it is somehow preferable with a notable remarked of pedestrianisation movements.

Cairo Governorate representative in May 2019 is aiming at attracting everyone into Downtown and not just the current population. They aim at making it an international destination. As for Ismailia Real Estate (June 2019), the firm is working towards attracting the creative class to Cairo's Downtown while maintaining the current social fabric and developing an inclusive city centre. The firm is attracting the creative class through supporting entrepreneurial initiatives through RISE UP Summit that is held on yearly basis and working with local entrepreneurs like up fuse on making Downtown go green through reusing plastic and decorating Kodiak Passage. As for

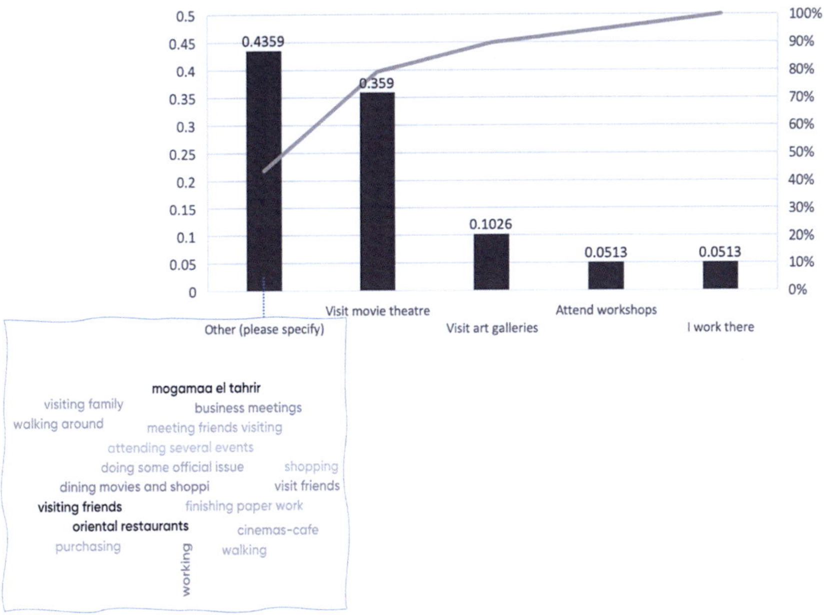

Fig. 30.4 Response to 'What are the usual activities that you do in Downtown?'

downtown, workers and visitors are not aware of the events and neither do they attend it.

The researchers here noticed a remark related to the land vale. Cairo Governorate representative in May 2019 stated that the current apartments are occupied with small companies and firms that pay a very low amount of rent that is not sufficing to maintain the building structure. The housing market and the concerns that it would drive the residents out of the city centre along withMaspero's redevelopment market would indeed place pressure on the housing market, but the governorate is leaving this aspect to the free market. The current rent in Downtown actually varies between 5000 and 12,000 LE per month.[1] When discussing the same concept of land value with the Ismailia's Real Estate in June 2019, the firm follows a business model where furbishing the building will raise its value. The artists (July 2018) have stated the rent is increasing, which coincides with the Cairo Governorate's statements.

The authors also noticed another remark regarding governance. Ismailia's Real Estate is part of the board of National Commission for the Development and Protection of Heritage Cairo. These local authorities and Cairo Governorate are part of it where the development of Downtown is discussed their efforts. However, they are yet to be effectively coordinated with the local governorates. Cairo Governorate Representative in May 2019 would be interested in implementing the same interventions as Ismailia's Real Estate. However, given the current capacity and the daily issues

[1]Which is almost equivalent to Dokki, Mohandessin and Heliopolis based on social media groups and Airbnb.

that face the governorate, this is still not feasible. The governorate would not object a collaboration where a win–win situation would be achieved. An important aspect of governance is that the artists are not present or aware that they are presented in this commission.

Finally, the current interventions in Cairo's Downtown are natural development method of any city and are the first step to place Cairo as an international cultural attraction competing with other cities in the Region. Gentrification is natural and inevitable processes that aim at enhancing the living environment, the economic and social situation.

The present research highlighted the pressures on cities of Global South to diversify their economy, along with the forces of competing internationally for positioning. In this, it is necessary to start fostering creative industries, and old central business districts make a viable place. It also needs to be pointed out that the private sector is merely attempting to benefit from existing initiatives and providing a space for them. It could be feasible for a private–public partnership to develop and work collectively on developing Downtown. This partnership could play a crucial role in stabilizing the housing market, in developing economically feasible projects that build on the existing creative industries and upscale them and maximize the reach of the arts and cultural events to reach other communities existing within Downtown or in neighbouring districts. Recently, international agencies like UN Women are working with local NGOs and artists to develop educational theatres that target refugees in Cairo and produce movies sheds light to harmful habits like female genital mutilation. This development needs to be investigated in further details to address any negative implications.

Finally, the results here confirmed the importance of documenting the ongoing process objectively and work on developing city and neighbourhood scale indicators and signs for gentrification and another development model. This exercise would result in documenting the process and its outcomes, allowing further analysis to upscale this form of development. For successful mapping, stakeholder consultation is essential, working with international agencies on data collection on the influence of arts and culture on Cairo's Downtown (i.e. propose and adopt the indicators and data developed by UN-Habitat Global Urban observatory). Attaining data would allow the stakeholders to provide an evidence-based strategic plan for development using arts and culture for Cairo's Downtown.

30.5 Conclusion

Cairo's Downtown is currently undergoing a culture-led development process through the private sector; targeting physical, social and economic aspects. The district is undergoing cultural gentrification process conducted by the private sector through using arts and culture as a tool for attracting national and international creative class and subsequently attracting culture-oriented tourists. Parallelly, the local authorities are implementing infrastructure projects to enhance the walkability of the neighbourhood, increasing its connectivity, investing in bike lanes and in public spaces like Al-Alfi, with the aim of making Downtown an international vibrant attraction. The efforts of the local government in enhancing the urban realm. The efforts of the private sector, that resulted in creating bubbles of arts and culture where the nuclei are the events/institutions; left Downtown with a great potential to be demographically gentrified.

The private sector development presence is exclusive in nature. This was determined through the interview with the private sector where it stated that it aims at attracting the creative class through its events and activities and through fostering further arts and culture events. This was further strengthened by the interviews with the artists that confirmed that upper-middle-class population could be observed these events. Through the field and online surveys, the visitors of Downtown did not attend or recognize the events fostered by Ismailia's Real Estate; because it does not match their interest in terms of the topic being presented through such activities. The visitors reported going to cinemas in Downtown to watch commercial Arabic films, have a walk and have coffee in a local café *'ahwa* in Arabic'.

Qualitatively, the research determines that the gentrified and gentrifying population is unaware of the actions needed from its end to cope up with the process. According to Maslow's needs, pyramid that depicts that the basic needs of the population need to be met before the population could show interest in arts and culture. The stakeholders need to collaborate to meet the needs of the population like investing in the infrastructure in neighbouring areas and finally collaborate on employing artists to change the cultural perception of the possibly gentrified population; to achieve an inclusive urban development of Downtown that does not cause any forms of displacement or create social exclusion.

Quantitively, it is important that future research aims at gathering further data on land value changes and the increase in arts and cultural business. This would assist in developing an evidence-based strategic development plan for Downtown. It is also important to develop social policies that target the vulnerable population inside Downtown and in neighbouring areas to reduce the consequences of gentrification, e.g. displacement and to minimize the effects of placing refurnished buildings onto the housing market. Along with that, it is important that the National Committee for the Preservation of Cairo's Downtown Board of trustees of Historic Cairo should have a stronger and coordinating role to the interventions of the different stakeholder and to develop a strategic plan for Downtown.

A future direction of research can be extended to investigate in depth the private sector interventions are forming pockets of arts and culture, attracting the creative class that will cause culture gentrification. It is recommended for future research to acquire this data and to further map the arts and culture in terms of kind, audience, outreach and impact.

References

Aboamer M, Abdelfattah D (2019) Cairo's Downtown in narratives: from singularity to hardship. In Abusaada H, Vellguth C, Elshater A (eds) Handbook of research on digital research methods and architectural tools in urban planning and design. IGI Global, pp 356–381. https://doi.org/10.4018/978-1-5225-9238-9.ch016

Abusaada H, Elshater A (2019) Effects of urban atmospheres on changing attitudes of crowded public places: an action plan. Int J Commun Well-Being, ahead-of-print. https://doi.org/10.1007/s42413-019-00042-w

Ahmed N, Elshater A, Afifi S (2019) Community participation in design process of livable streets: ethnographic analysis in Cairo. In: InterSol2019. Springer, Cairo

Afify H (2015) Downtown lost or reclaimed. Retrieved 14 Apr 2016, from Mada Masr website: https://www.madamasr.com/sections/politics/downtown-lost-or-reclaimed

Bigio AG, Licciardi G (2010) The urban rehabilitation of medinas: the World Bank experience in the Middle East and North Africa. World Bank 3

Binns L (2004) Capitalising on culture: an evaluation of culture-led urban regeneration policy. Futures Acad. Retrieved from https://arrow.dit.ie/futuresacart/5/

CLUSTER & British Council (2018) Creative initiatives: economic impact on Downtown Cairo

El-Gibaly L (2015) Creative cities_can downtown accommodate the many visions for its future. Retrieved 14 Apr 2016, from Mada Masr website: https://www.madamasr.com/sections/culture/creative-cities-can-downtown-accommodate-many-visions-its-future

Eldaidamony M, Shetawy AA, Serag Y, Elshater (2019) Adapting geographies of gentrification in egypt: lesson learned from fatimid cairo and heliopolis. In: Hawkes D, Bougdah H, Rosso F, Cavalagli N, Ghoneem MY, Alalouch C, Mohareb N (eds) Conservation of architectural heritage. Springer Nature Switzerland AG, pp 49–64. https://doi.org/10.1007/978-3-030-10871-7_5

Florida R, Mellander C, Stolarick K (2008) Inside the black box of regional development-human capital, the creative class and tolerance. J Econ Geogr 8(5):615–649. https://doi.org/10.1093/jeg/lbn023

Florida R (2014) The creative class and economic development. Econ Dev Q 28(3):196–205. https://doi.org/10.1177/0891242414541693

Franz Y (2011) Gentrification trends in Vienna. In: Urban sprawl and gentrification, pp 187–208

Hart KM, Hart RF (1989) Quantitative methods for quality improvement. ASQC Quality Press and Santosh, Milwaukee, WI (Pre Press)

Kingsley P, Berger M-J (2014) Why the battle for control of Cairo's Downtown is a fight for the future of Egypt. The Guardian. Retrieved from https://www.theguardian.com/cities/2014/nov/03/battle-control-downtown-cairo-fight-future-egypt?CMP=share_btn_fb

Lees L, Slater T, Wyly E (2008) Gentrification. Routledge

Lester D, Hvezda J, Sullivan S, Plourde R (1983) Maslow's hierarchy of needs and psychological health. J Gen Psychol 109(1):83–85. https://doi.org/10.1080/00221309.1983.9711513

Padilla C, Eastlick MA (2009) Exploring urban retailing and CBD revitalization strategies. Int J Retail Distrib Manag 37(1):7–23. https://doi.org/10.1108/09590550910927135

Rabie P (2015) The changing face of downtown. Retrieved 14 Apr 2016, from Mada Masr website: https://www.madamasr.com/sections/politics/changing-face-downtown

Roberts P, Sykes H (2000) Urban Regeneration-a handbook. In: Roberts P, Sykes H (eds) Sage

Shukrallah S (2015) Renovating Cairo's Downtown: contested visions of the heart of the city. Al Ahram

UN-Habitat (2016) Urbanization and development: emerging futures. In: UN Habitat world cities report 2016. Retrieved from https://wcr.unhabitat.org/main-report/

von Hoffman A (2008) The lost history of urban renewal. J Urbanism 1(3):281–301. https://doi.org/10.1080/17549170802532013

Wilkinson L (2006) Revising the pareto chart. Am Stat 60(4):332–334. https://doi.org/10.1198/000313006x152243

Part V
Climate-Smart Cities

Chapter 31
Spatial Morphology and Urban Heat Island: Comparative Case Studies

Esraa Elmarakby, Marwa Khalifa, Abeer Elshater, and Samy Afifi

Abstract Urbanisation is a process associated with the global industrial and technological development. Globally, this process causes a continuous change of the land cover and increases the urban areas. Scholars from various disciplines address these changes that affect the global climatic system, microclimate and the environment. Urban heat island (UHI) is considered as one of the climatic change phenomena which are associated with the urbanisation process. This phenomenon increases the air temperature in the urban areas compared by the surrounding rural ones. This makes up for the problem of overlook the local scale analysis of the UHI and focuses on the mesoscale. This paper aims to explore the relationship between the urban structures in residential areas of Cairo and the UHI. The research methods use remote sensing techniques and field monitoring devices to catch the variances in the air temperature. The key finding figured out the importance of considering the configuration of urban patterns in development plans. Future research in studying the phenomenon is needed for addressing the exact effect of each urban morphology parameters on UHI and analysing the spatial and temporal distribution of the phenomenon throughout various periods of the different the seasons.

Keywords UHI · Climatic change · Remote sensing · Geographic Information System (GIS) · Cairo

E. Elmarakby (✉) · M. Khalifa · A. Elshater · S. Afifi
Faculty of Engineering, Ain Shams University, Cairo, Egypt
e-mail: esraa.elmarakby@eng.asu.edu.eg

M. Khalifa
e-mail: marwa_khalifa@eng.asu.edu.eg

A. Elshater
e-mail: abeer.elshater@eng.asu.edu.eg

S. Afifi
e-mail: samy.afifi@eng.asu.edu.eg

Architecture and Urbanism: A Smart Outlook,
https://doi.org/10.1007/978-3-030-52584-2_31

31.1 Introduction

The rapid urbanization and the urban periodical growth exert a noticeable influence on the climate. Urban Heat Island (UHI) is a phenomenon resulting from increasing the temperature of the urban areas compared by the rural areas (Landsberg 1981). UHI exists in the urbanized areas and derives its intensity from the cities' urban morphology with their hard surface and arid nature which make them hotter than their surroundings (Moughtin 2005; Srivanit and Kazunori 2011; Zhou et al. 2017). UHI leads to high energy demand due to the cooling loads which needed to reduce its effect, especially in the hot weather events, and also effects on the human discomfort (Akbari et al. 1988). Most researches made on the Cairo context are targeting the spatial and temporal existence of the urban heat island at a regional scale using remote sensing techniques (Ghoniem 2009). UHI temporal and spatial distribution are also associated with the urban composition, characteristics and morphology of the cities which should be addressed at its local for better understanding and analysing of the phenomenon existence and intensity (Paravantis et al. 2015; Stewart 2011).

This research aims at studying the differences in the morphological patterns of the residential areas in Cairo and its relationship with UHI intensity at the local scale (Fig. 31.1). The research analyses the temporal and spatial change of UHI using remote sensing and GIS applications. The research uses here the field monitoring measurements to address the influence of the urban morphology of each zone on UHI and to find out the relation between the phenomenon and the urban composition.

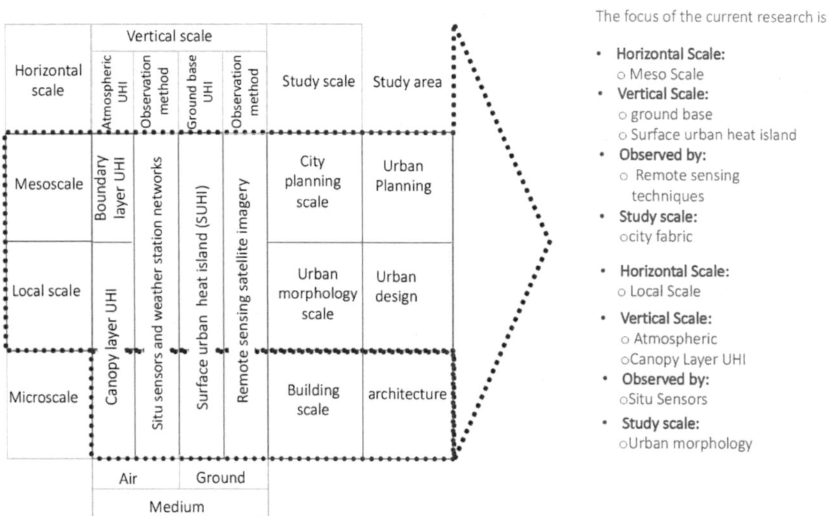

Fig. 31.1 Classifications of the Urban Heat Island (UHI)

31.2 Literature Review

31.2.1 *Classification of the Urban Heat Island*

The first detection of the urban heat island (UHI) was investigated by Luke Howard in 1820 (Mills 2008). From that time and until 1993, UHI was detected at its macro-scale and defined as the increase of the air temperature in the cities compared by the surrounding rural area (macro UHI) (Stewart 2002). In 1993, Michael and Aniello classified UHI into two basic types: macro-urban heat island and intra-urban heat island. They describe the intra UHI as hot spots existed in the city context (Aniello et al. 1995; Moughtin 2005). The intra-UHI various according to the scale it being observed at to vertical scale UHI and horizontal scale UHI (Zipper et al. 2016). The research detects each type using different methods and tools. The observation takes place through various mediums: Air-Ground (Stewart 2011; Williams et al. 2013).

The researchers define the horizontal scale of the urban heat island as the area at which the phenomenon observed (Fig. 31.1). The scale of observation includes three types: microscale, local scale and mesoscale. The microscale of UHI deals with the composition of a single street or covers an individual building (area less than one metre and extends to hundreds of metres). The local scale represents the neighbourhood scale. This scale relies on the urban morphology, characteristics and composition. Mesoscale represents the scale of the whole city and wide spreads to several kilometres (Oke 2006).

The vertical scale of UHI is sorted according to the medium through which the phenomenon detected to atmospheric type and ground type (Oke 1973; Weng 2012). The atmospheric UHI involves the boundary layer and Canopy layer UHI. The canopy layer urban heat island (CLUHI) begins from the ground surfaces and extends up to the mean building heights (Gartland 2008). Boundary-layer urban heat island (BLUHI) is located above the canopy layer, which represents the portion of the planetary layer (Oke 2006). The surface urban heat island is detected at the ground surface level and reflects the surface temperature (Fig. 31.2) (Weng 2012). Figure 31.3 summarizes the different types of the urban heat island with its horizontal and vertical scale besides highlighting the core type that the paper is going to study. This paper investigates the mesoscale of UHI, which exists at the urban scale and represents the city urban fabric. The researchers here measure UHI at the ground medium (surface layer) using remote sensing and GIS applications and study the local scale UHI at the air medium (canopy layer UHI) using field measurements.

31.2.2 *Urban Heat Island and the Urban Morphology*

Studies about urban morphology address the structure of the human settlements with analysing their formation and transformation process (Caliskan and Marshall 2011; Shane 2019). Urban morphology deals with the spatial structures and characteristics

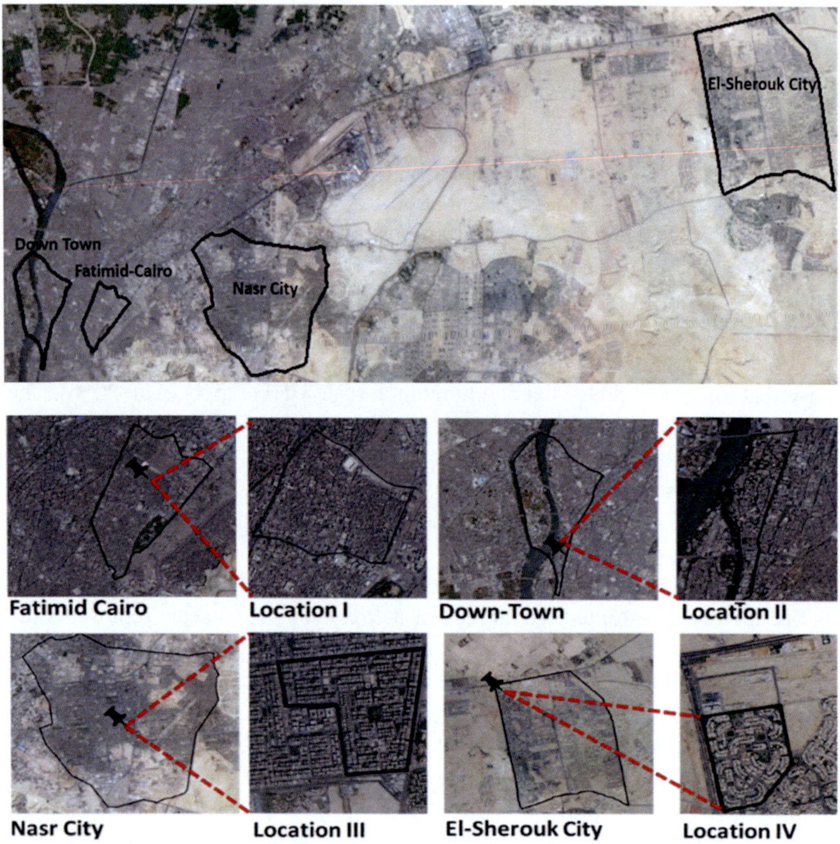

Fig. 31.2 Locations of the study areas

of metropolitan areas, cities, towns or villages. It studies the way through which cities grow, sprawl and changes their features (Kropf 2017; Farag et al. 2019). It opens the gate for studies related to different aspects of the urban environment such as urban microclimate, solar potential and the atmospheric environment where factors like built-up density, buildings heights, street pattern, land use and surface cover play a key role in controlling the urban climate and the heat island effects (Landsberg 1981; Oke 1973; Shahraiyni et al. 2016).The materials of the building like concrete, asphalt roads, huge areas of the pavements, narrow streets, the low urban permeabilities, tall buildings and more are shaping the urban morphology which suffers from the UHI existence. Accordingly, studying the influence of the urban morphology on UHI is an essential demand for better understanding of the relation between them (Boukhabla et al. 2013). This paper investigates the effect of the urban morphology on UHI phenomenon in Cairo's residential areas.

31.3 Materials and Methods

31.3.1 The Study Area

Cairo is the largest city in Africa and the Middle East and the 17th largest metropolitan area in the world, with a population of over than 20 million (CAPMAS 2019). It is located at 30° 2′ 39.4116″ N and 31° 20′ 24.0072″ E. Cairo shows a rapid urbanization process in the last decades with a highly increasing in the built-up areas. The urbanization processes in Cairo is associated with building many housing projects, which play a significant role in forming Cairo's urban fabric. Divergent and successive eras which passed through Cairo influenced on its planning. Each era reflected its own urban characteristics on Cairo urban fabric. Through analysing Cairo urban pattern, it is investigated that Cairo has unique dissimilar urban patterns formed during the distinct political and economic periods which passed through it. This research aims to study the effect of these different urban patterns on UHI. In this study, four locations were assessed by analysing their signature patterns and monitoring their different temperature variances. Fig. 31.2 shows these areas that comprise Fatimid Cairo (969–1798), Downtown (1798–1952), Nasr City (1952–1987) and El-Sherouk City 1987–now). The selected areas have a different location from the city business discotic (CBD). Downtown is considered as the CBD of Cairo. Fatimid Cairo is the nearest location to the CBD. Nasr City is at the border of Cairo, and El-Sherouk City is considered as Cairo's suburban. The methods used in this research follow the guidelines provided by Oke (1973) and Stewart (2002) using a three-step process.

31.3.2 Data Collection and Processing

This research depends on two methods to analyse UHI quantitatively. The methods vary according to the scale at which the phenomenon is observed and the techniques used in the observation. The first method is investigating UHI spatial distribution at the mesoscale (city scale) using remote sensing technique to evaluate UHI percentage in each study area and to provide the ability for the comparison between the four-study areas. Figure 31.3 shows the points of temperature monitoring in each zone. The second method depends on recording the temperature using field measurements at the local scale (neighbourhood scale) inside each study area and makes a matrices comparison between them to investigate UHI intensity.

In this paper, the quantitative analysis was used to investigate the relationship between UHI and the urban structure. It is represented in the statistical calculations, the records of the air temperature and UHI intensity calculations. To investigate the spatial distribution of UHI at the mesoscale, remote sensing techniques was used. Satellite images have many sources that can be downloaded from. The one which used in this analysis was downloaded from Landsat8 website channels (Earth Explorer).

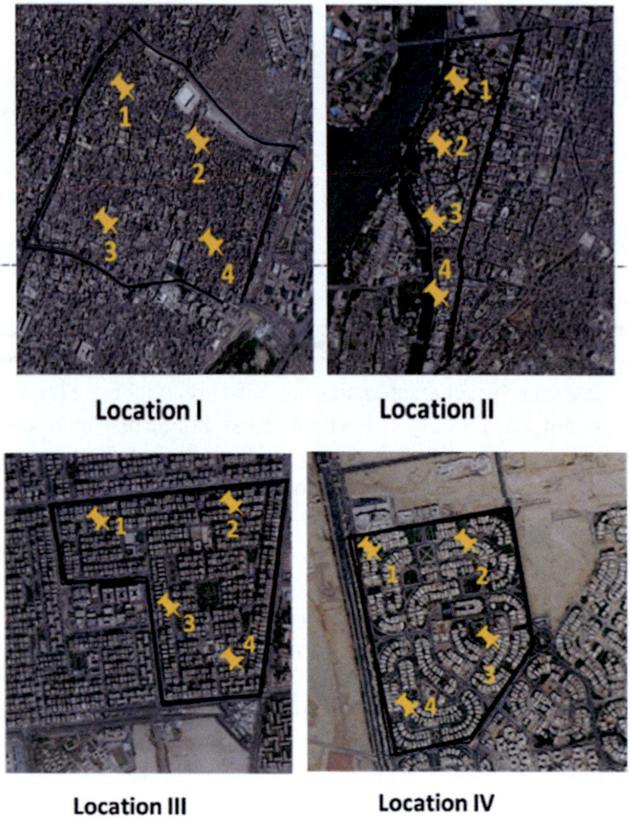

Fig. 31.3 Points of temperature monitoring in each location

GIS software was used to analyse the satellite image in order to obtain the land surface temperature and to investigate UHI percentage (Abusaada et al. 2019; Srivanit and Kazunori 2011). Table 31.1 shows the details of the satellite image, which are used in the analysis of case studies.

The remote sensing method analyses the satellite image through calculating the surface temperature (ST) and then measuring UHI percentage in each case study to compare between them. The present research follows equations to retrieve the ST (U.S. Geological Survey 2016):

Table 31.1 Details of the used Landsat 8 image

Satellite	Sensor	Acquisition time	Path and row	Spatial resolution	Cloud cover
Landsat 8	OLI-TIRS	5.5.2019	176/39	100[a]	2.85

[a]Resolution is resampled to 30

$$L_\lambda = M_L * Q_{\text{cal}} + A_L \tag{31.1}$$

where L_λ = spectral radiance [converting the digital numbers DN of the images to radiance values (W/(m^2 * sr * μm))]; M_L = radiance multiplicative scaling factor for the band from the metadata file attached with the image; A_L = radiance additive scaling factor for the band (from the metadata); Q_{cal} = Landsat band.

$$T = \frac{K_2}{\ln\left(\frac{K1}{L\lambda} + 1\right)} \tag{31.2}$$

where T = Top of atmospheric (TOA) brightness temperature, in kelvin; $L\lambda$ = Spectral radiance [W/(m^2 * sr * μm)]; $K1$ = thermal conversion constant for the band from the metadata; $K2$ = thermal conversion constant for the band from the metadata.

* These previous steps repeated twice representing the Landsat thermal Band 10 and Band 11.

$$\text{UHI} = T \geq \mu + \sigma \tag{31.3}$$

where UHI is the urban heat island temperature; T = brightness temperature; μ = the mean temperature; σ = standard deviation.

These equations measure UHI the areas where the temperature is one or more standard deviation above the mean temperature (CCVA 2015). The four-study areas: EL-Sherouk City, Nasr City, Fatimid Cairo and Garden City were compared according to UHI percentage existence in each area to correlate the intensity of the phenomenon to each type of urbanization of the four areas.

The second method depended on studying UHI at the local scale using the temperature detecting devices to catch the air temperature in the field and correlate the temperature variances to the urban structure of each study area. The local scale is defined as an area of 1 km * 1 km as a sample of each zone (see Fig. 31.3). The urban morphology of each sample was defined according to the Local Climate Zones (LCZ) approach. LCZs is a systematic classification produced by Oke (1973) and Stewart and Oke (2012) to be used in the urban heat island studies. It is an analytical method to categorize each urban form according to its surface cover and manufactured materials with human life that extends to several kilometres in the horizontal dimension (neighbourhood scale) (Weng 2012). Consequently, a sample is selected from each location (i.e. Fatimid Cairo, Downtown, Nasr City, El-Sherouk City). Each sample expends to the area about one km^2 which represents the neighbourhood scale.

The classification depends on sorting the urban form and its landscape to 17 standard classes to allow the comparison between the different urban fabric and their own climate). Observatory data sheets were used to identify the urban pattern of each location (Fig. 31.4). This sheet involves three sectors. The first is zone definition with a brief of the urban features in the studied area. The second is zone illustration using photographs of the area, either high angle or eye angle. The third is the urban properties which describe each physical urban components of the urban composition

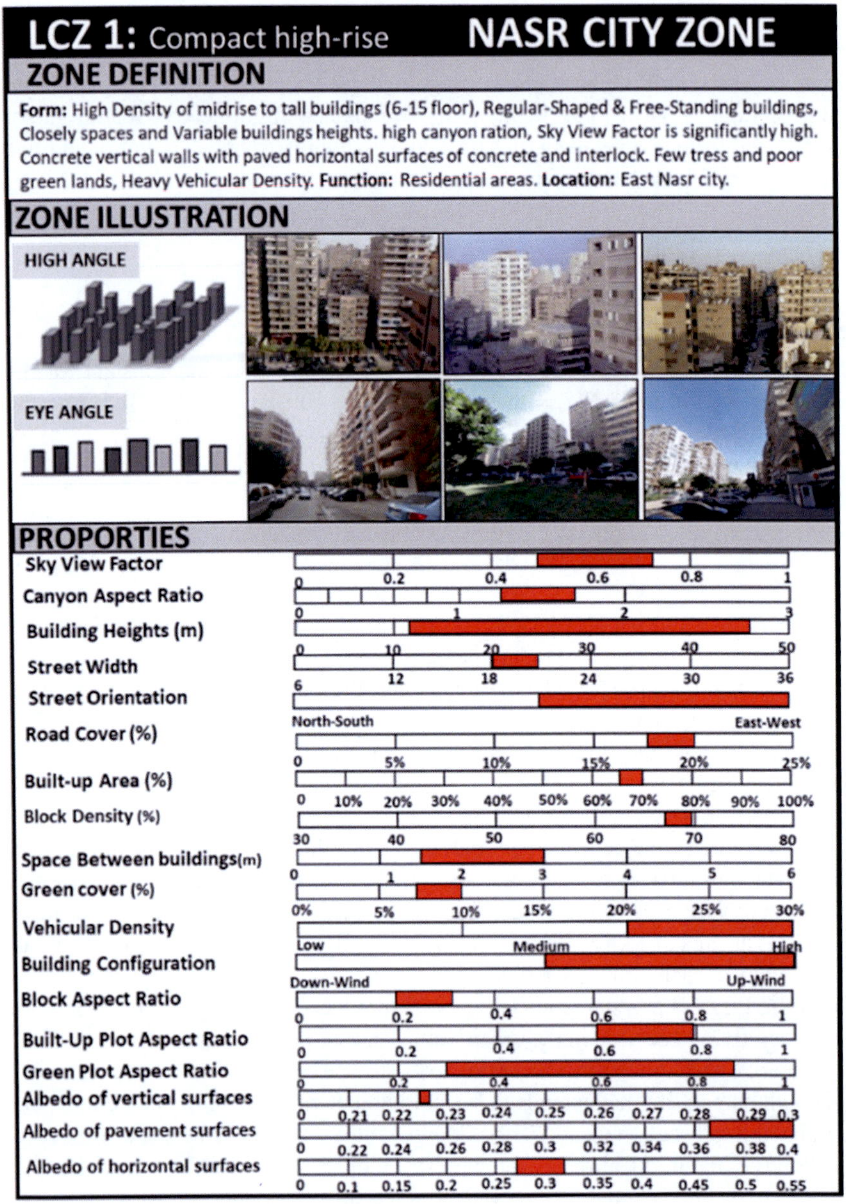

Fig. 31.4 Sample of the observatory datasheet

(Yvonne 2015; Kühl 2011). After defining the urban characteristics of each zone, each zone is classified according to the LCZ.

The air temperature was recorded in the site using temperature thermometer with LCD screen and accuracy one decimal with 0.001% error ratio. Four points in each location were detected, and their temperatures were observed in the same day of Landsat 8 image picking (5. 5. 2019), (Fig. 31.6). The researchers observed the temperatures at the same time in the four locations between 1:00 and 2:00 pm as it is the peak heat hour in the day. The intensity of the urban heat island is measured as the variances between the different local climate zones using the following formula:

$$\Delta T_{\mathrm{LCZ}} = T_{\mathrm{LCZ}\,X} - T_{\mathrm{LCZ}\,Y} \tag{4}$$

where ΔT is the temperature differences between the LCZ (X) and the LCZ (Y). As the more the ΔT increases, the more the effect of the urban form on UHI between the different zones is more significant (Stewart 2011; Stewart and Oke 2012; Yvonne 2015).

31.4 Results and Discussion

31.4.1 Addressing UHI

The land surface temperature (LST) for the four zones: Fatimid Cairo; Downtown; Nasr City, El-Sherouk City at their mesoscale which was obtained via remote sensing and GIS shows variance in UHI intensity (Figs. 31.5 and 31.6).

The investigation shows that Nasr City is the location that suffers from UHI strongly in comparison to the others due to the existence of the undeveloped lands which lifted over by their owner for financial concerns and the desert lands which exist in Almaza airport, Military areas, Alwafaa Wal Amal industrial zone, etc. Those undeveloped and desert lands are the main factors behind the high intensity of UHI in Nasr City. These also explain the existence of UHI percentage in El-Sherouk City which has a lot of the underdevelopment areas (areas under constructions). UHI in the Downtown zone is expressed mainly in its location in the heart of the built-up environment with its high density of vehicular capacities. In Fatimid Cairo, UHI zones are located in the highly condense urban form existed in Bab El-Nasr area with its highly condense urban fabric and very low urban permeability. Nasr City is the most significant zone that suffers from UHI effect, with a percentage of 30.3% while Downtown is the lowest zone suffering from UHI 23%. Fatimid Cairo suffers from 26.3% and El-Sherouk City agonises from 30.3% (see Fig. 31.5).

Remote sensing analysis cannot give a full understanding of the phenomenon, as shown in the map's analysis. The present research speculates that might be due to the effect of the existence of the desert lands within the built-up areas. This does seem to depend on the influence of the surrounding environment on the study areas. One

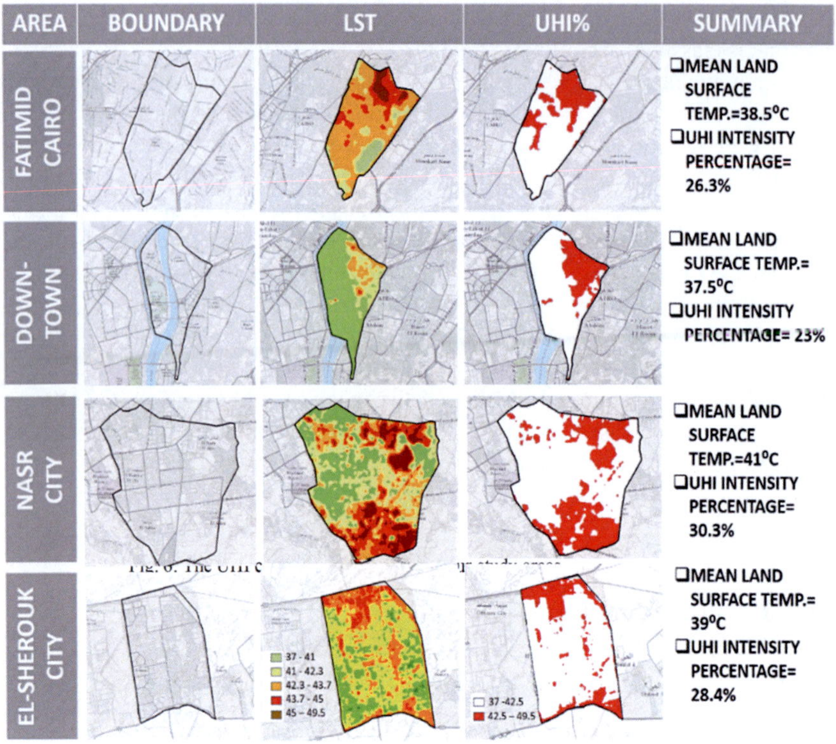

Fig. 31.5 UHI intensity in the four-study areas

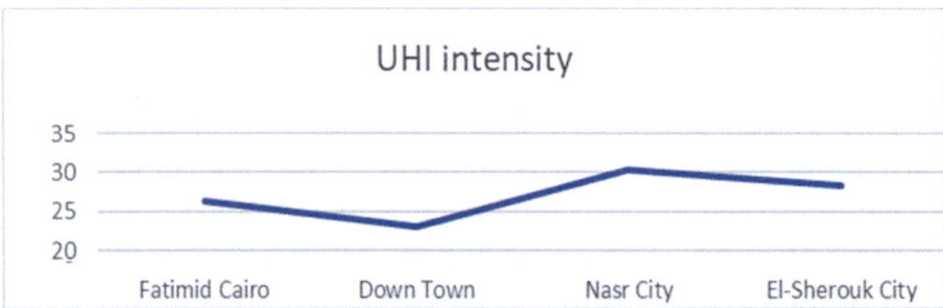

Fig. 31.6 The UHI comparison between the four-study areas

concern about the findings of using remote sensing was that it cannot give a proof of the real effect of the urban fabric on the UHI. It is by now generally accepted that there was a need to study the local scale of UHI (i.e. neighbourhood scale). This is important to illuminate the previous factor in investigating the real existence of the phenomenon.

31.4.2 *Urban Heat Island Intensity and the Spatial Morphology*

To classify the urban morphology of each zone, samples of the locations at the neighbourhood scale were observed (see Fig. 31.3). It is shown that the different urban forms of the four locations specify a certain local climate zone (Table 31.2). Fatimid Cairo location (I) is classified as a combat low rise (LCZ 3) which is characterized by its high-density bricks and stone-attached buildings, with no spaces. In this site, streets have a high canyon aspect ratio (H/W) of the interlocking pavement. The sky view factor (SVF)—the amount of the sky visible from a point at the street—is relatively low, no green lands and vehicular low density.

While Downtown is combat mid-rise (LCZ 2) with high-density concrete buildings, free-standing buildings with spaces of 1–2.5 m, high canyon aspect ratio of the asphalt-paved streets, concrete pavements and SVF are slightly low. It is obvious that there is a very low spread of green lands and high vehicular density. The form of Nasr City (LCZ1) has a high density of mid-rise to tall buildings (6–15 floors). This site is characterized by freestanding buildings with closely spaces. The canyon ratio and sky view factor are significantly high. It has concrete buildings, few trees, poor green lands and heavy vehicular density.

El-Sherouk City (LCZ 6), which is classified as open mid-low rise, is characterized by its low-density freestanding mid-rise concrete buildings. Theses building are ranging from three to four stories with spaces of 4–6 m with spacing between the building, low canyon ratio of the asphalt-paved streets with interlock and precast concrete pavements. The sky view factor is significantly high. In sum, each location shows temperature variance different from the others when the air temperature measured (Tables 31.3 and 31.4).

This study shows that UHI investigation is needed to be studied on the two levels (mesoscale and local scale) to get the full understanding of the phenomenon. On one side, remote sensing can be misleading as it translates all the underdevelopment and

Table 31.2 Local climate zone classifications according to the selected cases

Fatimid Cairo (Location I)	Downtown (Location II)	Nasr City (Location III)	El-Sherouk (Location IV)
LCZ3	LCZ 2	LCZ1	LCZ6

Table 31.3 Temperature records through the four locations

Fatimid Cairo (Location I) (°C)	Downtown (Location II) (°C)	Nasr City (Location III) (°C)	El-Sherouk City (Location VI) (°C)
31.0	34.7	35.2	32.9
31.2	34.2	35.1	32.8
31.3	34.5	35.1	33.0
31.0	34.7	35.0	32.9

Table 31.4 UHI intensity between LCZs

Location	LCZ	Description	Mean temperature (°C)	UHI using the mean air temperature difference of LCZ (ΔT LCZ $X-Y$) (°C)
Fatimid Cairo (Location I)	3	Compact low-rise	31.125	ΔT LCZ 3–2 = −3.400 ΔT LCZ 3–1 = −3.950 ΔT LCZ 3–6 = −1.775 ΔT LCZ 2–1 = −0.550 ΔT LCZ 2–6 = 1.625 ΔT LCZ 1–6 = 2.175
Downtown (Location II)	2	Compact mid-rise	34.525	
Nasr City (Location III)	1	Compact high-rise	35.075	
El-Sherouk (Location VI)	6	Open low-rise	32.900	

desert lands as the main core of UHI existence without giving the real proof about the effect of the urban composition on UHI. But on the other side, studying the local scale samples using the field monitoring measurements provides a suitable analysis for UHI existence. Fatimid Cairo (location I) is categorized as LCZ 3 and has the lowest air temperature in comparison with the three other locations.

The highest difference between the different LCZs is between LCZ 3 and LCZ 1, which means that the temperature in Nasr City location with its compact high-rise urban form is higher than Fatimid Cairo location by 3.950 °C. While the differences between Downtown location and the Fatimid Cairo location is the lowest with a variance of 0.550 °C, it is noticeable that Nasr City location with its condense urban form and high-rise buildings provides the worst-case example between all locations in storing heat and creating UHI effect around its forms.

Otherwise, Fatimid Cairo, with its combat low-rise pattern, shows a good example for providing a cool island effect in comparison with other locations. To some extent, El-Sherouk City, with its open low-rise form, also records a slightly low temperature. Downtown location still the nearest fabric of urban design which classified as a mid-rise compact and also suffered from UHI effect. Meanwhile, El-Sherouk City is considered a suburban area located beside the desert and is affected by it, but it has a lower UHI effect. However, Downtown and Fatimid Cairo are the nearest locations to the CBD; they have a temperature lower than Nasr City, which is at the broader of Cairo.

This research proofs that not only the location of the urban area is the key factor controlling the UHI, but also the urban morphology, which has the most significant influence on the existence of UHI. A number of questions regarding UHI remain to be addressed. A still unsolved question is the verification using measurements in different seasons in studying UHI. A more systematic and theoretical analysis is required for giving a full understanding of the phenomenon of spatial and temporal distribution.

31.5 Conclusion

The paper concludes by arguing that UHI is a significant phenomenon threatening the urban areas in contrast with the rural. This allows the conclusion that Cairo—as one of the biggest cities in Africa—suffers from UHI effects. The analysis of the case studies leads to the following outcomes. First, the divergent urban patterns in Cairo play a noteworthy role in creating UHI effects. Second, the contribution of this phenomena in the existence of the hot spots in the city context. Third, many methods are used to analyse UHI, but good research should not rely on one method. It is by now generally accepted that the reason behind using each method has a gap. In the current study, the research methods examined case studies using remote sensing and field measurement. As it is generally agreed that urban morphology of CBD is the most influencing factors affecting the absence of the UHI. However, the researchers here reported that there are considerable discussions among scholars to morphology characteristics of the urban areas. The key findings show that urban morphology is the key factor that influences on the UHI existence in the city.

Future research should consider the potential effects of the urban heat island (UHI) in wider cases in Cairo. This assumption might be addressed in future studies. In future work, it is needed to investigate the examination of each urban morphology parameter on the UHI intensity. The researchers of the current work believe that apart from looking for UHI, future research should look for expanding the implementation of remote sensing techniques through the different seasons. This is the key component in future attempts to overcome the phenomenon of UHI to be investigated during times through the seasons.

References

Abusaada H, Vellguth C, Elshater A (2019) Handbook of research on digital research methods and architectural tools in urban planning and design. IGI Global. https://www.doi.org/10.4018/978-1-5225-9238-9

Akbarl H, Huang J, Martlen P, Rainer L, Rosenfeld A, Taha H (1988) The impact of summer HI on cooling energy consumption and CO_2 emission. ACEEE summer study on the energy efficiency of the buildings. American Council, Washington, DC. pp 11–23

Aniello C, Morgan K, Busbey A, Newland L (1995) Micro-urban heat islands using Landsat TM and a GIS. Comput Gesci 21(8):965–969

Boukhabla M, Alkama D, Bouchair A (2013) The effect of urban morphology on urban heat island. Int J Ambient Energy 34(2):100–110

Caliskan O, Marshall S (2011) Urban morphology and design: introduction. Built Environ 37(4):381–392

CAPMAS (2019, June 6) Egypt statistics. Retrieved from Central Agency for Public Mobilization and Statistics. https://www.capmas.gov.eg/HomePage.aspx

CCVA (2015) Urban heat island protocol for mapping temperature projections. Climate Change Vulnerability Assessment, Cambridge

Farag AA, Badawi S, Doheim RM (2019) Digital simulation for the outdoor thermal comfort assessment. In: Abusaada H, Vellguth C, Elshater A (eds) Handbook of research on digital research methods and architectural tools in urban planning and design. IGI Global, pp 33–46

Gartland LM (2008) Heat islands: understanding and mitigating heat in urban areas. EarthScan, New York

Ghoniem E (2009) A remote sensing study of some impacts of global warming on the arab region. Boston University, AFED the Center for Remote Sensing

Kühl K (2011) Migropolis: spatial practices of migration case studies of migrant communities and their impact on local urban environments. Columbia University, Columbia

Kropf K (2017) The handbook of urban morphology. Wiley, Chichester

Landsberg HE (1981) The urban climate. Acadmic Press, London

Mills G (2008) Luke Howard and the climate of London. Weather, University College Dublin, Ireland, pp 57–153

Moughtin C (2005) Urban design: green dimension. Architectural Press, Oxford

Oke TR (1973) City size and the urban heat island. Atmospheric Environment Pergamon Press, Great Britain

Oke TR (2006) Initial guideline to obtain representative meteorogical observations at urban sites. World Meteorolgical Organization (WMO), Canada

Paravantis J, Tzavali A, Mihalakakou G, Aleksander F, Eleni S (2015) Urban heat island intensity: a literature review. Fresen Environmental Bulletin 24, 4537–4554

Shahraiyni HT, Sodoudi S, El-Zafarany A, Seoud TA, Ashraf H, Krone K (2016) A comprehensive statistical study on daytime surface urban heat island during summer in urban areas, case study: Cairo and its new towns. Remote Sensing 8:643–663

Shane DG (2019) Foreword. In: Abusaada H, Vellguth C, Elshater A (eds) Handbook of research on digital research methods and architectural tools in urban planning and design. IGI Global, pp xvii–xviii

Srivanit M, Kazunori H (2011, October) The influence of urban morphology indicators on summer diurnal range of urban climate in bangkok metropolitan area, Thailand. Int J Civil Environ Eng IJCEE-IJENS 11(5):34–46

Stewart DJ (2002) Changing Cairo: the political economy of urban form. Int J Urban Reg Res 23(1):103–127

Stewart, I. D. (2011, October). Redefining the urban heat Island. Columbia: UBC Library

Stewart ID, Oke TK (2012) Local climate zones for urban temperature studies. Bull Am Meteorol Soc 93(12):1879–1900

U.S. Geological Survey (2016) LandSat 8 DATA USERS HANDBOOK. Retrieved from Earthexplorer (29 Mar). https://www.usgs.gov/land-resources/nli/landsat/landsat-8-data-users-handbook

Weng Q (2012, November) Thermal infrared remote sensing for urban climate and environmental studies: methods, applications, and trends. In: Srivanit M, Hokao K (eds) ISPRS Journal of photogrammetry and remote sensing, 64(4):335–344. Retrieved from ResearchGate

Williams NS, Bosomworth RJ, Trundle A (2013) Planning for a cooler future: green infrastructure to reduce urban heat. Victorian Centre for Climate Change Adaptation Research: VCCCAR, Victoria

Yvonne (2015) A study of urban heat island using "Local Climate". Br J Environ Clim Change 5(2):116–133. https://doi.org/10.9734/bjecc/2015/13051

Zhou Y, Zhuang Z, Yang F, Yu Y, Xie X (2017) Urban morphology on heat island and building energy consumption. Proc Eng 205:2401–2406

Zipper SC, Schatz J, Singh A, Kucharik CJ, Townsend PA, Loheid SP (2016) Urban heat island impacts on plant phenology: intra-urban variability and response to land cover. Environ Res Lett 11(5):1–12

Chapter 32
Environmental Risk Assessment as an Approach of Resilience to Climate Change in Architectural Built Cultural Heritage

Noha A. Shaheen, Ayman A. E. Othman, and Mostafa Rifat Ismail

Abstract Built heritage symbolizes our culture, heritage, and identity, acting as testament defining who we are and declaring our past stories. Yet, the risk inherent by the global climate change exposes our heritage properties to increasing disaster risks, consequently impacting the country's tourism, economic state and social resilience. In addition to the growing recognition that an environmental risk management strategy is required to determine the damage of any potential danger on heritage buildings, the precautions required for risk mitigation and assessing the vulnerability of those buildings. Thus, prioritizing and defining any destructive factors. The purpose of this research is to investigate the appropriation of environmental risk assessment (ERA) toward providing guidance for eventual future efforts to promote resilience to climate change in heritage buildings. A comprehensive literature review is conducted to achieve this aim. It will investigate the built cultural heritage perception, classifications, characteristics and values in addition to understanding the risks imposed by climate change on built heritage and the potential impacts on the different built heritage aspects. As well as examining the current risk assessment, mitigation and adaptation strategies and approaches for effective implementation. The research findings will provide a climatic disaster's preparedness at heritage properties through a detailed matrix that was developed to combine the risk assessment techniques and mapping the climate change-related threats. Hence, acting as an assessment quantitative tool to ensure effective responses at all levels. In addition to investigating the resilience measure that would be taken by then based on the effective analysis.

N. A. Shaheen (✉) · A. A. E. Othman
Architectural Engineering Department, Faculty of Engineering, The British University in Egypt, Cairo, Egypt
e-mail: Noha.a.shaheen@gmail.com

A. A. E. Othman
e-mail: Ayman.Othman@bue.edu.eg

M. R. Ismail
Department of Architecture Engineering, Faculty of Engineering, Ain Shams University, Cairo, Egypt
e-mail: Mostafa_ismail@eng.asu.edu.eg

Architecture and Urbanism: A Smart Outlook,
https://doi.org/10.1007/978-3-030-52584-2_32

Keywords Heritage management · Risk assessment · Climate change · Cultural heritage · Climatic resilience

32.1 Introduction

Over the last years, climate change is perceived as one of the most severe problems posing additional risks on human settlements and buildings performance. Heritage sites represent a fundamental asset for the community's economic and social state. Yet, they are constantly exhibited to natural hazards such as earthquakes and volcanoes, climatic change consequences, and human-induced threats (Chmutina et al. 2016). According to a survey conducted by the world heritage center in 2005, around 72% of the 110 responses received agreed that the heritage is being affected by climate change (Forino et al. 2016), hence highlighting the importance of long-standing preservation against environmental threats (Croft 2013). Furthermore, climatic changes have a direct physical impact on the world cultural heritage. For instance, the underground where the archeological evidence is preserved is known for its balance between hydrological, chemical, and biological processes of the soil. Thus, any alterations in these parameters will definitely result in the non-survival of some materials. This can be proportionally related to the historic buildings, which are more intimate with the ground. Therefore, any increase in the soil moisture may result in bigger salt mobilization and loss of decorated surfaces through dryness from damaging crystallization (Colette 2007). This can be easily projected on Old Islamic Cairo, El Moez Street, where every day hundreds of stone layers are being lost from salt mobilization and crystallization (El-Sharief 2018). Accordingly, the previous studies emphasize on the importance of conducting a strategy relevant to the environmental crisis which would act as an early warning for heritage properties. Accordingly, this study will focus on the provision of an environmental risk assessment strategy as a tool for preparedness toward better knowledge, innovation, and resilience at built heritage properties.

32.2 Heritage Definition and Its Classifications

Heritage is perceived as anything that has a history or cultural value that can be conserved or inherited to others. However, according to the communities or governorates, the forms of heritages, which are cherished, regulated and protected in any conservation initiatives, entails tangible heritage such as buildings, places, archaeological sites, landscapes and intangible heritage such as languages, stories, traditional customs and beliefs (Tonkin 2011). It is stated that every tangible form of heritage is wrapped with an intangible form of heritage, which is manifested in social practices surrounding it or particular religion or even a language spoken by to describe it (Marmion 2012). Heritage can be also be viewed in terms of social and cultural form rather than only physical forms. According to Appadurai and Byrne (2008), communities use heritage to produce culture as known as production of locality. This view of heritage focuses on the role of heritage in the creation of collective memories among people that produce culture. It defines heritage as a something that is being created by people through their connections between themselves and other places and things.

As a way for classifying heritage, it is necessary to differentiate between cultural heritage, which is human-made and natural heritage which is not manufactured by humans. The United Nations Educational, Scientific and Cultural Organization (UNESCO) has categorized the cultural heritage in 2002 cited in Harrison (2010) into cultural and historical sites, natural scared sites, underwater cultural heritage, several museums and art galleries, arts and crafts, digital documentaries, oral traditions, festive events, religious rituals, different types of arts' performance, traditional sports and games, culinary traditions and traditional medicine such as cupping therapy.

32.3 Built Cultural Heritage Definition

The built cultural heritage (BCH) is classified under the human-made resources that are made up by people inhabiting the planet in order to preserve their cultural values, hence to be enjoyed by the present generations and passed on the future generations. It entails historic residential buildings, monuments, historic cities, landscapes, etc. According to Tweed and Sutherland (2007), BCH preservation is essential for the creation of communities' cultural identity and provision of a character for the place. The values of those buildings or monuments do not only lie in its appearance as an architectural heritage, but also in the integrity in all of its components that formed an exceptional outcome of a particular building at that time (Bento 2019). However, the vulnerability of those buildings to natural disasters and climate change effects has been recorded not only due to their fragility but also to their long exposure to all kinds of aging effects. Thus, the understanding and assessing those impacts is crucial for heritage preservation.

32.4 Valuing Heritage

According to Howard (2003) cited in Marmion (2012) states that the value of heritage is determined culturally rather than being lying within the stonework of historic buildings. He argues that the real value of heritage lies within the value of the communities engaging with it and according to their own lens of attractiveness, thus drawing the attention to the intangibility of heritage in terms of being associated with senses of heritage and their connections with it. According to Realdania (2015) built cultural heritage has a direct value which is generated from the architectural, environmental value and identity associated with it and an induced value, which is represented in the activities it attracts, for instance tourism which generates revenues for local businesses. In such a situation, there is a direct value for tourism while an induced value for the businesses benefiting from it with a greater revenue stream.

32.5 Climate Change as a Threat

According to Williams and Baldry (2015), the sustainability challenge is considered as the longest-term challenge facing the human race. Since we are all sharing the same atmosphere all over the globe, it is now the world concern to study the impact of global warming and carbon dioxide into the atmosphere. It was found that the past 50 years change in climate is due to the increase of greenhouse gas emissions by 31% since 1750 and an increase rate of 1.5 ppm; 0.4 % per year in CO_2 concentrations (Williams and Baldry 2015). This impact can be in the form of either meteorical impact such as storms, hurricanes, tornadoes or hydrological form such as rising in sea levels, floods, tides, inundation, or climatological form such as increase in temperature, wildfires and droughts, consequently affecting buildings and monuments stability, causing an alteration in their materials' characteristics and modifying the topography, landscaping and vegetation of sites. This analysis draws the attention to the importance of providing a risk perspective for the short- and long-term climate variations on built cultural heritage which is a recent topic to the literature and practices (Forino et al. 2016).

32.6 Potential Climate Change Risk Factors to Cultural Built Heritage

There are two types of events when it comes to the effect of climate on cultural built heritage: extreme events and slow events: Extreme events are characterized by being severe yet short lived, intensive, highly destroying, and uncontrollable such as strong winds, tornados, storms, flooding, heat waves and droughts. Those events' frequency is expected to be increased in the future, while slow events are characterized by being progressive, long-lived, eternal, less destroying and controllable. Some of them are linked to air pollution, for instance, deterioration in marble or limestone facades, stained stone surfaces, soiling and chemical leaching of stained glass and chemical erosion as clarified in Table 32.1. Others are linked to the geographical location, hence being affected in terms of salt crystallization in porous wall like stone, bricks, plaster, and wall paintings in addition to the rise of sea level that causes the submersion of monuments on coastal, instability of monuments due to swelling-shrinkage of clay minerals in soils (Schorlemer and Maus 2014).

Table 32.1 Impacts of climate change on heritage sites (Daly 2011)

Climate effect	Controlling parameters	Potential impact on heritage values		
		Cultural Landscape	Structures and Features	Buried deposits
TEMPERATURE • Increased annual temperatures • Reduction in freeze thaw events	Temperature Moisture	Change (i.e. loss and/or gain) of habitats and species Lengthening of growing season Changes in land use	Increased biological growth	Accelerated micro-biological activity Changes in land use
RAIN • Storms/heavy rain • Prolonged wet periods • Flooding • Altered water table • Prolonged dry periods	Rain intensity and duration Rain volume Catchment hydrology (i.e. flooding can be caused by rain elsewhere) Temperature	Soil erosion Silting of river beds Change/loss of habitats and species Loss of vegetation Deterioration of water quality Landslides Changes in irrigation practice* Increased risk of fires* Increased recreational use*	Mechanical erosion Change in humidity cycles (salts) Increase in time of wetness (leading to biological colonization) Rising damp Subsidence and collapse Changes in surface deposition and washing of pollutants Increased recreational use*	Mechanical erosion Changes in soil chemistry and pH Subsoil instability Eutrophication accelerating biological decay
WIND • Wind driven rain • Wind pressure • Wind driven particulates • Gusts and changes in wind direction	Wind speed Wind direction Rain intensity and duration	Soil erosion	Mechanical erosion and abrasion Increased penetration of water (leading to salt movement) Increased loading pressure Physical damage and collapse	Erosion of earthen monuments/soil cover

(continued)

Table 32.1 (continued)

		Additional coastal impacts		
INCREASED SEA TEMPERATURE • Sea Level Rise • Cyclones	Temperature Ocean currents	Coastal flooding Erosion Salt in soils and water table Migration of human population	Erosion Increase in salt damage Storm damage	Erosion of sites (exposure and/or loss) Sedimentation of sites Submersion Coastal protection engineering
COASTAL WINDS • Wind transported salts • Wind driven sand • Increased wave heights • Storm surge	Wind speed Wind direction Surface pressure	Erosion of sand dunes Coastal erosion Flooding	Increased penetration of salts and salt weathering Flooding Erosion of foundations	Exposure and/or erosion of sites in sand dunes, underwater and Flooding

* Refers to the requirement of both human and environmental influences assessment on the site

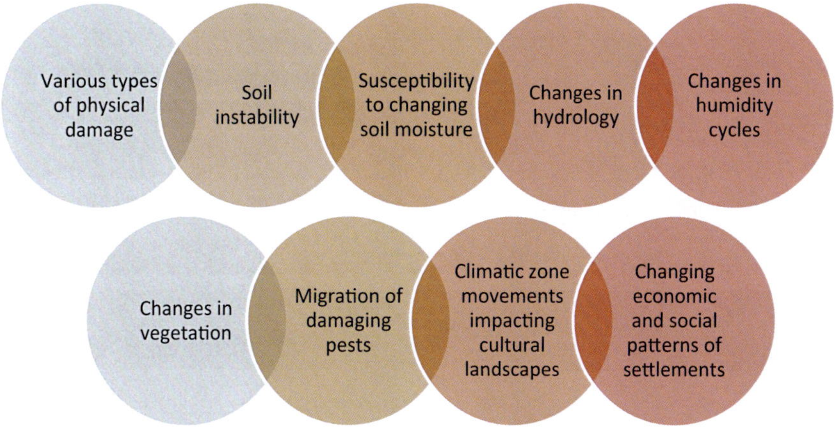

Fig. 32.1 Factors affecting cultural built heritage (Forino et al. 2016)

The previous table shows the relation between the climate parameter (temperature, rain, wind, rise in sea level) along with the factors associated with it in terms of moisture, rain, wind intensity and duration, wind speed and its impact on built cultural heritage, structures and architectural features. in addition to the consequences on the buried deposits (soil erosion, soil instability and sedimentation of sites).

In addition, Forino et al. (2016) identified nine listed categories as shown in Fig. 32.1., the factors affecting cultural built heritage (Forino et al. 2016) that might affect cultural built heritage which are:

All of which are posing new challenges that would face cultural built heritage, hence suggesting additional attention for heritage sites identification, mapping and documentation with a specific preference to those on coastal locations. In addition, it manifests the interlocking relationship between the nature, cultural and social systems. Hence, emphasizing on the importance with dealing with the system as a whole unit.

32.7 The Complexity of the Climate Change Dilemma

According to Schorlemer and Maus (2014), it was found that the general conventions for built cultural heritage protection does not properly consider the current aggressive climatic changes, consequently raising the why question where the following may act as possible explanations, therefore calling the attention to understanding the complexity of the climate change dilemma requirements.

Rules set by conventions can be generally applied to any country regardless to its degree of exposure to climate changes and its location on the globe. However, the fact is that weather and climate changes from a place (country) to another significantly.

Each country takes into consideration its own variations in annual weather and climate when implementing the conventions and its protection commitments.

Climate heritage properties includes materials that vary in its sensitivity to climate change which contradicts with the conventions that set out rules that can be generally applied to all cultural heritage properties that under lies their scope of application category.

The negotiation of some conventions was done when the climate change matter was not clearly identified as an un-predictable, distinct problem from the normal annual weather change.

32.8 The Nature of Built Cultural Heritage Risk Management and Risk Assessment Methodology

32.8.1 The Nature of Risk Management

Risks are considered as an integral part of our daily lives as humans. It represents the probability of harmful events' occurrence at a particular location and timing (Kohler et al. 2004). According to Forino et al. (2016), risk is defined as the product of vulnerability from a given event or a group/subgroup of a community and the possibility of occurrence of that event. In order to assess risks, there are two crucial elements as shown in Fig. 32.2, the concept of risk (Risk = hazard × vulnerability) (Kohler et al. 2004), which are hazards and vulnerability. Hazard is the probability of a harmful event of risk occurrence, while vulnerability is the injuries or the damage caused as a result to the risk that happened and the ability to provide protection against it, consequently providing the probability of risk occurrence and its magnitude of the damage.

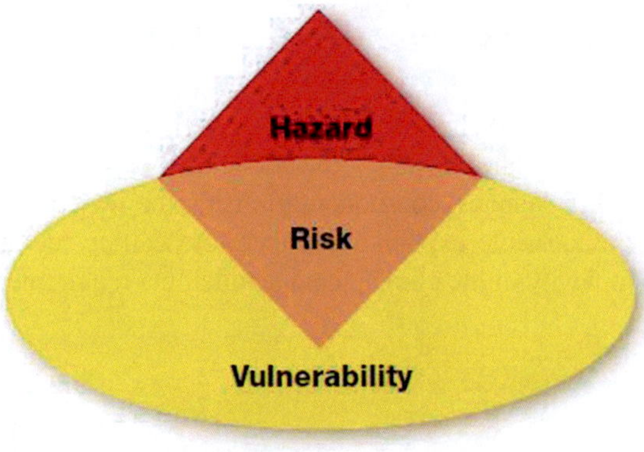

Fig. 32.2 Concept of risk (Risk = hazard × vulnerability) (Kohler et al. 2004)

32.8.2 *Built Cultural Heritage Risk Assessment Methodology*

As a matter of fact, climate change is expected to expand the risk imposed on built cultural heritage. Hence, a risk assessment mitigation plan should be developed as an initiative toward reducing the future impacts (Papathoma et al. 2016). Risk is related to events that have not occurred yet; however, there are solid reasons to believe that they will occur eventually. This should be assessed by both ways, qualitatively and quantitatively. The "EC Risk Assessment and Mapping Guidelines for Disaster Management" cited in Papathoma et al. (2016), identified the steps required for risk assessment as follows:

Risk Identification

The risk identification process entails identifying the context and the basis on which the risk will be assessed.

Risk Analysis

This part of the assessment has the highest demand of data, in addition to its being the most important part that consumes time and effort. It is the process to determine the nature and level of the risk. It may be qualitative or quantitative. Hence, it involves hazard and vulnerability analysis, all of which are then recorded on a raster risk map that identifies the likelihood × severity of its consequence as shown in Fig. 32.3. This map is used as an index to identify the risk level upon which the risk evaluation is made. For instance, whether it is an acceptable risk, shall be considered as a supplementary issue or non acceptable risk that requires a quick response development.

The risk matrix Fig. 32.3 combines the hazard occurrence probability, the consequences of the hazard and the risk criteria of the end user (high, medium, low risk) which is set by the working group in risk identification process.

Risk evaluation

On the basis of the criteria on which the risk is either accepted or has to be treated that was made in the beginning, the risk analysis and the criteria are compared. Hence, a decision is taken to deal with the risks.

Risk treatment

This phase considers and assesses measures in order to reduce risk levels. It involves identifying what are the appropriate decisions for risk management and evaluating the impact of their results after which the treatment plan is implemented that result in risk impact elimination or reduction.

Monitoring and reviewing

In order to ensure the effectiveness of the risk assessment process, it is crucial to establish monitoring and reviewing process that is responsible for reporting any changes or raising circumstances.

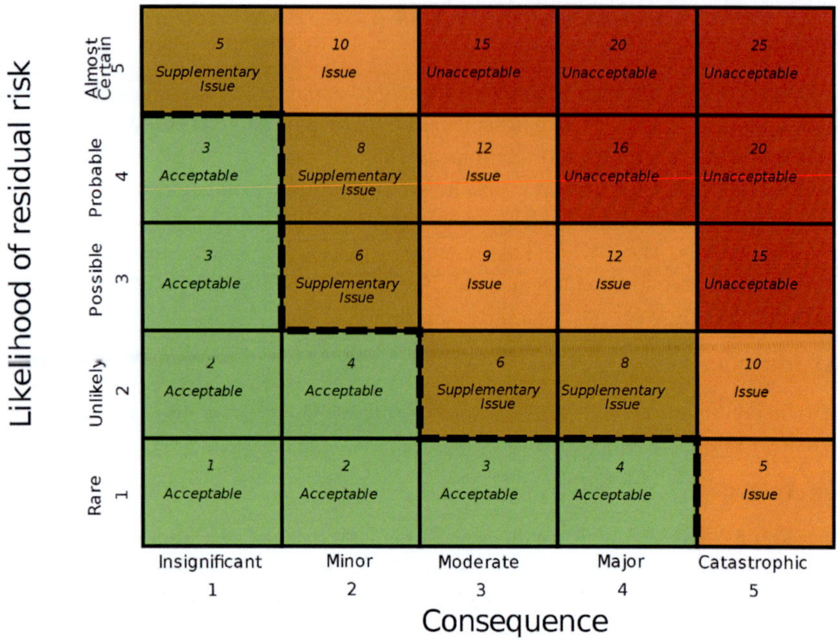

Fig. 32.3 Raster risk map sited from https://commons.wikimedia.org/wiki/File:Risk_Analysis_Chart.svg

32.9 Challenges Facing Governmental and Non-governmental Organizations Toward Built Cultural Heritage Preservation Against Climate Change

According to the Fawzi and El Quosy (2011), in order for developing countries including Egypt to achieve its strategy for climate change adaptation and risk reduction, the following principles have to be fulfilled:

1. Political will in terms of financial and natural resources provision, institutional frameworks reforming and adjustments, legislations and laws modification, enhancing the national system for the dissemination of information.
2. Monitoring, assessing, follow-up provision and identifying performance indicators, through the provision of a dynamic-based strategy.
3. Analyzing social and economic status based on a developed national model.

Fig. 32.4 Challenges for climate change considerations (Sesana et al., 2018)

This report has discussed only five sectors that were affected by climate change that includes the water resources sector, agricultural, health, tourism sector, urban areas, housing and roads sectors. However, the cultural built heritage sector was not discussed as one of the most vulnerable sectors that would be affected by climate change events. In addition, according to Sesana et al. (2018), interviews were carried out in the UK, Italy and Norway to assess how climate change is integrated in the preservation process and how the human resources involved in these sites act upon the information received on climate change considerations. All the results received were classified under five categories as shown in Fig. 32.4.

32.10 Conclusion

Climate change is a global phenomenon that requires policy and risk management making. This study explored the contribution of built cultural heritage to shaping the country's indemnity and provision of a sense of place to its community. However, it is now being threatened by the changing climate patterns, which impose its impacts on our heritage, thus leading to decays, partial and total loss in some areas. Despite the fact that the world organization such as the UNESCO world heritage launching events for public awareness and initiatives for a change yet, Egypt lacks a formal legal document for risk mitigation being imposed on its heritage which represents a main asset for Egypt's GDP. The study focuses on the importance of establishing a matrix as a tool for BCH assessment, hence providing an aiding tool for mitigation and adaptation measures that adopt a sustainable approach for responding to the climate influence on the BCH and as well as providing a strategy for providing a balance between conserving our BCH and climate change impacts. The following matrix was developed from a comprehensive literature research. This matrix will be used in a further research for case studies assessment and from which a framework would be developed as a quantitative tool for climate risks assessments (Table 32.2).

Table 32.2 Matrix developed for further case studies assessment of climate change impact on BCH (developed by the author, 2019)

State Building Type			**Identifying Affected Sectors**											
			Architecture					*Structure*				*Urban*		
Risk classification			Building appearance and finishes	Interior spaces	Materials durability and flexibility	Affecting indoor quality	Authentic features	Celling	Walls	Floors	Roof	Landscaping	Affecting urban density	Affecting urban environment
Risk identification	*Meteorological*	Storms												
		Hurricanes												
		Tornadoes												
		Heat Waves												
		Fire												
	Hydrological	Sea Level Rise												
		Floods												
		Tides												
		Inundation												
	Climatological	Soil Moisture Content Rise in Temperature												
		Relative Humidity Solar Radiation												
		Rainfall												
		Intensity												
		Duration												
		Wind												
		Intensity												
		Duration												
	Biological	Pests												
		Diseases												
Risk Analysis — Hazard analysis	*Geographical*	Location & Extent												
	Temporal analysis	Frequency												
		Duration												

(continued)

Table 32.2 (continued)

State Building Type				**Identifying Affected Sectors**											
				Architecture					*structure*				*Urban*		
Risk classification				Building appearance and finishes	Interior spaces	Materials durability and flexibility	Affecting indoor quality	Authentic features	Celling	Walls	Floors	Roof	Landscaping	Affecting urban density	Affecting urban environment
		Dimensional analysis (scale, intensity)	Local												
			Specific site												
			Regional												
			National												
		Risk occurrence probability	Certain												
			Likely												
			Unlikely												
			Rare												
			Extremely rare												
	Impact analysis	*Features Affected that contribute to the site value*	Physical												
			Biological												
			Geological												
		Risk Impact Significance	Insignificant												
			Minor												
			Moderate												
			Major												
			Catastrophic												

Risk Evaluation	***Integrity (Resistance to threat)***													
	Sensitivity (How easily the site can be affected by all the previous factors)	High												
		Medium												
		Low												
	Resilience measures (site ability to restore itself or recover to its original state)	Acceptable												
		Minor interventions												
		Major Interventions												

References

Appadurai A, Byrne D (2008) 'Heritage as social action' and "The globalisation of archaeology and heritage" (The Heritage, Reader edn. Routledge, New York

Bento R (2019) An interdisciplinary approach to the seismic assessment of built cultural heritage: case studies in lisbon and outskirts. Springer Link: Struct Anal Hist Constr 18:3–18

Chmutina K, Jigyasu R, Bosher L (2016) Understanding the impacts of climate change on cultural heritage buildings: a case of York, UK. The CIB World Building Congress: Intelligent Built environment for life, pp 188–198

Colette A (2007) Climate change and world heritage: report on predicting and managing the impacts of climate change on world heritage and strategy to assist States Parties to implement appropriate management responses. UNESCO World Heritage Centre, France

Croft A (2013) Assessment of heritage at risk from environmental threat. Atkins Heritage, Birmingham

Daly C (2011) Climate change and the conservation of archaeological sites: a review of impacts theory. School Surv Constr Manag 13(4):293–310

El-Sharief SA (2018) Cairo climate talks: digital climate education: open access to knowledge (13 May)

Fawzi MAE-R & El Quosy DED (2011) Egypt's national strategy for adaptation to climate change and disaster risk reduction. Egypt: The Egyptian Cabinet information & decision support centre, UNDP

Forino G, Mackee J, Meding J (2016) A proposed assessment index for climate change-related risk for cultural heritage protection in Newcastle (Australia). Int J Disaster Risk Reduction 19:235–248

Harrison R (2010) What is heritage? Manchester: Manchester University Press in association with the Open University, Manchester and Milton Keynes

Howard P (2003) Heritage: Management, interpretation and identity. London: Continuum

Kohler A, Jülich S, Bloemertz L (2004) Risk analysis—a basis for disaster risk management. Deutsche Gesellschaft für Technische Zusammenarbeit (GTZ) , Germany

Marmion MM (2012) Understanding heritage: multiple meanings and values. Bournemouth University, United Kingdom

Papathoma KM, Promper C, Glade T (2016) A common methodology for risk assessment and mapping of climate change related hazards—implications for climate change adaptation policies. J Clim Sci (2225-1154), 1–23

Realdania (2015) The value of the built heritage. Denmark

Schorlemer S, Maus S (2014) Climate change as a threat to peace: impacts on cultural heritage and cultural diversity. University of Dresden, Frankfurt

Sesana E, Gagnon AS, Bertolin C & Hughes JJ (2018) Adapting cultural heritage to climate change risks: perspectives of cultural heritage experts in Europe. Geosci 8:305

Tonkin S (2011) Essay: what is heritage? Retrieved 12 Jan 2019 from https://www.environment.gov.au/system/files/pages/f4d5ba7d-e4eb-4ced-9c0e-104471634fbb/files/essay-whatisheritage-tonkin.pdf

Tweed C, Sutherland M (2007) Built cultural heritage and sustainable urban development. ELSEVIER: Landscape Urban Plan 83, 62–69

Williams D, Baldry D (2015) Leveraging facilities management best practices in managing the effect of climate change on older buildings within the low carbon economy. University of Salford, UK

Chapter 33
Visual Pollution: An Approach to Reduce the Environmental Impact of Light Pollution in Egypt

Ashraf A. Nessim

Abstract Visual pollution can be contributed to many factors one of which is "Light pollution". Light pollution has been identified as a main environmental issue in urban areas during night-time due to the advancement in lighting and its related technologies. As part of Egypt's 2030 Vision, the government is currently executing mega projects to serve the public such as the national network of roads project, the national project for development of Sinai, building a number of major cities. All of the aforementioned projects can add to the light pollution if not addressed carefully. The aim of this research is to evaluate the environmental impact of light pollution in Egypt. In order to achieve the objective of this paper, an intensive literature review of recent papers discussing the problem has been performed, analysed and followed by assessment of the possible techniques that can be used to limit light pollution. The value of this research comes from the fact that light pollution is unavoidable due to the continuous increase of outdoor activities at night; thus, strict laws and proper regulations should be established. The detected findings of this paper discussed the technical and legal aspects, their objectives and proposed techniques to limit obtrusive light in Egypt which could help those involved in decision-making and setting legislations to regulate light pollution in Egypt.

Keywords Visual pollution · Obtrusive light · Light pollution

33.1 Introduction

Visual pollution is an expression that describes the amalgamated effect of excess, clutter and disorder of diverse graphics and objects in the landscape and the outdoor environment. There are various sources of visual pollution such as street furniture, lighting fixtures, building facades and abandoned buildings but the most typically is associated with excess and chaos imposition of outdoor advertisement and its associated lighting (Azeema and Nazuk 2016). In general, the harmful influence

A. A. Nessim (✉)
Ain Shams University, Cairo 11517, Egypt
e-mail: a.nessim@eng.asu.edu.eg

Architecture and Urbanism: A Smart Outlook,
https://doi.org/10.1007/978-3-030-52584-2_33

of oversaturation caused by the lighting of the outdoor advertisement may cause driver distraction, impacts planning strategies and affects negatively the outdoor environment visual quality. Unlike air and water pollution problems, for which there is a long track of research in the physical sciences, visual pollution is a physical and cultural study. The impact of this pollution is increasing remarkably day by day (Chmielowski et al. 2018).

In the meantime, light pollution can be identified as a visual problem that occurs at night in the outdoor environment and generated by unrestrained outdoor lighting hindering the viewing of the sky. This can be attributed to human behaviour caused by lighting their night-time environment for various reasons, including advertising, to facilitate transportation, orientation and way finding, to reduce perceptions of fear associated with night-time and to extend the usable hours of the outdoor environment after sunset.

The main causes for light pollution are the luminaires and types of outdoor lighting. Accordingly, light pollution is linked with street lighting, façade lighting, signage and advertisement lighting and security lighting, etc. (SLL 2012). The Institution of Lighting Engineers (ILE 2005) identified the following forms of light pollution: urban sky glow, glare and light trespass.

Consequently, light pollution has been recognized as one of those negative effects and can be seen as a major environmental problem in urban areas at night, and it is closely associated with the growth of urbanization (Jiang et al. 2017). Light pollution significantly impacts sustainable developments in terms of human health and well-being, ecological stability and energy conservation (Du et al. 2018). In spite of the growing interest among experts in different fields such as astronomy, ecology, health care and land-use planning, there is a lack of current quantification of its magnitude of light pollution on a global scale (Falchi et al. 2016).

33.2 Research Problem

Currently, and due to the vast urbanization and economic development worldwide, light pollution is increasing at an unprecedented pace and to an unmatched extent. In addition, major cities around the world are working towards establishing the concept of "City Beautification by Night" as a marketing strategy to the city. Consequently, attractive night lighting is increasing and humans progressively expand the scope of their night-time activities. As a result, many large cities around the globe have been threatened by light pollution due to the immense use of outdoor lighting.

According to Schroer and Hölker (2014), more than 60 % of the world population live under light-polluted skies in which the artificial sky brightness $\geq$10 % of the natural night sky brightness above 45° of elevation on the horizon. In addition, 99% of the inhabitants in Europe and the USA cannot distinguish the original night sky and almost 20% of the world's land is affected by increased artificial sky brightness. In 2016, Falchi et al. specified in the new world atlas of artificial night sky brightness

Fig. 33.1 Various forms of light pollution in Cairo, Egypt (*Source* Laila Nasser)

that 80% of the whole world is disturbed by light pollution and that 99% of sites within the USA suffer from some degree of light pollution.

As for Egypt, the uncontrolled outdoor lighting causes various forms of light pollution in existing major cities as seen in Fig. 33.1. Accordingly, clear codes should be established to prevent any further pollution resulting from the mega projects that Egypt is currently executing as part of its 2030 Vision which might lead to rapid urbanization expansion, economic growth and misuse of available technologies thus leading to additional light pollution.

By examining the photographs of Egypt provided by NASA Earth Observatory and comparing it to the photographs obtained from the light pollution map, which is a mapping application that displays Visible Infrared Imaging Radiometer Suite (VIIRS), U.S. Air Force Defence Meteorological Satellite Program (DMSP) and World Atlas to visualize the light pollution worldwide, it is clear that the light pollution, resulting from high illumination levels, is concentrated along Nile valley and Delta. Some of the brightest lights occur around Cairo. In addition, bright city lights also occur along the Suez Canal.

33.3 Research Aim

This paper aims at evaluating the environmental impact of light pollution in Egypt resulting from outdoor lighting as this area has not yet been widely documented due to the scarcity in the researches that discuss its causes in Egypt. In addition, the paper investigates the available methods to control light pollution. Thus, the objective of this paper is twofold;

First: To recognize and define light pollution, to identify its various types and to examine the light pollution map of Egypt.

Second: To reduce the environmental impact of outdoor lighting at night and to control light pollution.

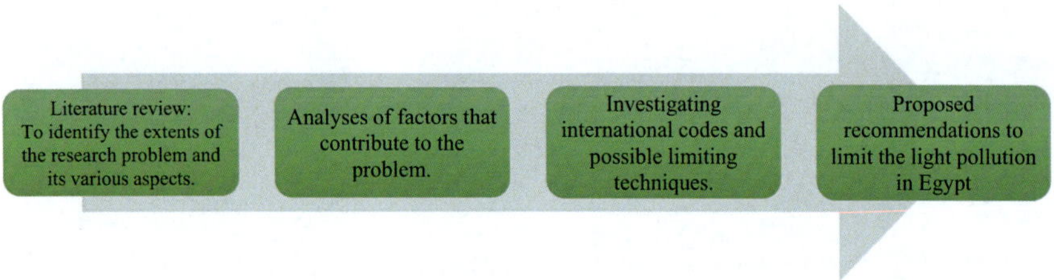

Fig. 33.2 Research methodologies (*Source* Author)

33.4 Methodology

In order to achieve the objective of this paper, an intensive literature review of recent papers discussing the problem internationally has been performed to understand the global challenge that light pollution represents. Afterwards, the various types of light pollution were analysed to identify the factors that contributes to this problem. Furthermore, the international codes and possible techniques that can be used to limit light pollution were examined. Finally, a set of recommendations were introduced to limit the light pollution in Egypt. Figure 33.2 summarizes the research methods and the aim of each.

33.5 Literature Review

The literature review is divided into three parts. The first part defines light as a pollutant. The second part focusses on the relevant methods for assessing the related light pollution, and the third part concentrates on strategies to reduce the resultant light pollution.

33.5.1 Light Pollution Recognized

The main aim of this part is to review the definitions of the light pollution and to classify its different types.

Defining Light as a Pollutant

Light pollution has become one of the key features of environmental pollution and can be seen as one of the most rapidly growing alterations to our environment. Light pollution was first discussed by astronomers. The diffuse urban lights resulted in a sky glow near cities which obstructed the visual observation of dim astronomical objects.

According to McGowan (2013), light pollution can be defined as "any annoying, unwanted, unnecessary or wasteful outdoor lighting at night which results in sky glow, light trespass, glare and or other harmful effects on people and the environment". The International Dark-Sky Association (IDA), which is a U.S. non-profit association to conserve and protect the night-time environment, defines light pollution as "any unfavourable effect caused in the outdoors as a result of the use of artificial light". This definition includes sky glow, glare, light trespass, light clutter, decreased visibility at night and energy waste (IDA 2019).

Types of Light Pollution

Obtrusive light, commonly referred to as light pollution, can be divided into three categories: sky glow, glare and light trespass as illustrated in Fig. 33.3 (ILE 2005). The first category of light pollution is sky glow which can be defined as the upward loss of light into the sky which causes the increase in the luminance of the night sky. It is a joint effect of reflected and refracted light from the atmosphere and produced by excessive artificial lighting. The negative effect of sky glow on the environment can be viewed clearly as a glowing effect above large-scale urban areas when the area is viewed from a darker external viewing point as seen in Fig. 33.4. The first measurements of urban sky glow were performed by astronomers (Kuechly et al. 2012).

The second category is light trespass (sometimes refereed to light intrusion or light spill), which can be considered as a localized light pollution phenomenon. It occurs when unwanted exterior light falls beyond the boundary of a property, illuminates an indoor space and cause nuisance to others. This type of obtrusive light is typically brought by light installations beyond the property boundary.

The third category is glare which is the uncomfortable brightness resulting from a large luminance contrast between the light source and the darker background causing

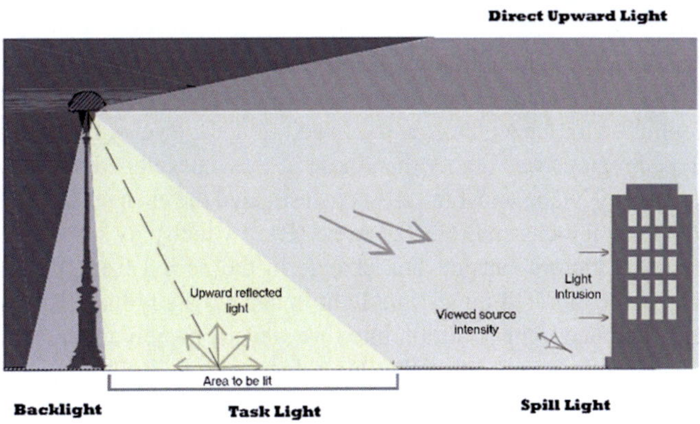

Fig. 33.3 Various types of obtrusive light including sky glow, glare and light trespass (*Source* Author after the Institution of Lighting Engineers (ILE), "Guidance Notes for the Reduction of Light Pollution")

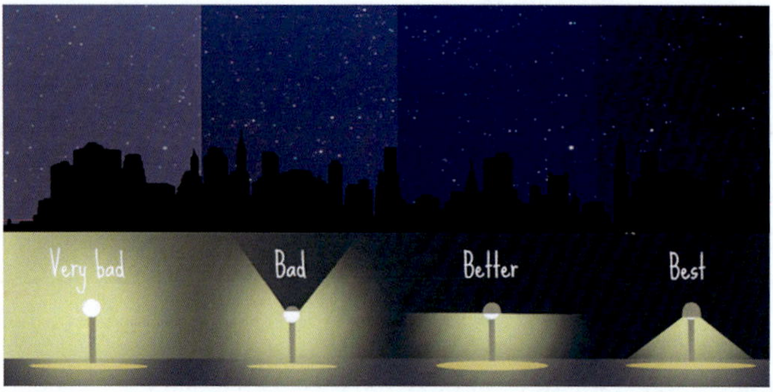

Fig. 33.4 Sky glow effect over urban areas as a result of using different types of luminaires (*Source* author after National Park Service)

discomfort or disability glare (SLL 2012). When compared to the previous two categories, glare could be more difficult to measure and control in urban areas at night.

33.5.2 The Relevant Literature Related to Assessing the Light Pollution can be Briefed as Follows

A paper by Jiang et al. (2017) assessed the light pollution in China based on night-time light imagery. The research focussed on studying the spatiotemporal patterns and the development of light pollution in China, during the period from 1992 to 2012, through analysing satellite observations. Satellite images give information on light pollution, intense regions of human activities and the level of economic and social activity. The researchers utilized a linear regression trend method to reveal evolvement of China's light pollution. The paper concluded that the light pollution was mainly focussed in areas of north and east China which accounted for over 50% of the light pollution for the whole country. In addition, the level of the light pollution in the developed regions was higher than those of the undeveloped regions.

Another paper by Xiang and Tan (2017) investigated the changes in China's light pollution with special focus on Protected Areas (PAs), which play key environmental functions, and the factors causing this change in the period from 1992 to 2012. The research used calibrated night-time light images to study the changes in light pollution and classified light pollution into three levels, namely moderate, medium and strong. The results showed that the PAs subject to light pollution increased by about 1.79 times, with a significant increase in the intensity of artificial light. The paper concluded that light pollution in PAs is shaped by both human activities and the PAs own characteristics.

A third paper by Kuechly et al. (2012) used aerial survey and spatial analysis of light pollution sources in Berlin, Germany. The researchers' produced a high-resolution (1 m) mosaic image of the city at night and spatially analysed a dataset to recognize the key sources of light pollution in the city. It was found that street lighting was the dominant source of directed light pollution by 31.6% than other land-use classes.

A paper by Osman et al. (2001) examined the light pollution effect from the city of Cairo at Kottamia Observatory. The research estimated light and electric energy losses for Cairo from ground observations and compared it with space results obtained from Defence Meteorological Satellite Program (DMSP) satellites. The paper compared between the light energy losses in 1980 and 1995. It was noticed that energy loss from Cairo is nearly equal to that of Paris and a little lower than that of London and much higher than Amsterdam and Bruxelles.

33.5.3 The Relevant Literature Concerned with Reducing the Light Pollution Can be Briefed as Follows

A paper by Elsahragty and Kim (2015) investigated the strategies to reduce light pollution using Geographic Information Systems (GIS). The paper investigated light pollution at the Port of Long Beach. A GIS model is generated to verify the level of light pollution at the study area. Relevant lighting power densities (ASHRAE 90.1-2007) were applied to find solutions to reduce the level of light pollution. The effect of light pollution generated from the Port area is assessed by comparing it against the sky glow generated from the Port and other areas throughout the coast of Southern California, as well as comparing how intense the sky glow penetrates the ocean. The outcomes were authenticated by comparing against the lighting requirements used in the area of study. The lighting strategies proposed include adjustments to the height and spacing of light poles.

A thesis by Pedithep (2015) used comparative law studies to recognize the strengths and weaknesses of light pollution laws in different areas to limit light pollution from outdoor light fixtures. The thesis aimed at improving light pollution regulatory frameworks in the UK. The thesis compared regulatory measures in the English legal system with other international and European light pollution laws. The thesis proposed a set of essential modifications to regulatory standards to limit light pollution.

33.6 Light Pollution Reduction: Methods to Reduce the Environmental Impact of Artificial Light at Night

Several studies have been conducted to control light pollution. According to Schroer and Hölker (2014), simple solutions as shielding luminaires can be used to prevent obtrusive and trespass. Moreover, using shades is an efficient means to reduce trespass from interior lights of glazed facades thus reducing sky glow. Furthermore, artificial light should be turned off whenever it is not needed and smart, flexible lighting systems can aid in the use of artificial light with accuracy. In addition to avoiding glare, light should be guided to where it is needed and dimmed to the lowest possible level possible. Finally, all details related to the lighting installation such as the location, mounting height and aim of exterior luminaries need to be taken into consideration for efficient use of lighting energy.

33.6.1 International Outdoor Lighting Codes

Since 2005, several codes have been developed to reduce light pollution from exterior lighting luminaires. The Illuminating Engineering Society (IES) proposed a metric for assessing the luminaire's 3D photometric distribution and its impact on light trespass, sky glow and glare. This metric was named Backlight, Uplight and Glare (BUG) ratings (IES 2019).

Meanwhile, the International Commission on Illumination (CIE) adopted the Upward Light Ratio (ULR) for exterior luminaires aiming to limit the sky glow. According to CIE, the ULR is defined as "the proportion of the flux of a luminaire or installation that is emitted, at and above the horizontal, when the luminaire is mounted in its installed position". Figure 33.5 illustrates the ULR.

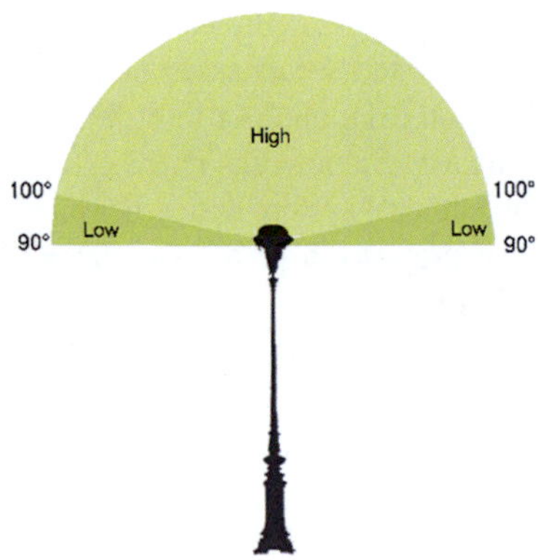

Fig. 33.5 Critical luminaire angles for reducing sky glow. Critical angles for ULR are between 90° and 100° (*Source* Author after Lighting Analysts: Illumination engineering software)

33.6.2 Environmental Zones for Outdoor Lighting

Environmental zoning is a qualitative approach that is used to organize the use of exterior lighting installations and to reduce obtrusive lighting in various areas. In 2003, the CIE outlined a system consisting of four zones based on the "night brightness" of the local environment. The first zone is E1, and it includes intrinsically dark areas, E2 includes low district brightness areas, E3 includes medium district brightness areas, and E4 includes high district brightness areas. (ILE 2005)

Recent modifications in the CIE guide (CIE 150-2003) related to obtrusive light added lighting zone E0 which is only applied in 'intrinsically dark' areas in which permanent illumination is not allowed to be fitted. E0 includes protected wildlife areas and corridors, IDA Dark-sky Parks and main optical observatories (Du et al. 2018).

33.6.3 Design Guidelines for Limiting Obtrusive Light

The Institution of Lighting Engineers [2005] established obtrusive light limitations for exterior lighting installations including the environmental zone, sky glow maximum ULR, illuminance into windows, light source intensity and building luminance. The environmental zone E0 was introduced in which no lighting is accepted. Values for other environmental zones were specified (ILE 2005).

33.6.4 Requirements of Green Buildings Rating Systems

As per Ahmed et al. (2016), two of the leading rating systems are BREEAM (UK) and LEED (USA). The U.S. Green Building Council (USGBC) has included a potential credit for reducing the amount of light trespass and sky glow into the USGBC's green building rating system. The LEED v4 Sustainable Sites Credit 8 (SS8) provides guidelines for reducing interior and exterior lighting levels, awards up to 1 point if this requirement is met, and aims at: U.S. Green Building Council (USGBC) (2019).

- Reducing the light trespass from the site.
- Minimizing sky glow thus improving sky visibility at night.
- Decreasing glare probability at night-time.
- Reducing harmful effect on nocturnal environments.

Similarly, the Building Research Establishment (BRE) has incorporated a credit for the control of light pollution in the Building Research Establishment Environmental Assessment Method (BREEAM). The "Pol 04 Reduction of Night time Light Pollution" credit aims to ensure that outdoor lighting is focussed in the proper areas and that upward lighting is minimized thus reducing light pollution, energy consumption and annoyance to neighbours (BRE 2019).

33.7 Findings

Egypt can work on reducing, if not minimized, the light pollution in its different forms and their associated negative impacts on the environment by working on two levels: the first level is to develop the technical aspects related to this issue, and second level is to establish a rigorous legal framework that addresses, specifically the light pollution problem. Accordingly, it is important to assess all environmental issues related to the light schemes of the mega projects currently being executed as part of Egypt's 2030 Vision. Table 33.1 summarizes the technical aspects and legal framework objectives and methods to address each.

33.8 Conclusions

This paper discussed the effect of light pollution as part of the visual pollution at night. Light pollution is a global issue, and its most notable effect is mainly aesthetic. In addition, it is an ecological issue, mainly in developed countries, with negative effect on both humans and nocturnal living organisms. To overcome the light pollution problem in Egypt, the following recommendations should be followed:

- Developing the related technical aspects by:
  - i. Considering the light pollution when carrying out the environmental impact assessment reports;
  - ii. Analysing the aesthetic lighting appearance of the projects so that the used lighting systems provide citizens with a subtle and adequate outdoor lighting themes;
  - iii. Using computer software to provide simulations of the proposed outdoor lighting systems so that the quality and quantity of lighting needed is specifically designed to meet the project's requirements without any straying light that cause environmental problems; and
  - iv. Utilizing appropriate technologies to minimize light pollution including full cut-off luminaires, glare reduction accessories and narrow beam angle spotlights.
- Expanding the governmental legal framework by optimizing the laws and regulations to prohibit high levels of outdoor lighting, use of extreme contrast, and establishing strict control measures and mechanisms to reduce the light pollution. Accordingly, the following needs to be addressed:
  - i. Establishing rigorous Egyptian light pollution prevention standards and regulations;
  - ii. Setting a clear classification of the Egyptian light pollution zones;
  - iii. Creating specific exterior lighting design guidelines;

Table 33.1 Summary of the different aspects, objectives and techniques to limit obtrusive light in Egypt

Level	Objective	Method
• Technical aspect	• Reducing the light trespass from the site	• Using narrow beam angle spotlights
		• Checking the aiming of the luminaires
		• Testing aiming of luminaires on site
	• Minimizing sky glow thus improving sky visibility at night	• Using full cut-off luminaires
		• Adopting Upward Light Ratio (ULR)
	• Decreasing glare probability at night-time	• Using glare reduction accessories
		• Adopting Backlight, Uplight and Glare (BUG) ratings
		• Examining contrast levels
	• Carrying out outdoor lighting calculations	• Using computer simulation software
		• Establishing appropriate illumination levels for different environmental zones
	• Preparing environmental impact assessment reports	• Carrying out visual analysis
		• Setting appropriate illumination levels
		• Evaluating exterior contrast levels
• Legal framework	• Creating appropriate lighting themes	• Analysing outdoor environmental zones
		• Setting environmental zones for outdoor lighting
	• Establishing outdoor lighting codes	• Analysing international lighting codes
		• Adopting appropriate items that matches the requirements of the Egyptian context
		• Adding additional items according to the requirements of the Egyptian context/local environment
	• Design guidelines for limiting obtrusive light	• Specifying environmental zones
		• Setting maximum ULR for the installation to prevent sky glow
		• Setting maximum vertical point illuminance (in lux) to prevent light intrusion into windows

(continued)

Table 33.1 (continued)

Level	Objective	Method
		• Specifying maximum light source intensity I (in kcd)
		• Identifying the building luminance before curfew
	• Applying of regulations	• Including the environmental impact assessment reports as part of the project permits
		• Following-up the application of the regulations through governmental authorities

Source Author

iv. Implementing the regulations as part of the license procedure; and
v. Founding official authorities that follow-up the application of the established regulations.

- Spreading the Egyptian's knowledge and awareness about this type of pollution.
- Replacing old lighting systems that does not meet the required specifications, regulations, and standards with new ones to reduce the negative environmental effect caused by exterior lighting.
- Encouraging the implementing of the green building's rating systems as they have specified credits dedicated to address the light pollution.

Further research work includes:

i. Establishing a database to identify the light pollution zones in Egypt;
ii. Determining the outdoor lighting environmental zones in Egypt; and
iii. Executing a spatial analysis of causing factors of light pollution in the major Egyptian cities.

References

Ahmed M, Abul Hasan M, Mallick J (2016) World Green building rating systems: a comparative study. In: International conference CUM exhibition on building utilities, New Delhi, India. https://www.researchgate.net/publication/315516234_World_Green_Building_Rating_Systems_A_Comparative_Study. Retrieved 08 Apr 2019

Azeema N, Nazuk A (2016) Is billboard a visual pollution in Pakistan? Int J Sci Eng Res 7(7):862–877

Building Research Establishment (BRE) (2019) Building Research Establishment Environmental Assessment Method (BREEAM), "Pol 04 reduction of night time light pollution". Retrieved 08 Apr 2019, from BRE Website: https://www.breeam.com/BREEAM2011SchemeDocument/Content/12_pollution/pol04.htm

Chmielowski S, Samulowska M, Lupa M, Lee D, Zagajewski B (2018) Citizen science and WebGIS for outdoor advertisement visual pollution assessment. Comput Environ Urban Syst 67:97–109

Du J, Zhang X, King D (2018) An investigation into the risk of night light pollution in a glazed office building: the effect of shading solutions. Build Environ 145:243–259

Elsahragty M, Kim J (2015) "Assessment and strategies to reduce light pollution using geographic information systems", International conference on sustainable design, engineering and construction. Proc Eng 118:479–488

Falchi F, Cinzano P, Duriscoe D, Kyba C, Elvidge C, Baugh K, Portnov B, Rybnikova N, Furgoni R (2016) The new world atlas of artificial night sky brightness. Am Assoc Adv Sci Res Art Sci Adv. http://advances.sciencemag.org/content/2/6/e1600377. Retrieved 31 Dec 201

Illuminating Engineering Society (IES) (2019) Addendum A for IES TM-15-11: Backlight, Uplight, and Glare (BUG) ratings. https://www.ies.org/wp-content/uploads/2017/03/TM-15-11BUGRatingsAddendum.pdf. Retrieved 31 Dec 2018

Institution of Lighting Engineers (ILE) (2005) The outdoor lighting guide. Taylor and Francis, London, UK, p 16

International Commission on Illumination (CIE), Upward Light Ratio (ULR). http://www.cie.co.at/eilv/1381. Retrieved 30 Dec 2018

International Dark-Sky Association (IDA) (2019) Retrieved 30 Dec 2018, from IDA Website: http://www.darksky.org/

Jiang W, He G, Long T, Wang C, Ni Y, Ma R (2017) "Assessing light pollution in china based on nighttime light imagery", Remote sensing. MDPI Open Access J 9:135

Kuechly H, Kyba C, Ruhtz T, Lindemann C, Wolter C, Fischer J, Hölker F (2012) Aerial survey and spatial analysis of sources of light pollution in Berlin, Germany. Remote Sens Environ 126:39–50

Lighting Analysts: Illumination engineering software. https://lightinganalysts.com/swatting-bugs/. Last accessed 07 Nov 2019

McGowan T (2013) "Light pollution". Encyclopedia of color science and technology. Springer Science + Business Media, New York

National Park Service. https://www.nps.gov/grca/learn/nature/night-skies.htm. Last accessed 07 Nov 2019

Osman A, Isobe S, Nawar S, Morcos A (2001) "Light pollution and energy loss from Cairo", Preserving the astronomical sky, international astronomical union symposium, vol 196, pp 107–110 (provided by the NASA Astrophysics Data System)

Pedithep Y (2015) Comparative environmental and planning law relating to light pollution control in England and other jurisdictions. De Montfort University, Ph.D. Thesis. https://ethos.bl.uk/OrderDetails.do;jsessionid=6CBC142693B0F55EADFB256E51AFA510?uin=uk.bl.ethos.679030. Retrieved 31 Dec 2018

Schroer S, Hölker F (2014) Light pollution reduction: methods to reduce the environmental impact of artificial light at night. Springer International Publishing, Switzerland, Handbook of Advanced Lighting Technology

Society of Light and Lighting (SLL), Part of the Chartered Institution of Building Services Engineers (CIBSE) (2012) Guide to limiting obtrusive light. Lavenham, Suffolk, England

U.S. Green Building Council (USGBC) (2019) Retrieved 08 Apr 2019, from USGBC Website, "Light pollution reduction (SS8)". https://www.usgbc.org/credits/ss8?view=language

Xiang W, Tan M (2017) "Changes in Light Pollution and the Causing Factors in China's Protected Areas, 1992–2012", Remote Sens MDPI Open Access J 9:1026

Chapter 34
Artificial Intelligence Shaping Sustainable Cities for Climate Change Mitigation: A Review of Literature

Ahmed K. Abd El-Hameed

Abstract Among the processes that regulate the stability of the Earth, Climate Change has crossed the boundary levels. Cities are responsible for high energy consumption, significantly causing Climate Change with the CO_2 emissions they trigger. Besides the growth of populations living in cities globally, the complexity of modern cities and the pressure they place on resources mean that Climate Change drivers and threats will ever increase in concentration. Through integrating established knowledge, design strategies and innovative technologies based on the rapid advancement in Artificial Intelligence, the Fourth Industrial Revolution offers many opportunities to mitigate Climate Change. The paper highlights the role of Artificial Intelligence in mitigating Climate Change in cities, including applications in transportation, urban energy, water use and waste management. The review proves that these applications can impact city planning through reshaping the urban planning and design of the built environment. The paper also shows that implementing such applications will improve the sustainability of future cities.

Keywords Artificial Intelligence · Climate Change · Mitigation · City · Sustainability

34.1 Introduction

The progressed human development during previous industrial revolutions has been largely achieved at the expense of the Earth. Nine systems that regulate the planet stability and resilience have been identified by scientists, where Climate Change is considered one of the four systems that have currently crossed the boundary levels due to human activities (Steffen and Rockström 2015). And accordingly, combating Climate Change is declared as one of the 17 United Nations Sustainable Development Goals.

A. K. Abd El-Hameed (✉)
Faculty of Engineering, Ain Shams University, Cairo, Egypt
e-mail: ahmed.khaled@eng.asu.edu.eg

Architecture and Urbanism: A Smart Outlook,
https://doi.org/10.1007/978-3-030-52584-2_34

Focusing on 100 cities worldwide in a recent report published, the mean annual temperatures are projected to rise by 0.7–1.5 °C in the 2020s, 1.3–3.0 °C in the 2050s and 1.7–4.9 °C in the 2080s. Moreover, the mean annual precipitation is projected to remarkably change by −7 to +10% in the 2020s, −9 to +15% in the 2050s and −11 to +21% in the 2080s. These projections were based on 35 global climate models and 2 representative concentration pathways, RCP4.5 and RCP 8.5 (Rosenzweig et al. 2015).

Keeping the Paris Agreement pledges, the world average temperature in 2100 is still projected to be 3 °C above the pre-industrial levels, above the defined targets to avoid the worst impacts of Climate Change (UNEP 2017). The fresh water required to support the world economy may fall 40% short in amount by 2030, as Climate Change and pollution impact the global water cycle (WWAP 2015). Moreover, recently, around 92% of the global population lives in areas that fail to meet the air quality guidelines of the World Health Organization (WHO 2018). And as the global population is expected to grow to 9.8 billion by 2050, the demand for food, transport and energy will accelerate, causing further environmental degradation that affect public health, livelihoods and security (Combes et al. 2018).

The United Nations Framework Convention on Climate Change (UNFCCC) focuses on two key terms: "mitigation" which aims at reducing greenhouse gas (GHG) emissions to minimize induced global warming or "avoiding the unmanageable", and "adaptation" which aims at coping with a changed climate or "managing the unavoidable". A simpler distinction between mitigation and adaptation has been made by Baum et al., where mitigation or "climate friendly development" helps reduce GHG emissions, while adaptation or "climate safe development" reduces the vulnerability to direct and indirect impacts of Climate Change (Baum et al. 2009). However, the interrelationships between both terms are significant.

As most of the accelerating world population currently living in cities is projected to rise in the coming decades (Rosenzweig et al. 2015), these large urban populations are projected to suffer more frequently from severe events induced by Climate Change that require specific attention on mitigation and resilience building. The United Nations adopted a new 11th Sustainable Development Goal in 2015, targeted to "Make cities and human settlements inclusive, safe, resilient and sustainable" (Rosenzweig et al. 2015). This goal cannot be achieved without regarding Climate Change as a key factor. Likewise, the effective responses to Climate Change cannot advance without understanding sustainability with its larger context. Cities are widely recognized as possible main implementers of targeted climate resiliency. As cities can mitigate Climate Change while adapting to changed conditions, thorough changes in land uses, urban energy, water use, transportation, growth patterns, consumption and lifestyles will be required.

Trends in current global energy consumption show an increase in using fossil fuels, particularly in parts with rapid urbanization worldwide. Key challenges include lowering the urban heat island effect, air pollution and GHG emissions as well as ensuring resilience in changing climates. Cities need access to clean, affordable and safe mobility; with co-benefits including improved reduced congestion, better air quality, mass transit development and sustainable infrastructure towards low-carbon transportation. Cities account for more than 70% of GHG emissions, with the transportation sector directly accounting for about 30% of total end-use energy-related CO_2 emissions (Rosenzweig et al. 2015).

Climate Change impacts cause changes in storm water and rainfall patterns, rising sea levels and changes in both surface water and groundwater, increasing the pressure in urban areas. Higher demands for water will cause competition for the limited resources as cities expand. Waste management is linked to increasing urbanization and Climate Change, as improper waste management cause about 3–5% of global GHG emissions (Rosenzweig et al. 2015). Key challenges include effective collection and treatment systems to reduce emissions, waste recovery through both reuse and recycling, as well as preventing waste at the source through promoting upstream interventions.

34.1.1 Research Aim

The paper analyses and highlights the role of Artificial Intelligence (AI) in mitigating Climate Change (CC) in cities, as well as impacting city sustainability and planning.

34.1.2 Research Methodology

A literature review was conducted for recent AI applications used to improve air quality, energy management, water management, urban management, transportation and waste management. Later, the categorized AI applications were analysed and summarized in a table according to their detailed impacts on CC mitigation, sustainability and city planning as followed:

- CC mitigation: Managing GHG emissions, less GHG emissions and zero GHG emissions.
- Sustainability: Less resource use (less pollution), less water use (high efficiency), less energy use (high efficiency), using/improving renewable energy and less waste.
- City planning: Optimal density (population), land uses (zoning), horizontal urban expansion, decentralized/off-grid utilities and building design (orientation).

34.2 Literature Review

The diverse definitions of AI have helped the applications to grow and advance at an accelerating pace. Nils J. Nilsson provided a definition of AI as: "Artificial intelligence is that activity devoted to making machines intelligent, and intelligence is that quality that enables an entity to function appropriately and with foresight in its environment" (Nilsson 2010). Many factors have induced the AI revolution, including the maturing of machine learning supported by web-based data gathering and cloud computing. Machine learning has advanced using "deep learning", an adaptive form of artificial neural networks that are trained using the backpropagation method (Stone et al. 2016). Such advanced performance of developed information processing algorithms has also been accompanied by an advanced progress in hardware technology for basic operations such as sensing, perception and object/pattern recognition.

Among the Industry 4.0 technologies, AI is predicted to have the most impact on all industries and daily life (Combes et al. 2018). Being often combined with other Industry 4.0 technologies, there are many opportunities to apply AI to current pressing challenges. Smart cities should be more immediately concerned due to the current complexity of modern cities and the pressure they place on all resources. As much of the global population currently lives in cities through natural growth and migration from rural areas, this means that challenges will occur in ever-accelerating concentrations. First, AI can support rural populations and help cities hold their populations at sustainable levels by slowing down the migration flow, or even encouraging people to move from cities (Hoar et al. 2017). Second, some reports and papers foresee AI applications for managing and monitoring air quality, energy, water, built environment, transportation, manufacturing and waste that can mitigate Climate Change in cities.

34.2.1 Air Quality

AI applications are advancing real-time monitoring of air quality, with the example of using smartphone cameras by AirTick as proxies for air pollution sensing that harness machine algorithms and image recognition to analyse images across cities at low cost (Rutkin 2016). The Green Horizons initiative of IBM combines IoT and machine learning to develop predictive analytics for 7–10 day forecasts by harnessing data from air quality stations and more sources such as weather stations and satellites, topographic maps, traffic systems, industrial activity and even social media (IBM 2017). Such applications can help monitor and manage urban air quality. Concerning real-time indoor air pollution source detection that enables management scenarios, smart monitors use neural network algorithms to relate a pollutant to a source in a given environment (Combes et al. 2018).

Transportation pollution intensity can be predicted in urban areas using AI that leverage pollutants and environmental data (solar irradiation, humidity and temperature), enabling management responses to reduce pollution impacts such as congestion charge and traffic restrictions (Combes et al. 2018). Moreover, AI used in new connected platforms can harness data from cameras, vehicles and radar sensors to optimize urban traffic flows and improve air quality by reducing stop-start driving and stationary vehicles (Bell et al. 2014).

34.2.2 Energy Management

AI can be used to optimize energy use in cities, utilizing deep learning and machine learning analysis of energy consumption patterns to make real-time intelligent decisions that maximize the energy efficiency. Machine learning is being used for real-time matching of energy generation and demand in "smart grids", reducing unpredictability while increasing power balancing, efficient use and storage of renewable energy (Global Agenda Council on the Future of Cities 2015). Agder Energi is also using the cloud and AI to predict and manage changing energy demands in Norway (Agder Energi 2017), such applications provide more reliable, effective and autonomous energy grids.

Neural networks are being developed to improve the efficiency and reliability of renewable energy. DNV GL uses sensors attached to wind and solar power plants to provide data for improving machine learning monitoring capabilities, enabling predictive maintenance, remote inspection of sites and energy resource forecasting (DNV GL 2017). This application increases both control and maintenance efficiency while reducing costs of wind and solar energy. AI applications can also help face the intermittency of integrating renewables such as wind and solar energy into urban energy networks due to weather variations. Creayla et al. developed a model for forecasting solar irradiance to calculate next-day generation of solar energy from PV systems by using site-specific meteorological and solar irradiance parameters as inputs to a machine learning algorithm. The model had a 5% increase in accuracy when other parameters such as precipitation and wind direction were added (Creayla et al. 2017).

Integrating AI in energy distribution systems is critical for overall decarbonizing of grids, increasing energy efficiency and expanding the usage of renewables. Using AI in Virtual Power Plants (VPPs) can aggregate, integrate and optimize the use of microgrids, solar panels, energy storage systems and other facilities (Combes et al. 2018). Energy distribution grids can also be further extended to allow the integration of AI-enabled solar roads to connect, optimize and expand the grid, while using AI to adjust the traffic lanes of roads based on real-time vehicle flow.

34.2.3 Urban Management

AI can be used to optimize energy use at building and urban scales, using machine learning algorithms (e.g. the Energy-Plus model) to guide energy efficiency through modelling efficient building layouts at the design phase for further optimisation of energy use at the operation phase (AI.Business 2017). AI is also used to provide acoustic cue responsive heating/cooling and lighting in buildings and streets to optimize energy use (Combes et al. 2018).

AI deep learning can also improve urban planning, leading to resilient cities with minimal environmental impacts. AI can be used to both simulate and automate the creation of zoning laws and building ordinances, as well as being used along with smart metres and the IoT to help predict and optimize urban energy production and demand at both building and city scales (Combes et al. 2018). Real-time optimized energy efficiency using AI can have a substantial impact on building energy consumption, with the example of Google using DeepMind's learning algorithms to optimize cooling in its data centres that cut down energy use by 40% (Evans and Gao 2016).

Combining real-time data on water and energy availability and consumption, people flows, traffic flows and weather at city scale can create an "urban dashboard" that can be AI-enabled to optimize urban energy and water use, potentially avoiding costly additional infrastructure as well as lowering the city's environmental footprint by reducing pollution and congestion (Combes et al. 2018).

34.2.4 Transportation

The large-scale availability of data has made modern transportation an ideal development domain for AI applications. Optimal search algorithms in applications such as Google Maps and MapQuest have been used widely for using public transportation, routing trips, receiving real-time traffic information with predictions for its conditions, as well as locating services (Stone et al. 2016). Sensing algorithms will also achieve driving capabilities with super-human performance, as automated vision and perception are already near or at the level of human performance for tasks such as tracking and recognition (Stone et al. 2016). Such advances will be followed by more algorithmic improvements in high-level reasoning and planning capabilities. Self-driving vehicles are predicted to be adopted by 2020 in large numbers (Greenough 2016). Beyond self-driving applications in personal, public and peer-to-peer transportation, advances in robotics can facilitate the adoption of other autonomous vehicles such as robots and drones.

Smart mobility can provide safe transportation to children, elders and people with disabilities, as well as reducing traffic congestion, noise and accidents. In addition, it can use energy more efficiently and reduce emissions. Kaabachi et al. developed a bi-objective green version of the classical "Travelling Salesman Problem with Time Windows" that considers more parameters aiming at not only finding the shortest routes but also reducing fuel consumption and CO_2 emissions (Kaabachi et al. 2017). Meneguette et al. also proposed a method for detecting, classifying and managing traffic congestions using inter-vehicle communications based on an artificial neural network, which achieved reductions in fuel consumption and CO_2 emissions by avoiding congested roads (Meneguette et al. 2016).

Street lights can use AI algorithms to gather data from cameras and radar sensors with the aim of detecting traffic and building a timing plan for street lights that increases the traffic flow efficiency (e.g. Surtrac) or proposes optimal traffic navigation (e.g. Nexar), which eases congestions and reduces air emissions (Combes et al. 2018). AI can also analyse data such as user behaviour and weather to generate management insights for city transportation networks, enabling a more energy-efficient mobility service.

AI-enabled autonomous vehicles (AVs) can use deep neural network techniques and machine vision algorithms that will enable a future transition to on-demand mobility (World Economic Forum 2015). Connected AVs can unlock substantial GHG reductions through reducing congestion and driving miles using route optimization applications, prioritizing energy efficiency using eco-driving algorithms, as well as minimizing car ownership and miles travelled using autonomous ride-sharing services (Park 2017). Minimizing car ownership can further lower manufacturing and operational emissions of vehicles. AI will also accelerate the widespread of autonomous connected electric vehicles (EVs), which will ultimately cut GHG emissions through the communication between vehicles and infrastructure that optimizes network and navigation efficiency (Combes et al. 2018).

34.2.5 Water Management

AI can simulate reservoir performance and use weather forecasts to project water usage for a specific geographical area, informing better policy decisions (Combes et al. 2018). Valor Water Analytics combines AI with operational interactions and industry intelligence to detect anomalies or inaccuracies in smart water metres using machine learning algorithms, further enabling leak detection and understanding real-time water flows (Valor Water Analytics 2017). Flo Technologies and Pluto AI use cloud-based machine learning algorithms for real-time water management for households, industrial users and utilities, ensuring water efficiency and optimizing energy consumption (Combes et al. 2018). Utilis uses also AI to detect underground leaks-prone pipes in water supply networks using machine learning that analyses satellite imagery (Combes et al. 2018).

Water Planet's IntelliFlux uses AI to analyse data from water pressure sensors and determine optimal filtration performance in case influent water quality varies, minimizing water loss through self-adaptive water filtration (Water Planet 2017). AI can use satellite data to analyse surface water and soil conditions and predict weather patterns in order to predict drought conditions, helping communities affected to optimize reservoir management and allocation of water resources (Tan and Perkowski 2015). AI has also the potential to manage distributed "off-grid" water resources, further combining machine learning, the IoT and blockchain to establish decentralised water systems where local water resources and closed-loop recycling gain more value (Combes et al. 2018).

34.2.6 *Manufacturing and Waste Management*

AI processing tools for natural languages (e.g. eRevalue) can analyse social, environmental and governance data about supply chains to improve their efficiency through monitoring suppliers and informing better management conditions (Combes et al. 2018). Recycling stations can also use neural networks to gather real-time feeds to select and sort recyclable waste (Combes et al. 2018).

34.3 Discussion

Concerning Climate Change (CC) mitigation, sustainability and city planning, the roles of AI applications previously mentioned can be accordingly analysed by category as followed in Table 34.1.

Table 34.1 shows that implementing AI applications can help cities improve their sustainability and mitigate Climate Change (CC). The table also demonstrates that AI applications can impact city planning through reshaping the urban planning and design of the built environment. Referring to the AI applications summarized in Table 34.1, the prominent impacts on CC mitigation, sustainability and city planning can be categorized as followed:

- Improving rural communities: AI applications can reduce rural–urban migration. Such lower migrations will help cities reach their optimal populations and

Table 34.1 Roles of AI applications in Climate Change mitigation, sustainability and city planning

		CC mitigation			Sustainability in cities					City planning				
AI applications		Managing GHG emissions	Less GHG emissions	Zero GHG emissions	Less resource use (Less pollution)	Less water use (High efficiency)	Less energy use (High efficiency)	Using /Improving renewable energy	Less waste	Optimal density (Population)	Land uses (Zoning)	Horizontal urban expansion	Decentralized /Off-grid utilities	Building design (Orientation)
Improving rural communities (Lower rural-urban migration)			■		■	■	■		■	■				
Air quality	Air pollution sensing (AirTick)	■			■									
	Air quality forecasting (Green Horizons initiative)	■			■									
	Indoor air pollution source detection	■			■									
	Transportation pollution forecasting	■			■									

(continued)

Table 34.1 (continued)

		CC mitigation			Sustainability in cities					City planning				
AI applications		Managing GHG emissions	Less GHG emissions	Zero GHG emissions	Less resource use (Less pollution)	Less water use (High efficiency)	Less energy use (High efficiency)	Using /Improving renewable energy	Less waste	Optimal density (Population)	Land uses (Zoning)	Horizontal urban expansion	Decentralized /Off-grid utilities	Building design (Orientation)
Energy management	Matching generation and demand patterns	■	■				■	■						
	Forecasting and managing demand (Agder Energi)	■	■				■							
	Remote inspection and predictive maintenance of solar/wind plants (DNV GL)			■			■	■				■	■	
	Forecasting solar irradiance for PV systems			■			■	■						
	Integration in distribution systems	■	■				■	■						
	Integration in Virtual Power Plants	■	■				■	■				■	■	
Transportation	Smart mobility		■		■		■							
	Green Travelling Salesman Problem	■	■		■		■							
	Detecting and managing traffic congestions	■	■		■		■							
	Detecting traffic and timing street lights (Surtrac & Nexar)	■	■		■		■							
	Generating management insights for transportation	■	■		■		■			■	■	■		
	Autonomous on-demand mobility		■		■		■							
	Autonomous ride-sharing mobility		■		■		■							
	Autonomous connected electric vehicles		■		■		■							

(continued)

Table 34.1 (continued)

AI applications		CC mitigation			Sustainability in cities					City planning				
		Managing GHG emissions	Less GHG emissions	Zero GHG emissions	Less resource use (Less pollution)	Less water use (High efficiency)	Less energy use (High efficiency)	Using /Improving renewable energy	Less waste	Optimal density (Population)	Land uses (Zoning)	Horizontal urban expansion	Decentralized /Off-grid utilities	Building design (Orientation)
Urban management	Modelling efficient building layouts (Energy-Plus model)	■	■				■							■
	Optimized heating/cooling and lighting (DeepMind)	■	■				■							
	Auto-creation of zoning laws and building codes	■	■		■	■	■	■	■	■	■		■	■
	"Urban dashboard" for optimal city operation	■	■		■	■	■	■	■					
Water management	Projecting water usage for a specific geographical area		■			■	■					■	■	
	Detecting inaccuracies in water meters (Valor Water)		■			■	■		■					
	Detecting leaks-prone pipes in supply networks (Utilis)		■			■	■		■				■	
	Water management for households, industrial users and utilities (Flo & Pluto)	■	■			■	■							
	Optimal filtration of influent water (IntelliFlux)		■			■	■		■					
	Managing distributed "off-grid" water resources	■	■			■	■		■			■	■	
Waste management	Improving the efficiency of supply chains (eRevalue)	■	■		■	■	■		■		■			
	Selecting and sorting recyclable waste		■		■				■					

urban densities, which will help reduce GHG emissions to mitigate CC as well as decreasing resource use, water use, energy use and waste production.

- Air quality: AI applications used for sensing and forecasting air pollution help manage GHG emissions for better CC mitigation decision-making.
- Energy management: AI applications mainly manage GHG emissions that help inform high efficiency practices and encourage higher dependence on renewable energy that reduce GHG emissions. AI applications used in remote inspection and predictive maintenance of solar/wind plants as well as forecasting solar irradiance

for PV systems help improve renewable energy production with zero GHG emissions. Concerning the impacts on city planning, AI applications used in remote inspection and predictive maintenance of solar/wind plants as well as integrating AI in Virtual Power Plants (VPPs) can encourage more horizontal urban expansion using decentralized and off-grid power utilities.

- Urban management: AI applications used for modelling energy-efficient building layouts and optimizing heating/cooling and lighting help increase the energy efficiency that reduces GHG emissions. Using AI-enabled "urban dashboard" for optimal city operation and AI-enabled creation of zoning laws and building codes help achieve lower waste production, higher dependence on renewable energy and higher efficient use of resources, water and energy that reduce GHG emissions. Concerning city planning, using AI for modelling energy-efficient building layouts help inform optimal building design and orientation. Moreover, using AI-enabled creation of zoning laws and building codes can inform optimal urban density, land uses, building design and decentralized/off-grid utilities.
- Transportation: AI applications can increase the efficiency of energy used and manage the resulted emissions, which in turn help reduce overall energy consumption, pollution and GHG emissions. Concerning city planning, using AI to generate management insights for transportation networks can help inform optimal urban densities and land uses while encouraging further horizontal expansion. Concerning the indirect impacts, applications such as AI-enabled vehicles used for shared and on-demand mobility can reduce car ownership and manufacturing that can help reduce overall energy use and GHG emissions.
- Water management: AI applications can increase the efficiency of using both water and energy that help reduce GHG emissions. Moreover, such applications like detecting inaccuracies in water metres, detecting leakage in supply pipes, optimal filtration of influents and managing distributed off-grid water resources can also help reduce GHG emissions through minimizing water losses and related energy consumption. Concerning city planning, using AI to manage distributed off-grid water resources can encourage further horizontal urban expansion using decentralized and off-grid water utilities.
- Waste management: AI applications can help save resources and reduce GHG emissions through minimizing the waste produced. Concerning other sustainability and city planning impacts, using AI to improve the efficiency of manufacturing supply chains can increase the efficiency of using both water and energy, thus informing optimal land uses and further reducing GHG emissions.

34.4 Conclusion

In this paper, AI was proved to have a promising role in shaping cities in terms of city planning as well as improving their sustainability in order to mitigate Climate Change. The paper showed that various AI applications can help manage and reduce GHG emissions in cities, and even encourage and improve clean practices with zero emissions. This can be achieved by using AI applications that manage and improve sustainable practices such as using resources, water and energy more efficiently while reducing pollution and waste production.

The paper summarized some AI applications that can be used for improving air quality, energy management, urban management, transportation, water management, manufacturing and waste management. The paper also demonstrated that these applications can impact city planning in terms of encouraging further horizontal expansion with improved decentralized/off-grid utilities, informing optimal urban densities, land uses and building layouts while improving the operational performance of buildings and utilities.

Further investigation is required for the role AI applications can play to further mitigate Climate Change through decarbonizing cities, especially in sectors with high energy use and GHG emissions such as product manufacturing and building construction. Such sectors require improving the energy efficiency of their processes and closing the loops for product and building life cycles. The impacts of such applications on city planning should be further investigated.

References

AI.Business (2017) Machine learning and energy efficient building design. Retrieved from http://ai.business/2017/03/23/machinelearning-%0Aand-energy-efficient-building-design/

Baum S, Horton S, Choy Low D, Gleeson B (2009) Climate change, health impacts and urban adaptability: case study of Gold Coast City

Bell M, Goodman P, O' Brien J, Namdeo A (2014) A congestion sensitive approach to modelling road networks for air quality management. Int J Environ Pollut 54(2/3/4), 213–221. https://doi.org/10.1504/ijep.2014.065122

Combes B, Herweijer C, Ramchandani P, Sidhu J (2018) Fourth industrial revolution for the earth: harnessing artificial intelligence for the earth. In: World economic forum. Retrieved from https://www.pwc.com/gx/en/sustainability/assets/ai-for-the-earth-jan-2018.pdf

Creayla CMC, Garcia FCC, Macabebe EQB (2017) Next day power forecast model using smart hybrid energy monitoring system and meteorological data. Proc Comput Sci 105:256–263. https://doi.org/10.1016/j.procs.2017.01.219

DNV GL (2017) Making renewables smarter: the benefits, risks, and future of artificial intelligence in solar and wind. Retrieved from https://www.dnvgl.com/publications/making-renewables-smarter-104362

Energi A (2017) Retrieved from https://www.ae.no/konsernet/om/english/

Evans R, Gao J (2016) DeepMind AI reduces Google data centre cooling bill by 40%. Retrieved from Deepmind Blog website: https://deepmind.com/blog/deepmindai-%0Areduces-google-data-centre-cooling-bill-40/

Global Agenda Council on the Future of Cities (2015) Top ten urban innovations. Retrieved from http://www3.weforum.org/docs/Top_10_Emerging_Urban_Innovations_report_2010_20.10.pdf

Greenough J (2016) 10 million self-driving cars will be on the road by 2020. In: Business insider. Retrieved from http://www.businessinsider.com/report-10-million-self-driving-cars-will-be-on-the-road-by-2020-2015-5-6?IR=T&r=US&IR=T

Hoar C, Atkin B, King K (2017) Artificial intelligence: what it means for the built environment. Retrieved from http://www.rics.org/Global/RICS_Insight_AI_in_the_built_environment_2017.pdf

IBM (2017) Green horizons: harnessing the power of cognitive computing and IoT to help fight pollution and climate change. Retrieved from http://www.research.ibm.com/green-horizons/#fbid=dQJLQ99TYMS

Kaabachi I, Jriji D, Madany F, Krichen S (2017) ScienceDirect a bi-criteria ant colony optimization for minimizing fuel consumption and cost of the traveling salesman problem with time. Proc Comput Sci 112:886–895. https://doi.org/10.1016/j.procs.2017.08.105

Meneguette RI, Filho GPR, Guidoni DL, Pessin G, Villas LA, Ueyama J (2016) Increasing intelligence in inter-vehicle communications to reduce traffic congestions: experiments in urban and highway environments. PLoS ONE 11(8). https://doi.org/10.1371/journal.pone.0159110

Nilsson NJ (2010) The quest for artificial intelligence: a history of ideas and achievements. Cambridge University Press, Cambridge

Park J (2017) Sensemaking/what will autonomous vehicles mean for sustainability? Retrieved from https://thefuturescentre.org/articles/11010/what-will-autonomous-vehicles-mean-sustainability

Rosenzweig C, Solecki W, Romero-Lankao P, Mehrothra S, Shakal S, Bowman T, Ali Ibrahim S (2015) ARC3.2. summary for city leaders. In: Climate change and cities, second assessment report. Urban Climate Research Network (UCCRN), Columbia University. https://doi.org/10.1080/23298758.1994.10685604

Rutkin A (2016) Pic-scanning AI estimates city air pollution from mass of photos. Retrieved from New Scientist website. https://www.newscientist.com/article/2076562-pic-scanning-ai-estimates-city-air-pollution-from-mass-of-photos/

Steffen BW, Rockström J (2015) Planetary boundaries: guiding human development on a changing planet. Science 347(6223):1–11. https://doi.org/10.1126/science.1259855

Stone P, Brynjolfsson E, Calo R, Etzioni O, Hager G, Hirschberg J, ... Brooks R (2016) Artificial intelligence and life in 2030. In: One hundred year study on artificial intelligence: report of the 2015–2016 study panel. https://www.ai100.stanford.edu

Tan R, Perkowski M (2015) Wavelet-coupled machine learning methods for drought forecast utilizing hybrid meteorological and remotely-sensed data. 2015 Int Conf Data Min 50–56

UNEP (2017) The emissions gap report 2017: A UN environment synthesis report. Retrieved from https://wedocs.unep.org/bitstream/handle/20.500.11822/22070/EGR_2017.pdf?sequence=1&isAllowed=y

Valor Water Analytics (2017) Valor product overview. Retrieved from https://www.valorwater.com/solutions-overview/

Water Planet (2017). Kevin Costner, Water planet team up to advance sustainable water reuse with smart membrane products. Retrieved from PR Newswire website: https://www.prnewswire.com/news-releases/kevin-costner-water-planet-team-up-to-advance-sustainable-water-reusewith-%0Asmart-membrane-products-300427906.html

WHO (2018) Ambient (outdoor) air quality and health. ISBN 9789241511353

World Economic Forum (2015) The driverless car revolution. Retrieved from http://reports.weforum.org/digital-transformation/the-driverless-car-revolution/

WWAP (2015) World water development report 2015: Water for a sustainable world. https://doi.org/10.1016/S1366-7017(02)00004-1

Index

A

Ain Shams University, 49, 54, 55, 61
Alexandria, 159, 161, 162, 166, 167, 169–171, 247, 258, 262, 402
Algorithms, 65–72, 74–77, 79, 80, 85, 86, 143, 486–489
Al Rehab City, 396
Ambience, 105, 106, 109–113, 116
Architectural design, 216, 217, 233, 305, 413
Architecture, 3, 5, 8, 21, 22, 24, 26, 33–37, 41, 46, 47, 49–52, 54–56, 61, 77–79, 82, 86, 89, 90, 92, 94–96, 99, 101, 102, 105–107, 115, 116, 134, 135, 174, 185, 186, 232, 234, 235, 284, 325, 331, 341, 359, 373, 375, 380, 383, 428, 434, 466, 467
Artificial intelligence, 65, 66, 75, 78, 483, 485–494
Aswan, 159, 161, 162, 164, 166, 167, 169–171, 243, 248
Atmospheres, 109, 311, 458, 473
Azhar Park, 270, 271, 274

B

Behavioural patterns, 321
Beirut, 357–359, 361, 366, 367
Building Information Modelling (BIM), 185–187, 189–191, 196, 231, 233
Bioinspired design, 20
Biomimetic design, 19–21, 23, 24, 26–30, 90, 94
Biomimicry, 19, 20, 89–94, 96, 99–102
British University in Egypt (BUE), 33, 35, 37, 39
Building envelope, 93, 163, 168, 169
Building information modelling, 231, 232, 235, 237
Built environment, 34, 50–52, 55, 105–109, 113, 116, 147, 174, 215, 217, 219, 221, 269, 295, 305, 322, 330, 341–343, 353, 358, 361, 363, 367, 403, 483, 486, 490

C

CAD, 233
Cairo, 38, 122, 147–150, 156, 158, 159, 161–164, 166, 167, 169–171, 185, 186, 215, 222, 259, 269, 270, 272–274, 277–280, 282–285, 289–291, 293, 294, 296, 300, 353, 377, 394, 395, 397, 401–403, 405, 406, 408, 425–427, 429–433, 435–437, 441, 442, 445, 447, 449, 451–453, 456, 471, 475
Children, 185–188, 191–194, 196, 201–205, 209, 212, 227, 254, 263, 265, 323, 396, 488
Cittaslow, 371–382, 385
Cittaslow Charter, 371, 372, 377, 378, 381, 382
City branding, 387, 388, 391, 396, 398
City centre, 427, 434, 435
Climate change, 122, 304, 317, 455–459, 461–466, 483–486, 490, 493, 494
Climatic change, 159, 160, 441, 456
Cultural identity, 266, 457
Culture-based economy, 425

Architecture and Urbanism: A Smart Outlook,
https://doi.org/10.1007/978-3-030-52584-2

D
Design education, 19, 20, 23, 26, 30, 46, 50–52, 55
Developing countries, 185, 186, 188, 189, 196, 254, 282, 290, 321, 322, 326, 427, 464
Downtown Cairo, 289–291, 293, 296, 300, 425, 430–432

E
Ecological architecture, 375
Ecological footprint, 307, 308, 310–312, 317
Ecological urbanism, 357
Egypt, 33, 35, 37, 39, 49, 52, 54, 55, 111, 121, 122, 130, 137, 147–149, 157, 160–164, 170, 185, 186, 200, 206–208, 213, 216, 241, 242, 245, 253, 254, 258, 259, 264, 279, 280, 289–291, 296, 298, 371, 372, 377, 379, 380, 387–390, 394, 401, 402, 408, 409, 426, 428, 429, 432, 464, 469, 471, 472, 478–480
El-Fayoum City, 371, 377, 379
El Shorouk City, 387, 389, 394–398
Energy consumption, 51, 121–123, 126, 127, 129, 130, 147, 148, 150, 152, 153, 155–159, 161–164, 166, 169–171, 293, 311, 477, 483, 484, 487–489, 493
Energy simulation, 148, 149, 159, 161, 162, 169
Environmental psychology, 176
Evidence-based approach, 425, 436, 437

F
Facility management, 185–187
Fortified city, 328
Fourth industrial revolution, 483

G
Gated communities, 147–150, 157, 158, 269–272, 278–283, 285, 396, 398, 403
Gentrification, 367, 426, 428, 430, 436, 437
Geographic information systems, 235, 475
Geospatial Information Systems (GIS), 185–187, 189–191, 196, 235, 411–413, 415, 419–422, 442, 443, 446, 449, 475
Giza, 215, 222, 289, 294
Green architecture, 37, 89
Green spaces, 124, 270, 272, 278, 305, 364

H
Healing environment, 51
Healthcare architecture, 215, 217
Healthcare buildings, 218, 219
Healthcare facilities, 215, 222
Heliopolis, 277, 395, 432, 435
Heritage, 231, 232, 235–239, 246, 248, 253, 254–260, 262, 264–267, 293, 296, 300, 328, 331, 332, 337, 359, 367, 379, 418, 427, 428, 432, 435, 455–459, 461–465
Historical buildings, 232, 234
Historic urban landscape (HUL), 253–258, 264–267
Housing and Building National Research Centre (HBRC), 164
Housing unit, 52, 79, 85, 345, 347, 349, 350, 353, 397
Human behaviour, 34, 321, 324, 330, 336, 470
Human life, 282, 447
Human needs, 304, 342, 426, 428, 430
Human resource, 189, 191, 465

I
Indoor environmental quality, 173, 175
Indoor environments, 86, 122, 174–176
Industrial heritage, 427
International Park, The, 432
Interview, 105, 109, 110, 116, 254, 263–265, 267, 289, 290, 296, 297, 332, 345, 353, 395–398, 425, 426, 430–433, 437, 465

L
Landmark, 232, 235, 329
Landscape, 38–40, 51, 55, 99, 205, 208, 209, 213, 241–243, 245, 248, 250, 251, 253–258, 262–266, 270, 357–363, 367, 447, 456, 457, 459, 461, 469
Landscape management, 55
London, 49, 54, 55, 160, 199–201, 206–208, 213, 219, 254, 428, 465, 475, 477
Low-income neighbourhoods, 341, 343, 344, 351, 353

M
Maslow, 351, 352, 430, 437

Mediterranean Sea, 361
Mobility, 284, 312, 313, 378, 382, 427, 432, 484, 488, 489, 491, 493
Motion parameters, 136, 138

N
Natural environment, 51, 201, 203, 205, 267, 372, 377
Neighbourhood, 164, 185–187, 189, 190, 193, 195, 264, 281–283, 289–292, 294, 300, 318, 341, 343–346, 349–353, 357, 359–361, 363, 365, 366, 397, 403, 408, 421, 426, 428, 429, 436, 437, 443, 445, 447, 450, 451
Net-Zero Energy Building (NZEB), 147, 148, 161, 166
New Cairo City, 401–403, 405, 408
Newly planned cities, 406
Night life, 477
Nile Delta, 402
Nile River, 243, 401, 402
Notre Dame University in Lebanon, 231, 232, 235, 236, 238

O
Occupants comfort, 94, 173

P
Paolo Saturnini, 371, 373
Paris, 231, 235–237, 326, 373, 374, 475, 484
Physical environment, 16, 112, 113, 219, 255, 342
Placemaking, 387
Post-Occupancy Evaluation (POE), 215, 217, 219, 220, 223, 224, 226, 341
Psychological needs, 352, 353
Public parks, 270, 364
Public realm, 321–324, 331
Public spaces, 269, 270, 272, 273, 277–283, 285, 321–325, 332, 339, 372, 389, 411–419, 422, 423, 429, 432, 437
Public sustainability, 304, 373
PV, 147–150, 154–160, 166, 169–171, 487, 491, 493

Q
Qualitative method, 215, 217, 223, 430
Quality of life, 173, 255, 258, 264, 265, 267, 303, 304, 312, 318, 341–344, 350, 352, 353, 371, 373, 376, 377, 388, 390
Quantitative method, 215, 217, 223, 430
Questionnaire, 3, 5, 7, 11, 13–16, 24, 33, 35, 40–44, 46, 57, 175, 345, 350, 352, 353, 395

R
Regenerative cities, 303–307, 312–314, 317, 318
Rehabilitation, 286, 425, 427, 430
Renewable energy, 147–150, 155–162, 164, 166, 170, 304, 305, 314, 485, 487, 490–493
Residential satisfaction, 341, 344, 345, 349, 352, 353
Rosetta, 253, 254, 258–267

S
Sense of belonging, 282, 312, 352
Smart cities, 313, 486
Smart technology, 182
Social attributes, 341, 352
Social cohesion, 269, 283, 285
Social events, 207, 411–414, 416, 422
Social policies, 437
Social resilience, 455
Social sustainability, 52
Socio-economy, 262, 272, 278, 284, 311
Space of belonging, 401
Space syntax, 401, 404, 405, 415
Spatial configuration, 404, 406
Stakeholders, 54, 55, 185–187, 189–193, 196, 215, 216, 224, 225, 264, 266, 290, 300, 358, 372, 425–427, 429, 432, 436, 437
Street activities, 281
Street barriers, 269, 274, 275, 280, 282, 285
Street networks, 408
Sustainability, 256, 257, 291, 303–305, 308, 372–375, 458, 483–484, 490–493
Sustainable architecture, 34, 49, 54, 56, 61, 89
Sustainable cities, 483
Sustainable design, 33, 34, 36, 40, 46, 50–52, 54, 55, 61, 90
Sustainable development, 188, 255, 264, 292, 293, 279, 379, 388, 470, 484
Swarm intelligence, 77–79, 81–86

T
Thermal performance, 123
Tunis village, 371, 372, 380, 381, 385

U
UNESCO, 52, 186, 253, 255–258, 457, 465
University of Stuttgart, 19–21, 135
Urban catalysts, 257, 264
Urban design, 77, 78, 111, 269, 270, 281, 283, 339, 358, 366, 452
Urban fabric, 427, 443, 445, 447, 449, 450
Urban Heat Island (UHI), 441–447, 449–453, 484
Urban landscape, 253–257, 262, 264–266, 270, 357, 367
Urban metabolism, 304, 307, 357, 359, 388
Urban quality of life, 255, 284, 378
Urban reality, 416
Urban regeneration, 253–255, 257, 292, 293, 357, 358, 367, 427
Urban sprawl, 402

V
Vivid experiences, 105–111, 113, 116
Voids, 322, 363

W
Well-being, 174, 308, 342
Workers performance, 173, 174